Lecture Notes in Computer Science 1309

Edited by G. Goos, J. Hartmanis and J. van Leeuwen

Springer
Berlin
Heidelberg
New York
Barcelona
Budapest
Hong Kong
London
Milan
Paris
Santa Clara
Singapore
Tokyo

Ralf Steinmetz   Lars C. Wolf (Eds.)

Interactive Distributed Multimedia Systems and Telecommunication Services

4th International Workshop, IDMS '97
Darmstadt, Germany, September 10-12, 1997
Proceedings

 Springer

Series Editors

Gerhard Goos, Karlsruhe University, Germany

Juris Hartmanis, Cornell University, NY, USA

Jan van Leeuwen, Utrecht University, The Netherlands

Volume Editors

Ralf Steinmetz
Darmstadt University of Technology
Department of Electrical Engineering and Information Technology
Merckstrasse 25, D-64283 Darmstadt, Germany
and GMD IPSI
Dolivostrasse 15, D-64293 Darmstadt, Germany
E-mail: Ralf.Steinmetz@kom.th-darmstadt.de

Lars C. Wolf
Darmstadt University of Technology
Department of Electrical Engineering and Information Technology
Merckstrasse 25, D-64283 Darmstadt, Germany
E-mail: Lars.Wolf@kom.th-darmstadt.de

Cataloging-in-Publication data applied for

Die Deutsche Bibliothek - CIP-Einheitsaufnahme

**Interactive distributed multimedia systems and
telecommunication services** : 4th international workshop ;
proceedings / IDMS '97, Darmstadt, Germany, September 10 - 12
1997. Ralf Steinmetz ; Lars C. Wolf (ed.). - Berlin ; Heidelberg ;
New York ; Barcelona ; Budapest ; Hong Kong ; London ; Milan ;
Paris ; Santa Clara ; Singapore ; Tokyo : Springer, 1997
   (Lecture notes in computer science ; Vol. 1309)
   ISBN 3-540-63519-X

CR Subject Classification (1991): H.5.1, H.4.3, C.2, B.4.1, K.6.5, D.4.6, J.1

ISSN 0302-9743
ISBN 3-540-63519-X Springer-Verlag Berlin Heidelberg New York

© Springer-Verlag Berlin Heidelberg 1997
Printed in Germany

Typesetting: Camera-ready by author
SPIN 10545866   06/3142 – 5 4 3 2 1 0   Printed on acid-free paper

Preface

This International Workshop on Interactive Distributed Multimedia Systems and Telecommunication Services (IDMS) is the fourth in a series which started in 1992. The first workshop was organized by Prof. K. Rothermel and Prof. W. Effelsberg, and took place in Stuttgart in 1992. It had the form of a national forum for discussion on multimedia issues related to communications. The succeeding event was "attached" as a workshop to the German computer science conference (GI Jahrestagung) 1994 in Hamburg organized by Prof. W. Lamersdorf. The chairs of the third IDMS, Dr. E. Moeller and B. Butscher, enhanced the event to become a very successful international meeting in Berlin in March 1996 (the proceedings were published by Springer-Verlag as Lecture Notes in Computer Science, Volume 1045). At that time it was decided to hold the fourth IDMS in Darmstadt under the responsibility of both of us (Prof. R. Steinmetz as general chair and Dr. L. Wolf as program chair). We organized this event at the Darmstadt University of Technology where we started in spring 1996 to build up from scratch a new professorship and related research groups.

For many years we have known that the convergence of multimedia and communications provides for a large variety of challenging research issues. Furthermore, applications and systems resulting from research and development in this area will find, and partly have found already, wide interest in the society in general. The goal of the workshop is to provide a forum for discussion of current work and future research areas between the related fields of multimedia systems, computer communications, and telecommunications – for example, the role of quality of service with respect to resource management and adaptation policies, protocols, and mechanisms. Hence we attracted contributions from the following areas:

- High-speed and multimedia networks
- ATM networks and applications
- Mobile multimedia systems
- Multimedia communication protocols
- Compression algorithms
- Quality of service and media scaling
- Resource management
- Multimedia operating systems
- Synchronization
- Multimedia database and storage
- Video-on-demand systems, components and architectures
- Multimedia programming languages, abstractions, and APIs
- Development tools for distributed multimedia applications
- Multimedia-specific intelligent agents
- Multimedia/hypermedia applications and tools, production, and authoring
- Conferencing
- Computer supported collaborative work
- Digital libraries
- Interactive television
- Virtual reality systems

We received over 100 submissions, 94 of which were considered to comply with the areas and style of the workshop. The submissions came from 20 countries all over the world, most from Europe, but many from America and Asia as well. These papers were distributed to the international program committee for review. In total, we requested a large number of reviews from the program committee and additional reviewers and, thanks to the very good work of all reviewers, we received at least three to four reviews per submission. Since the overall quality of the submissions was high, the program committee had the difficult task of deciding which papers to accept. During the PC meeting 41 papers were selected for presentation at the workshop and for publication in the proceedings. All these selected papers are included in this book. The PC meeting also decided to have a best paper award. The work on the MBone VCR performed by W. Holfelder under the guidance of Prof. W. Effelsberg at the University of Mannheim was selected as the best paper of IDMS'97.

Due to our strong commitment to the scientific world and the professional societies, IDMS'97 is held in cooperation with GI, VDE ITG, ACM SIGMM, and IEEE Germany Section, and with technical co-sponsorship by IEEE Communications Society. Thanks to grants and help from Deutsche Telekom (Darmstadt), Digital Equipment CEC (Karlsruhe), Ericsson Eurolab (Herzogenrath), IBM Stiftungsfond (Essen), NEC Europe (Berlin), o.tel.o (Essen), and Panasonic European Laboratories (Langen), we have moderated the fees such that researchers in academia as well as in industry could afford to participate.

The next IDMS will be chaired by Prof. T. Plagemann, University of Oslo, Norway, and will take place in Oslo in late 1998 or early 1999. If you as reader and/or participant at IDMS'97 have any suggestions as to how to improve or what to change, please let us know.

Finally we would like to thank Martin Karsten and his team from the Industrial Process and System Communications (KOM) group at Darmstadt University of Technology for their exceptional work in organizing all issues around our successful workshop.

Darmstadt, September 1997   Ralf Steinmetz and Lars Wolf

Address from the Minister of Science and Arts

of the Federal State Hessen, *Dr. Christine Hohmann-Dennhardt*, to the "4th European Workshop on Interactive Distributed Multimedia Systems and Telecommunication Services (IDMS'97)", September 10th–12th, at Darmstadt University of Technology.

The Federal State Government of Hessen and I are delighted to extend our cordial greetings and best wishes for a successful workshop to the participants of IDMS'97. I am glad to see so many experts with international reputation in the field of information technology congregated at Darmstadt University of Technology. This provides the opportunity to present interesting and deep research results to a broad professional audience.

Multimedia technology has made the worldwide exchange of information, ideas and cultural entities possible to new heights. In an early stage, Hessen has recognized that this innovative technology has also an economical impact. Therefore universities have been supported in corresponding activities. The upcoming start of the "Multimedia-Initiative Hessen" is expected to give this research area additional stimulus. For these reasons I am convinced that this workshop will raise major interest here in Hessen.

I would like to wish all participants an interesting and pleasant stay in Darmstadt.

Dr. Christine Hohmann-Dennhardt
(Minister of Science and Arts of the Federal State Hessen)

Supporting/Sponsoring Societies

ACM SIGMM
Gesellschaft für Informatik (GI)
GMD
IEEE Germany Section
IEEE Communication Society, TC on Multimedia Communication
Informationstechnische Gesellschaft im VDE (ITG)

Supporting/Sponsoring Companies

Deutsche Telekom
Digital Equipment CEC
Ericsson
IBM
NEC
o.tel.o
Panasonic

General Chair

R. Steinmetz GMD IPSI and Darmstadt University of Technology, Germany

Program Committee

Berthold Butscher	T-Berkom, Germany
Andre Danthine	University of Liege, Belgium
Luca Delgrossi	Universita Cattolica, Italy
Jörg Eberspächer	Technical University of Munich, Germany
Wolfgang Effelsberg	University of Mannheim, Germany
Jose Encarnação	FhG-IGD, Germany
Domenico Ferrari	Universita Cattolica, Italy
Borko Furht	Florida Atlantic University, USA
Nicolas Georganas	University of Ottawa, Canada
Simon Gibbs	GMD, Germany
Wendy Hall	University of Southampton, UK
Ralf Guido Herrtwich	o.tel.o, Germany
Andy Hopper	Cambridge University / ORL, UK
Jean-Pierre Hubaux	EPFL, Switzerland
David Hutchison	Lancaster University, UK
Bob Ip	Siemens AG, Germany
Winfried Kalfa	TU Chemnitz, Germany
Tom D.C. Little	Boston University, USA
Friedemann Mattern	Darmstadt University of Technology, Germany
Eckhard Möller	GMD FOKUS, Germany
Klara Nahrstedt	University of Illinois, USA
Erich Neuhold	GMD IPSI, Germany
Martin Paterok	IBM ENC, Germany
Stephen Pink	Swedish Institute of Computer Science, Sweden
Thomas Plagemann	University of Oslo, Norway
Radu Popescu-Zeletin	Technical University Berlin, Germany
Venkat Rangan	University of California, USA
Kurt Rothermel	University of Stuttgart, Germany
Jean Schweitzer	Siemens AG, Germany
Hide Tokuda	Keio University, Japan
Michael Weber	University of Ulm, Germany
Fiona Williams	Ericsson, Germany
Lars Wolf	Darmstadt University of Technology, Germany (Chair)

Local Organization

Martin Karsten Darmstadt University of Technology, Germany (Chair)

Table of Contents

Embedding Data in 3D Models

Ryutarou Ohbuchi, Hiroshi Masuda, Masaki Aono

ohbuchi@acm.org, masuda@trl.ibm.co.jp, aono@acm.org

IBM Japan Tokyo Research Laboratory

1623-14 Shimo-tsuruma, Yamato-shi, Kanagawa, 242, Japan

Abstract. The recent popularity of digital media such as CD-ROM and the Internet has prompted exploration of techniques for embedding data, either visibly or invisibly, into text, image, and audio objects. Applications of such data embedding include copyright identification, theft deterrence, and inventory.

This paper discusses techniques for embedding data into 3D models. Given objects consisting of points, lines, (connected) polygons, or curved surfaces, the algorithms described in this paper produce polygonal models with data embedded into either their vertex coordinates, their vertex topology (connectivity), or both.

A description of the background and requirements is followed by a discussion of where, and by what fundamental methods, data can be embedded into 3D polygonal models. The paper then presents several data-embedding algorithms, with examples, based on these fundamental methods.

Additional Keywords: 3D graphics, data hiding, digital watermarking, steganography, geometric modeling, copyright protection.

1. Introduction

The advent of digital media, such as CD-ROMs and the Internet, has made possible rapid dissemination of various kinds of data objects, including, images, text, movies, audio data, and recently 3D models. The primary advantage of digital data lie in the ease with which they can be duplicated, distributed, and modified. These advantages, however, have prompted unauthorized duplication and distribution of data.

One way of addressing this problem is to add invisible structures, or *(digital) watermark*, to the data objects. The structures convey information, such as owner identifications or copyright notices. Watermarks can be used, for example, to deter theft, to notify users of how to contact the copyright owner for payment of licensing fees, to discourage unauthorized copying, or to take inventory. The technology associated with adding watermarks is called *steganography, (digital) watermarking, data embedding*, or *fingerprinting*. In this paper, the act of adding watermark is called *embedding*, and retrieving the information encoded in the watermark for perusal is called *extraction*. Following [Pfitzmann96], the information to be embedded is called *embedded<datatype>*, the object in which the information is embedded is called *cover<datatype>*, and the object with watermark is called *stego<datatype>*. Suffix "<datatype>" varies with data types such as image or text. For example, embedding text data into a still image will be termed as, "embedded-text is embedded into cover-image, producing stego-image".

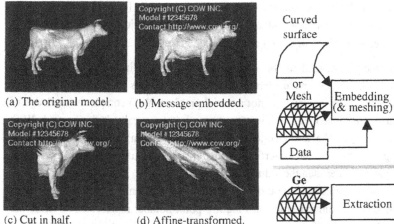

(a) The original model. (b) Message embedded.

(c) Cut in half. (d) Affine-transformed.

Curved surface

or Mesh

Data

Embedding (& meshing)

Polygonal model **Ge**

with watermark

Ge

Extraction → Data

Figure 1. (a) The original model of a cow (5804 triangles). (b) A message is embedded. The message, which is displayed on demand, survives (c) resection or (d) affine transformation.

Figure 2. Data are embedded into 3D polygonal models. Embedded data travels with the polygonal model.

Data embedding has been studied for cover-data types of still image, movie image, audio and text data (for example, [Tanaka90, Cox95, Walton95, Berghel96, Braudway96, O'Ruanaidh96, Zhao96]). To the author's knowledge, however, no study has been published on data-embedding techniques for 3D geometrical objects, such as a model described by using Virtual Reality Modeling Language (VRML) [ISO96].

In the past, the comment and annotation capabilities of scene description formats have been the primary means for adding information to 3D models. However, these comments and annotations can be easily removed, either intentionally or unintentionally. For example, programs for converting between 3D model formats often remove comments and annotations. As a result, comments and annotations cannot meet most of the requirements of watermarks.

This paper discusses techniques for embedding data into 3D models, specifically, 3D polygonal models. Inputs of embedding may include curved surfaces, in addition to polygonal geometric models. Watermarks are added to polygonal models by altering their geometry (the coordinates of their vertices) and/or their topology (connectivity of vertices). Figure 1 shows an example of such embedding into a polygonal model of a cow using an algorithm described in Section 3.2. The embedded message, "Copyright (C) COW INC. <CR> Model #12345678 <CR> Contact http://www.cow.org/.", can be extracted and displayed by clicking a button while using a browser enhanced with the extraction capability. The watermark withstands such alterations as Affine transformation and resection.

Data-embedding techniques for 3D models have the following required characteristics, whose priorities vary according to the intended applications:

Unobtrusive: The embedding must not interfere with the intended use of a model, such as viewing. One published example of image watermarking [Braudway96] used the visibility of the watermarks as an advantage, but in most applications, the watermarks must be unnoticeable in terms of the model's intended use.

Robust: Robustness is crucial to the success of data embedding. Using conventional cryptography, data can be made unreadable to third parties [Schneier96]. Making watermarks indestructible, however, is not a trivial problem. With complete knowledge of how watermarks are embedded, any watermarks can theoretically bed removed. With partial knowledge (e.g., the knowledge of the basic algorithm), the removal must be difficult enough so that it either interferes with the intended use of the models or the effort of removal is greater than the value of the model.

What kind of modifications should a watermark in a 3D model resistant to? Data format conversion is a common practice. Floating-point-number representation errors are often introduced during file format conversions. Models are (geometrically) transformed. While transformations are sometimes limited to rotation, uniform-scaling, and translation, more general affine transformations are common. Local deformation is occasionally used to reshape a model. Topological alterations, such as resection of a desired part of a model, remeshing, and polygon simplification, may also be performed. Assuming that the degree of modification is limited so that the utility of the model is not compromised, watermarks in 3D models should *ideally* withstand these and other possible alterations, regardless of whether they are intentional or otherwise.

Space efficient: A data-embedding method should be able to embed a non-trivial amount of information into models whose complexity and value warrants the effort involved.

It is important to note that these three requirements are at odds. For example, in general, more robust watermarks are able to carry less amounts of information. Note also that, depending on applications of watermarks, there could be other requirements, such as reliable identification of source or intended recipient.

2. Fundamental methods of data embedding

2.1. Objects types in 3D models for embedding

A 3D scene model may contain many types of data objects, such as geometry, vertex color, images for texture mapping, vertex normal vectors and even Universal Resource Locator links. We argue that geometry is the best candidate as a target for data embedding since it is the least likely to be removed. While there are many possible representations of 3D geometry, we chose *polygonal models* as the output of embedding for the study reported in this paper (see Figure 2). A "polygonal model" in this paper may include one or more of the following geometrical primitives: points, lines, polygons, connected polygons (e.g., an indexed face set), polyhedrons, and connected polyhedrons. Some data embedding algorithms require topology (connectivity) among points, while others do not. Topology can be added - for example by Delaunay triangulation [O'rourke94] - to polygonal models.

As inputs, embedding algorithms in this paper may accept curved surfaces (e.g., NURBS patches) in addition to polygonal models. The curved surfaces are tessellated into polygonal meshes as they are being embedded with stego-messages.

While we do not discuss details in this paper, components of 3D scenes other than polygonal models of geometry can become targets of data embedding. For example, various other representations of geometry, such as control points of curved surfaces or voxels, can be used as targets of embedding. Depending on the intended use of the model, non-geometrical objects such as per-vertex texture coordinates and per-vertex normal vectors, and to a lesser degree, per-vertex colors or face colors, are viable candidates for data embedding. Image texture, movie texture, and audio data are natural targets for data embedding by using "conventional" data embedding techniques. Note, however, that many of these non-geometrical components of 3D scenes are in general less suitable for data embedding since they can be removed or altered more easily than geometry.

2.2. Embedding primitives

There are two kinds of attributes in a polygonal model that can be modified in order to add watermarks. One is *geometry* of geometrical primitives (e.g., points or triangles) and the other is *topology* of these primitives. Units of modification, either geometrical or topological, are called *embedding primitives* in this paper.

2.2.1. Geometrical primitives

Geometrical values - specifically, the coordinates of points and vertices - can be modified to embed data. Modifications of coordinates change scalar or vector quantities, some of which are classified below according to the transformations to which they are invariant. The *invariance* property is quite useful in constructing data embedding algorithms that are robust against a given class of transformations.

- Altered by all the transformations listed below
  - Coordinates of a point
- Invariant to translation and rotation
  - Length of a line
  - Area of a polygon
- Invariant to rotation, uniform-scaling, and translation
  - Two quantities that define similar triangles (e.g., two angles)
- Invariant to affine transformation
  - Ratio of the lengths of two segments of a straight line
  - Ratio of the volumes of two polyhedrons
- Invariant to projection transformation
  - Cross-ratio of four points on a straight line [Farin96]

Embedding modifies these quantities in such a manner that the modification do not affect the intended uses (e.g., viewing by a browser) of models. For geometrical primitives, this usually means that the amount of a coordinate displacement must remain small. However, this is not always the case. If curved surfaces, instead of polygonal surfaces, are input to an embedding algorithm, the algorithm could have a large degree-of-freedom in number and positions of vertices in output polygonal meshes. The algorithm could exploit this freedom of vertex placement and features of the surfaces (e.g., curvatures) to produce robust watermarks.

2.2.2. Topological primitives

Watermarks can be embedded in the topology of a model. Modifications in this class involve changes in topology, although the geometry might also be changed as a result. Simple examples of topological embedding primitives include encoding of a binary symbol by using two alternative ways of triangulating a quadrilateral, ☑ and ◩, or two different mesh sizes, ☐ and ⊞.

2.3. Arranging embedding primitives

A lone embedding primitive usually can not encode meaningful amount of data. By arranging a set of embedding primitives into an ordered arrangement, a significant amount of information can be embedded into the arrangement of primitives. Examples of arrangements are 1D sequences generated by sorting triangles according to their areas, and 2D arrangements of embedding primitives based on the connectivity of triangles in an irregularly tessellated triangular mesh. A set of arranged primitives could be used to encode an ordered sequence of symbols, such as a character string and binary-number digits, or geometrical patterns, for example, shapes of letters. Note that, conventional targets of data embedding, such as audio and image data, fortunately have implicit ordering. For example, a 2D image has pixels arranged in a regular 2D array. To embed data into 3D polygonal models, however, primitives in the models must be arranged explicitly by embedding algorithms.

Arrangement of primitives can be achieved by the following two methods:

a. Topological arrangement employs topology (e.g., connectivity of vertices) to arrange embedding primitives. Topological arrangement requires topology among the embedding primitives. This arrangement is applicable to both topological and geometrical primitives. This arrangement can survive almost any geometrical transformation.

b. Quantitative arrangement sorts primitives into arrangement by using inequality relations among quantities associated with embedding primitives. This arrangement is applicable only to geometrical primitives, since it requires quantities for comparison. In order for the arrangement (and thus the watermark) to survive, the quantity used must withstand expected disturbances, such as a class of geometrical transformation.

It is often critical to find an *initial condition* - for example, the first primitive of a one-dimensional arrangement - in order to start an arrangement. The initial condition must be invariant to expected disturbances (e.g., affine transformation.); if the initial condition is changed irrecoverably, watermarks are lost.

Arrangements of embedding primitives can also be classified by their locality into *global*, *local*, and *subscript arrangements*.

a. Global arrangement arranges a set of all the embedding primitives.

b. Local arrangement subdivides the embedding primitives into disjoint subsets, and arranges each subset.

c. Subscript arrangement is similar to local arrangement with very small subsets (e.g., a few primitives per subset). Each subset in this case is called a *macro-embedding-primitive*, which embeds {data-symbol, subscript} pair so that the subscript identifies data symbol's position in an arrangement.

In the latter two arrangement methods, a set of primitives are grouped into a subset by proximity, either topological or quantitative, of their members.

3. Embedding algorithms

In the previous section, we have discussed fundamental methods for embedding data into 3D polygonal models. These fundamental methods can be combined to generate practical data-embedding algorithms. This section describes two such algorithms.

Both algorithms are implemented by using a kernel for a non-manifold modeler [Masuda96]. Using *radial edge* structure [Weiler86] to represent topological relationships among vertices, edges, faces, and regions, the modeler efficiently performs such operations as computation of the spanning tree of vertices on a triangular mesh.

3.1. Triangle similarity quadruple embedding

A pair of dimensionless quantities, for example, {b/a, c/a}, {S/{a*a}, b/c}, or {θ_1, θ_2} in Figure 3a, defines a set of similar triangles. One of these pairs can be used as a primitive to watermark triangular meshes so that the watermarks are robust against translation, rotation, and uniform-scaling transformations. Combining this primitive with subscript arrangement and repeated embedding of a message, the watermark becomes resistant to resection or local deformation of the stego-3D-models.

In order to realize subscript ordering, a quadruple of adjacent triangles that share edges (Figure 3b) is used as a macro embedding primitive that stores a pair of data symbols, a marker, and a subscript together. A marker is a special value (or values) that identifies macro embedding primitives. This algorithm, which is called Triangle Similarity Quadruple (TSQ) algorithm, embeds a message according to the following steps.

(1) Find a macro embedding units on the input triangular mesh. In doing so, avoid triangles that have already been used for the watermark. Avoid triangles that are unfit for stable embedding, e.g., if its two dimension-less quantities are either too small.

(2) For each macro embedding unit, embed *{subscript, mark, data1, data2}* quadruple by displacing vertices by small amount.

(3) Repeat (1) and (2) above until all the data symbols of a message are embedded.

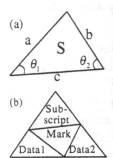

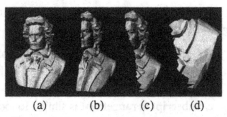

(a) (b) (c) (d)

Figure 5. A model of a Beethoven's bust (4889 triangles) resected repeatedly.

Figure 3. (a) A pair of dimension-less quantities defines a set of similar triangles. (b) A macro-embedding-primitive.

Figure 4. Triangles in dark gray are the macro-embedding-primitives.

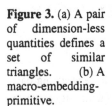

	No. of triangles	Data remained intact
(a)	4889	6 copies, 132 bytes each
(b)	2443	132/132 bytes
(c)	1192	102/132 bytes
(d)	399	85/132 bytes

Table 1. Data loss due to resection in the example shown in Figure 5.

In order to embed multiple copies of a message, steps (1) to (3) are repeated many times. Each repetition of a message is embedded in a topological proximity.

The TSQ extraction algorithm does not require the original cover-3D-model for extracting embedded message. The algorithm traverses all the triangles in the model to find triangles with markers. Each marker identifies a macro-embedding-primitive, which contains two data symbols and a subscript. The subscript puts the symbol pair at a proper place in the sequence of extracted message symbols.

Figure 5a shows a model of Beethoven's bust (4889 triangles, 2655 vertices) in which six identical copies of a message, each of which consists of 132 bytes, have been embedded by using the TSQ algorithm. Figure 5b-c shows the result of resection by arbitrary planes (the first one just happened to cut the model in half at the medial line.) As shown in Table 1, cutting the model in half left the message intact. After the model has been roughly quartered 102 bytes out of 132 bytes remained. Note that, since a subscript arrangement was used, intact characters still tended to be in the correct positions within the message string.

3.2. Tetrahedral volume ratio embedding

A ratio of volumes of a pair of tetrahedrons is the embedding primitive for the Tetrahedral Volume Ratio (TVR) embedding algorithm described in this section. The algorithm is designed to accept triangular meshes as input and embedding primitives are ordered topologically into a global one-dimensional arrangement to embed a sequence of symbols. The algorithm does not require cover-3D-model for extraction, and the watermark survives affine transformation. The TVR algorithm, however, is not robust with respect to topological modifications such as resection and remeshing.

The crux of the TVR algorithm is establishing global one-dimensional ordering of embedding primitives. This is accomplished in accordance with the following steps;

(1) Find a spanning tree of vertices *Vt*, called *vertex tree*, on the input triangular mesh *M*, given an initial condition *Ivt* for *Vt*. Convert *Vt* into a sequence of triangles *Tris*, called a *triangle sequence*.

(2) Convert *Tris* into a sequence of tetrahedrons *Tets*, called a *tetrahedron sequence*. To do this, compute a common apex as the centroid of the coordinates of a few triangles selected from the triangle sequence. The selected triangles are removed from the triangle sequence so that their coordinates are not modified by embedding.

(3) Convert *Tets* into a sequence of ratios of volumes *Vrs*. To do this, a volume of a tetrahedron (e.g., the first one) in *Tets* is selected as a common denominator of all the ratios, and volumes of the remaining tetrahedrons are used for numerators.

(4) Embed a symbol into each ratio by displacing vertices of numerator-tetrahedrons. The vertex displacements for the current symbol must not interfere with modifications of the previously embedded symbols. (In Figure 8, triangles that are used for embedding, which are colored dark gray, do not share edges because of this constraint.)

We now explain details of the first step, starting with the method to create a triangle sequence, and later come back to explain how to find an initial condition *Ivt*.

Generating vertex tree *Vt* from an input triangular mesh requires that the input mesh is an orientable manifold. To generate *Vt*, traverse vertices from a given initial vertex in the initial traverse direction, starting with the *Vt* initialized to empty. At each vertex, by scanning the edges in counter-clockwise order, find an edge that is not a member of *Vt* and does not loop back to any of the vertices covered by *Vt*. If such an edge is found, add it to *Vt*. Figure 6 shows an example of a vertex tree, in which vertex 1 is the initial vertex.

The vertex tree *Vt* is converted into the triangle sequence *Tris* as a set of edges *Tbe*, called a Triangle Bounding Edge (TBE) set is constructed. The *Tbe* is initialized to a set

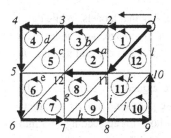

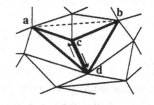

Figure 6. An example of a vertex tree (vertices are numbered), a triangle bounding edge set (edges are numbered as ⬈), and a triangle sequence (triangles are numbered as ①) for a triangular mesh.

Figure 7. Volume of a tetrahedron a-b-c-d, which is subtended by two triangles a-d-c and b-c-d that are adjacent to the edge c-d, is computed. (Two arrows indicate possible alternative initial traverse directions given an initial edge c-d.)

Figure 8. Triangles used for embedding by the TVR algorithm are shown in dark gray.

of edges that connect vertices in the *Vt*. To add an edge to the *Tbe*, vertices are traversed according to the *Vt*, starting from the root. At each vertex, all the edges adjacent to the vertex are scanned clockwise, and the scanned edge is added to the *Tbe* if it is not a member of the *Tbe*. A new triangle is added to the *Tris*, which started as an empty sequence, if all three edges of the triangle are in the *Tbe* for the first time, and the triangle is not already in the *Tris*. In the example shown in Figure 6, edges (except the initial entries of the *Tbe*) are marked by alphabets in the order of addition to the *Tbe* using alphabetical ordering. In the figure, members of the *Tris* are marked by numbers in circles.

The algorithm selects an initial *edge*, instead of an initial vertex and an initial traverse direction, using the method described below. Using an initial edge as initial condition allows two possible alternatives to start traverse of vertices to construct a vertex tree. This ambiguity is resolved by a trial-and-error method. The algorithm extracts sequence of symbols by using both alternatives, and choose the direction that yielded correct predetermined lead-in symbol sequence.

To select an initial edge, the algorithm computes, for every edge in the model, the volume of the tetrahedron subtended by the two triangles that are adjacent to the edge (Figure 7). The algorithm selects, as the initial edge, the edge for which tetrahedron's volume is the largest. (Note that these tetrahedrons are different from the ones used to embed symbols.) Since the initial edge with the largest volume may be incorrect due to noise and other reasons, the TVR algorithm employs the trial-and-error method again. The algorithm tries multiple candidate edges until it finds the correct lead-in sequence.

The TVR algorithm can be made resistant to resection and local deformation by using a local or subscript arrangement method combined with repeated embedding of a message. Examples shown in Figure 1 used a strengthened version of the TVR algorithm by using a local arrangement and repeated embedding of a message.

4. Summary and Future Work

We have applied the concept of data embedding to 3D polygonal models. While data-embedding techniques have previously been studied for text, image, and audio data objects, to the author's knowledge, our work is the first to study data embedding into 3D geometrical models.

After describing the background and requirements, we argued that geometry is the best data type in a 3D model for embedding. We then chose polygonal models as targets of embedding for this study. We presented fundamental methods for embedding data into polygonal models, namely, geometrical and topological modification primitives and methods for introducing ordering into a set of modification primitives. Finally, we presented two data-embedding algorithms and gave examples of their execution, thus demonstrating that data embedding into 3D polygonal models is a practical technique.

Much work in this area remains to be done. Our embedding algorithms are meant to be examples of simply embedding data into 3D polygonal models. Ideas such as spread spectrum communication [Cox96, Smith96], scrambling of symbol sequences by pseudo-random number sequence, and others may prove valuable in order to improve security, reliability, and robustness of embedding algorithm. We would also like to investigate data embedding into targets other than polygonal models, for example, control parameters of curved surfaces. In doing so, we need to understand requirements of applications other than viewing; for example, a manufacturing CAD system should have different criteria for unobtrusiveness than a simple browser of 3D polygonal models.

References

[Berghel96] H. Berghel and L. O'Gorman, Protecting Ownership Rights Through Digital Watermarking, *IEEE Computer*, July 1996, pp101-103.

[Braudway96] G. Braudway, K. Magerlein, and F. Mintzer, Protecting Publicly-Available Images with a Visible Image Watermark, *IBM Research Report*, TC-20336 (89918), January 15, 1996.

[Cox95] I. J. Cox, J. Kilian, T. Leighton, and T. Shamoon, Secure Spread Spectrum Watermarking for Multimedia, *Technical Report 95-10*, NEC Research Institute, 1995.

[Farin96] G. Farin, *Curves and Surfaces for CAGD: a Practical guide*, 4th edition, Academic Press, 1996.

[ISO96] ISO/IEC JTC1 SC24/N1596 CD #14772 Virtual Reality Model Language (VRML 2.0)

[Masuda96] H. Masuda, Topological operations for non-manifold geometric modeling and their applications, *Ph. D dissertation*, Department of Precision Machinery Engineering, University of Tokyo, 1996 (in Japanese).

[O'Rourke94] J. O'Rourke, *Computational Geometry in C*, Cambridge University Press, 1994.

[O'Ruanaidh96] J. J. K. O'Ruanaidh, W. J. Dowling, and F. M. Boland, Watermarking digital images for copyright protection, *IEE Proceedings on Vision, Image, and Signal Processing*, Vol. 143, No. 4, pp.250-256.

[Pfitzmann96] B. Pfitzmann, Information Hiding Terminology, in R. Anderson, Ed., Lecture Notes in Computer Science No.1174, pp347-350, Springer-Verlag, 1996.

[Schneier96] B. Schneier, *Applied Cryptography*, Second edition, John Wiley & Sons, Inc., New York, 1996.

[Smith96] J. R. Smith and B. O. Comiskey, Modulation and Information Hiding in Images, in R. Anderson, Ed., Lecture Notes in Computer Science No.1174, pp207-296, Springer-Verlag, 1996.

[Tanaka90] K. Tanaka, Y. Nakamura, and K. Matsui, Embedding Secret Information into a Dithered Multilevel Image, *Proc. 1990 IEEE Military Communications Conference*, pp216-220, 1990.

[Walton95] S. Walton, Image Authentication for a Slippery New Age, *Dr. Dobb's Journal*, April 1995.

[Weiler86] K. Weiler, The Radial Edge Structure: A Topological Representation for Non-Manifold Geometric Boundary Modeling, *Geometric Modeling for CAD Applications*, North Holland, pp3-36, May 1986.

[Zhao96] J. Zhao and E. Koch, Embedding Robust Labels into Images for Copyright Protection, *Proc. of the Int. Congress on Intellectual Property Rights for Specialized Information, Knowledge, and New Technologies*, Vienna, August 1995.

Estimation of Motion Parameters of a Rigid Body from a Monocular Image Sequence for MPEG-4 Applications

A. Smolic[1,2], B. Makai[1], G. Lin[1,3] and T. Sikora[1]

[1] Heinrich-Hertz-Institute Berlin GmbH (HHI)
[2] RTL plus Deutschland Fernsehen GmbH & Co. Betriebs-KG
[3] Department of Electronics Engineering, Xiamen University, PR China

Abstract

In this paper we present a method for the estimation of rigid body motion parameters from a monocular image sequence for MPEG-4 applications, such as SNHC face animation. Based on feature extractions in every frame the motion parameters of a human face are estimated with an extended Kalman filter that performs a prediction and correction loop at every timestep. With this recursive structure of the estimation process the temporal redundancies of the motion are taken into account. The non-linear motion equation is linearized at every timestep within the extended Kalman filter and therefore the rotation is not restricted to be small and the motion model can be based on the first frame and must not describe the frame to frame motion. Results are presented which demonstrate the accuracy of our estimation method on synthetic data as well as on a real image sequence where we estimated the motion parameters of a human face.

1 Introduction

The estimation of motion parameters of a rigid body from a monocular image sequence is an important task in a variety of image processing applications. Our focus is on model-based coding and visually controlled face/body animation [1-9] in the content of MPEG-4 SNHC and we use a 3D-wire-grid-model of the human face that is approximating the facial anatomy and physiology [12, 1-7]. In a model-based coding system the estimated parameters are used to make a prediction of the actual image and in an animation system they are used to animate a virtual face based on the motion of a real person.

The SNHC group of MPEG-4 is working on the description of synthetic human faces and bodies and their animation. They are also working in the field of text-to-speech synthesis and its possible interface to facial animation. This part of the standard shall provide the possibility to create virtual characters or talking heads which can be used in a variety of Multimedia and Internet applications like information systems, video games, tv and movie productions and so on. In this paper we present a method that can be used to estimate the global motion parameters of a human face from a monocular image sequence. The estimated parameters can be used for the animation of virtual characters in a realistic and natural way.

We assume that there is a moving 3D-object in the "real world" and that the image sequence is the 2D-projection of it. Based on the 2D-observation, an object model, a motion model and a camera projection model the motion parameters of the 3D-object are estimated with an extended Kalman filter that performs a prediction-and-correction loop at every timestep. In our formulation the rotation is not restricted to be small and the global motion parameters describe the continuous motion over the whole sequence relatively to the first image. This overcomes the problem of error accumulation that appears when the motion parameters describe the frame to frame motion.

A person's face is not a rigid body but the nonrigid motion can be divided into the global motion of the head and the local changes caused by facial expressions. Especially the eyes and the mouth region are concerned by the local changes which can be parametrized using the FACS action units [11]. In this work we deal only with the global motion of the human head which we assume to be rigid. Knowing the motion parameters and the underlying models the images can be synthesised at the decoder or a virtual face can be animated.

2 Estimation Method

For the estimation of the global motion parameters the person's face is regarded approximately as a rigid body. The motion of a rigid body has six degrees of freedom and can be described by six parameters. The co-ordinates of every point $\mathbf{P} = (x, y, z)^T$ can be calculated by

$$\mathbf{P} = \mathbf{R} \cdot \overline{\mathbf{P}} + \mathbf{T} \tag{1}$$

where $\mathbf{R} = f(\varphi_x, \varphi_y, \varphi_z)^T$ is the rotation matrix and $\mathbf{T} = (t_x, t_y, t_z)^T$ is the translation vector. The angles φ_x, φ_y and φ_z denote the rotations around the corresponding axes and t_x, t_y and t_z denote the translations (see fig. 1). The vector $\overline{\mathbf{P}} = (\overline{x}, \overline{y}, \overline{z})^T$ contains the co-ordinates of the reference point for the actual point obtained from the first picture. The rotation matrix is defined as

$$\mathbf{R} = \begin{pmatrix} r_{11} & r_{12} & r_{13} \\ r_{21} & r_{22} & r_{23} \\ r_{31} & r_{32} & r_{33} \end{pmatrix} \tag{2}$$

with

$$r_{11} = \cos \varphi_z \cos \varphi_y$$
$$r_{12} = \sin \varphi_z \cos \varphi_y$$
$$r_{13} = -\sin \varphi_y$$
$$r_{21} = -\sin \varphi_z \cos \varphi_x + \cos \varphi_z \sin \varphi_y \sin \varphi_x$$
$$r_{22} = \cos \varphi_z \cos \varphi_x + \sin \varphi_z \sin \varphi_y \sin \varphi_x$$
$$r_{23} = \cos \varphi_y \sin \varphi_x$$
$$r_{31} = \sin \varphi_z \sin \varphi_x + \cos \varphi_z \sin \varphi_y \cos \varphi_x$$
$$r_{32} = -\cos \varphi_z \sin \varphi_x + \sin \varphi_z \sin \varphi_y \cos \varphi_x$$
$$r_{33} = \cos \varphi_y \cos \varphi_z.$$

In the work of others these equations are approximated for small angles φ with $\sin\varphi \approx \varphi$ and $\cos\varphi \approx 1$ and the estimated motion parameters describe the frame to frame motion where the rotation can be assumed to be small [1-5]. We do not linearize in this way and so our motion parameters are reliable for arbitrary rotations. The advantage is that we can base our motion model relatively to the first picture and our motion parameters describe the continuous motion over the whole sequence. Long term motion parameters relatively to the first image are needed if a texture which is taken from a real image sequence shall be mapped on the wire-grid. In this case the texture values are only known for a discrete number of points and the mapping over the sequence requires the interpolation of subpixel values. Mapping from frame to frame would therefore result in a low pass effect reducing the image quality. The accumulation of frame to frame motion parameters to long term motion parameters would lead to an error accumulation. The direct estimation of long term motion parameters overcomes these problems.

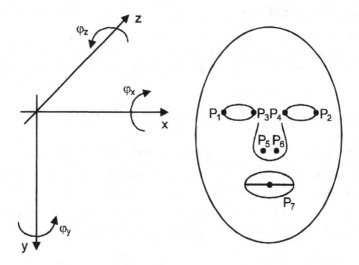

Fig. 1. Control points used for the estimation of the global motion parameters

The co-ordinates of the point in the image plane are denoted as (u, v). Under the assumption of parallel projection (1) can be rewritten as

$$\begin{pmatrix} u \\ v \end{pmatrix} = \begin{pmatrix} r_{11} & r_{12} & r_{13} \\ r_{21} & r_{22} & r_{23} \end{pmatrix} \cdot \begin{pmatrix} \overline{u} \\ \overline{v} \\ \overline{z} \end{pmatrix} + \begin{pmatrix} t_x \\ t_y \end{pmatrix}. \tag{3}$$

The values (u, v) are measured at every frame for a set of points $P_1...P_7$ (see fig. 1) and the values $(\overline{u}, \overline{v})$ are measured at the first frame. The influence of the facial expressions on the selected control points is small and so we can assume that these points participate only in the global motion. In the first frame we adapt a generic wire-grid (we use the a modified CANDIDE model [12]) to the actual face. This is done automatically based on feature extractions. The depth of the reference point $\overline{Z}$ is obtained from the adapted wire-grid at the first frame. We have 14 equations for the 5 unknowns. This oversizing of the estimation problem results in a better stability and robustness against noise.

We define a state vector X is which consists of the motion parameters which have to be estimated

$$X = \left(t_x, \ t_y, \ \varphi_x, \ \varphi_y, \ \varphi_z \right) . \tag{4}$$

Estimations for the real state vector are calculated with the EKF at each timestep. It is assumed that the system dynamics can be described by

$$X_n = X_{n-1} + \mathbf{w}_n . \tag{5}$$

This is an identity transform plus a random vector $\mathbf{w}_n$ that describes the stochastic input to the observed system. The non linear measurement equation is obtained by combining (3) for the seven control points. Formally the measurement process can be described by

$$Y_n = C\left(X_n \right) + \mathbf{v}_n . \tag{6}$$

The vector Y_n consists of the co-ordinates of the seven control points. The operation $C(\ )$ describes a non linear function of the current state vector and the vector $\mathbf{v}_n$ describes the measurement noise. The estimated state vector $\tilde{X}_n$ is calculated at each timestep n by

$$\tilde{X}_n = \tilde{X}_{n-1} + \mathbf{K}_n \cdot \left[Y_n - C\left(\tilde{X}_{n-1} \right) \right] . \tag{7}$$

The Kalman-gain-matrix $\mathbf{K}$ has to be updated at each timestep based on the system dynamics (5), the measurement (6) and a priori knowledge about the statistics of the stochastic variables $\mathbf{w}$ and $\mathbf{v}$. Here the operation $C(\ )$ has to be linearized [10], but it is done around the actual value of $\tilde{X}_n$ and not around $\mathbf{0}$ like in the classical linearizations described above. Therefore arbitrary rotation angles can be allowed. The EKF performs a one step ahead prediction of the system state which is corrected by a term coming from the measurement. In this way temporal redundancies can be exploited and time consistent motion parameters can be obtained.

3 Experiments on synthetic data

Before the experiments on real imagery the EKF was tested with synthetic data to prove the estimation accuracy. The real motion parameters are not known in experiments with real imagery and the estimation error can not be analysed. Therefore motion parameters were generated. We took initial co-ordinates for the control points from the extracted feature points and the adapted wire-grid of the first frame of the sequence "Claire" (see section 5). Thus we used the same geometry in both types of experiments. Then we generated sequences of 1000 sets of control points with equation (3) and the motion parameters shown in fig. 3 and fig. 4. The generated motion parameters are sinusoidals describing a motion with rotations up to 360 degrees and Translations up to more than 70 pels. To simulate the measurement error we added zero mean gaussian noise to the generated control point co-ordinates. The resulting trajectory of one of the noisy measurement points is shown in fig. 2.

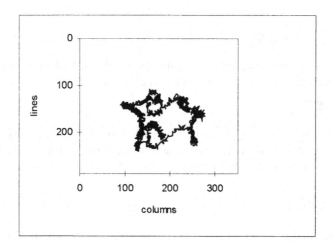

Fig. 2. Trajectory of the generated motion of the control point P_1 (see fig. 1) with noise added

Then we used the EKF-algorithm described in section 2 to recover the motion parameters from the noisy measurement of the control points. The results are shown in fig. 3 and fig. 4 where the generated and the estimated translation and rotation parameters are compared. Both types of motion are tracked very well. The estimation of the large rotation angles is very accurate.

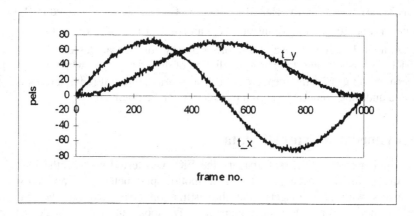

Fig. 3. Generated (smooth line) and estimated translation parameters in the experiments on synthetic data

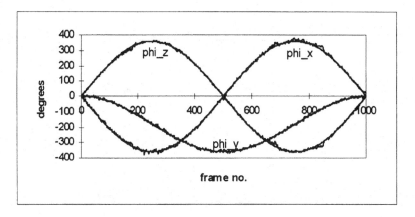

Fig. 4. Generated (smooth line) and estimated rotation parameters in the experiments on synthetic data

4 Experiments on real Imagery

Our EKF-estimator was tested on the sequence "Claire", a typical head-and-shoulder sequence. The co-ordinates of the control points were measured automatically for the first 30 frames of the sequence and the global motion parameters were estimated with the EKF-algorithm described in section 2. The estimated translation and rotation parameters are presented in fig. 5 and fig. 6 respectively and they show a good correspondence to the real motion of the 3D-object in the scene (see fig. 7). E.g. the person "Claire" lifts her head at the beginning of the sequence and then she drops it again. This is described by the parameter φ_x in fig. 6. At the end of the sequence she bends her head to the side which is described by the parameter φ_z in fig. 6.

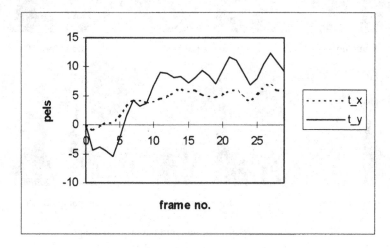

Fig. 5. Estimated translation parameters for the test sequence "Claire"

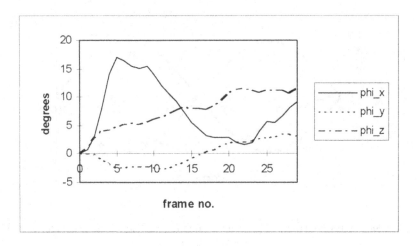

Fig. 6. Estimated rotation parameters for the test sequence "Claire"

In the next step a general wire grid was adapted to the face based on the estimated parameters for every frame by applying (3) to every vertex point of the adapted wire-grid from the first frame. Fig. 7 shows the animated wire grid drawn into the picture for selected frames of the sequence. It can be seen that the transformed wire-grids fit well to the actual faces and that the motion is thus estimated accurately.

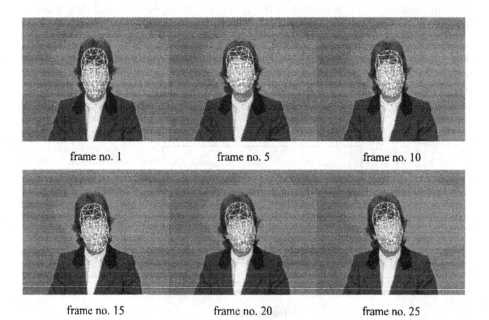

frame no. 1   frame no. 5   frame no. 10

frame no. 15   frame no. 20   frame no. 25

Fig. 7. Adapted wire-grids drawn into the original images

5 Conclusion

We have shown that the proposed EKF-implementation provides good estimations of the motion parameters of a rigid body and can be readily used for applications such as model-based coding or for the animation of synthetic faces in the context of MPEG-4 SNHC. The method is, however, not restricted to face parameters. It could be used for any problem where 3-D motion parameters of a rigid body have to be estimated from a monocular image sequence. Our estimation method is able to deal with large rotation angles and it can exploit the temporal redundancy of the motion due to it's recursive prediction-and-correction structure.

Further extensions could be the simultaneous estimation of the depth structure of the moving objects and the estimation of the translation along the z-axis using a camera model with perspective projection.

References

[1] K. Aizawa, H. Harashima and T. Saito, "Model-Based Analysis Synthesis Image Coding (MBASIC) System for a Person's face", Signal Processing : Image Communication 1, 1989, Elsevier Science Publishers B. V.

[2] C. S. Choi, K. Aizawa, H. Harashima and T. Takebe, "Analysis and Synthesis of Facial Image Sequences in Model-Based Image Coding", IEEE Transactions on Circuits and Systems for Video Technology, Vol. 4, No. 3, June 1994.

[3] H. Li, P. Roivainen and R. Forchheimer, "3-D Motion Estimation in Model-Based Facial Image 'Coding", IEEE Transactions on Pattern Analysis and Machine Intelligence, Vol. 15, No. 6, June 1993.

[4] H. Li and R. Forchheimer, "Two-View Facial Movement Estimation", IEEE Transactions on Circuits and Systems for Video Technology, Vol. 4, No. 3, June 1994.

[5] G. Bozdagi, M. Tekalp and L. Onural, "3-D Motion Estimation and Wireframe Adaptation Including Photometric Effects for Model-Based Coding of Facial Image Sequences", IEEE Transactions on Circuits and Systems for Video Technology, Vol. 4, No. 3, June 1994.

[6] D. Terzopoulos and K. Waters, "Analysis and Synthesis of Facial Image Sequences Using Physical and Anatomical Models", IEEE Transactions on Pattern Analysis and Machine Intelligence, Vol. 15, No. 6, June 1993.

[7] I. Essa, T. Darrel and A. Pentland, "Tracking Facial Motion", Proc. of the IEEE Workshop on Nonrigid and Articulate Motion, Austin, Texas, November 1994.

[8] B. Moghaddam and A. Pentland, "An Automatic System for Model-Based Coding of Faces", Proc. of the IEEE Data Compression Conference, Snowbird, Utah, March 1995.

[9] A. Azarbayejani, T. Starner, B. Horowitz and A. Pentland, "Visually Controlled Graphics", IEEE Transactions on Pattern Analysis and Machine Intelligence, Vol. 15, No. 6, June 1993.

[10] V. Krebs, "Nichtlineare Filterung", R. Oldenbourg Verlag, München, Wien, 1980.

[11] P. Ekman and W. Friesen, "Facial Action Coding System", Consulting Psychologists Press, Palo Alto, California, 1977.

[12] M. Rydfalk, "CANDIDE : A Parametrized Face", Dep. Elec. Eng. Rep. LiTH-ISY-I-0866, Linköping Univ., October 1987.

System for Screening Objectionable Images Using Daubechies' Wavelets and Color Histograms *

James Ze Wang   Gio Wiederhold   Oscar Firschein

Department of Computer Science, Stanford University, Stanford, CA 94305

Abstract. This paper describes WIPE$_{TM}$ (Wavelet Image Pornography Elimination), an algorithm capable of classifying an image as objectionable or benign. The algorithm uses a combination of Daubechies' wavelets, normalized central moments, and color histograms to provide semantically-meaningful feature vector matching so that comparisons between the query image and images in a pre-marked training set can be performed efficiently and effectively. The system is practical for real-world applications, processing queries at the speed of less than 10 seconds each, including the time to compute the feature vector for the query. Besides its exceptional speed, it has demonstrated 97.5% recall over a test set of 437 images found from objectionable news groups. It wrongly classified 18.4% of a set of 10,809 benign images obtained from various sources. For different application needs, the algorithm can be adjusted to show 95.2% recall while wrongly classifying only 10.7% of the benign images.

I   Introduction

Every day, large numbers of adults and children use the internet for searching and browsing through different multimedia documents and databases. Convenience in accessing a wide range of information is making the internet and the world-wide web part of the everyday life of ordinary people. To protect the freedom of speech, people are allowed to publish various types of material or conduct different types of business on the internet. However, due to this policy, there is currently a large amount of domestic and foreign objectionable images and video sequences available for free download on the world-wide web and usenet newsgroups. Accessing objectionable media by under-aged "netters" is increasingly a problem that many parents are concerned about.

I.1   Related Work in Industry

There are many attempts to solve the problem of objectionable images in the software industry. Pornography-free web sites such as the *Yahoo! Web Guides for Kids* have been set up for protecting those children too young to know how to use the web browser to get to other sites. However, there is still a possibility that children may be able to gain access to an objectionable site starting from such a site.

Programs such as *NetNanny* or *CyberSitter* have been put on market for parents to prevent their children from accessing objectionable documents. However, the algorithms used in this software are still in an elementary stage. Some software stores more than 10,000 IP addresses and blocks access to objectionable

---

* *Correspondence to:* wangz@cs.stanford.edu

sites by matching the site addresses. Other software blocks all unsupervised image access. Apparently, there are problems with this software. The internet is so dynamic that more and more new sites and personal pages are added to it everyday. Memorizing many sites is neither efficient in speed nor feasible. Eliminating all images is not a feasible solution since the internet will not be as useful and attractive to children if we do not allow children to view images.

I.2   Related Work in Academia

Academic researchers are also actively investigating alternative algorithms to screen and block objectionable media. Many recent developments in shape detection, object representation and recognition, people recognition, face recognition, and content-based image and video database retrieval are being considered by researchers for use in this problem. A detailed summary of previous work can be found in [8].

To make such algorithms practical for our purposes, extremely high recall with reasonably high speed and precision are necessary. In this application, *recall* is defined as the ratio of the number of objectionable images identified, to the total number of objectionable images in the database; *precision* is defined for a set of images labeled as objectionable as the ratio of images that are actually objectionable to the number of images in the set. A perfect system would identify all objectionable images and not mislabel any benign images, and would therefore have a recall and precision of 1. In the present application, a high recall is desirable, i.e., the correct identification of almost every image likely to be objectionable even though this may result in some benign images being mislabeled. As one can imagine, even a few exposures of objectionable images would be harmful enough for children.

The following properties of the internet objectionable images make the problem extremely difficult:

- mostly contain non-uniform image background;
- foreground may contain textual noise such as phone numbers, URLs, etc;
- may range from grey-scale to 24-bit color;
- may be of very low quality (sharpness);
- taken by all possible camera positions.
- may be an indexing image containing many small icons;
- may contain more than one person;
- persons in the picture may be of different skin colors;
- may contain both people and animals;
- may contain only some parts of a person;
- person in the picture may be partially dressed;
- person in the picture may be fully dressed, but the facial expression is objectionable;
- and so on.

Forsyth's research group [7, 8, 9] has designed and implemented an algorithm to screen images of naked people. Their algorithms involve a skin filter and a human figure grouper. As indicated in [8], 43% recall and 57% precision have been obtained for a test set of 565 images with naked people and 4289 assorted benign images. However, it takes about 6 minutes on a workstation for the grouper in their algorithm to process a suspect image passed by the skin filter.

I.3 Overview of Our Work

Our approach is different from previous approaches. Instead of carrying out a detailed analysis of an image, we match it against a small number of feature vectors obtained from a training database of 500 objectionable images and 8,000 benign images. If the image is close in content to a threshold number of pornographic images, e.g., matching two or more of the marked objectionable images in the training database within the closest 15 matches, it is considered objectionable. To accomplish this, we attempt to effectively code images based on image content and match the query with statistical information on the feature indexes of the training database. The foundation of this approach is the content-based feature vector indexing and matching developed in our multimedia database research.

Image feature vector indexing has been developed and implemented in several multimedia database systems such as the IBM QBIC System [5, 16] developed at the IBM Almaden Research Center. Readers are referred to [10, 13, 17, 18, 21] for details on this subject.

In the WIPE project, we developed a new algorithm to efficiently and accurately compare the semantic content of images mainly consisting of objects such as the human body. Using Daubechies' wavelets, moment analysis, and histogram indexing, the algorithm produces feature vectors that provide excellent accuracy in matching images of relatively isolated objects such as the human body. We use a novel multi-step metric to compute the distance between two given images. A training database of about 500 objectionable images and about 8,000 benign images has been indexed using such an algorithm. When a query comes in, we compute the feature vector and use it to match with the training database. If it matches with objectionable images in the training database, we classify it as an objectionable image. If it does not match with objectionable images in the training database, we classify it as a benign image. Promising results have been obtained in experiments using a test set of 437 objectionable images and 10,809 benign images.

II Related Background

II.1 Content Based Image Indexing and Retrieval

There are several ways to index the images so that queries can be performed by comparing the indexes. The color histogram is one of the many ways to index color images and it preserves the color information contained in images very well. However, a global histogram does not preserve the color locational information within the images. Using this measure, two images may be considered to be very similar to each other even though they have completely unrelated semantics. Shape and texture-based detection and coding algorithms are other techniques for indexing images. They both have substantial limitations for general-purpose image databases. For example, current shape detection algorithms such as the *snake* algorithm only work effectively on images with relatively uniform backgrounds. Texture coding is not appropriate for non-textural images.

Storing color layout information is another way to describe the image content. A wavelet and statistical analysis-based layout indexing technique developed at the Stanford WBIIS project [21] performs best when the images contain a lot of high frequency information such as sharp color changes.

II.2 Daubechies' Wavelets and Fast Wavelet Transform

Wavelets, developed in mathematics, quantum physics, and statistics, are functions that decompose signals into different frequency components and analyze each component with a resolution matching its scale. Applications of wavelets to signal denoising, image compression, image smoothing, fractal analysis and turbulence characterization are active research topics.

Theoretical details on wavelet analysis, wavelet basis and Daubechies' wavelets can be found in [6, 15, 4, 14, 21]. Daubechies' wavelets give remarkable results in image analysis and synthesis due to its mathematical properties.

Daubechies' wavelets transform is more like a weighted averaging which better preserves the trend information stored in the signals if we consider only the low-pass filter part. Various experiments and studies have shown that Daubechies' wavelets are better than other wavelet forms for dealing with general-purpose images.

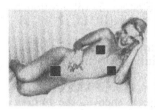  

    Original image   *Traditional edge detection* *Wavelet edge detection*
                      (with post-processing) *(without post-processing)*

Fig. 1. Comparison of the traditional edge detection and our wavelet algorithm. The amount of background noise can be controlled by using different Daubechies' wavelets.

In shape-based image indexing, we want to represent the object shape in the image as exactly as possible in the coefficients of the feature vector. When using the Haar wavelet, we get too much noise in the high-pass bands. Traditional edge detection algorithms have the same problem, as illustrated in Figure 1. Daubechies' wavelets offer (1) a multiresolution analysis, which has the potential for high speed algorithm design, and (2) a wide range of flexibility. For example, we may select the appropriate wavelet basis to obtain the exact amount of fluctuation we desire in the high-frequency bands to represent the object shape.

II.3 Moments

Moments are descriptors widely used in shape and regional coding [11]. For a 2-D continuous surface $f(x, y)$ embedded on the xy-plane, the *moment of order* $(p + q)$ is defined as

$$m_{pq} = \int_{-\infty}^{\infty} \int_{-\infty}^{\infty} x^p y^q f(x, y) dx dy \qquad (1)$$

for $p, q \in \mathbb{N} \bigcup \{0\}$. Theory of moments has shown that the moment sequence $\{m_{pq}\}$ is uniquely determined by $f(x, y)$ and vice versa.

The *central moment* is defined as

$$\mu_{pq} = \int_{-\infty}^{\infty} \int_{-\infty}^{\infty} \left(x - \frac{m_{10}}{m_{00}} \right)^p \left(y - \frac{m_{01}}{m_{00}} \right)^q f(x, y) dx dy. \qquad (2)$$

For discrete cases such as a digitized image, we define the *central moment* as

$$\mu_{pq} = \sum_x \sum_y \left(x - \frac{m_{10}}{m_{00}} \right)^p \left(y - \frac{m_{01}}{m_{00}} \right)^q f(x, y). \tag{3}$$

Then the *normalized central moments* are defined as

$$\eta_{pq} = \frac{\mu_{pq}}{\mu_{00}^\gamma} \quad where \quad \gamma = \frac{p + q + 2}{2} \tag{4}$$

for $p + q = 2, 3, 4, ...$

A set of seven *translation, rotation, and scale invariant moments* can be derived from the 2nd and 3rd moments. A detailed introduction to these moments can be found in [12, 11]. These moments can be used to match two objectionable images containing people having the same posture but taken from different camera angles.

III   Screening Algorithm in WIPE

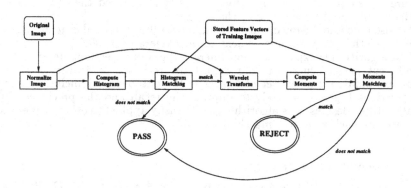

Fig. 2. Basic structure of the algorithm.

III.1   Overview

We have developed a new shape-based indexing scheme using forward and backward Daubechies' wavelet transforms, variant and invariant normalized central moments, and color histogram analysis that is able to capture the object posture. The screening algorithm uses several major steps. Figure 2 shows the basic structure.

We apply a 1-level fast wavelet transform (FWT) with Daubechies-3 wavelet to each image in the training set of objectionable and benign images. Then we perform an inverse wavelet transform on each of the three high frequency blocks obtained from the wavelet transform using the same wavelet basis. Multidirectional edge detection is then applied to the three inverse transforms to obtain fluctuation information stored in the original image. The normalized central moments of the edge image, invariant moments of the edge image and the color histogram of the rescaled (R, G, B)-space image are stored as a feature vector.

Given a query image, the search is carried out in three steps after the feature vector for the query is computed. In the first step, the color histogram of the query is used. If the amount of yellow in the query image is lower than a threshold, we classify it as a benign image since we have found that objectionable images contain large areas of yellow or colors close to yellow. In the second step, a crude selection based on the individual moments stored is carried out. Then a weighted version of the Euclidean distance between the moments of an image in the training set selected in the first step and those of the querying image is calculated, and the images with the smallest distances are selected and sorted. In the third step, a color histogram is again used to perform the final match so that images with both human body shape and skin color histogram will be selected as matching images to the query. If one or more objectionable images is found in the best matching images, we mark the query as an objectionable image. Otherwise, the query is marked as a benign image.

Our design has several immediate advantages.

1. It does not rely too much on color when detecting sharp edges. That means that naked people of different races can be detected without bias. It also has the potential for shape-based matching of benign images. Image background does not affect the querying results unless the background is not reasonably smooth. Also, the query image can be of different color quality.

2. We used Daubechies' wavelet rather than a traditional edge detector to capture the shape information in the images. This reduced the dependence on the quality or the sharpness of the images.

3. We used a combination of variant and invariant normalized central moments to make the querying independent of the camera position.

4. Our algorithm is expandable for different types of objectionable images and different needs. Queries in our algorithm are based on feature indexes of an expandable training set of images in the database. By including images of various types, it is likely to improve the recall and the precision of the algorithm. Besides, this algorithm can be easily applied to other shape-based searching such as object matching.

III.2   Normalize the Images

Many color image formats are currently in use, e.g., GIF, JPEG, PPM and TIFF are the most widely used formats. Because images in an image database can have different formats and different sizes, we must first normalize the data for histogram computation. For the wavelet computation and moment analysis parts of our algorithm, any image size is acceptable.

A rescaled thumbnail consisting of 128×128 pixels in Red-Green-Blue (i.e., RGB) color space is adequate for the purpose of computing the histogram feature vectors.

III.3   Color Histogram Indexing and Matching

It is clear to us that objectionable images contain mostly large areas of yellow or colors close to yellow. This is not the case for benign images.

The color histogram indexing and searching techniques used in the WIPE system are fairly standard when compared to systems such as the IBM QBIC System. We use a total of 512 bins to compute the histogram.

A much more efficient approach is implemented. We manually define a certain color range in the color spectrum as yellow. If we define

$$yellow(r, g, b) : [0, 255] \times [0, 255] \times [0, 255] \rightarrow [0, 1] \qquad (5)$$

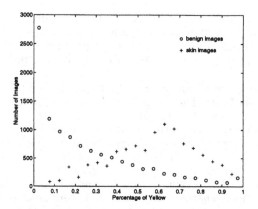

Fig. 3. Percentage of yellow in benign images and skin images.

as the weight of yellow for a certain color (r, g, b), then $yellow(r, g, b) = 1$ means the color is the most yellow color and $yellow(r, g, b) = 0$ means the color is the least yellow color. Denote

$$histogram(r, g, b) : [0, 255] \times [0, 255] \times [0, 255] \to \mathbb{R}+ \qquad (6)$$

as the histogram of a given image. Then the function

$$YLW(image) = \sum_{r=0}^{255} \sum_{g=0}^{255} \sum_{b=0}^{255} \left(yellow(r, g, b) \times histogram(r, g, b) \right) \qquad (7)$$

indicates the weighted amount of yellow that the given image contains.

For simplicity, we set

$$yellow(r, g, b) = \begin{cases} 1 & (r, g, b) \text{ close to yellow} \\ 0 & \text{otherwise} \end{cases} \qquad (8)$$

Figure 3 shows histograms of the values of $YLW(image)$ for benign images and skin images. If we set a threshold of, say, 0.15, about one half of the benign images are then classified correctly, while only a small amount of skin images are classified incorrectly.

Using this approach, the run time of the histogram matching step for skin images has been reduced from $O(n)$ to $O(1)$, where n is the number of images in the training database.

III.4 Edge and Shape Detection and Matching Using Wavelets and Moments

Clearly the color histogram approach alone is not sufficient. Sometimes two images may be considered very close to each other using this measure when in actuality they have completely unrelated semantics.

We apply the wavelet transform to perform multidirectional and multiscale edge detection. Readers are referred to [1] for the theoretical arguments on the effectiveness of a similar algorithm. Our purpose is not to obtain a high quality edge detection algorithm for this application. Rather, since the goal here is to effectively extract the main shape information for objects and textural information for areas from the image, it is not necessary to produce a perceptually pleasant edge image. Consequently, we try to keep the algorithm simple to achieve a fast computation speed.

We start the edge detection process by transforming the image using the Daubechies-3 wavelet basis. The image is decomposed into four frequency bands with corresponding names LL, HL, LH and HH. The notation is borrowed from the filtering literature [20]. The letter 'L' stands for low frequency and the letter 'H' stands for high frequency. The left upper band is called 'LL' band because it contains low frequency information in both the row and column directions. The details of the filtering terminologies are given in [20]. An even number of columns and rows in the querying image is required due to the downsampling process of the wavelet transform. However, if the dimensions of the image are odd, we simply delete one column or one row of pixels from the boundaries.

The LH frequency band is sensitive to the horizontal edges, the HL band is sensitive to the vertical edges, and the HH band is sensitive to the diagonal edges [4, 1]. This property has been used in a textual information detection and elimination system designed for secure medical image distribution [22].

We detect the three types of edges separately and combine them at the end to construct a complete edge image. To detect the horizontal edges, we perform an inverse Daubechies-3 wavelet transform on a matrix containing only the wavelet coefficients in the LH band. Then we apply a zero-crossing detector in vertical direction to find the edges in the horizontal direction. The mechanism for using zero-crossing detector to find the edges can be found in [1]. Similar operations are applied to the HL and HH band. The difference lies with the zero-crossing detector. For the HL band, we use a zero-crossing detector in the horizontal direction to find vertical edges and for the HH band, we use zero-crossing detector in the diagonal direction to find diagonal edges.

After we get the three edge maps, we combine them to get the final edge image. To numerically show the combination, let us denote[2] the three edge maps by $E_1[1:m, 1:n], E_2[1:m, 1:n]$ and $E_3[1:m, 1:n]$. The image size is $m \times n$. Then the final edge image, denoted by $E[1:m, 1:n]$, can be obtained from

$$E[i,j] = \left(E_1[i,j]^2 + E_2[i,j]^2 + E_3[i,j]^2\right)^{\frac{1}{2}}. \qquad (9)$$

Once the edge image is computed, we compute the normalized central moments up to order five and the translation, rotation, and scale invariant moments based on the gray scale edge image using the definitions in Section II.3. A feature vector containing these $21 + 7 = 28$ moments is computed and stored for each image in the training database. When a query comes in that has passed the histogram matching step, a moment feature vector is computed and a weighted Euclidean distance is used to measure the distance of the query and an image in the training database. The weights are determined so that matching of the 21 normalized central moments has higher priority than the matching of the 7 invariant moments. In fact, many objectionable images are of similar orientation.

If the query matches with objectionable images in the training database, we classify it as an objectionable image. If it does not match with objectionable images in the training database, we classify it as a benign image.

---

[2] Here we use MATLAB notation. That is, $A(m_1 : n_1, m_2 : n_2)$ denotes the submatrix with opposite corners $A(m_1, m_2)$ and $A(n_1, n_2)$.

IV   Experimental Results

Type of Images (total)	Eliminated (percentage)	Passed WIPE (percentage)	Eliminated (percentage)	Passed WIPE (percentage)
Pornographic (437)	426 (97.5%)	11 (2.5%)	416 (95.2%)	21 (4.8%)
Benign (10809)	1993 (18.4%)	8816 (81.6%)	1155 (10.7%)	9654 (89.3%)
	Configuration A		Configuration B	

Table 1. Overall Classification Performance of the WIPE System.

This algorithm has been implemented on a Sparc-20 workstation. We selected about 500 objectionable images from news groups and 8,000 benign images from various sources such as the Corel Photo CD-ROM series for our training database. When we downloaded the objectionable images, we tried to eliminate those from the same source, i.e., those of extremely similar content. To compute the feature vectors for the 8,000 color images of size 640 × 480 in our database requires approximately 3 hours of CPU time.

We have also selected 437 objectionable images and 10,809 benign images as our queries. The matching speed is very fast. Using a SUN Sparc-20 workstation, it takes less than 10 seconds to process a query and select the best 100 matching images from the 8,500 image database using our similarity measure. Once the matching is done, it takes almost no extra CPU time to determine the final answer, i.e., if the query is objectionable or benign.

Besides the fast speed, the algorithm has achieved remarkable accuracy. Table 1 shows the overall classification performance of two system configurations of the WIPE system. In System Configuration A, we set the yellow color threshold to 10% in the color matching and the the minimum number of objectionable images within the top 15 best matching images for the shape matching stage to 1. That is, the query passes the WIPE system only if it contains less than 10% pre-defined yellow or else it does not match any objectionable image within the top 15 best matching images. Otherwise, it is rejected by the WIPE system. In System Configuration B, we set the two thresholds to 10% and 3. This allows many more benign images to pass the WIPE system correctly, but somewhat more objectionable images are mistakenly allowed. Choosing a more constraining filter, i.e., more protective for children, will cause a greater loss of valid images, but that trade-off may still be worthwhile for some parents. The algorithm can be set up so that the parent user can set the level of protection.

Figure 5 and 6 show typical images being mistakenly marked by the WIPE system. Most of the few failures for marking objectionable images happen when the query contains textual noise and/or a frame, as in Figure 6.a.

V   Conclusions and Future Work

In this paper, we have demonstrated an efficient shape-based indexing and matching system using Daubechies' wavelets developed by us for screening objectionable images.

It is possible to improve the search accuracy by fine-tuning the algorithm, e.g., using neural network, using a perceptually-comparable color space, adjusting weights for different matching steps, or adding more complicated preprocessing to eliminate textual noise or frames. The accuracy may also be improved by fine-tuning the training database.

Fig. 4. Animal images are sometimes mistaken for objectionable images. Some areas of objectionable images are blackened and blurred.

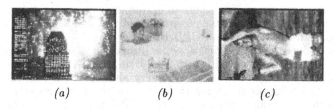

(a)　　　　　　　(b)　　　　　　　(c)

Fig. 5. Typical benign images being marked mistakenly as objectionable images by WIPE. (a) too much yellow (b) hard to tell (bathing) (c) partially undressed.

It is also possible to make the searching faster by developing a better algorithm for storing and matching the feature vectors. Significant speed-up is also possible if a more extensive statistical analysis can be utilized. The algorithm can also be modified to execute in parallel on multi-processor systems. Experiments with our algorithm on a video database system could be another interesting study.

Finally, we are working on applying this technique to shape-based search of regular image databases containing, for example, merchandise catalog images and museum art images. It is also possible to extend this technique to texture-based search.

(a)　　　　　　　(b)　　　　　　　(c)

Fig. 6. Typical objectionable images being marked mistakenly as benign images by WIPE. (a) frame and textual noise (b) dressed but objectionable (c) image too dark and of extremely low contrast. Some areas of objectionable images are blackened and blurred.

VI Acknowledgements

We would like to thank Jia Li of Stanford University Electrical Engineering Department for useful discussions and valuable help in implementing the algorithm. We would also like to thank the Stanford University Libraries and Academic Information Resources (SULAIR) for providing computer equipment during the development and testing process.

References

1. Turgut Aydin et al, Multidirectional and Multiscale Edge Detection via M-Band Wavelet Transform, *IEEE Transactions on Image Processing*, Vol. 5, No. 9, pp. 1370-1377, 1996.
2. Dana H. Ballard, Christopher M. Brown, *Computer Vision*, Prentice-Hall, Inc., New Jersey, 1982.
3. Ingrid Daubechies, Orthonormal bases of compactly supported wavelets, *Communications on Pure and Applied Mathematics*, 41(7):909-996, October 1988.
4. Ingrid Daubechies, *Ten Lectures on Wavelets*, CBMS-NSF Regional Conference Series in Applied Mathematics, 1992.
5. C. Faloutsos et al, Efficient and Effective Querying by Image Content, *J. of Intelligent Information Systems*, 3:231-262, 1994.
6. G. B. Folland, *Fourier Analysis and Its Applications*, Pacific Grove, Calif., 1992.
7. Margaret Fleck, David A. Forsyth, Chris Bregler, Finding Naked People, *Proc. 4'th European Conf on Computer Vision*, 1996.
8. David A. Forsyth and Margaret Fleck, Finding Naked People, *journal reviewing*, 1996.
9. David A. Forsyth et al, Finding Pictures of Objects in Large Collections of Images, *Proceedings, International Workshop on Object Recognition*, Cambridge, 1996.
10. Amarnath Gupta and Ramesh Jain, Visual Information Retrieval, Communications of the ACM, vol.40 no.5, pp 69-79, 1997.
11. Rafael C. Gonzalez, Richard E. Woods, *Digital Image Processing*, Addison-Wesley Publishing Co., 1993.
12. M. K. Hu, Visual Pattern Recognition by Moment Invariants, *IRE (IEEE) Trans. Info. Theory*, Vol. IT-8, pp. 179-187.
13. C. E. Jacobs, A. Finkelstein, D. H. Salesin, Fast Multiresolution Image Querying, *Proceedings of SIGGAPH 95, in Computer Graphics Proceedings, Annual Conference Series*, pp.277-286, August 1995.
14. Gerald Kaiser, *A Friendly Guide to Wavelets*, Birkhauser, Boston, 1994.
15. Yves Meyer, *Wavelets: Algorithms & Applications*, SIAM, Philadelphia, 1993.
16. W. Niblack et al, The QBIC project: Query image by content using color, texture and shape, *Storage and Retrieval for Image and Video Databases*, pages 173-187, San Jose, 1993. SPIE.
17. R. W. Picard, T. Kabir, Finding Similar Patterns in Large Image Databases, *IEEE ICASSP*, Minneapolis, Vol, V., pp.161-164, 1993.
18. A. Pentland, R. W. Picard, S. Sclaroff, Photobook: Content-Based Manipulation of Image Databases, *SPIE Storage and Retrieval Image and Video Databases II*, San Jose, 1995.
19. M.J. Swain and D. H. Ballard, Color Indexing, *Int. Journal of Computer Vision*, 7(1):11-32, 1991.
20. Martin Vetterli, *Wavelets and Subband Coding*, Prentice Hall, N.J., 1995.
21. James Ze Wang, Gio Wiederhold, Oscar Firschein, Sha Xin Wei, Wavelet-Based Image Indexing Techniques with Partial Sketch Retrieval Capability, *Proceedings of the Fourth Forum on Research and Technology Advances in Digital Libraries (ADL'97)*, Washington D.C., May 1997.
22. James Ze Wang, Michel Bilello, Gio Wiederhold, Textual Information Detection and Elimination System for Secure Medical Image Distribution, *submitted for conference publication*, March 1997.
23. Gio Wiederhold, Digital Libraries, Value, and Productivity, *Com. ACM*, Vol.38 No.4, April 1995, pages 85-96.

MUSIC: an interactive MUltimedia ServIce Composition environment for distributed systems.

Piergiorgio Bosco, Giovanni Martini, Giovanni Reteuna

CSELT - Via G.Reiss Romoli 274, 10148 - Torino - Italy

{Piergiorgio.Bosco,Giovanni.Martini}@cselt.it

Keywords: Distributed Multimedia Applications Development, IMA, CORBA, TINA

Abstract - The objective of MUSIC is to devise a development environment for the composition of distributed multimedia applications over CORBA platforms, that uses a graphical model to support their specification. A movie-like approach has been adopted to assemble both processing service objects and physical networked devices.

1. Introduction

The distributed multimedia systems combine in an integrated way the advantages of the distributed elaboration [13] and the possibility of manipulating discrete multimedia information (text, images) with continuous information (audio, video) [3]. It is reality an Intranet/Internet scenario where the components of a service, the processing objects and the physical devices, located in different nodes, are shared among different applications and users. Conversely, this fact imposes a composition process that is more complex than in the case of stand-alone system [16]. It is required to go further the actual multimedia "microworlds" [3], i.e. self-contained and platform dependent development systems, whose design and evolution is under the control of a single vendor, and that act as a proprietary middleware between applications and low-level system functions. In these situations, processing objects and multimedia physical peripherals can be located on distinct nodes. Proper abstractions are also required for supporting platform independent media devices, and the work carried out in this direction by the Interactive Multimedia Association (IMA) [1] and by the ISO PREMO standard [2] is of the most valuable significance. In this scenario, applications are combined by selecting and plugging-in components from a pool of already developed elements, with no problems of generality and portability conflicts. Additional efforts are also undertaken in this direction, but at a low infrastructural level by OMG [11].

The objective of the MUSIC (MUltimedia Service Composition) project is to devise a multimedia services composition prototype that by means of an intuitive graphical approach, allows to define the main necessary steps for the composition and the customisation of multimedia systems. The distributed platform referenced by MUSIC is OMG CORBA [10], although the processing elements are defined according to the TINA[1] computing architecture [17]. For their definition the TINA ODL (Object Definition Language) [18] has been adopted. The services composed by MUSIC allow dynamic binding at runtime, and make use of platform services (e.g. Trader, Event

---

[1] *TINA is an effort carried by a world-wide consortium to overcome current network limitations in the provisioning of a service (http://www.tinac.com).*

Service) in order to exchange interface references and for synchronisation activities. The operations among the objects composing a new service are describe by means of a sequence of snapshots, i.e. visual representation of the state of the multimedia flows at a certain time. Proper synchronisation rules apply in order to control the execution of the sequence of snapshots. The composition of a new service is basically made up of the creation and management of a proper sequence of snapshots, composing a final storyboard [8]. A movie-like approach has been adopted, because of its expressiveness, of easy comprehension for the user and because it allows coordination of the "time" dimension. The generated code is a JAVA applet [15], as well as the MUSIC prototype.

2. Related Works

The issue of (distributed) multimedia service composition is relatively new in literature, being many proposals more oriented to standalone multimedia authoring systems [3]. In MUSIC it has been adopted the same approach proposed for the IMA Multimedia System Services (MSS) [1] and being adopted in PREMO [2]: all the control interfaces of all the service components are defined in CORBA IDL. MSS define a synchronisation model of the actions based on the timeline diagrams, very effective but loosely coupled with the graphical composition model. Another interesting research project is CINEMA [2]. It adopts component and nested objects that are conceptually similar to the correspondent entities defined in the TINA computational model [17]. Moreover, in CINEMA a script is eventually generated as a result of the service composition activity. Similar issues are also considered in SAW (Software Assembly Workbench) [7]. SAW is an ATM-based platform for rapid generation of multimedia applications that adopts a graphical model. The expressiveness of the visual programming approach is thereby illustrated, with the possibility of simulating the single service components.

3. MUSIC: the project

One of the main issue in literature is the lack of the adoption for a standard solution for the representation of the objects and for their retrieval, in favour of different proprietary solutions. In addition, nevertheless the usage of graphical formalisms and of visual programming [14] approaches, the definition in temporal diagrams of the object interactions and connection set-ups is not completely tackled. In MUSIC the sequence of events is expressed by means of a proper graphical formalism (the movie-like paradigm, section 4), while the composition issues are coped with embracing well defined and adopted standards: CORBA [10] for the distributed aspects, JAVA [15] for the code mobility and TINA [17] for the definition of the single components and for an overall open architecture. The envisaged user of MUSIC is a business consultant who specifies new multimedia services by assembling existing (processing and device) objects, in order to deploy them on an open system.

3.1 The reference object model

In MUSIC the service components are represented by objects defined according to the TINA computational model [17]. The objects support two types of multiple dynamic interfaces: computational and stream. The computational are simply mapped onto CORBA interfaces, and CORBA IDL is used for their specification; while the stream interfaces are used for abstracting , and for specifying flows of data. They contain a certain number of end-points identifying the multimedia stream flows (sinks and sources).

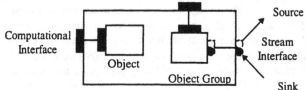

Figure 1: TINA objects and object group.

The objects and their interfaces are described by means of TINA ODL (Object Definition Language) [18] (regarded as an extension of IDL), and proposed to the MUSIC user by means of a graphical representation (Figure 1).

3.2 The basic multimedia components

In MUSIC the TINA (CORBA) objects represent the processing components of a service, whereas the physical devices are depicted as predefined terminal objects. A particular set of icons allows the user to identify the endpoints of a multimedia connection (see Figure 2). They do not intend to be exhaustive, but only placeholders for more precise and detailed hierarchical definitions (we refer to the MSS definitions as in IMA [1] and in PREMO [2]). The concept of port [1], through which a device receives/transmits media data is related to the one of stream endpoint (sink/source) of TINA objects.

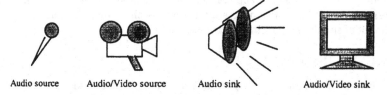

Audio source   Audio/Video source   Audio sink   Audio/Video sink

Figure 2: Terminal multimedia objects in MUSIC.

3.3 Code generation

The output of the composition activity is the generation of code that is able: (1) to dynamically retrieve the interface references of each TINA object and for the physical devices located on the network nodes (e.g. by means of the Trader); (2) to execute the operations indicated for the proper customisation and configuration of the service components; (3) to invoke operations offered by a platform API for the setting-up of the connections among the stream end-points of the participating objects. The generated code is a JAVA applet (in order to provide mobility and portability) although it is only a skeletal file, customisable by the user. The availability of JAVA CORBA brokers (e.g. OrbixWeb [12]) and the availability of CORBA-enabled browsers is therefore a basic factor for devising open distributed multimedia systems.

4. Actions synchronisation by means of snapshots

In MUSIC, the approach for interactively representing the whole composition of a multimedia service is based on the movie-like paradigm [8]. By means of a "storyboard" the user (intended as the business consultant, and not the final user of the service) illustrates all the necessary steps to be executed in order to set-up a new multimedia service.

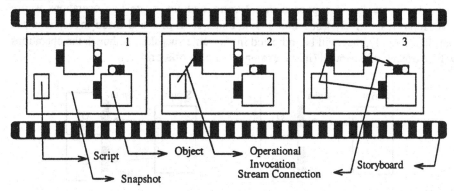

Figure 3: The movie-like composition paradigm in MUSIC.

Eventually, a sequence of snapshots for the new service will indicate the proper phases, e.g. starting from the retrieval of the composing objects, and indicating the proper (operational and stream) dependencies among them.

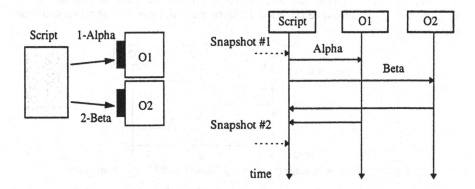

Figure 4: Snapshot #1 (parallel case) and corresponding temporal diagram.

The advantage of such a paradigm relies on the expressive merging of spatial relationships (e.g. representing connections to be set-up) with temporal dependencies (expressed by the sequence of snapshots) occurring among objects. The temporal information is large-grained, regarding only interactions among objects. It doesn't deal with fine-grained temporal relationships and synchronisations, that are already defined within the processing objects and that are better represented with other approaches (e.g. [4],[5]). The semantics of the underlying adopted model is the following one: all the operations represented within the same snapshot (numbered for the sake of clarity) are intended to be executed in parallel, whenever it's possible(Figure 4). The (n+1)th operation will be activated after the start of the n-th operation, but not necessarily before its completion. In case a strict succession has to be followed (i.e. a certain operation is invoked only after the completion of another one), multiple snapshots must be used. The operations within a snapshot will be started only when all the invocations started in the previous snapshot are completed.

In Figure 4 it is supposed that Alpha and Beta are two operations whose invocations respectively last 5 and 3 time units. If Alpha and Beta invocations can start indifferently, they will be indicated in the same snapshot, although the operations will be executed in parallel (in the example, Alpha has priority).

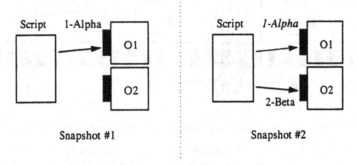

Figure 5: Snapshots #1 and #2 (sequential case).

A different situation can be devised in another scenario (sequential). If the invocation of Alpha must be completed before the invocation of Beta two snapshots must be used.

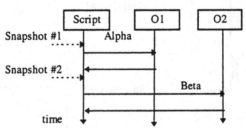

Figure 6: Temporal diagram for snapshots #1 and #2 (sequential case).

From the previous diagrams (Figure 5, where the invocation of Alpha is in italics to indicate that it has been previously defined) the Figure 6 temporal diagram can be drawn, where it is highlighted that Beta can be invoked only after the completion of Alpha. MUSIC offers the possibility of defining, for each snapshot, auxiliary synchronisation conditions, and once that they are satisfied, the invocation of the snapshot can start. These conditions (showed as clock-shaped icons) are expressed as boolean predicates, referring to local variables declared in a previous snapshots or to any application event. A possible usage is to synchronise the start of the actions of a snapshot with any application generated event (e.g. Start of the multimedia flow). The synchronisation conditions of the initial snapshot are the start-up ones of the service.

5. MUSIC: the composition model

The MUSIC composition model makes intensely use of the features of the visual programming [14]. The components (software modules and devices) are represented by icons and the links among them identify physical connections and object interactions.

5.1 The connections types

Three types of connections are used for modelling object interactions: operation invocations, parameters passing and stream connections.

Operation invocations

This type of connection represents an invocation of an operation offered by a computational interface of an object. The types of invocations have been divided in two additional categories: descriptive and executive invocations.

Descriptive: it's a normal operation invocation between two service components. The user knows of such an interaction by consulting the textual behavioural descriptions for each object (from the ODL specifications) and graphically depicts such kind of interaction for the sake of the comprehension of the service. This kind of representation doesn't generate final code, being the corresponding interaction already defined within the object behaviour. This type of representation is mainly due to the actual lack of a formalism for expressing the basic semantics of an object behaviour.

Executive: it's a normal operation invocation, comprising all the interactions that have to be carried out by the final user in order to properly control the composition of a service. The interface references are retrieved at run time (i.e. when eventually the service is being executed) from the Trader. MUSIC generates in the final script the proper code for these operations. Operation invocations are depicted by continuos arrowhead links, with associated labels identifying the progressive number of the operation invocation and the name of the target interface and operation. The arrow that characterises an executive invocation starts from a particular icon representing the final script.

An example of a simple interaction schema is hereby reported. It is supposed that two objects (A and B) offer an operational interface each one. At the time of the service execution, after having retrieved (e.g. from the Trader) the references to their, it must be invoked the operation "Start" of the object A. In turns the operation A invokes the operation "Service" of the B object: it is supposed that this behaviour is textually described in the ODL of the object A.

```
Object A {                          Object B {
   interface i1 {                      interface i2 {
   //operations:                       // operations:
     Start ()                            Service()
     behaviour:  {                    }
     type-Bobject.service();  }
     ....
   }
 }
 }
```

Figure 7: ODL specifications for class A and B.

In the MUSIC model, the graphical representation corresponding to the situation of Figure 7 is reported in Figure 8, where the invocation of the Start operation is an execution, and it is necessary to indicate it because it allows the definition of the service. On the contrary, the invocation of the operation Service from A to B is descriptive, and its purpose is to illustrate which operations the object A will execute after having been activated. The generated code contains only the Start[2] operation: ...;i1.Start();...

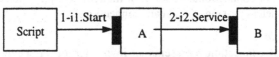

Figure 8: Example of operation invocations.

The code of all the examples is automatically generated by MUSIC: JAVA primitives (OrbixWeb[12]), with invocations of CORBA interfaces.

Parameters passing

The parameters passing completes the scenarios of the possible operational interactions.

Descriptive parameters passing: the user wants to represent that a given reference is passed to another as a parameter (e.g. due to an operation invocation). Anyway, MUSIC doesn't generate code in this case, as this kind of interaction is implicitly defined within the objects' behaviour. The parameters passing is indicated by dotted arrowhead lines. In the example of Figure 9 the MUSIC user has also specified that a reference to a B's interface is being passed as parameter to the object A. It is possible to add textual information to the dotted line, to indicate that the B instance value will be requested at runtime from A (e.g. using a dialog box). The code is: ...;i1.Start();...

Executive parameters passing: it is a special case, when a reference is passed as parameter to an operation invoked from the outside in order to compose the service. The operation invocation is usually executive and also the corresponding parameters passing must be made explicit in the final generated script. This type of parameters passing will therefore produce generated code.

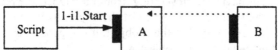

Figure 9: example of generic parameters exchange.

---

2 *The code for retrieving the interface offered by A and B will be also produced; for each interface, an Import() invocation will be generated.*

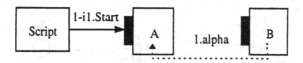

Figure 10: example of parameters exchange in an operation.

The arrow labelled "1-i1.Start" indicates that the operation Start must be invoked on the interface of type i1 offered by the object A, while the arrow directed from B towards A (labelled "1.alpha") indicates that the reference to B is being passed as the actual parameter "alpha" to the operation number 1 invoked on A ("alpha" is the name of one of the formal arguments of Start of interface i1).The code is: ...;A.Start(B);...

Stream connections

The endpoints offered by the stream interfaces may be bound together only if they are compatible (i.e. an MPEG source may be connected only with an MPEG sink, not with an audio source). In MUSIC, the compatibility among endpoints follows the rules of subtyping defined in [17], and this means that two endpoints are compatible when: (1) they have the same type and (2) they have two different types, but subtypes of a common type. The type of an endpoint is obtained from the ODL specifications. A connection between two endpoints doesn't imply that the data stream flows from one object to the other, but only that this can happen at runtime as a consequence of the connections specified by the users. The effective control of the stream (e.g. by means of Start, Stop, operations) is left to the user of the final application. MUSIC, supports only the composition of a new service, and not the service specific details of its usage (that are contained in the processing objects). The code generated for a stream connection is mapped onto an invocation of a proper binding operation. The references of the interfaces are passed as parameters. The binding operation is provided by the connection management service of the underlying platform, responsible for the set-up of a physical connection between endpoints (terminal points of the network) with the proper QoS values.

Figure 11 illustrates an audio data flow starting from a physical source (represented by a microphone icon) and reaching a sink endpoint of a stream interface of object A. This object will process the input flow (e.g. compress the signal) and will generate an output flow that will transit in the connection A-B, that in turn, will do its own processing and eventually the result is transmitted into an audio sink. As indicated, the stream connections may be drawn from sources and sinks of TINA objects, likewise terminal device objects can take part in the connection. The final code for the example of Figure 11 is:

```
StreamBinding(device-mike,objA-name_itf-sink);
StreamBinding(objA-name_itf-source,objBname_itf-sink);
StreamBinding(objB-name_itf-source,device-speakers);
```

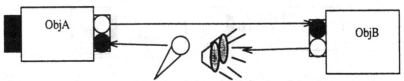

Figure 11: Example of audio flows connections.

The StreamBinding() operation is offered by the Connection Management service of the underlying platform. It is out of the scope of this work to deep into the functional details of a connection management service. The StreamBinding() operation is hereby indicated as an abstraction of the functionalities offered by the platform for setting up proper network connections. It is possible to take into account QoS parameters associated to each stream (extracted from the ODL specifications). The QoS values may be passed to the StreamBinding() function in the generated code.

5.2 The generated code

The user can generate a JAVA applet with skeleton definitions (to be refined) for the methods that are broadly defined in this way:

- "`init`": This method (executed by the browser the first time the applet is downloaded) contains the operations (a sequence of `Import()` to the Trader), for the retrieval at run time of the involved objects (processing and physical devices).

- "`start`": In this method (executed each time that the page containing the applet is exposed) all the operations related to the connections among objects are reported, and in particular: (1) invoking operations on interfaces; (2) expressing the actual values of the formal parameters passed to the invoked operations; (3) invoking the Connection Management `StreamBinding()` operation (5.1). Moreover, for each synchronisation condition, a proper waiting instruction is generated.

- "`stop`", "`destroy`": They are not generated, although the user can add any proper additional operations related to the termination of the service.

6. Future Works and Conclusions

The illustrated concepts and ideas constitute part of our current activity, and a first prototype (integrated with OrbixWeb[12]) has been produced, although some additional work is still needed. Next evolution will comprise a tight integration with tools supporting behavioural specifications (e.g. ACE [9]), as well the usage of platform services (e.g. Event Service).

7. References

[1]IMA, *Multimedia System Services*, Draft Recommended Practice, 1995

[2]ISO, *Presentation Environments for Multimedia Objects:Part 1*, 1996

[3]M.Mühlhäuser et al., *Services, Frameworks, and Paradigms for Distributed Multimedia Applications*, IEEE Multimedia, Fall 1996

[4]S.Gibbs, *Composite Multimedia and Active Objects*, OOPSLA'91, 1991

[5]S.Gibbs et a., *Multimedia Programming Objects*, Addison-Wesley,1994

[6]K.Rothermel, I.Barth, T.Helbig, *CINEMA-An Architecture for Configurable Distribuited Multimedia Applications*, Universität Stuttgart, 1994

[7]E.R.Beyler et al., *An ATM-Based Platform for Rapid Generation of Multimedia Applications*, AT&T Technical Journal, 1995

[8]P.G.Bosco, *Requirements for Multimedia Service Creation, CSELT*, 1996

[9]P.G.Bosco, G.Martini, C.Moiso, *ACE: An environment for specifying, developing and generating TINA services*, IEEE/IM'97, May 1997

[10]OMG, *The Common Object Request Broker: Architecture and Specification*, 1995

[11]OMG,*Control and Management of A/V Streams RFP*, OMG n. 96-08-01

[12]IONA, *OrbixWeb Programming Guide*, 1996

[13]R. Orfali et al., *The Essential Distribuited Objects*, Wiley, 1996

[14]M.M.Burnett et al., *Visual Object-Oriented Programming*, Manning, 1995

[15]J.Gosling, H.Mc Gilton, *The JAVA Language Environment*, Sun, 1996

[16]TINA-C, *Service Composition Problem Statement*, 1996

[17]TINA-C, *Computational Modelling Concepts, vers.3.2*, 1996

[18]TINA-C, *Object Definition Language Manual*, 1996

Achieving Networked Virtual Environments Interoperability

Michel Soto and Hubert Lê Văn Gông

Laboratoire LIP6 – Université Pierre et Marie Curie
75252 Paris Cedex 05, FRANCE

Abstract. The interoperability between heterogeneous distributed virtual environments is one of the main problem Virtual Reality will have to face in the very close future. This article presents the main concepts we propose in NOVÆ (which stands for Networked Open VirtuAl Environment). These concepts are presented as a set of paradigms based on the semantic of a virtual world and its virtual entities.

Key-words : Networked Virtual Environments, Interoperability, Cooperative Work, Object Modeling.

1 Introduction

As the range of existent applications for synthetic environment (SE) is rapidly growing, the will of being able to interact between different users on different SEs will dramatically increase.

It is now generally accepted that multi-user networked SEs constitute a major area of interest. SEs allow users located in different geographical areas to jointly perform complex tasks. Thus, networked SEs can be applied to almost all the human domains of activities.

Today, there are several existing networked SEs and numerous other ones will appear in the close future. Current existing networked SEs can be considered as closed in the way that users of two distinct SEs are not able to work together. This situation is no longer desirable and constitutes a paradox while the World is becoming a global village where information highways will link the main places.

New networked SEs must allow users to cooperate freely or, in a more general way, to interoperate with other networked SEs. Meanwhile already existing SEs have to evolve taking into account this new requirement. Some efforts have already been done in this direction. The DIS protocol is a good example of what have been obtained in this topic of research [DIS 1993]. Nevertheless, to be successful on the way of interoperability, we believe in the need of new paradigms and in the consolidation of existing ones. Taking into account the interoperability requirement immediately raises the problem of heterogeneity. We sometimes can restrict heterogeneity but never avoid it. Therefore, successful paradigms aiming at interoperability must take it into account. Heterogeneity appears at any level between distinct SEs: hardware, software, operating system, communication stack protocols, input/output devices. Solutions already exist (e.g., ISO and TCP/IP protocols, ODP, CORBA) to deal with these levels of heterogeneity and thus allowing interoperability. The point is that heterogeneity also appears at the application level which is the most important for networked SE. At this level, interoperability paradigms must allow to answer questions such as:

- Having users working together from their respective synthetic environments, what does their interaction capabilities become ?
- What does the behavior of a virtual entity migrating from its native synthetic environment to a new one become ?

In this paper we propose architectural paradigms to achieve interoperability between networked SEs. Section 2 depicts the proposed paradigms (virtual entity, metaphor, virtual system) and highlight their major benefits. Finally we discuss the compatibility evaluation of SE systems.

2 Architectural Paradigms

The paradigms we propose aim at providing an open architecture for virtual environments. The starting point of our reflexion is the semantic associated to a virtual entity. This semantic represents the purpose the entity has been designed for. For instance, a virtual meeting desktop may be designed to automatically duplicate the document to the meeting participants. If we immerse this entity in a virtual world created to provides meeting points, the entity semantic is adequate to the virtual world's one.

If we now consider two different virtual worlds, they possibly have two different semantic. Is it possible to use the virtual meeting desktop in both environment in order to work together ? What can I expect to do whith an object imported from another environment ?

We do believe that a solution to provide virtual environments openness relies on adressing the semantic and the computing aspects of the problem. The former one having consequences on the solutions to the latter one.

Our solution relies on the distinction we make between an action and the consequences of this action. In order to do that we use the Influence/Reaction model developped in multi-agents systems by Ferber [Ferber 1995]. This distinction allows us to address the fact that it is impossible to define all the possible consequences of an action because we cannot know by advance in which virtual world the entity will be immersed.

2.1 The virtual entities

The *virtual entity* is undoubtedly a central notion in a SE. Unfortunately, its definition can slightly vary from a system to another one. For example, AVIARY [Snowdon et West 1994] implements autonomous and communicating objects whereas NPSNET [Pratt 1993] provides semi-automated forces (SAF) and autonomous Agents (AA). We represent virtual entities as a set of modules, each one corresponding to a particular aspect of objects in virtual worlds (figure 1).

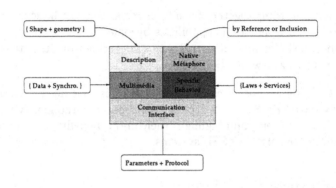

Fig. 1. A virtual Entity

• **The Appearance** corresponds to the description of the geometry of the entity. It also refers to any information related to the appearance of the entity (e.g., graphical, haptic, audio). The description of these data is performed using well-known languages (e.g. Autodesk's DXF, JPEG, VRML...) and thus can be processed by engines.

• **The Multimedia** component plays a specific role in the interactivity capacity of the entity. Embedded multimedia is already sensed as a key feature of future virtual environment [Zyda *et al.* 1993] because of the number of potential virtual applications which would benefit of multimedia capacities (e.g., distant learning, telemedicine). Several standards such as MHEG ([ISO 1994, ISO 1996]), HyTime [HyT 1992] and HyperODA [ISO 1992] enable designers to describe multimedia documents. All those formats also need specific engines to exploit their data and synchronize the flows of the different media. Some SEs already implement similar possibilities, such as NPSNET which provides *anchor* as a deposit of multimedia information [Zyda *et al.* 1993].

• **The Communication Interface** is a repository for the necessary parameters to enable the communication between the virtual entity and the rest of the virtual world. The interface also provides a communication protocol such as DIS [DIS 1993] to ensure communication possibility between virtual entities and with the environment.

• **The Specific Behavior** gives the opportunity to a virtual entity designer to modify the default behavior of the entity with another pattern. This behavior represents the peculiar actions of the entity, the rules it is the only one to follow. We represent the behavior by a set of laws using a formalism based on the Influence/Reaction model. Entities designed with language such as VRML 2.0 may have scripts containing behaviors defined with our formalism.

Some SEs already provide specific behavior to virtual entities. For instance, AVIARY [West *et al.* 1993] provides objects whith specific behavior called *volitional behavior*. The benefit is to ease the definition of autonomous actors to populate the virtual worlds and to increase the quality of the simulation.

• **The Native Metaphor** is a reference to the metaphor the entity has been designed for (see definition of the term metaphor below 2.3). That means the entity was thought to fully behave according to the laws contained in the mentioned metaphor. Indeed, an entity is always designed to be used in a certain way even if our architectural paradigms intend to make possible interaction with any entity.

2.2   The laws

Two types of laws are needed to distinguish effectively between an action and the consequences of the action.

- The *behavioral law* type is used to specify the actions performed by any virtual entity.
- The *reaction law* type allows to specify how the virtual world changes (i.e., reacts) according to the current state of the virtual world and according to **all** the actions of the virtual entities. The reaction laws are included in the metaphor rulling the virtual world. So, specifying the behavior of a virtual entity, we do not have to describe all the possible consequences on the state of the virtual world.

For example, if a single virtual robot pushes a virtual table, this action will modify the current state of the virtual world since the location of the moved table will change. The final location of the table will be different if there are, now, several robots pushing simultaneously the table in different directions. The distinguishing between action and reaction allows us to focus on what is the push action in the behavior of a robot without the burden of envisaging an infinity of scenarios when the action is performed. The new state of the virtual world (i.e., the final location of the table in our example) will be calculated by the virtual system using the reaction laws of the metaphor.

The dynamic of a virtual world is based on a two phases schema: the production of influences followed by the reaction of the virtual world. The figure 2 presents this schema.

The application of both a behavioral law and a reaction law depends on conditions to be fulfilled. A pre-state defines a condition on the state of   the

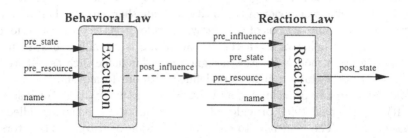

Fig. 2. The Dymamic of the Virtual World

virtual world. For example a pre-state for performing a push action on the virtual table is that the robot and the table must be in contact. The pre-resource is a special pre-state as detailled in section 2.3. When the pre-conditions of a behavioral law are satisfied, one or more post-influences are produced. A post-influence indicates an attempt to modify the state of the virtual world. For instance, a post-influence in the behavior of a robot will indicate its attempt to push the table in a given direction. The post-influences become pre-influences for reaction laws.

Thus a pre-influence is a condition on an attempt of modification of the virtual world. Arbitrary complex scenarios of interaction between virtual entities can be adressed in this manner. When the pre-conditions of a reaction law are satisfied one or more post-states are produced. A post-state thus defines an actual modification of the virtual world. The phases of influences production and reaction are reapeted infinitly.

2.3 The metaphor

In our model, a metaphor is the expression of the virtual world's semantic. It's an intention, a goal related to a virtual world. This intention induces the specification of the world's reactions.

Hence, a metaphor is a set of laws defining the reactions of the environment to the influences generated by the entities. Those reactions may vary from an environment to another, and it's the metaphor's job to implement those characteristics. For instance, the lift up of a rocket might not produce the same result if we're in Space or in Paris.

The previous example shows that the metaphor also have to express impossibilities in order to achieve modelling of consistent virtual worlds : the space environment should forbid the propagation of soundwaves. However, it would be wrong to prevent a rocket from making noise : the engine is still functionning in the same way, but we cannot hear it. In order to deal with this problem, we propose to determine the execution of a law by the presence of resources. These resources represent a part of the world's semantic.

We define two kind of resources in a metaphor : positive and negative. The former one means that the resource exists in the virtual world. This resource represents a property of the virtual world. For example, a virtual application designed for the simulation of an ecosystem might need an *atmosphere* resource to distinguish baterium who can't live in air from thos who can.

A negative resource means that this resource doesn't exist in the virtual world and that it is forbidden in order to maintain the world consistent. If a virtual world contains such a resource, this is to indicate that the creator of the world designed it with this incompatibility in mind.

The absence of a resource in a metaphore means that the virtual world has been modelled to be used without this resource, but that the presence of it shouldn't lead to inconsistency (at least, there's no explicit incompatibility).

To summarize, we propose the following definition :

> **A metaphor is the representation of a virtual world's reactions to the influences produced by its virtual entities. It is composed of a set of resources and a set of laws.**

The benefits As we explain it in [Gông *et al.* 1994], the notion of metaphor is one of the key point of our architecture because it provides important benefits :

- **Consistency**: As a metaphor gathers all the laws composing the environment's reaction, the designer of a virtual world has a better overview of the global behavior of its virtual applications. Using the formalism we propose to design the laws, it would even be possible to use some formal methods to determine the consistency of the virtual world.

- **Simplicity**: Without the model we propose, if we want an object to have a consistent behavior in two different virtual worlds, we have to define the result of an action in both virtual world. Therefore, it is impossible to predict all the different consequences of an action among the virtual world an entity is immersed in. Using our model, the designer defining a particular behavior has no longer to define all the possible consequences of these actions. Obviously, the use of the metaphor represents a benefit in time (for the design of new virtual entity) and in performance.

- **Interoperability**: The main benefit for interoperability resides in the fact that the virtual world's reactions to entities influences are defined in the metaphor. This allows us to re-use entities while avoiding inconsistency problems thanks to the resource notion.

Also, the evaluation of the needs in resources (network, system and harware ones) is more accurate using the semantic content of the metaphor. For instance, the needed network bandwidth to maintain the coherence of a virtual world populated by rapidly moving entities (e.g., planes, data stream) will be higher than if the world was populated by static virtual entities (e.g., humans) [Council 1995, Macedonia et Zyda 1995].

Using our metaphor model also eased us to develop the needed mechanisms (1) to determine levels of compatibility between different virtual environment to interoperate, (2) to calculate the resulting behavior of each virtual entity involved in an interoperability process.

2.4   The underlying virtual system

The virtual system is the base of our architecture. It contains the different managers in order to deal with the internal mechanisms. We do not impose the content of those manager, but we give their external definition.

We defined the following managers as necessary in the virtual system :

- **the entities manager** is in charge of the system's structures related to the virtual entities. It is responsible for the instanciation and the destruction of entities. For instance, when an entityis imported using the migration service, the entities manager has to be informed of this new entity in order to perform its initialisation. For distributed virtual environments, the architecture of the entities database may vary from a system to another.
- **the law manager** is responsible for the behavioral laws applications. It gets the influences generated by the entities laws analyses them and finally launch the reactions. As the architecture of our virtual system lay be distributed, each site have at least one law manager. On a single site, one may need to have more than one law manager because there's no official language for the modeling of a virtual entity's management. Another reason to duplicate the law manager is the fact that it represents a potential bottleneck.
- **the interoperability manager** has to deal with interoperability operations. It is able to determine the compatibility level between an entity and a metaphor or between two metaphors. The other part of the Interoperability manager (IM for short) is to decide which mechanism is to be used in order to increase the compatibility level when this level is too low. These mechanisms may be the use of virtual RPC (called VRPC) or the use of a metaphors server.
- **the SIP** (**S**ervice **I**nformation **P**rovider) has been introduced in our model in order to inform new entities. The informations it provides are related to the services the entity may request. Indeed, for a new entity, it is important to be able to get some information concerning the different services available. This mean of discovering the world will allow a better adaptation of an incoming or a new entity. The SIP is very similar to the ODP trader plus the dynamicity of the CORBA ORB. We'll note that HLA (the High Level Architecture) also admits the importance of such mechanism because it uses some elements of a CORBA implementation.

The following figure (fig 3) represent the general architecture of a virtual environment.

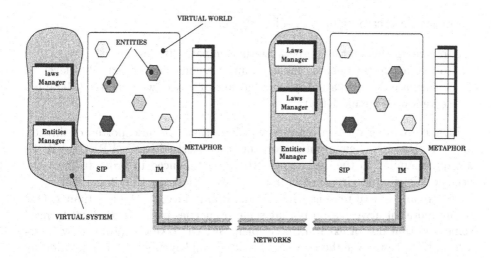

Fig. 3. The General Architecture

3 The Compatibility Evaluation

Compatibility evaluation is necessary to determine the potential of interoperability between different SEs. There are three kinds of compatibility depending on the protagonists involved:

- **Metaphor-Virtual entity**: This level of compatibility is necessary for the migration service.
- **Metaphor-Metaphor**: for sharing services.
- **Virtual entity-Virtual entity**: if the goal is to enable the interaction of two virtual entities (for instance, a virtual application that wants to deal with a resource on another virtual environment using direct sharing service).

Both compatibilities require such an evaluation to determine the quality of future interactions between the virtual entity/metaphor and the metaphor of the receiving SE. This compatibility level ranges from 0 (interoperability is impossible) to 1 (complete interoperability).

The calculus of the compatibility level is made possible by the use of the formalism we introduced to describe the laws. Using this formalism and the resources mentionned in the metaphor, we are able to determine a compatibility level and finally to predict the potential behavior of an entity in a virtual world.

Once a compatibility level has been calculated, several possibilities exist depending on the value of this compatibility level:

1. *level = 0* : the interoperability is impossible.
2. *level = 1* : the interoperability is full.

3. $0 < level < 1$: several possibilities exist, we describe them below.

In the third case, the protagonists may decide :

1. to give up the interaction because the compatibility level is too low,
2. to interact in the debased mode obtained among the compatibility level,
3. to increase the compatibility using the fourth service we proposed (Metaphor servers with virtual RPC).

A metaphor server is a repository of laws that can be exported or directly applyed in a virtual environment via the use of virtual RPC (VRPC). VRPCs are quite similar to the well-known RPC but enhanced in order to meet the needs of distributed virtual environments (see [Zelesko et Cheriton 1996]).

There are several reasons why a compatibility level can be less than 1. One of the more intuitive reason is that some properties of the virtual entity need some laws that do not exist in the recipient SE. We solve this problem by transferring those laws over the network. BrickNet implements a similar solution by transferring the methods to generate the entity's behavior [Singh *et al.* 1995])

However, for complex entities such as big applications over wide network bandwidth, performance will gain using virtual RPCs. They consist in asking the remote environment (or a metaphor server) to apply those laws on the local virtual world.

4 Conclusion and Further Works

We proposed architectural paradigms to take into account the unavoidable heterogeneity of networked SEs. They rely on the central concepts of metaphor defined as a set of laws and resources, and virtual entity. These concepts allow us to calculate the potential degree of interoperability between involved SE systems. The metaphor is also used during the interoperability session to calculate the resulting behavior of virtual entities.

To enable interoperability, we exhibit in [Bréant *et al.* 1994], [Gông *et al.* 1994] the need of 3 types of services: migration, cooperative sharing and direct sharing which cover all the possible VR applications. Our concept and services lead to achieve interoperability between heterogeneous networked SEs.

Those paradigms allow us to propose :

- A formalism to represent virtual environment and entities semantic. We do believe this is a rather new possibility.
- A methodology to develop open virtual environment.
- A way to predict the behavior of an entity in another environment.

We currently implement the architecture we propose. The kernel of the virtual system, the law manager has been developped in C. It implements the Influence/Reaction model and use the formalism we built in order to represent behavioral and reaction laws.

A VRML 2.0 parser is also being developped in order to load VRML2-compatible worlds. This parser is part of the entities manager. The next step is to develop the compatibility evaluation and the interoperability manager.

In the close future, we will use the semantic to extract criteria for an efficient load balancing of the distributed virtual system.

References

[Bréant et al. 1994] F. Bréant, H. Lê Van Gông, et M. Soto. Towards open virtual environments interoperability. Tokyo, Japan, July 1994. International Conference on Artificial Reality and Tele-Existence.

[Council 1995] National Research Council. *Virtual Reality, Scientific and Technological Challenges*. National Academy Press, 1995.

[DIS 1993] *IEEE Standard for Information Technology – Protocols for Distributed Interactive Simulation Applications, Entity Information and Interaction* . Technical Report ISBN 1-55937-305-9, The Institute of Electrical and Electronics Engineers, 1993.

[Ferber 1995] J. Ferber. *Les Systemes Multi-Agents, vers une intelligence collective*. 1995.

[Gông et al. 1994] H. Lê Van Gông, F. Bréant, et M. Soto. Architecture for virtual environments cooperation. San Antonio, Texas, USA, October 1994. IEEE International Conference on Systems, Mans and Cybernetics.

[HyT 1992] Hytime - hypermedia time based structuring language. (10744), 1992.

[ISO 1992] ISO. Hyperoda - hypermedia office document architecture. (8613/PDAM 7-10), October 1992.

[ISO 1994] ISO. Mheg: Information technology - coding of multimedia and hypermedia information, part 1: Mheg object representation, base notation (asn.1). 1994.

[ISO 1996] ISO. *MHEG: Information Technology - Coding of Multimedia and Hypermedia Information, Part 2: alternate notation (SGML)*. Technical report, 1996.

[Macedonia et Zyda 1995] M. Macedonia et M. Zyda. A taxonomy for networked virtual environments. In *Workshop on Networked Realities*, Boston, MA, October 1995.

[Pratt 1993] D.R. Pratt. *A Software Architecture for the Construction and Management of Real-Time Virtual Worlds*. PhD thesis, Naval Postgraduate School, Monterey, California, June 1993.

[Singh et al. 1995] G. Singh, L. Serra, W. Png, A. Wong, et H. Ng. Bricknet: Sharing objects behaviors on the net. In IEEE Computer Society Press, editor, *Virtual Reality Annual International Symposium'95*, March 1995.

[Snowdon et West 1994] D. Snowdon et A. West. The aviary distributed virtual environment. In Robin Hollands, editor, *Proceedings of the 2nd UK VR-SIG Conference*, pages 39–54, 1994.

[West et al. 1993] A.J. West, T.L.J. Howard, R.J. Hubbold, A.D. Murta, D.N. Snowdon, et D.A. Butler. Aviary - a generic virtual reality interface for real applications. In R.A. Earnshaw, M.A. Gigante, et H. Jones, editors, *Virtual Reality Systems*, chapter 15, pages 213–226. Academic Press, March 1993.

[Zelesko et Cheriton 1996] M. Zelesko et D. Cheriton. Specializing object-oriented rpc for functionality and performance. In *International Conference on Distributed Computing Systems, ICDCS*, 1996.

[Zyda et al. 1993] M.J. Zyda, C. Lombardo, et D.R. Pratt. Hypermedia and networking in the development of large-scale environments. In *International Conference on Artificial Reality and Tele-existence*, pages 33–39, Tokyo, Japan, July 1993.

A Redundant Hierarchical Structure for a Distributed Continuous Media Server*

Cyrus Shahabi, Mohammad H. Alshayeji, and Shimeng Wang

Integrated Media Systems Center and
Computer Science Department
University of Southern California
Los Angeles, California 90089
[cshahabi, alshayej, shimeng]@cs.usc.edu

Abstract. The growing number of digital audio and video repositories has resulted in a desperate need for effective techniques to deliver data to users in a timely manner. Due to geographical distribution of users, it is not cost effective to have a centralized media server. In this paper, we investigate issues involved in the design of a distributed video server (DVS) to support movie-on-demand (MOD) application. We propose a redundant hierarchical (RedHi) architecture for DVS where the nodes are continuous media servers and the edges are dedicated network lines. With RedHi, each node has two or more parents. We show that the redundant links in RedHi yield a more reliable and efficient system. Our simulation results demonstrate that RedHi can tolerate a single link failure with no degradation in performance while with pure hierarchy almost 2.5% of requests are rejected due to the failure. In normal mode of operation, RedHi outperforms pure hierarchy significantly (160% improvement on the average when counting the number of rejections). In the context of RedHi, we also propose and evaluate alternative object management policies, and load balancing heuristics.

1 Introduction

The growing number of digital audio and video repositories has resulted in a desperate need for effective techniques to deliver data to users in a timely manner. Due to geographical distribution of users, it is not cost effective to have a centralized media server. Distributed media servers can be employed by numerous application domains, e.g., digital libraries, health-care information systems, and educational applications to name a few.

In this paper, we investigate issues involved in the design of a distributed video server (DVS) to support movie-on-demand (MOD) application. With MOD, the request of a user to watch a (say 2 hour) movie should be served within a

---

* This research has been funded by the Integrated Media Systems Center, a National Science Foundation Engineering Research Center (NSF grant EEC-9529152) with additional support from the Annenberg Center for Communication at the University of Southern California and the California Trade and Commerce Agency.

reasonable amount of delay (say 2 minutes). DVS consists of a number of continuous media servers, such as *Mitra* [GZS+97] or *Fellini* [MNO+97], connected to each other via dedicated network lines. Unlike the customers of a video rental store, the users of DVS do not have to drive to the store to rent a movie nor need to worry about the availability of their desired movies. As compared to *pay-per-view* TV, the users of DVS are provided with a larger selection and can watch their selected movies at any desired time.

There are two alternative designs to support MOD as opposed to the DVS design: 1) a single large centralized server, and 2) a number of independent servers. The geographical disbursement of users results in a high communication cost for the first approach. The bandwidth requirement for this approach is estimated to be as high as 1.54 Pb/s (Peta-bit per second) for the continental United States [NPSS95]. With the second approach, each server must be large enough to accommodate all the movies. This requires local servers with large storage space and imposes an extra overhead for maintaining each server (i.e. insertion of new movies and deletion of older ones). Although DVS still requires large national servers that can store a rich set of movies, objects can be cached on smaller local servers closer to the users in order to reduce the communication cost. Moreover, the number of large servers in this design is much less than that of the independent servers design.

To strike a compromise between the communication cost and the server cost, some studies [BP96,NPSS95] have proposed a hierarchical network topology for a DVS. In this paper, however, we propose a redundant hierarchy (RedHi) architecture. With RedHi, each node has two or more parents. This not only makes the system more fault-tolerant (if a connection fails, the client will not become isolated), but also improves load balancing (each node has a choice to where to obtain the data in order to balance the load). In Sec. 7, we argue that the improvement in the performance and reliability of the system justifies the extra cost of the redundant links.

Assuming the RedHi architecture, we address the following issues in this paper:

- Distributed information management: Each node of RedHi (i.e., a server) stores *static* information about all the objects such as object size and display format. Nodes do not, however, keep track of *dynamic* information such as the locations of the objects. We show that an object can be located and retrieved in a distributed manner. Distributed information management eliminates the possibility of bottlenecks and makes the system more fault-tolerant.
- Load balancing: Since nodes of RedHi have more than one parent, each node might be able to obtain an object through multiple paths. We propose a number of heuristics to find the *best* path to balance the load. We show that our heuristics can be implemented in RedHi with low complexity.
- Performance evaluation via a simulation study: We compared RedHi with pure hierarchical architecture in both normal mode of operation and in the presence of failure. Our simulation studies demonstrates an average of 160%

improvement with RedHi in normal mode of operation. When the system is not over-utilized (i.e., load $\leq$ 100%), a single link failure resulted in an average of 2.5% rejections with pure hierarchy while RedHi rejected no request. In addition, we evaluated the performance of our replacement policies and path finder heuristics.

The rest of this paper is organized as follows. Sec. 2 covers some related work on distributed server architectures. The RedHi architecture is formally defined in Sec. 3. In Sec. 4, our object caching and replacement policies are explained. The heuristics for balancing the load in RedHi are discussed in Sec. 5. In Sec. 6, our simulation experiments are described. Finally, Sec. 7 concludes the paper and provides an overview of our future plans.

2   Related Work

A significant amount of research was conducted on centralized multimedia servers (e.g., [GZS+97,GVK+95,MNO+97,HLLD97]), as well as high-speed networks (e.g.,[Pry95,RRM+93]). The challenge, however, is to marry these two technologies to achieve cost effective delivery of multimedia objects to users.

A number of distributed server architectures were proposed in [BP96,NPSS95,LLQW96,DVV94]. Frequently, the proposed network architecture in these studies is assumed to be a d-ary tree with each node representing a switch (see Fig. 1(a)). Moreover, at each node (switch) there exists a multimedia server which is employed to cache a subset of the objects. At the root of the tree sits what is referred to as a regional or national server. A national server should be large enough to store all the available objects. A regional server, however, is connected via another network to other regional servers which collectively store all the objects available in the system. At the lowest level of the hierarchy are the head-end nodes to which users are connected. Asymmetric digital subscriber line (ADSL) seems to be the strongest candidate for connecting users to these head-end nodes [LV94]. The higher levels of the hierarchy, however, are typically based on the asynchronous transfer mode (ATM) technology.

In this paper, we are extending the above architecture into a redundant hierarchical architecture to achieve a better fault-tolerance and load balancing. The work in [BP96,NPSS95,LLQW96,DVV94] concentrated on the communication aspects of this distributed server architecture while our major focus is to develop methods and policies for the distributed management of objects in this architecture. It is also important to point out that although DVS has some similarities to traditional distributed information systems (e.g. [Sim96]), it is mainly designed for uninterrupted delivery of continuous media objects, something that has not been considered in traditional distributed information systems.

3   System Architecture

The main motivation behind proposing hierarchical DVS (pure or redundant), as mentioned earlier, is to reduce the communication cost. Communication cost

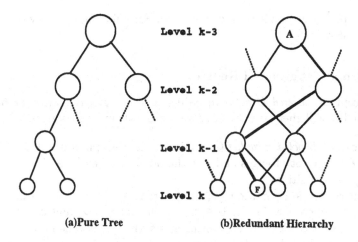

Level k-3

Level k-2

Level k-1

Level k

A

F

(a)Pure Tree   (b)Redundant Hierarchy

Fig. 1. Illustration of Pure Hierarchy & RedHi

can be defined as the cumulative bandwidth requirement of the system. Allowing users to connect to local head-ends, as opposed to having them connected to a centralized server, can significantly reduce the cumulative bandwidth requirement of the system.

In hierarchical DVS, uninterrupted delivery of continuous media object is not a trivial task. In this paper, we assume a pipelined object delivery. That is, the first block of data reaches the user after an initial delay; all consequent blocks will then arrive in a timely fashion with no interruptions. To illustrate, suppose that a user connected to node F (see Fig. 1b) requests an object O_i (a 2 hour MPEG-2 movie with a bandwidth requirement of 2 Mb/s). Node F needs to locate O_i and select a path to retrieve it (locating an object and selecting a path will be discussed in Sec. 5). Suppose that the highlighted path in Fig. 1b was selected to retrieve O_i from node A. Prior to the initiation of the display, the system needs to reserve the required bandwidth (2 Mb/s) at node A and all the links participating in the path. Note that it is not essential (but possible) to materialize O_i at any of the nodes through the path (including node F). After reserving the resources along the path, a pipeline is established. That is, the display of O_i starts after the arrival of its first block to node F and continues with no interruptions for 2 hours. In [SAW97], we illustrate how to use client side buffering techniques to relax the resource reservation requirements and reserve less or more bandwidth than what is required for object display.

In addition to eliminating the one parent restriction in RedHi, as opposed to pure hierarchy in [BP96,NPSS95], we do not assume that nodes at the same level have the same amount of resources (server bandwidth and storage capacity). Nor do we assume that links connecting nodes of level k to level k-1 of the hierarchy have an identical bandwidth. Therefore, when the number of users of a head-end increases, the system can be expanded by adding more resources at the head-end (e.g., more disk space), adding a new link, increasing the bandwidth of existing links or applying a combination of the three. For guidelines on how to configure RedHi see [SAW97].

3.1 Formal definition of RedHi

The concept of *Leveled Graph* can be defined as a generalization of Bipartite graph. RedHi will then be defined as a special case of leveled graph.

Definition 1. *Leveled Graph* $\mathcal{LG} = (\mathcal{V}, \mathcal{E}, n)$ is a graph where $\mathcal{V}$ is the set of vertices, $\mathcal{E}$ is the set of edges and n is the number of levels, and it satisfies the following two conditions:

1). There exists a partition $\{\mathcal{V}_i \mid i \in 1..n, n \geq 1\}$ of $\mathcal{V}$, such that if $n > 1$ then $\forall e \in \mathcal{E}$, where $e = (u, w), \exists j \in 2..n$, such that $u \in \mathcal{V}_j$ and $w \in \mathcal{V}_{j-1}$. That is, edges connect vertices of neighboring levels. Here, w is called the *parent* node of u, and u is called the *child* node of w.

2). If $n > 1$ then $\forall v \in \mathcal{V}_i$, where $i \in 2..n, \exists e = (u, w) \in \mathcal{E}$ such that $v = u$. $\forall v \in \mathcal{V}_1, \exists e = (u, w) \in \mathcal{E}$ such that $v = w$. That is, every vertex in $(\mathcal{V} - \mathcal{V}_1)$ has a parent and every vertex in $\mathcal{V}_1$ has a child. ∎

Level $\mathcal{V}_1$ contains vertices that are called *root* or *top level* nodes. Nodes that have no children (i.e. leaf nodes) are termed *head-ends*.

Pure Hierarchy (or Tree) is an $\mathcal{LG}$ that has one and only one root and every node except the root has one and only one parent. *Redundant Hierarchy* is an $\mathcal{LG}$ that has at least two roots and every node except the roots have at least two parents.

Definition 2. *Distributed Server Architecture* $\mathcal{DSA}$ is an $\mathcal{LG} = (\mathcal{V}, \mathcal{E}, n)$ where

1). A *link* of $\mathcal{DSA}$ is an edge of its corresponding $\mathcal{LG}$ with two parameters: $MaxB$ and B, which are the maximum and available bandwidth of the link, respectively.

2). A *server* of $\mathcal{DSA}$ is a vertex of its corresponding $\mathcal{LG}$ with five parameters: $MaxB$ (maximum bandwidth), B (available bandwidth), $MaxC$ (maximum capacity), C (available capacity), and $ObjInfo$ (objects information); where each object has six parameters: $Name$ (object name), $Time$ (recent access time for the object), $Freq$ (accumulated access frequencies), $Flag$ (if the object exists in the server), $Cont$ (content of the stored object), and $Size$ (object size). ∎

Notation: $\mathcal{DSA}$ can either be a Pure Hierarchy or a Redundant Hierarchy ($RedHi$). A *path* in $\mathcal{DSA} = (\mathcal{V}, \mathcal{E}, n)$ is a sequence $\langle v_0, v_1, ..., v_k \rangle$ of servers, where k is the length of the path, and $(v_{i-1}, v_i) \in \mathcal{E}$ for $i = 1, 2, .., k$. Furthermore, for a path to be able to deliver the requested object in $\mathcal{DSA}$, termed *delivery path*, the following condition must be satisfied: $\forall j \in 0..(k-1), v_j$ be the child of v_{j+1}.

3.2 Distributed information management

Each node of a DSA stores a table that contains *static* information about all the objects such as object size and display format. In addition, each node keeps track of the *local access frequency* of each object (i.e. the number of instances that the children of this node have requested each object). These frequencies will be used by the replacement policies that will be discussed in Sec. 4. Nodes do not, however, keep track of *dynamic* information such as the locations of the objects. When a node receives a request for an object, it either responds to the requester indicating that it has the object and able to deliver it, or forward this request to its parents. The request will continue propagating up in the hierarchy until the object is found. One alternative is to keep the information about all objects at a centralized location. In this case, every request and update should visit the centralized server and hence it might become a bottleneck. Another alternative is that each node maintains information about the location of every object in the system. This results in a higher maintenance overhead and may suffer from inconsistency problems.

4   Objects Management

The system strives to minimize communication cost by caching objects closer to users. As mentioned earlier, head-end servers can accommodate only a small subset of the rich selection of objects in the system. Therefore, a replacement policy is needed when the storage space of head-ends is exhausted. The problem is very similar to the memory hierarchy problem. The popularity of the object is usually the main factor to consider when it is cached. A semi-dynamic approach for object replacement was proposed in [BP96]. They assumed that it is possible to anticipate the popularity of some objects before their insertion to the system. This anticipated popularity is used to decide the residence of the object. The actual popularity is then checked periodically (e.g. daily) and whenever the popularity drops, the object is replaced by a more popular one.

We, however, propose a dynamic approach that is similar to the least frequently used (LFU) replacement policy used in memory hierarchy. In this policy, objects are not assumed to have a predicted popularity, and are always inserted into the root nodes. Instead of examining the popularity periodically, we check it every time an absent object is requested by the children of a node. To illustrate, assume that object Oi is requested by a user connected to node F (Fig. 1). Suppose we somehow located Oi in node A and selected the highlighted path to display the object to the user (see Sec. 5). Node F, as well as every other nodes in the path from A to F, should compare the popularity of object Oi with the popularity of all the objects resident at that node. A replacement occurs if three conditions are met; 1) the popularity of Oi exceeds the popularity of one or more object(s) available at that node ($Freq$(the old objects) < $Freq$(the new object) if LFU) 2) the removal of the less popular object(s) frees enough space to accommodate the new object, and 3) the server has enough bandwidth (B) to store the new object. If any of these conditions is violated, the object is forwarded without caching (i.e., a copy of the object is not materialized at the server). The use of a dynamic replacement policy makes the system more adaptive to sudden changes in objects popularity which is typically the case in the movie industry.

It is important to point out that when a node decides to cache Oi, it will follow the PDF dataflow paradigm introduced in [GDS95]. That is, the caching and the delivery of Oi to the lower node are performed in parallel.

New objects are inserted into the system by storing them at the root nodes. As the popularity of these objects increases, the system replacement policy will cache them closer to the users. In Sec. 6, the performance of LFU is compared with that of LRU. The main difference between these two replacement policies is the way object popularity is calculated. LFU uses the frequency of access as its popularity indicator where LRU uses the latest time that the object was referenced as its popularity indicator.

5 Load Balancing

The existence of redundant links in RedHi introduces a new problem that was not there with the pure hierarchy (PHi) architecture. With PHi, when a node receives a request for an object not residing at that node, it propagates the request to its parent. Since each node has only one parent, this request continues to traverse up the tree until it reaches a node that contains the object, possibly the root. With RedHi, each node has two or more parents and hence there might be more than one source to retrieve the object. There might even be alternative paths to each of these sources. Therefore, a policy is required for choosing the *best* source-path combination for better load balancing.

5.1 Cost Functions

To balance the load, one can choose the least loaded path among all the candidate paths. There are many ways to quantify the load of a path. One method is to take the summation of the loads (i.e., reserved bandwidth) of all the participating nodes and links in the path. In this case longer paths, with more links, are more likely to have higher loads. Trivially with this approach longer paths are penalized.

An alternative method would be to take the average of the loads of all participating nodes and links in the path. This method does not provide shorter paths with any advantage. However, if a path has a number of lightly loaded links and nodes but a single nearly saturated node, this method would choose this path over an averagely loaded path. This may saturate the heavily loaded node and prevent it from serving other requests while the nodes in the other path are less loaded. Therefore, we have decided to choose the cost of the most loaded component in the path as the cost of the path (i.e. the node or the link which is most loaded determines the load of its corresponding path).

We propose three cost functions corresponding to our three heuristics to measure the load of a node or a link in order to select the least loaded path. The first cost function $FreeBW$ uses the available bandwidth (B) of servers or links as the load indicator[1]. The disadvantage of *FreeBW* is that it tends to select nodes with higher bandwidth in the higher levels of the hierarchy over nodes at the lower levels with less bandwidth. Therefore, a large node operating at 50% of its maximum bandwidth is going to be selected over a smaller node, lower in the hierarchy, with the same load percentage. This can unfairly saturate higher nodes and might yield a higher communication cost since most of the objects will be retrieved from higher levels of the hierarchy.

To overcome this drawback, a second cost function *RatioBW* is proposed. Instead of using available bandwidth as load indicator, *RatioBW* uses the available bandwidth ratio ($RatioBW = B/MaxB$) to measure the load.

The third cost function *UserBW* employs an alternative method to eliminate the disadvantage of *FreeBW*. *UserBW* divides the available bandwidth of a node by the number of users served collectively by the node and its children ($UserBW = B$/total number of users served by the node). Similarly the *UserBW* of a link $e = (u, w)$ where $u \in V_j$ and $w \in V_{j-1}$ is defined as the available bandwidth of e divided by the number of user served collectively by u. These three cost functions are compared in Sec. 6, and the result shows the superiority of *RatioBW*.

5.2   Object Locating Methods

As mentioned earlier, the goal of object locating methods is to find a node that contains the requested object and is capable of delivering it. We investigate two methods to locate objects in RedHi: *Local* and *Global*. With *Local*, when a node Si receives a request for an object Oi, it first examines if Oi is available locally and if it has the required server bandwidth to deliver Oi; if both are true, then the node responds to the requester declaring itself as a possible source for Oi. Otherwise, the request is forwarded to Si's parents.

Instead, with *Global*, a node Si forwards the request to its parents even if Oi is available locally. By doing so, *Global* hopes to find a node, higher in the hierarchy, which can deliver Oi at a lower cost.

By stopping the search after locating the first copy of the object in a path, *Local* attempts to utilize objects that are closer to the user. Retrieving closer objects can result in a significant reduction in communication cost. On the other hand, if we keep on selecting objects in the lower levels of the hierarchy, we run the risk of overloading these nodes while leaving nodes in the higher level underutilized. *Global* attempts to balance the load between nodes at the higher and lower levels of the hierarchy.

We have developed efficient heuristics to calculate the costs of the paths, compare these costs and select the least expensive path. The main structure of the heuristics is identical and can be divided into two phases: upward and downward. The two phases of *Local* and *Global* are identical: compute the cost of the entire path (the maximum costs of its components utilizing one of the cost functions), then select the least expensive path among the candidate paths

---

[1] In this case, the cost function is $-B$.

as the *designated* path for object retrieval. The difference is in the way they decide on stopping the upward phase and starting the downward phase. *Local* stops after finding the first source in the path capable of delivering the object, where *Global* stops when it either detects a saturated link or reaches a root. For details on these two heuristic see [SAW97].

Although it seems similar to the shortest path problem in graph theory [GJ79], finding the least loaded path in RedHi differs in the way the path cost is computed. The cost of a path here is the maximum cost of all the nodes and links participating in that path rather that the summation of these costs. This makes the traditional shortest path algorithms inapplicable in this case.

The complexity of the upward and downward phases of these heuristics are $O(n * p)$ where n is the number of levels in the hierarchy and p is the number of candidate paths from head-end to roots. Since n is not expected to be high and since the communication cost forces p to be low [SAW97], the complexity of these heuristics is not high.

5.3 Resource Reservation Policies

When a node is exercising its option to find the best possible path for obtaining an object, it might reserve all the resources along all candidate paths so that other requests cannot grab the resources (a *Pessimistic* approach). By doing so, this approach commits extra resources to the node preventing other nodes from using these resources. Alternatively, a node might not reserve the resources hoping that the selected path will not be occupied by other requests (*Optimistic* approach). In this case, the node is running the risk of finding the resources reserved by other nodes when the path is selected.

To avoid the disadvantages of both *Pessimistic* and *Optimistic* strategies, the *Server-Lock* strategy is proposed. With this strategy, each participating server exclusively processes one request at a time (in sequence), until either a path is chosen or the request is rejected. Only then the server can process the next request. Details of these strategies are eliminated due to lack of space (see [SAW97]).

6 Performance Evaluation

We conducted a number of simulation experiments to obtain some insights on RedHi. Our main objective is to compare RedHi to a pure hierarchy with respect to fault-tolerance and load balancing. We also compared the performance of our different heuristics and replacement policies.

6.1 Simulation Model

In our simulation, the Distributed Server Architecture $\mathcal{DSA}$ has 4 levels. Similarly pure hierarchy is assumed to be a 4-level 3-ary tree. The parameters of these hierarchies are detailed in the Tabs. 1 and 2. It is important to point out

Level	Vertices ($\mathcal{V}_{level}$)	Servers		Links	
		MaxB(Mb/s)	MaxC(GB)	edges	MaxB(Mb/s)
1	{0}	384	583.2	$\emptyset$	
2	{1,2,3}	96	194.4	$\{(i,0) \mid i \in \mathcal{V}_2\}$	240
3	{4,..,12}	48	64.8	$\{(i,(i-1)/3) \mid i \in \mathcal{V}_3\}$	144
4	{13,..,39}	24	21.6	$\{(i,(i-1)/3) \mid i \in \mathcal{V}_4\}$	64

Table 1. Pure Hierarchy Parameters

Level	Vertices ($\mathcal{V}_{level}$)	Servers		Links	
		MaxB(Mb/s)	MaxC(GB)	edges	MaxB(Mb/s)
1	{0,40}	192	583.2	$\emptyset$	
2	{1,2,3}	96	194.4	$\{(i,j) \mid i \in \mathcal{V}_2, j \in \mathcal{V}_1\}$	120
3	{4,..,12}	48	64.8	$\{(i,j) \mid i \in \mathcal{V}_3, j = (i-1)/3$ or $j = (i-1)/3 + 1,$ if $j > 3$ then $j = 1\}$	72
4	{13,..,39}	24	21.6	$\{(i,j) \mid i \in \mathcal{V}_4, j = (i-1)/3$ or $j = (i-1)/3 + 1,$ if $j > 12$ then $j = 4\}$	32

Table 2. RedHi Parameters

that the total amount of resources (server bandwidth and capacity as well as links bandwidth) assigned to these two architectures was identical. For example, although with RedHi the number of links is twice that of pure hierarchy, the capacity of each link of RedHi is half the capacity of the corresponding link in the pure hierarchy.

A static movie population was assumed: all movies were initially stored at the root node(s), and no new movies were inserted during the execution of the simulation. Each root node contained a total of 324 movies.

For each simulation experiment, 18000 user requests were generated. Poisson distribution was used to simulate the inter-arrival period of requests per each head-end. To simulate movie selection, Zipf's law [NPSS95] was employed. All the movies were assumed to have identical lengths and bandwidth requirements. Without loss of generality, each movie was a 2 hour MPEG-2 compressed with a 2 Mb/s bandwidth requirements.

6.2   Simulation Results

A request only arrives at the head-end nodes. Subsequently, a request is either served immediately[2], or rejected if its required resources cannot be allocated. The number of rejected requests was used as the performance metric. Each experiment was repeated multiple times with different random number generator seeds. The system behavior was recorded under different system loads with all head-end nodes observing identical arrival rates. Hence, for all the reported

---

[2] We ignored the time required to select and reserve a path as well as the time elapsed until the first block of the requested object reaches the head-end node. We expect that due to fast speed of network links, this delay is in order of seconds. We plan to quantify this factor as part of our future research.

graphs, the x-axis is the percentage of the load on each head-end (varied from 40% up to 240%), and the y-axis is the number of rejected requests. Each experiment had a warm-up period of 2000 requests during which the objects were replicated from the root(s) to other levels of the hierarchy dynamically. The results are only reported for the post warm-up period (i.e., total of 2000 requests were generated).

The following four sets of simulation results will be detailed: 1) Architecture: Pure Hierarchy vs. RedHi, 2) Cost functions: FreeBW, RatioBW and UserBW, 3) Object locating method: *Local* vs. *Global*, 4) Replacement policy: LRU vs. LFU. Unless stated otherwise, the default parameters for all of these simulation experiments were: *Optimistic* (resource reservation policy), *FreeBW* (cost function), LFU (replacement policy), *Local* (object locating method).

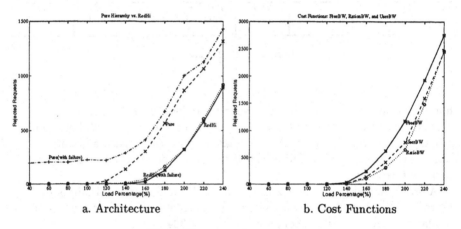

a. Architecture   b. Cost Functions

Fig. 2. Two sets of experiments

Pure Hierarchy vs. RedHi In normal mode of operation, fewer requests were rejected by RedHi than by the pure hierarchy (see Fig. 2a). Trivially, under low system load ($\leq 120\%$), the number of rejected requests is very low with both architectures. Under higher load ($> 120\%$), however, the superiority of RedHi is obvious. For example, when the system load was 200%, pure hierarchy rejected 168% more requests than RedHi. The load balancing capability of RedHi enabled it to function much better under higher system loads.

To simulate a link failure, we disabled a link (selected at random) connecting a head-end to a node one level higher after serving 9000 requests. The impact of this failure was analyzed by comparing the performance of the system with and without the failure. Fig. 2a shows that RedHi performed almost identically with and without a failure. The performance of pure hierarchy, however, degraded significantly due to the link failure. An interesting observation is that RedHi did not reject any request when $load \leq 100\%$ while pure hierarchy rejected 2.5% of the 9000 requests arriving after the link failure. We also compared RedHi with Pure hierarchy in the presence of two failures. In the presence of two link failures (not shown in the graph) the performance of RedHi still remained almost unchanged, while with *load* ≤ 100% pure hierarchy rejected 5% of the 9000 requests.

Cost Functions Cost functions are an integral part of our object locating and retrieval heuristics. The three cost functions presented in Sec. 5 were compared. Fig. 2b shows that *RatioBW* and *UserBW* consistently outperformed *FreeBW*. For example, when *load* = 200%, *RatioBW* rejected 21% and 80% less requests as compared to *UserBW* and *FreeBW*, respectively. As mentioned earlier, *FreeBW* has a tendency to retrieve objects from the higher levels of the hierarchy. Therefore, more links are occupied resulting in bad load balancing.

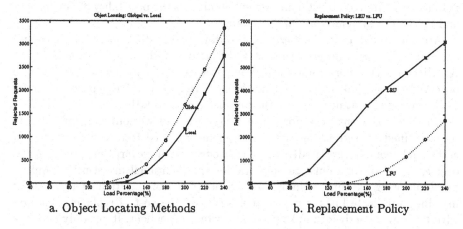

a. Object Locating Methods   b. Replacement Policy

Fig. 3. Other two sets of experiments

Object Locating Methods: *Local* vs. *Global* Fig. 3a depicts the comparison between *Global* and *Local*. We expected to see a better performance from *Global* since it attempts to find the least expensive path globally. Instead, our simulation results show the inverse to be true. This is because *Global* by selecting more higher level (closer to *root*) nodes to deliver objects, occupies more links (similar to the problem observed by *FreeBW*). This means that the system will use more resources to deliver the same number of objects. Although these resources were less loaded (according to the cost function) at the time that the request arrived, using more resources to deliver objects in the long run yields more rejections. As a result, *Local* outperformed *Global* (e.g., 47% less rejection with *Local* as compared to *Global* when *load* = 180%).

Replacement Policy: LRU & LFU Finally, we compared the performance of the replacement policies. A large margin of difference between the performance of LRU and LFU was observed (see Fig. 3b). This is because LFU is consistent with the employment of Zipf's law.

We did not compare the performance of the three resource reservation strategies discussed in Sec. 5.3. We were not able to investigate the performance of these strategies due to the fact that we assumed a zero message delay in our simulation model. Currently, we are modifying our simulation model to incorporate these message delays.

7 Conclusions

In this paper, we proposed a redundant hierarchical (RedHi) architecture for distributed continuous media servers. We showed that RedHi has a better fault-tolerance and load balancing capabilities. With RedHi, the task of locating an object and the decision of from where to retrieve it are performed in a distributed manner. Distributed object management eliminates the possibility of bottlenecks and makes the system more fault-tolerant.

We conducted a set of experiments to illustrate RedHi's fault-tolerance and load balancing capabilities. We observed that under the same load conditions, RedHi can serve more requests as compared to the pure hierarchy. We also observed that a link failure degraded the performance of the pure hierarchy while having almost no effect on the performance of RedHi.

The improvement in performance both with and without link failures can be quantified by estimating the extra income resulted from the admission of additional requests. For RedHi to be cost-effective, however, this extra income over a period of time must compensate for the additional investment required for duplicating the links. Unfortunately, accurate computation of the cost of installing duplicate links is not straightforward. The cost of replacing one link with two lower bandwidth links is highly system dependent. It not only depends on the geographical distances that the duplicated links have to cover, but also the geographical locations of these links. For example, the installation of an ATM link in a suburban area with a well established infrastructure is likely to cost less than that of the same link in a rural area. Our initial estimations indicate that duplicating the links should not increase the cost of the system substantially, making RedHi cost-effective.

Other experiments were performed to gain some insights about RedHi. When comparing object locating methods, it was observed that *Local* consistently outperformed *Global*. Moreover, LFU replacement policy was found to be superior to LRU. Out of the three cost functions presented in this paper, *RatioBW* showed the best performance while *FreeBW* was the worst.

This study can be extended in many ways. As short term plans, we intend to consider the network delay. Consequently, we can compare the impact of our resource reservation strategies (*Server-Lock, Optimistic*, and *Pessimistic*). Once network delay is introduced, we intend to add a new cost function which takes into account the length of a path. We also plan to design and evaluate policies to study the impact of employing adaptive buffering mechanisms on the system performance. Finally, we intend to investigate partial object replication and migration as well as dynamic request migration. As a long term plan, we want to investigate new challenges introduced by other applications when assuming RedHi. For example, the temporal relationships introduced by digital editing applications [CGS95] or query scripts introduced by MM-DBMS [SDG97] applications.

References

[BP96]   C. Bisdikian and B. Patel. Cost-Based Program Allocation for Distributed Multimedia-on-Demand Systems. *IEEE MultiMedia*, pages 62–72, Fall 1996.

[CGS95]   S. Chaudhuri, S. Ghandeharizadeh, and C. Shahabi. Avoiding retrieval contention for composite multimedia objects. In *Proceedings of the International Conference on Very Large Databases*, 1995.

[DVV94]   D. Deloddere, W. Verbiest, and H. Verhille. Interactive Video on Demand. *IEEE Communication Magazine*, pages 82–88, May 1994.

[GDS95]   S. Ghandeharizadeh, A. Dashti, and C. Shahabi. A Pipelining Mechanism to Minimize the Latency Time in Hierarchical Multimedia Storage Managers. *Computer Communications*, March 1995.

[GJ79]   M. Garey and D. Johnson. *Computers and Intractability: A Guide to the Theory of NP-Completeness*. W.H. Freeman and Company, New York, 1979.

[GVK+95]   D. J. Gemmell, H. M. Vin, D. D. Kandlur, P. V. Rangan, and L. A. Rowe. Multimedia Storage Servers: A Tutorial. *IEEE Computer*, May 1995.

[GZS+97]   S. Ghandeharizadeh, R. Zimmermann, W. Shi, R. Rejaie, D. Ierardi, and T.W. Li. Mitra: A Scalable Continuous Media Server. *Kluwer Multimedia Tools and Applications*, January 1997.

[HLLD97]   Jenwei Hsieh, Mengjou Lin, Jonathan C.L. Liu, and David H.C. Du. Performance of a mass storage system for video-on-demand. *To appear on Journal of Parallel and Distributed Computing on Multimedia Processing and Technology*, 1997.

[LLQW96]   Victor O. K. Li, Wanjiun Liao, Xiaoxin Qiu, and Eric W. M. Wong. Performance Model of Interactive Video-on-Demand systems. *IEEE Journal on Selected Areas in Communications*, August 1996.

[LV94]   T. Little and D. Venkatesh. Prospects for intractive video-on-demand. *IEEE MultiMedia*, pages 14–24, Fall 1994.

[MNO+97]   Martin, P.S. Narayan, B. Ozden, R. Rastogi, and A. Silberschatz. The Fellini multimedia storage. *to appear Journal of Digital Libraries*, 1997.

[NPSS95]   J. Nussbaumer, B. Patel, F. Schaffa, and J. Sterbenz. Network Requirements for Interactive Video on Demand. *IEEE Jornal on Selected Areas in Communications*, 13(5):779–787, June 1995.

[Pry95]   M. Prycker. *Asynchronous Transfer Mode :Solution for Broadband ISDN*. Prentice Hall International UK, 1995.

[RRM+93]   D. Reiniger, D. Raychaudhuri, B. Melamed, B. Sengupta, and J. Hill. Statical Multiplexing of VBR MPEG Compressed Video on ATM Network. In *IEEE INFOCOM*, March 1993.

[SAW97]   C. Shahabi, M. Alshayeji, and S. Wang. A redundant hierarchical structure for a distributed continuous media server. Technical Report USC-CS-TR97-647, University of Southern California, 1997. URL http://cwis.usc.edu/dept/cs/technical_reports.html.

[SDG97]   C. Shahabi, A. Dashti, and S. Ghandeharizadeh. Profile aware retrieval optimizer for continuous media. *submitted to ACM CIKM*, 1997.

[Sim96]   Errol Simon. *Distributed information systems : from client/server to distributed multimedia*. McGraw-Hill, 1996.

Program Caching and Multicasting Techniques in VoD Networks

Giacinto Dammicco and Ugo Mocci

Fondazione Ugo Bordoni
via B. Castiglione, 59 – 00142 Roma, Italy
Tel: +39 6 5480 3440; Fax: +39 6 5480 4404
E-mail: dammicco@fub.it; mocci@fub.it

ABSTRACT

This paper analyses a tree hierarchical network architecture employing pooled cache memories and multicasting network capability to deliver Interactive Video on Demand (I-VoD) services. The cache memories are located close to the customer and are able to store segments of the most popular programs in a dynamic way. They can reduce both transmission bandwidth and storage capacity required in the network, thus reinforcing the substantial network resource savings achievable through the multicasting feature allowed in the ATM switches. In the present paper, we analyse this type of VoD network with the aim to provide optimal dimensioning of the cache memories and to evaluate the load reduction produced in the servers and in the links as a function of the main system parameters (cache memory dimensions, number and popularity of programs, number of user clusters involved, etc.).

1. INTRODUCTION

With the emergence of B-ISDN, large information networking bandwidths would become available to the end-users, enabling the provision of several new services with very high bandwidth requirements. Moreover, in the recent years, new technologies and standards have been developed to exploit video compression and high-speed local access to the customer premises via copper (ADSL), coaxial cable or optical fibre [1, 2]. In this framework, Video on Demand (VoD) is believed to become one of the most promising service for the emerging broadband networks in metropolitan areas. There are a variety of ways in which the video programs, generally compressed resorting to coding algorithms offering MPEG1 standard (2 Mbit/s) [3], can be delivered to the customers, but two extreme forms are represented by the Near VoD (N-VoD) and the Interactive VoD (I-VoD) (Figure 1).

With N-VoD, the video streams of a given program starts at fixed instants (e.g., every 30 minutes) and customers are restricted to view at the beginning of these predetermined intervals [4]. In this case, several customers may share the same video stream on a path connecting the viewer locations to a network server, where a copy of the requested program is stored, thus greatly reducing the amount of resources needed to implement VoD services on a large scale. As drawbacks, the new program request has to wait for the fixed instants, and this waiting time on the average is simply half of the interval; moreover, no interactivity is allowed for this kind of service.

In I-VoD service implementation, a dedicated path interconnects the viewer locations to the network server [5]. This solution is able to provide instantaneous

Work carried out in the framework of the agreement between Telecom Italia and Fondazione Ugo Bordoni processing of viewer commands and high level of interactivity for the service, supporting typical VCR-like controls (pause, search, fast forward/rewind). With this approach the complexity of the transport network is minimal; however, two main drawbacks occur, related to the huge transmission flows and server access capacities required.

To avoid or at least to reduce these drawbacks, and at the same time to take as much as possible the advantages of both the solutions, the I-VoD service can be offered based upon a more sophisticated implementation, called Staggered VoD (S-VoD) (Figure 1). This implementation is based on the fact that many users can independently require the same program with small differences in time (seconds or minutes); in this case, programs are transmitted in segments of length L, (usually few minutes) and pooled cache memories (CM), close to the customers, are used to store the same program segments [4, 6]. In this way, all viewers requiring the same program within an interval of L minutes can access to those cache memories, still maintaining the instantaneous starting of their vision but sharing the same video stream from the server; in this case both the access rate to the server and the network load are significantly reduced. Resource saving can be still higher whether the network switches are able to provide multicasting features, as it happens in ATM networks [7]. In fact, the probability of having in the same time period a higher rate of requests addressing the same program increases with the number of subscribers connected to the same server and program multicasting can interest a wider geographical area.

One of the main problems implied by the S-VoD network implementation, concerns the way to assure the instantaneous response of the system to the user requests that must be compatible with the utilisation of the mentioned program caching and multicasting techniques. When a program requested is received by the service manager, the system verifies whether the required information is present in the cache memory located close to the customer or, if negative, whether it is multicasted at that time from the server. If both controls are negative a new program stream is generated by the server and downloaded in the cache memory serving the cluster to which the subscriber belongs. According to this S-VoD service implementation, the transfer of information from the server to the cache memories can be realised in different ways, depending on the requested program and on the network status, that is determined by the previously requested programs. The following transfer modes can be involved [5, 8]:

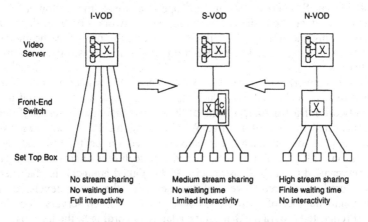

Figure 1 - VoD service logical implementation

(i)   a continuous transfer at a rate equal to the bandwidth required by the single video program (e.g., 2 Mbit/s);

(ii)   a temporary initial fast download, at a peak rate (e.g., 150 Mbit/s), needed to transfer the first L minutes of the program, followed by a continuous transmission at lower rate (e.g., 2 Mbit/s);

(iii)   a sequence of fast downloads used to transfer all the program segments.

Alternative (i) allows only the sharing of the cache memory among the viewers belonging to the same user cluster. Alternative (ii) allows a program request to hook on a multicasted flow of a previous request for the same program arrived within L minutes before. Alternative (iii) allows to adapt the transmission rate to the server output rate and to achieve high transmission link utilisation, in case of programs transmitted at regular and staggered intervals.

The actual advantages achievable by each alternative depend on several factors: the concentration of the requests on a limited number of programs (program popularity), the extension of the user cluster connected to a server area, and the specific operational rules chosen to manage the S-VoD network.

With reference to the described context, the main purposes of the present paper are:
- to analyse a S-VoD system and give criteria to dimension the cache memories;
- to evaluate the reduction on the server and link load determined by a cache memory system as a function of many system parameters, like cache memory dimensions, number and popularity of programs, number of user clusters involved;
- to estimate the sensitivity of the system with respect to the inherent uncertainty about user requests (number, type, duration) and the actual program popularity.

As to the outline, the paper is organised as follows. In section 2 we discuss general aspects of a distributed VoD architecture using a cache memory system. In section 3, the operating modes employed in the S-VoD service to satisfy the user requests are examined with more detail. In section 4 we analyse the main aspects involved in the system modelling, including demand model and program popularity, and report some preliminary numerical results, achieved by simulation. Finally, in section 5 some conclusions are summarised.

2. NETWORK ARCHITECTURE

For delivering S-VoD service in metropolitan areas, the following items are particularly important: the choice of an appropriate network structure; the location of the servers storing the library of available programs; the use of cache memories to store segments of programs; the exploitation of the multicasting capability of the ATM networks.

As to the network structure, the convenience of considering tree hierarchical networks is related to the need to reduce the costs through at least partial use of the existing network infrastructure, in spite of the lower reliability, in terms of number of users disconnected in case of failure. Within hierarchical network structures, two approaches can be distinguished, respectively based on centralised and distributed server location [9, 10, 11]. A fully centralised architecture is characterised by a single massive server used by all the customers. A distributed approach is characterised by programs stored at different levels of the network: the lower levels, close to a relatively small number of customers, consist in local video servers, containing the set of most requested programs, that can be adapted according to the local user requests [12, 13]; at the upper level of the network, a central video server contains all the other programs, or eventually the whole library for sake of security. The distributed approach is more effective in terms of reliability, manageability, user quality of service and especially bandwidth saving [4, 6].

In a distributed architecture, the lower level nodes can be equipped with pooled cache memories [14, 15], which can be accessed by different viewers watching the same program with small differences in time (seconds or minutes). Differently from a local video server, the cache memories are able to store segments of the most popular programs in a dynamic way, actually on the basis of customer requests, reducing the bandwidth required in the transport network, as well as improving the round trip time response to user requests, thus providing true I-VoD service.

The resource savings can be still higher whether the network switches are able to provide multicasting features. In fact, the probability of having in the same time period a higher rate of requests addressing the same program increases with the number of subscriber clusters connected to the same server; in this case, program multicasting can interest a wider geographical area. So, the relevant dimensioning of cache memories can lead to an optimised use of the bandwidth resources and storage capacity in the network.

For the purpose of our evaluation, a basic reference network architecture can be adopted, with a single central video server connected to different front-end switches (Figure 2). The main components of this architecture are the following:

- a central video server, where programs are stored on hard disk array according to different schemes, one movie per disk or on several disks in a striped format [16, 17];
- an ATM core network, connecting the video server to front-end switches (FE$_i$) and capable to carry traffic streams up to several Mbit/s;
- the front-end switches (FE$_i$), connecting given user clusters (CL$_i$) to the network;
- the access network, connecting the STB of the end-user to the front-end switches; it can be realised adopting different technological solutions (radio, coaxial cable, twisted pairs with ADSL technology, optical fibre and PON);
- the user set-top-box (STB), including decompression and decoding devices, for allowing the video signal to be displayed at the user television set.

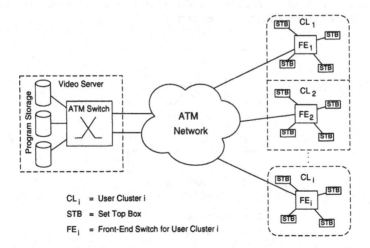

Figure 2 - VoD network reference architecture

The front-end switches (Figure 3) are equipped with local cache memory (CM$_i$), consisting of video buffers (M$_i$) of size L (measured in minutes) to store a program segment with the same duration L. These video buffers are not assigned to a single user, but can be pooled over all the users connected to the same front-end switch.

3. SERVICE IMPLEMENTATION

To describe the functionalities involved in the implementation of S-VoD service, we consider what happens when a new request for a program P is generated from a user belonging to a given cluster CL$_i$ connected to a front-end switch FE$_i$ (Figure 4). This request will be processed following different "data transfer schedules"; in this respect, we can distinguish between an initialisation phase, describing the transmission of the first minutes of the requested program, and a continuous feeding phase, regarding the transfer of the remaining and longer part of the same program.

Initialisation phase

When a new request arrives at the front-end switch, two events can occur:
- the first minutes of the requested program P are already stored in the local cache memory CM$_i$; in this case the service is provided by a primary rate (e.g., 2 Mbit/s) connection set up between the front-end switch FE$_i$ and the user (CMPR = *Cache Memory Primary Rate transfer schedule*);
- the program is not found in the local cache memory CM$_i$; in this case two alternatives can be envisaged:
  - empty space is available in the local cache memory CM$_i$: a fast load request will be launched to the central video server in order to transfer the first minutes of the program at high bit rate (e.g., 150 Mbit/s) (VSHR = *Video Server High Rate transfer schedule*);
  - no space is available in the local cache memory CM$_i$: the program will be transmitted from the central video server to the user via a dedicated point-to-point connection (DSHR = *Dedicated Server Primary Rate transfer schedule*).

Continuous feeding phase

After the initialisation, the next segments of the program are provided feeding the cache memory at the same speed (e.g., 2 Mbit/s) used by the customer to read out it; this feeding can be realised in two different ways:

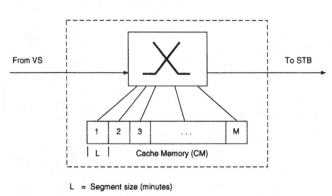

L = Segment size (minutes)

M = No. of segments in the CM

Figure 3 - Front-end switch with cache memory

- the program is already available in another cache memory CM_j ($j=1,2,...,S$; $j\neq i$) in the network: the same video stream from the central server to the relevant front-end switch FE_j can be reused via multipoint connections resorting to the multicasting capability of the ATM network (MCPR = *Multi Casting Primary Rate transfer schedule*)
- the program is not available in no one cache memory CM_j ($j=1,2,...,S$; $j\neq i$) in the network: a new video stream has to be generated from the central server (SSPR = *Shared Server Primary Rate transfer schedule*)

According to this processing mechanism, all the requests generated by the users can be satisfied properly combining the previously defined transfer schedules CMPR, VSHR, MCPR and SSPR into *transfer schedule combinations*. Four alternatives (C1, C2, C3 and C4) can be envisaged (Figure 5):

- C1 = requests are satisfied by the connection to the local cache memory CM_i (only transfer schedule CMPR is involved);
- C2 = a fast download between the central server and the local cache memory CM_i is launched in order to transfer the first segment of the program; the next parts of the program are provided feeding the buffer at a lower speed, resorting to the multicasting capability of the ATM network (transfer schedules CMPR+VSHR+MCPR);
- C3 = the whole program is transferred through a connection between the central server and the local cache memory CM_i: a fast download is provided for the first segment of the program and a lower bit rate transfer for the next parts (transfer schedules CMPR+VSHR+SSPR);
- C4 = requests are satisfied by a dedicated connection between the central server and the user (only transfer schedule DSPR is involved).

4. MODELLING AND EVALUATIONS

In the previous sections we have given a qualitative description of the S-VoD service. The development of analytic models for quantitative performance evaluations and dimensioning results rather difficult due to the complexity of network functionality. For the purpose of our analysis and to achieve some general guidelines about cache memory dimensioning, we have developed a simulation tool, on the basis of the modelling assumptions which are detailed in the following of this section.

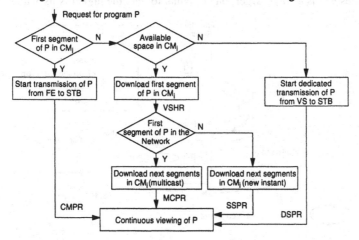

Figure 4 - Processing of a new request from each user cluster

Demand model

In general, a comprehensive demand model should include the most representative parameters describing the user affinity for certain programs, the user command rate (number of stop, pause, fast forward/reverse commands for viewed program), the number of programs and their popularity, the probability that a program is watched during certain time periods of the day. The average duration of a program is also an important input parameter. The average viewing time of a program depends not only by its duration, but also on the class of program and on the user behaviour in terms of number of VCR-like commands per program.

As VoD is still in the experimental/introductory phase the available demand/traffic data are not very representative [18], due to the particular conditions in which the service is offered to the customers (often free or at very low tariffs). However, in this paper we are mainly interested in the transport network dimensioning and not in signalling network. So we assumed an average program duration equal to 100 minutes and that a user watches the selected movie in its entirety and no VCR-like operations are requested. We also assumed that customer requests to a front-end switch for a new program occur randomly at a rate λ (Poisson process).

Considering a stationary user request distribution, a program request probability can be associated to every program stored in the network. The overall program request probabilities distribution is a parameter strongly affecting the performances of a VoD service. Given N_P programs, ordered according to a decreasing value of their popularity, different user request distribution functions, like Zipf's law, Rectangular distribution and Exponential distribution, are often addressed in literature [3, 19]. Such user request distribution function could be adopted in the simulation. For sake of simplicity, in the present analysis we have adopted the exponential distribution, in which the probability associated to the program i is recursively defined by a constant probability ratio D_{HP}, where:

$$D_{HP} = \frac{p_{i-1}}{p_i} \qquad \text{and} \qquad p_i = \frac{1 - (1/D_{HP})}{1 - (1/D_{HP})^{N_P}}$$

The parameter D_{HP} is the ratio between the $(i-1)$-th and the i-th program request probabilities and is always larger than or equal to one: the higher is its value, the less

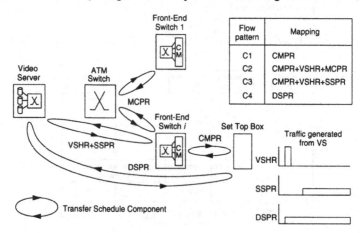

Figure 5 - Mapping program requests on network and server

probable is the request probability of programs with a higher index, while $D_{HP} = 1$ means an identical request probability for all the programs.

Numerical evaluations

In the evaluations, we adopt the two levels network architecture described in Figure 2: a single central video server is connected to different front-end switches, each serving 10,000 users; the number of these front-end switches has been varied between 1 and 3.

In our analysis, we focused on the following main aspects:
- cache memory effectiveness;
- impact of the number of user clusters on the transport network;
- sensitivity of the solution with respect to the number of programs in the program library.

Cache memory effectiveness

The first group of results regards the cache memory effectiveness and is relevant to a single user cluster. The following parameters have been assumed: number of programs, $N_P = 100$; exponential distribution ratio $D_{HP} = 1.1$; segment size $L = 4$ min. The percentage of requests completely satisfied by the access to the local cache memory (transfer schedule combination C1) is reported in Figure 6 as a function of the request rate λ and of a different cache memory size, CM (number of segments).

The share of requests satisfied by C1 increases with the request rate λ and with the

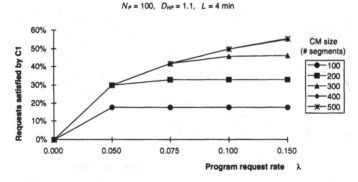

Figure 6 - Cache memory effectiveness vs. cache memory size

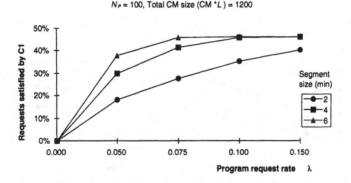

Figure 7 - Cache memory effectiveness vs. segment size

cache memory size, up to a maximum value, that we can consider as a "saturation" level. Over this value, a larger cache memory size is not able to collect more requests. A "saturation level" can also be achieved if we vary the segment size instead of the number of segments, as shown in Figure 7, where we change the segment size keeping constant the dimension of the cache memory. These results lead to dimension the cache memory at a maximum total size, which depends on the number of programs and on their popularity distribution.

Impact of the number of user clusters

To analyse the impact of the number of user clusters on the transport network, we can consider how the requests are satisfied according to the different transfer schedule combinations C1, C2, C3, C4. As shown in Figure 8, for fixed values of number of programs, cache memory size and program request rate, when the number of user clusters increases, the percentage of requests satisfied by C2 also increases and at the same time the percentage of requests satisfied by C3 complementary reduces. This reduction can be taken as a measure of the advantages for the transport network and server load achievable through the multicasting capability of the network. The percentage of requests satisfied by C1 and C4 remains obviously constant.

For any fixed number of user clusters, we can also evaluate the request-distribution percentage among the different transfer schedule combinations for an increasing value of cache memory size. In particular, as shown in Figure 9 for $N_p = 50$, $\lambda = 0.1$ and

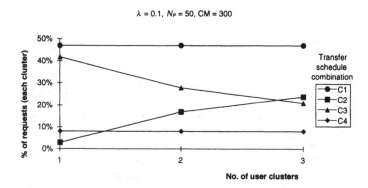

Figure 8 - Impact of the number of user clusters on the transport network

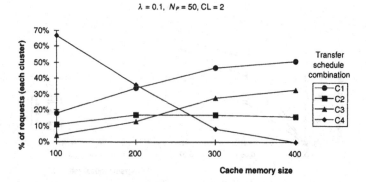

Figure 9 - Impact of the cache memory size on the transport network

CL = 2, when the cache memory size reaches the value of 400 segments, the percentage of requests satisfied locally by C1 increases up to 51%, and the percentage of requests exploiting multicasting capability (C3) grows up to 33%; at the same time only 16% of the requests imply a new video stream from the server (C2), while no one dedicated connection between the central server and the users is set up (C4 = 0%).

Sensitivity to the number of programs
    The last group of results analyses the robustness of the solution with respect to the variation of the number of offered programs. They are reported in Figures 10a and 10b for two different values of exponential distribution ratio D_{HP}. As shown in Figure 10a, if $D_{HP} = 1.1$, i.e., the requests mainly concentrate on the most popular programs, the share of requests satisfied by C1 remains substantially unchanged when the number of programs increases above a given threshold (e.g., $N_P = 50$ in figure). On the other side, as shown in Figure 10b, if $D_{HP} = 1.01$, corresponding to a more smoothed distribution of program popularity, the best advantages to enlarge the cache memory size can still be obtained for the most popular programs.
    In conclusion, we can also observe that the convenience to use cache memories with large size depends not only on the reduction needed in the transport network, but also on the required robustness characteristics of the implementation.

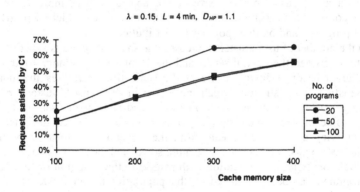

Figure 10a - Sensitivity to the number of programs

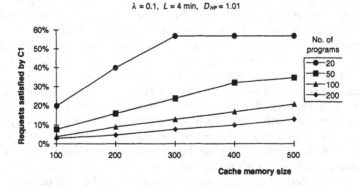

Figure 10b - Sensitivity to the number of programs

5. CONCLUSIONS AND FUTURE WORKS

In the previous sections we have discussed and analysed S-VoD operating mode to deliver I-VoD services, with the aim to evaluate the reduction on the server and link load achievable by resorting to the use of cache memories and multicasting network capabilities. The cache memories, located close to the customer, are able to store segments of the most popular programs in a dynamic way and can reduce both transmission bandwidth and storage capacity required in the network, as a function of cache memory dimensions, number and popularity of programs, number of involved user clusters.

For the evaluation we have adopted a VoD reference network architecture consisting of two hierarchical levels where a single central video server is connected to different front-end switches, each one equipped with a cache memory. A simulation tool has been developed to make an initial estimate of convenient cache memory size and number of segments in the cache memory, so to reduce the impact on the transport network and on video server, by the variations of the number of programs and of the number of user clusters. The results of the analysis have led to some general guidelines for cache memory dimensioning.

The cache memory effectiveness for a single user cluster, i.e., the share of requests satisfied by the local cache memory, increases with higher values of request rate λ and with larger cache memory size (number of segments and/or segment size) as well. This number reaches a maximum value, that we can consider as a "saturation" level, over which a larger cache memory size is not able to collect more requests, thus leading to consider a maximum total cache memory size, which depends on the number of programs and on their popularity distribution.

When the number of user clusters increases, the share of requests satisfied through a connection between the central server and the local cache memory to transfer the whole program becomes lower, while at the same time increases the number of requests satisfied using streams which are active in other user clusters, exploiting multicasting capability of the network. For any fixed number of user clusters and an increasing value of cache memory size, it is possible to reduce the share of the requests implying a new video stream from the server to the local cache memory, while dedicated connection between the central server and the users can be eliminated.

Also the sensitivity of the solution with respect to the variation of the number of offered programs has been evaluated. If the popularity is very diversified and the requests mainly concentrate on the most popular programs, the share of requests completely satisfied by cache memory is substantially unchanged when the number of programs increases above a given threshold. On the other side, if there is a more smoothed distribution of program popularity, the best advantages to enlarge the cache memory size can still be obtained for the most popular of programs.

As a conclusion, we can observe that the convenience to use cache memories with large size depends not only on the reduction needed in the transport network, but also on the required robustness characteristics of the implementation.

The use of a simulation tool, even though useful to achieve preliminary guidelines about the cache memory dimensioning, can result too much heavy in terms of processing time, especially when the problem dimensions (cache memory, program library, network, etc.) increase. To extend the analysis of VoD networks with cache memories, further researches [20] are being carried out mainly addressed to: the development of analytic models for replacing, at least partially, simulation methods in cache memory analysis; the development of analytic models for backbone network dimensioning.

REFERENCES

[1] P.V. Rangan, H.M. Vin and S. Ramanathan, "Designing an On-Demand Multimedia Service", IEEE Communications Magazine, July 1992

[2] J. Sutherland and L. Litteral, "Residential Video Services", IEEE Communications Magazine, July 1992

[3] L. De Giovanni, A.M. Langellotti, L.M. Patitucci and L. Petrini, "Dimensioning of Hierarchical Storage for Video on Demand Services", ICC '94, New Orleans, May 1-5, 1994

[4] G.H. Petit and D. Deloddere, "A Video On Demand Network Architecture Optimizing Bandwidth and Buffer Storage Resources", XV International Switching Symposium, April 1995

[5] J.P. Nussbaumer, B.V. Patel, F. Schaffa and J.P.G. Sterbenz, "Networking Requirements for Interactive Video on Demand", IEEE JSAC, vol. 13, No. 5, June 1995

[6] D. Deloddere, W. Verbiest and H. Verhille, "Interactive Video On Demand", IEEE Communications Magazine, May 1994

[7] K.C. Almeroth and M.H. Ammar, "The Use of Multicast Delivery to Provide a Scalable and Interactive VoD Service", IEEE JSAC, vol. 14, No. 6, August 1996

[8] G. Thomas, "A Novel Architecture for a Video on Demand Network", 37th MW Symposium on Circuits and Systems, 1995

[9] G. Bianchi, R. Melen and A. Rainoni, "Performance of Video Storage Hierarchy in Interactive Video Services Networks", Globecom '95, Singapore, November 13-17, 1995

[10] J. Hayes and M. Kerner, "Network Architecture for Multimedia Services", SPIE Advanced Networks and Services Symposium, Amsterdam, March 20-23, 1995

[11] S. Ramanathan and P.V. Rangan, "Architectures for Personalized Multimedia", IEEE Multimedia, February 1994

[12] C.C. Bisdikian and B.V. Patel, "Issue on Movie Allocation in Distributed Video on Demand Systems", ICC '95, Seattle, June 18-22, 1995

[13] F. Schaffa and J.P. Nussbaumer, "On Bandwidth and Storage Tradeoffs in Multimedia Distribution Networks", INFOCOM '95, Toronto, June 12-16, 1995

[14] S.A Barnett, G.J. Anido and H.W.P. Beadle, "Caching Policies in a Distributed and Video on Demand System", ATNAC '95, Sydney, December 11-15, 1995

[15] C.H. Papadimitriou, S. Ramanathan, P.V. Rangan and S.S. Kumar, "Multimedia Information Caching for Personalized Video on Demand", Computer Communications, vol. 18, No. 3, March 1995

[16] Y.N. Doganata and A.N. Tantawi, "A Cost/Performance Study of Video Servers with Hierarchical Storage", Intern. Conf. on Multimedia Computing and Systems '94, Boston, May 14-19, 1994

[17] A. Kovalick, D. Coggins and J. Burgin, "The Fundamental Concepts of Media Servers", ATNAC '94, Melbourne, December 5-7, 1994

[18] M. Naldi, "Traffic Characteristics of Multimedia Services and their Impact on Access Network Dimensioning", 13th ENPW '97, Les Arcs, March 9-15, 1997

[19] H. Ghafir and H. Chadwick, "Multimedia Servers – Design and Performance", Globecom '94, San Francisco, November 28-December 2, 1994

[20] V. Eramo and U. Mocci, "Dimensioning of Staggered VOD Networks", submitted to 2nd IFIP Workshop on Traffic Management and Synthesis of ATM Networks, Montreal, September 24-26, 1997

PENGUIN: DAVIC and the WWW in Coexistence

Reinhard Baier, Christian Gran, Petra Hoepner, Klaus Hofrichter, Angela Scheller
GMD FOKUS, Berlin, Germany
{baier,gran,hoepner,hofrichter,scheller}@fokus.gmd.de

Abstract: In parallel to the widely known WWW with HTML and Java, ISO and ITU-T in conjunction with DAVIC have developed standards such as MHEG-5 and DSM-CC that support interactive multimedia services-on-demand. Potential users and content providers are now faced with the decision, which of the two platforms to use. Rather than arguing which development better suits the needs of the users, this article presents an architecture and corresponding implementation efforts supporting the coexistence and interworking between both environments. This includes native MHEG-5 clients and servers with HTML-MHEG gateways as well as WWW browsers using Java applets or plug-ins for the presentation of MHEG-5 objects.

Keywords: interactive multimedia services-on-demand, DAVIC, WWW, Java, MHEG, interworking

1 Introduction

Today, potential users of interactive services and the corresponding content providers are faced with two major platforms for such services-on-demand. On the one hand there is the widely adopted Internet/WWW with HTML and Java [1] and on the other hand the fairly new Digital Audio Visual Council (DAVIC) [2] with MHEG-5 [3], [4] and DSM-CC [5]. Depending on who is asked, these two "worlds" are either classified as competing or complementary to each other. In order to better understand the current situation, a step back to the roots of both might help.

The development of MHEG was initiated by the International Standardization Organization (ISO) in 1992. Yet at that time nobody talked about Java and the functionality of HTML was limited. The scope of MHEG was to provide a portable, standardized interchange format for multimedia and hypermedia objects. MHEG-1 was finally published in 1995. In parallel to that the work on MHEG-5 started in 1994. Originally, it was meant to become a subset of MHEG-1 designed for endsystems with limited resources such as set-top-boxes (STB), but by now it also includes some features which are not part of MHEG-1 (partly due to DAVIC). The final text of the MHEG-5 standard was published in 1997.

DAVIC is an industry consortium exploring the potential of digital technologies applied to audio and video. As far as possible, DAVIC uses international standards and only fills the gaps with its own specifications. The first set of specifications, DAVIC 1.0, was published in December 1995. It concentrates on TV distribution, near-video-on-demand and video-on-demand addressing residential users with STB access. The specifications do not only include protocols and interchange formats, but also interfaces for STBs and cable modems. DAVIC 1.0 refers to international standards such as MHEG-5, MPEG-2 and DSM-CC. The next version, DAVIC 1.1, addresses additional applications and Internet access.

The history of the WWW is widely known. In 1992, when the development of MHEG-1 started, the WWW was not taken up by commercial content providers yet. At that time its use was still restricted to the scientific community. HTML has been in use by the WWW since 1990, but the HTML+ Specification was published in late 1993. The functionality of the HTML version at that time was quite limited and far away from the scope of MHEG. Even today, HTML 3.2 imposes too many restrictions on applications which are not acceptable for a commercial environment. This has changed since Java arrived on the stage. Java originates from internal projects of Sun Microsystems. The development already started in 1990. Finally in 1995 it was recognized that Java could be used in the WWW and might be suitable to overcome many of the HTML restrictions. As such it was made publicly available in late 1995.

So, all these developments have in common that their current scope goes far beyond the original intention. This phenomena is quite common in standardization. Various activities start in parallel without any overlap but later on the scope is broadened. When this is finally realized by a larger community, it is too late to stop either of them because of the huge investments. Rather than arguing which one is better and which one should not have been defined in the first place, joint forces shall be used to support the content providers and users by harmonizing both approaches and providing some interworking strategies.

As a first step to overcome these problems, this paper describes an architecture and the corresponding implementation activities that combine both environments. The so-called *Penguin architecture* includes native MHEG-5 clients and servers with HTML-MHEG gateways as well as WWW browsers which use Java applets or plug-ins to enable the presentation of MHEG-5 objects.

Section 2 shortly introduces the state of the art with regard to interactive services environments. The Penguin architecture will be introduced in Section 3. In Section 4 the various options for MHEG-enabled clients are explained in detail. Section 5 introduces the corresponding server alternatives and Section 6 briefly discusses some communication issues. Finally, Section 7 concludes and outlines future R&D in the Penguin project.

2 State of the Art

2.1 JAVA

The fast development of the WWW in the Internet established WWW browsers as the standard clients for Internet service access. To handle specific content types, WWW browsers are often extended by platform-dependent special programs called plug-ins. The disadvantage of platform-dependency is now overcome by the introduction of Java applets, which extend WWW browsers with platform-independent executable programs.

Java is an object-oriented programming language developed by Sun Microsystems. A comprehensive introduction into Java can be found in [1]. Even not originally developed for the Internet, Java applets now provide the features for the access of the browser resources. The WWW traditional server side activity is thereby transferred to the client.

2.2 MHEG

MHEG is an ISO and ITU standard for exchanging interactive multimedia applications in a platform-independent way. The idea of MHEG-based interchange is, that independent of the production environment the resulting objects may be imported and used by different runtime systems. The interchange process itself is not defined by MHEG. Other standards like DSM-CC are applied.

MHEG-1 specifies features for the extension of the standard to fit to the needs of particular application areas. This is used by MHEG-5 for the application domain of video-on-demand and similar applications. MHEG-5 provides a framework for the distribution of interactive multimedia applications across minimal resource platforms of different types.

The major extension of MHEG-5 in relation to MHEG-1 is the support of page-oriented application models, thus reducing encoding overhead. Within MHEG-5 a so-called 'Scene Object' represents a single page. An MHEG-5 application is made up of scenes and objects that are common to all scenes. A scene contains a group of objects, called 'Ingredients', that represent information and an associated behavior based on event firing (for example the 'Left' button being pushed activates a sound). At most one scene is active at any one time. Navigation is accomplished by transitioning between scenes. Several object classes describe the content, the behavior and the possible user input. Content is either included directly in an object or the object contains a reference to external data. MHEG-5 scenes and applications can be specified in a textual or in a binary notation [ASN.1 ISO/IEC 8825].

A complete MHEG-5 application resides on a server; as needed, portions will be downloaded to the client. MHEG-5 applications can run in arbitrary types of endsystems, but are specifically designed for the STB environment, where the data flow is asymmetric, i.e. small bandwidth from the client to the server and large bandwidth from the server to the clients.

2.3 DSM-CC

The Digital Storage Media Command and Control (DSM-CC) specification is a set of protocols providing control functions and operations for managing MPEG-1 and MPEG-2 bitstreams. The MPEG-2 Systems Standard [ITU-T Rec. H.222 / ISO/IEC 13818-1] was developed to specify coding formats for multiplexing audio, video and other data into a form suitable for transmission or storage, i.e. MPEG streams may also contain MHEG objects and other data. The goal of the standard is to enable information providers to load content into services and client applications to retrieve that content.

DSM-CC defines two types of information flow: User-to-User (U-U) and User-to-Network (U-N). The U-U primitives set enables multimedia applications to run using the MPEG delivery system in heterogeneous environments. DSM-CC supports a synchronous deferred pipelining of requests. This will enable applications to prefetch information and prepare services in advance of the time needed. For accessing the various objects on the server side, the DSM-CC Directory, File and Stream objects are used. Each scene object, application object and content data is stored in a separate file. U-N primitives are used to control sessions and network resources.

DSM-CC can be implemented using any RPC mechanism. The U-U API primitives that use the RPC are defined in terms of an Interface Definition Language (IDL). Functions are specified in OMG IDL [ISO/IEC 14750] to permit the use in multiple RPC and language environments, including, but not limited to UNO, ONC, DCE, C++ and C. Operation in a CORBA system environment is intended.

3 Penguin Architecture

Figure 1 shows the logical components of the Penguin architecture and their relationships. The upper part comprises the information sources and server systems, whereas the lower part shows the various client systems for accessing the information servers. The user has the choice to use the same interactive services-on-demand either through WWW or MHEG clients. A number of native MHEG clients is available for PCs and UNIX workstations. Clients for STBs are under development.

Available WWW browsers can be enabled to present MHEG objects in two different ways: either by means of a plug-in which provides the complete functionality of an MHEG client for the corresponding platform or by means of a Java applet which provides the functionality of the MHEG presentation system in a platform-independent way and communicates with the MHEG core on the server.

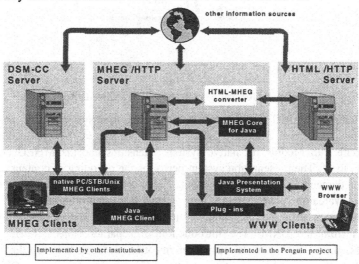

Figure 1: Penguin architecture

In addition to native MHEG objects, MHEG users can also retrieve HTML pages by means of an HTML-MHEG converter which creates MHEG objects on-the-fly from HTML sources. The user of a WWW client is restricted to the use of HTTP servers, whereas MHEG clients are provided for DSM-CC as well as for HTTP servers. Additional information from other sources can be obtained by CGI-like mechanisms regardless which kind of server is used.

4 Penguin Clients

In addition to native MHEG clients, the Penguin architecture offers a number of MHEG-enabled clients with corresponding servers which allow the user to chose the best-fitting configuration. All these solutions are structured according to the general architecture shown in Figure 2, but they differ with regard to the organization into client and server components and the protocols among them.

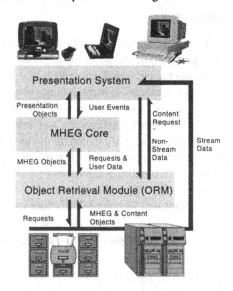

Figure 2: General Architecture of MHEG-enabled Clients

A complete MHEG system consists of a presentation system for the user interface, the Object Retrieval Module (ORM) for the request of the application data, the Procedure Call Module for native system extensions and the MHEG core (MHEG engine) itself. The presentation system performs the decoding of content data, renders them on demand and receives user input. The user input is send to the MHEG core for further evaluation. The MHEG core parses and interprets the MHEG objects, communicates with the presentation environment and requests additional MHEG objects from the ORM. The introduction of separate Object Retrieval Modules (ORMs) allows the use of various communication protocols between the MHEG core as well as the presentation system and the different types of servers. Objects can be retrieved either by file transfer, HTTP or DSM-CC.

The MHEG standard defines only the possible range of MHEG objects and their encoding; the design of a complete MHEG system as outlined above, is outside the scope of the standard and up to the individual implementation.

4.1 Native MHEG Clients

In order to verify the MHEG standard as early as possible, GMD FOKUS has developed a number of MHEG tools in parallel to the standardization process. These tools are used in several national and European projects [6],[7]. Among them are native MHEG implementations for PCs (Windows 95/NT) and UNIX Workstations (Sun/SPARC). A STB implementation is planned for the near future. The current MHEG clients are completely programmed in C++.

In all these implementations, the MHEG core as well as the presentation system reside on the client and communicate with the server through the ORM. However, as explained in section 4.2 the presentation system could also be located separately from the MHEG core requiring communication features among them.

The clients are able to read MHEG-5 scenes and applications in ASN.1 [ISO/IEC 8825] as well as in the textual notation. Since the MHEG-5 standard does not describe content types, a profile was defined based on DAVIC requirements and expected requirements for the interworking tests and user trials. A wide range of audio, video and bitmap content formats are supported by this profile.

4.2 Java Applet as MHEG Presentation System

The aim of the Maja project (MHEG Applications utilizing Java Applets) is the integration of the DAVIC and Internet environments by supporting MHEG-5 applications in WWW systems. A contrary but not competitive approach is proposed by the DAVIC 1.1 standard, which features the Internet access by DAVIC-conforming applications and platforms.

In Maja the presentation system is developed as a platform-independent Java applet that is located in a Java-enabled browser. Its task is to retrieve and to present the objects as requested by the server-located MHEG core and to propagate the user input. The applet connects to the MHEG/HTTP server with a specific URL. There an MHEG-5 engine is spawned as a CORBA object on demand of the client. The applet binds to this object. The engine runs the MHEG-5 application as specified by the parameters of the applet and controls the rendering of content while interpreting MHEG-5 objects. The applet utilizes a CORBA 2.0 compliant ORB using the Internet Inter-ORB Protocol (IIOP) [8]. In addition, due to the currently insufficient support of video within Java a video plug-in was developed.

The defined class hierarchy of the Maja presentation system strongly relates to the MHEG class hierarchy and is therefore especially suited for MHEG application presentation. It is implemented in two separate packages: mheg-5 and the main package. The classes within the main package represent the objects that are displayed to the user. The classes within the mheg-5 package are applet-independent. They contain basic variables and methods that can be used by inherited classes in the main package. In the future they will be used also for the implementation of a full MHEG client in Java (see 4.4 below).

4.3 WWW Browser with MHEG Plug-in

Plug-ins for available WWW browsers are another alternative for accessing MHEG objects within a WWW client. In contrast to the Maja approach the user depends on the availability of the plug-in for his preferred working environment whereas Maja can be used by any Java-enabled browser. A Netscape MHEG plug-in for Solaris and Windows95/NT was developed for comparison purposes.

4.4 Java MHEG Client

A full Java MHEG implementation is currently under development which includes the MHEG core as well as the presentation system. However, in order to be able to support the full range of potential MHEG applications a number of Java extensions regarding multimedia features (e.g. video) are needed.

5 Servers used in Penguin

MHEG-enabled servers provide the clients with MHEG objects and the associated content. They are also responsible for authorization and accounting. The actual content may be stored separately from the MHEG objects on special content servers (in particular in the case of stream content). Depending on the quality of service requirements the same content might be stored on different content servers and/or in different formats. Different server solutions are introduced below.

5.1 HTML/HTTP Server

In the Penguin architecture MHEG-enabled clients can access all the information on traditional HTML/HTTP servers using an HTML-MHEG converter (see below) for converting HTML3.2 documents to MHEG-5 objects (in real-time or non real-time).

5.2 MHEG/HTTP Server

The MHEG/HTTP server used within the Penguin architecture is basically designed like a traditional HTML/HTTP server. The main difference is, that the accessing clients receive MHEG objects instead of HTML documents. The origin of the MHEG objects is thereby wide-ranging: native MHEG objects, converted HTML documents or MHEG objects produced by external services and CGI-like processes.

For the transmission of video data a video pump to the client is used. In contrast to the WWW, each client usually connects to a particular MHEG/HTTP server that also resolves all requests to objects which are stored on other servers. This provides an advantage for low-resource environments, because certain services such as email do not have to be provided on the client and other services such as accounting, billing and authorization can be provided easier.

Special components implemented in the Penguin architecture on the MHEG/HTTP server are described below.

HTML-MHEG converter

A converter for HTML3.2 documents to MHEG-5 objects and the corresponding content objects was developed by VTT Finland. It can be used twofold: for the non-real time, i.e. batch mode conversion of self-contained HTML documents and for the real-time, i.e. on-the-fly resolution of URLs referenced within MHEG-5 applications.

External Services and CGI-like processes

External Services and CGI-like processes allow the access to non-MHEG services, to present non-MHEG multimedia objects and to update time-variant data for MHEG applications by the required access methods/protocols. The returned information objects as well as content objects are converted to MHEG.

During the creation phase external information is forwarded to the MHEG object generator together with some context information. This information is used to retrieve the template description for the MHEG page to be created. The template specifies the general structure and layout of a page, e.g. default buttons for navigation to other pages, the background image, sounds, etc. The content information which has been created dynamically is added to the layout specification of the page. It is up to the MHEG object generator to decide whether all the content is embedded into the MHEG object directly or stored in separate content objects.

MHEG Core for Maja

Within Maja, an MHEG core on the MHEG/HTTP server is needed for the Java presentation system described in section 4.2. After the Java presentation system connects to the server with a specific URL, a loader process on the server is initiated. The loader process creates a unique ID for the identification of the communication connection using the hostname and an internal reference. This ID is returned to the applet and a new MHEG-5 engine with the name of the scene as parameter is invoked on the server. The newly created MHEG-5 engine exports an MHEG-5 CORBA object with the given ID. The applet binds to this object. Afterwards the applet is able to transmit user events to the engine and to request commands (like 'draw a bitmap') from it. All user events are transferred from the applet back to the MHEG-5 engine. The engine interprets the events and transfers a bundle of new commands (for example: the layout of a new scene) to the applet.

The consequence of this design is, that for each client a separate engine process has to be instantiated. In order to avoid heavy load on the server, appropriate monitoring and resource allocation will become quite important. The loader process may be enhanced to identify the client, check the authorization of the client and provide billing information.

5.3 DSM-CC Server

In order to ensure interoperability, DAVIC selected DSM-CC for the access and retrieval of MHEG objects and their associated content objects. An MHEG client has to connect via DSM-CC protocol elements to a DSM-CC server. To differentiate the type of information, DAVIC defines content-information and control-information.

Content-information flow thereby is defined as a uni-directional flow from a server to the STB carrying encoded video/audio content and associated data, binary objects to be used by the STB, and other information types. This flow does not alter the behavior of the information source and destination objects, e.g. audio, video, or data transferred. Objects can be transmitted in reply to the DSM-CC File read function with the Universal Network Object (UNO) data representation and transport mechanism. Real-time streams are mapped directly to an MPEG-2 Transport Stream without using DSM-CC, but stream control is supported by DSM-CC control functions such as start, stop, pause etc. Control-information flows from an Application Service Layer source object to a peer destination object. The flow is transparent to any intermediate object through which the flow passes. The behavior may change as a result of the flow. In order to ensure interworking, DAVIC has organized interoperability meetings where content- and control-information flows were tested.

Additional DAVIC streams handle the User-to-Network functionality for session, connection and management information. Future DSM-CC server products will therefore support not only information transmission but also all functionality regarding service access, access control and resource management. User-to-Network primitives have to be implemented in the next step. These primitives will complete the full DSM-CC stack.

6 Communication Issues

The MHEG format is independent of a specific communication system. However, in order to realize the Penguin platform a number of protocols have to be specified between the Penguin presentation system, the MHEG core (and associated modules) and the various kinds of servers (application server, content server, gateways).

The introduction of separate Object Retrieval Modules (ORMs) allows the use of alternative communication protocols between the MHEG core and different types of servers. Objects can be read either by FTP, HTTP or DSM-CC. The same applies to the communication between the presentation system on the one hand and the ORM and servers on the other hand.

Within Maja, the communication between the applet (i.e. the presentation system) and the MHEG core is realized with CORBA. The motivation for the use of CORBA within Penguin is to enable a later extension of the Penguin system to other distribution concepts and programming languages. The IDL definition covers all necessary functionality for a client (presentation system) to communicate with the server (MHEG core) using standard CORBA-mechanisms. The IDL defines two objects to which the client connects - the loader object and the MH5 object. The client uses the start-function to identify itself to the loader and to force the loader to invoke a new incarnation of the MHEG core. The client receives a unique ID which it uses to bind to its personal MHEG core. From now on the client requests in certain time-intervals new commands. The get_command function can return a whole group of commands from MHEG core in one packet. On the other side the client transfers all user events with the send_command function to the MHEG core, which reacts on these events.

7 Conclusion

WWW/Java and MHEG are the two main candidates for the future development of interactive applications which address the needs of business as well as residential users. Java is a programming language. MHEG is an interchange format. As such they cannot be directly compared. The interpretation of the interchange format can certainly be implemented with the programming language. However, this is not as straightforward as Java advocates would like it to be. The support for media such as audio and video in Java for example is still quite limited.

Within the Penguin architecture and implementation client and server components have been developed that suit the needs of business and residential users without excluding one of the candidates.

Maja shows that the combination of Java applets and CORBA technology can lead to a client-server model with a sufficient turn-around time for real-time communication. A WWW browser with the Maja applet can be used as a client to display MHEG-5 objects which enable the integration of MHEG into the WWW. This provides the opportunity to use DAVIC-conforming MHEG applications on different platforms and in different environments.

On the other hand, a number of approaches are discussed in the WWW community to introduce certain new features to HTML which allow an advanced layout control. The major browser developers Microsoft and Netscape suggest their own (i.e. incompatible) extensions of HTML, whereas the W3C pushes the standardization process through a working group. This development is crucial for both MHEG and HTML, since interoperability is the main issue for both environments.

Acknowledgements

The work discussed in this paper was partly performed in the context of the GLUE Project funded by Deutsche Telekom/T-Berkom and the ACTS Project IMMP. The authors would like to acknowledge especially Andreas Kraft, GMD FOKUS, for ideas and comments on this paper. The HTML-MHEG converter which is mentioned in this paper as part of the overall Penguin architecture was developed by Markku Savela from VTT in Finland in the course of the IMMP Project.

References

[1]   Sun Microsystems, Inc., *Java Home Page,* http://java.sun.com/

[2]   DAVIC Consortium, *DAVIC Home Page,* http://www.davic.org

[3]   ISO/IEC DIS 13522-5, Information technology — Coding of Multimedia and Hypermedia information — Part 5: MHEG Subset for Base Level Implementation.

[4]   K.Hofrichter, *MHEG 5 - Standardized Presentation Objects for the Set Top Unit Environment,* in Proceedings of the European Workshop on Interactive Distributed Multimedia Systems and Services, Berlin, Germany, 4-6 March 1996, Springer Lecture Notes in Computer Science, 1996

[5]   ISO/IEC DIS 13818-6, Information technology — Generic Coding of Moving Pictures and Associated Audio - Part 6: Digital Storage Media Command and Control (DSM-CC), 1995

[6]   *IMMP Home Page,* http://www.nokia.com/projects/IMMP/ and http://www.fokus.gmd.de/ovma/immp/

[7]   *GLUE Home Page,* http://www.fokus.gmd.de/ovma/glue/

[8]   OMG , *CORBA 2.0/IIOP Specification,* OMG Technical Document PTC/96-03-04, http://www.omg.org/corba/corbiiop.htm

mTunnel: A Multicast Tunneling System with a User-Based Quality-of-Service Model

Peter Parnes, Kåre Synnes and Dick Schefström

Luleå University of Technology,
Department of Computer Science/Centre for Distance-spanning Technology,
971 87 Luleå, Sweden,
{Peter.Parnes,Kare.Synnes,Dick.Schefstrom}@cdt.luth.se

Abstract. This paper presents a system, called mTunnel, for application level tunneling of multicast traffic in a lightweight manner, where the end-user is responsible for deciding which MBone-sessions and multicast groups to tunnel. mTunnel is primarily designed for easy deployment and easy-to-manage tunneling. Therefore it runs as an user application and does not need access to restricted system resources.

Information about currently tunneled sessions and control of mTunnel is provided through a Web-interface. To allow the user to easily start tunneling of announced MBone-sessions, mTunnel listens for announcements and presents this information through the Web-interface.

To save bandwidth, tunneled streams can be translated in four ways: audio streams can be recoded to an encoding that requires lower bandwidth, several audio streams can be mixed together, streams can be switched based on activity in another stream and streams can be scaled by dropping a certain percent of the traffic.

1 Introduction

The Multicast Backbone, MBone [6] has existed for several years and is slowly being deployed in production networks. When a host sends a packet on the MBone it does not send it to any specific host, but instead to a so called *multicast group*. The sending host does not need to know which hosts are members of this group, but that information is stored in and kept up to date by the network. Packets sent to a group are only sent to the parts of the Internet, where there are listeners. Local routers detect if there are any local listeners to a specific group using the *Internet Group Management Protocol - IGMP* [1]. IGMP is also used by hosts to signal that they want join or leave a multicast group.

To ease the development of the MBone it was first developed as a virtual network on the Internet consisting of tunnels between machines that acted as virtual routers, sending multicast packets encapsulated in unicast packets. The software used is called *MRouted* [4]. Today, multicast routing functionality is

---

* This work was supported by the Centre for Distance-spanning Technology (CDT), Luleå, Sweden

being built into standard Internet routers, but there is still no general support for multicast over point-to-point links (such as analog modems or ISDN-links).

MRouted uses the normal routing tables and does real multicast routing, which means that a lot of routing information is also sent over the tunnels. On low bandwidth links, this can cause a problem because much of the bandwidth is used for control traffic and not the data itself.

MRouted also use IGMP to detect members of multicast groups and decide on which groups to tunnel based on the group membership information. This means that if a user on the local network joins a group, traffic to and from that group will automatically be tunneled as long as the users tool is running. This automatic tunneling can be a problem on low bandwidth links, as local users do not see which groups are currently being tunneled and may make the total bandwidth requirement too high by joining to many groups. If this happens, all currently tunneled groups get affected as packets will be dropped at random.

Another reason for using an unicast tunnel is firewalls. At some company's it has proven to be very hard to convince the system administrators to open the companies firewall for multicast traffic.

mTunnel is an application for tunneling of multicast traffic based on user requests rather than based on IGMP. This means that users explicitly have to choose which groups to tunnel. mTunnel, also allows the users to specify how different groups should be prioritized, which is useful if a group of users are currently using mTunnel to tunnel an electronic meeting and do not want to be disturbed by another user that, for instance, want to watch the current NASA mission in Space[2]. mTunnel uses the World Wide Web as its main user interface.

The first design goal of mTunnel was to easily tunnel multicast traffic through non-multicast enabled parts of the Internet or within an intranet. The second design goal was to let the end users easily make the decision on which groups to tunnel. The end users should also be able to easily prioritize different parts of the traffic. The third design goal was that it should waste as little as possible of the available bandwidth on control data. The fourth and last design goal was that it should be as platform independent as possible.

The rest of this paper is divided into: Sect. 2 that presents the general architecture of mTunnel and its features, Sect. 3 that presents the user based Quality-of-Service model used by mTunnel, Sect. 4 that presents the implementation and current status and Sect. 5 that gives a summary and conclusion about the work.

2   The mTunnel application

mTunnel consists of four main parts, the *tunneler* which performs the tunneling itself, the *controller* that keeps both ends of the tunnel synchronized and updates control-clients, the *Web-interface* which present a user interface to mTunnel using the World Wide Web, and finally the *translator* which can translate streams in various ways.

---

[2] NASA currently multicasts most of their shuttle missions.

As with all tunnels it has two end-points (see Fig. 1), between which the traffic is tunneled. For that reason there must always be two copies of mTunnel running, one at each side of the tunnel.

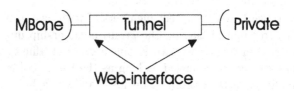

Fig. 1. A tunnel connecting a private network to the MBone.

mTunnel runs as a user process to allow for easy deployment of new and perhaps temporary tunnels.

Sessions

The term *session* is used as a name for one tunneled stream. A session consists of: a multicast group, a base port which is the lowest port number to tunnel, the number of consecutive ports[3] including the base port to tunnel, the Time To Live of the outgoing packets (see Sect. 2.6), and the name of the session.

Statistical information about the number of bytes and packets sent and received is also stored within a session.

A session can also include information about prioritization, if different from the default value (see Sect. 3).

2.1 The tunneler

The *tunneler*, is the part of mTunnel that does the actual tunneling. It listens on a number of multicast sockets (based on which groups to tunnel), encapsulates the incoming data, and sends it through the tunnel (see Fig. 2). The mTunnel on the other side, receives the encapsulated unicast packet, decapsulates it and resends it locally using multicast.

The tunneled data is sent over a unicast "connection" between the two end-points and all data is sent over the same port number. This allows for easy tunneling through firewalls (as the systems administrator only has to open one port in the firewall).

---

[3] Most multicast streams on the MBone use the Real-time Transfer Protocol [10], that uses more than one port, one for data and one for control.

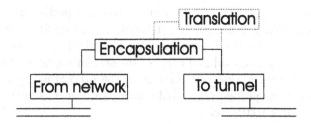

Fig. 2. Path of an incoming packet with optional translation.

Encapsulation of data

Each multicast packet is encapsulated by adding the multicast address of the group from which the packet was received, and the port, to the end of the packet (see Fig. 3). Because mTunnel runs as a user process, it does not have access to raw IP packets or the kernel buffers used to receive the packet by the operating system. Therefore, to minimize the number of copy-operations needed, the tunneler uses a memory buffer that is larger than needed to receive the incoming packets. This allows the tunneler to add the encapsulation information after the data in the same buffer as the packet was received in. On the other side of the tunnel, the same packet is recent without any extra copy-operation, by stripping of the earlier added encapsulation data.

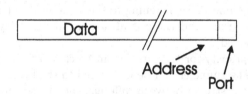

Fig. 3. An encapsulated packet where 'Data' is the original multicast packet.

Transmission loops

mTunnel is designed to connect a network or a single host that currently is isolated from the MBone, to the MBone. But, if both ends of a tunnel is started within the MBone, a transmission loop can occur if the tunneled MBone-sessions and TTL values are not chosen carefully. mTunnel therefore does not forward packets trough the tunnel if the sender matches the other end of the tunnel. Unfortunately, if two separate tunnels are deployed that together create a loop, packets will be forwarded over and over again.

If mTunnel suspects that a loop has occurred (the packet rate through the tunnel suddenly raises dramatically), it sends out a special probe-packet and waits for the probe-packet to be received again. If the probe-packet *is* received all current tunneling is stoped and users of the system are notified by the Web-interface. If the probe-packet *is not* received, the process is repeated a couple of times, because the probe-packet could have been lost on the way due the to best-effort nature of UDP-packets.

2.2 The controller

The *controller*, is the part of mTunnel that keeps the two endpoints synchronized. That is, if a new session is added on one end of the tunnel, the controller sends a message to the other end telling it to add the same sessions. The controllers communicate with each other over a TCP-connection for reliable messaging. If one end of the tunnel is restarted, the other end automatically updates the first end with information about current sessions. The controller also allows for connections from clients who wants a simpler and faster (than through the Web-interface) way of communicating with the server.

Using the controller each session can be controlled in various ways:

- **Pause/Continue**: The tunneler can be instructed to temporarily pause tunneling of a session. If the pause lasts for more than one minute the tunneler also leaves the multicast group by sending an IGMP leave message for that session, which makes traffic to that group stop. This has the advantage of the traffic to the end of the tunnel connected to the MBone lowers, but unfortunately this also makes the delay longer when the tunneling of the session is continued and no one else is listening to the same session (since the tunneler must rejoin the group and there is always an initial delay before the first packets are received).
- **Priority**: The priority for the session can be changed (see Sect. 3).
- **Translation**: The tunneler can be instructed to start a translator. A translator can recode a stream between different encodings; mix several streams into one stream; switch the traffic, meaning that only packets from a certain source is forwarded based on another stream (for instance, video packets are only tunneled for the member that currently is speaking; and streams can be scaled, meaning that parts of the traffic is just dropped (see Sect. 2.4).

2.3 The Web-interface

As one of the design goals was to allow for easy deployment, there is a small and minimal Web-server built into mTunnel. The alternative was to use a separate Web-server and the CGI-interface [7], but the built in Web-server allows for faster access than through a CGI-program (it also allows for a simpler implementation).

The Web-interface lets the user watch information about current sessions and statistics about the total number of packets and bytes sent through the tunnel. It also allows the user to control the current sessions and create new sessions.

Creation of a new tunnel session

Tunneling of new sessions can be specified in two ways: 1) manually entered or 2) chosen from a list of earlier announced MBone-sessions:

1. **Manually:** The user first chooses how many sessions to start (normally one session per available media). Second, the user specifies a name that is common for all the new sessions, the TTL with which outgoing packets should be sent, and for each session the user specifies the media-type, the multicast group, the base port and the number of ports to tunnel.
2. **Chosen:** The user first chooses a MBone-session to tunnel from a list over announced sessions. The information about announced sessions is gathered using a built in version of the multicast Session Directory, mSD [8], which is an application for displaying information about currently announced MBone-sessions. Second, the user chooses which sessions to create (an announced MBone-session can include several different media which results is several tunnel sessions).

2.4 The translator

The last part of mTunnel is the *translator*, that translates the traffic in various ways. Currently the translator includes a *recoder*, that translates between different encodings; a *mixer*, that mixes several streams into one stream; a *switch*, which only forwards packets from a certain source based on another stream (for instance, video packets are only tunneled for the member that currently is speaking); and a *scaler*, that rescales the traffic by dropping packets in predefined ways.

- The *recoder*, parses incoming traffic and translates it to another format. Currently, translation from PCM to GSM is supported, which for a normal MBone audio session would result in a reduction in bandwidth of about 75% (from 78 to 17 Kbps). LPC support is currently under development and would result in a reduction of about 88% if translating from PCM.
- The *mixer*, mixes several streams into a single stream. This allows for a bandwidth reduction, if several sources are active at the same time, but has the drawback of that the final receiver can not choose only to play out data from one sender. Currently mixing of PCM and GSM is supported.
- The *switch*, selects which packets to forward through the tunnel based on the activity in another stream, e.g. it can be instructed to only forward video packets from the source that currently is also sending audio packets.
- The *scaler*, rescales traffic by dropping certain parts of a traffic flow. This is currently only applicable on video traffic where the scaler drops packets randomly or drops complete frames depending on the type of traffic. This rescaling is possible due to the nature of video encodings on the MBone today, which are designed to tolerate packet-loss.

The different modes of the translator can of course be stacked after each other, e.g. an audio session can be both mixed and recoded.

2.5 Security issues

There are three main aspects of security in mTunnel: access to the interface, access to the tunneled data and firewalls.

Access to the Web-interface and the controller can either be public or based on simple user/password authentication. If the access is public, anyone that has access to the Web-interface can get information about current sessions and start new sessions, but only the host that created a session can alternate it (that is pause, continue, stop and/or change the priority).

When tunneling data over public networks, there is sometimes a need to hinder vicious persons from reading the tunneled data. This is solved using encryption of the tunneled traffic.

A third concern regarding security, is how to tunnel data through firewalls where it has proven administratively hard to open the firewall for regular multi-cast traffic. Here, mTunnel makes it easier to tunnel traffic as all tunneled data always have the same connection, i.e. its source and destination host and port is always the same. By this, it is easier to convince security administrators to open the firewall based on that connection pattern.

2.6 Time To Live

When packets are sent on the MBone today, their reach is limited by a so called *Time To Live (TTL)* value. For instance, if a user wants to send multicast packets to the local network only, he sends them with a TTL of 1.

In the current standard version of the sockets interface[4] under Unix and Windows, there is no way for a user application to get information about the TTL of an incoming packet. Due to the socket interface, and the fact that mTunnel runs as a user-application, mTunnel can only forward packets based on the TTL value specified when the session was created. Unfortunately, this means that if a user sends traffic in an announced MBone-session with a lower TTL than the announced TTL, the local packets will be "amplified" and retransmitted with a higher TTL than intended by the original sender.

The only available solution to this problem today is to make the users of mTunnel aware of the problem.

2.7 MBone Session Announcements

mTunnel includes an option for tunneling of MBone session announcements. These announcements are multicasted using the Session Announcement Proto-col [5] over a known group-address and port-number. When mTunnel is started, it can be instructed to automatically tunnel these announcements. This allows for users on the side that is normally not connected to get information about current MBone-sessions. This SAP-tunneler, can also be instructed to only tun-nel information about MBone-sessions that are currently tunneled. It can also

---

[4] The way an application speaks with the operating system and the network.

create new announcements for sessions that are manually entered (that is, they are not announced in the normal MBone manner).

To allow for faster update, known MBone-sessions are cached locally and are read from the cache and re-announced when mTunnel is restarted.

3 User based Quality-of-Service and session prioritization

The MRouted application (as presented in Sect. 1) connects networks to the MBone and makes them "true" members of the MBone, by tunneling all requested traffic. The decision on which sessions to tunnel, is based on requests made by the multicast aware applications that the users on the other side of the tunnel starts.

This model works very good if the bandwidth is not limited, as several sessions can be tunneled at the same time. But, if the bandwidth is limited (like over analog modems or ISDN-links) the users have to quit their MBone applications to stop a session from being tunneled (i.e. if an application "wants" multicast traffic for a special multicast group, usually the only way is to quit the application to stop the traffic). Also, if several users share the same narrow link, it might complicated or even impossible to coordinate which sessions to tunnel (several users join different sessions at the same time).

mTunnel instead uses a *user based Quality-of-Service model* where the local users explicitly have to choose which sessions to tunnel. This has several advantages, such as: making the end users aware of other currently tunneled sessions and removing the need for users to quit their MBone tools to stop the tunneling of specific multicast groups.

Another disadvantage with using MRouted for tunneling over dial-up links is that it uses the same link continuously for exchanging router information, even if no actual traffic is currently being tunneled. This means that links that have automatic set-up and tear-down will be kept dialed up as long as the MRouted program is running. mTunnel does not exchange this router information, as it only tunnels multicast traffic and is not a full router implementation (other multicast routers will not see the mTunnel as a router, but only as simple end-host in the network).

3.1 Prioritization

If the total currently needed bandwidth exceeds the available bandwidth, the tunneler can throw away packets (not send them through the tunnel) based on user defined priorities.

By default, all sessions have the same priority, but using the controller and the Web-interface, the priorities can be changed. Priorities for one or several sessions can also temporally be locked, meaning that no other session can get a higher priority than that/those sessions. This is useful if an important electronic meeting is conducted over the tunnel and the participants do not want to be disturbed by another user who wants to watch some other MBone-session.

Priorities, can also be configured in mTunnel, based on a number of different variables in the session: the media-type, the multicast address and port, the used bandwidth, the name and the description. This allows for advanced selection of priority schemes, that enables a user to participate in sessions even if the total needed bandwidth is not available.

4 Implementation and Status

The current prototype is implemented in the platform independent Java language (version 1.1), except some parts of the audio recoder that is implemented in C for efficiency reasons. The audio recode functionality is currently only available on Sun/Solaris, but the rest has been tested to work under both Unix and Windows95/NT4.

mTunnel is currently being used in three different ways: to connect different parts of a large software company's intranet, to connect computers at users homes, and to connect industry networks to the MBone.

More information about the current version and status of mTunnel can be found at [9].

4.1 Further issues

Important issues currently not addressed within the mTunnel development includes the ideas of *header compression* [2, 3], which together with stream based flow labels would result in a reduction of the bandwidth requirement. This means that not the full header would be transmitted through the tunnel, but common parts between packets in a stream would only transmitted with regular intervals and removed from the packets inbetween.

The encryption of the security part is currently completely missing and will be implemented in the future.

Another issue that requires further examination is compression of data sent over the tunnel. Several tunneled packets could be grouped together and then be compressed before sent through the tunnel.

The translator should also be extended to include a larger variety of encodings.

5 Summary and Conclusions

This paper presents a system for allowing users to easily connect to the MBone infrastructure and to connect different isolated multicast capable networks.

mTunnel allows users easy access to information about current tunneled sessions through a Web-interface, which also allows for easy configuration of existing and future sessions.

mTunnel does not start tunneling of MBone-sessions based on current multicast group activity, but instead makes the user responsible for deciding which

MBone-sessions to tunnel. This allows for a *user based Quality-of-Service model* where service decisions are left to the user.

To save bandwidth, data streams can be translated in four different ways: audio can be recoded to an encoding that requires lower bandwidth, several simultaneous audio streams can be mixed into a single stream, streams can be switched based on another stream (for instance, video packets could be forwarded, based on which audio packets are currently being received), and streams can be scaled by dropping certain parts of the traffic.

The system is currently being used to connect different parts of a large software company's intranet, to connect computers at users homes to the MBone and to connect industry networks to the MBone.

The usage of mTunnel has shown and proven that it is useful and that there is a need for this kind of applications.

Acknowledgments

Thanks to Mattias Mattsson, Fredrik Johansson, Håkan Lennestål, Johnny Widén and Ulrika Wiss, CDT, for interesting comments, encouragement and feedback.

This work was done within Esprit project 20598 MATES, which is supported by the Information technology part of the 4:th Framework Program of the European Union. Support was also provided by the Centre for Distance-spanning Technology (CDT).

References

1. S. Deering. Internet Group Management Protocol - IGMP. IETF RFC1112.
2. M. Degermark, M. Engan, B. Nordgren, and S. Pink. Low-loss TCP/IP header compression for wireless networks. In *Proceedings from MobiCom*, 1996.
3. M. Degermark and S. Pink. Soft state header compression for wireless networks. In *Proceedings from 6th International Workshop on Network and Operating System Support for Digital Audio and Video (NOSSDAV)*, 1996.
4. B. Fenner. MRouted. <URL:ftp://ftp.parc.xerox.com/pub/net-research/ipmulti />.
5. M. Handley. Session Announcement Protocol - SAP. work in progress.
6. V. Kumar. The MBone information web. <URL:http://www.mbone.com/>.
7. NCSA. Common Gateway Interface - CGI. <URL:http://hoohoo.ncsa.uiuc.edu/ cgi/>.
8. P. Parnes. The multicast Session Directory - mSD. <URL:http://www.cdt. luth.se/~peppar/progs/mSD/>.
9. P. Parnes. The multicast Tunnel system - mTunnel. <URL:http://www.cdt. luth.se/~peppar/progs/mTunnel/>.
10. H. Schulzrinne, S. Casner, R. Frederick, and V. Jacobson. RTP: A transport protocol for real-time applications. IETF RFC1889.

A Distributed Delay-Constrained Dynamic Multicast Routing Algorithm

Quan Sun and Horst Langendörfer

Institute of Operating Systems and Computer Networks
Technical University of Braunschweig
Bültenweg 74/75, 38106 Braunschweig, Germany

Abstract. Many new distributed multimedia applications involve dynamic multiple participants, have stringent end-to-end delay requirement and consume large amount of network resources. In this paper, we propose a distributed delay-constrained dynamic multicast routing algorithm (DCDMR) to support these applications. DCDMR scales well because the source of the multicast tree needs only limited computation or may even not be involved in the route computation. When group membership changes, the existing multicast tree is perturbed as little as possible and the average resulting tree cost is very satisfactory.

1 Introduction

Many new distributed real-time applications of computer networks such as distance education and videoconferencing involve multiple participants, have stringent end-to-end delay requirement and consume large amount of network resources. In order to support these new applications efficiently, multicast routing algorithms computing least cost multicast trees that satisfy a given end-to-end delay constraint are desirable. Because the membership of the multicast group changes frequently in some applications (e.g. teleconferencing), the routing algorithms should also have the ability to alter an existing multicast tree to accommodate membership changes. Although a number of delay-constrained multicast routing algorithms (e.g. [1] [2] [3]) have been proposed in the past few years, there exist few delay-constrained multicast routing algorithms that explicitly support dynamic multicast groups. Some existing multicast routing protocols used in the Internet (MBone) can very well support dynamic multicast groups. However, the underlying multicast routing algorithms have been designed only for best-effort delivery.

In this paper, we propose a distributed delay-constrained dynamic multicast routing algorithm which perturbs the existing multicast tree as little as possible when group membership changes. The proposed algorithm can be easily incorporated into the distance vector based routing protocol [12]. The benefits of this algorithm are that a given end-to-end delay constraint is satisfied and efficient network resource utilization can be achieved. In the next section, we give our network model and the formal definition of the Delay-Constrained Dynamic Multicast routing problem (DCDM). In section 3 and 4, we give our solution for the DCDM and the simulation results. Section 5 concludes the paper.

2   Network Model and Problem Definition

A network is modeled as a connected, directed graph $G=(V,E)$ where V is the set of nodes and E is the set of links. Two non-negative real value functions are associated with each link e ($e \in E$): delay $D(e)$ and cost $C(e)$. The link delay $D(e)$ is the delay a data packet experiences on the corresponding link, the link cost $C(e)$ is a measure of the utilization of the corresponding link's resources. Links are asymmetrical, namely the cost and delay for the link $e=(i,j)$ and the link $e' = (j,i)$ may not be the same.

Given a source node s ($s \in V$), a set of destination nodes S ($S \subseteq V - s$), a tree T ($T \subseteq G$) rooted at s and spanning all of the nodes in S (with all leaf nodes $\subseteq S$) is called a *multicast tree*. The cost of tree T is defined as follows:

$$C_T = \sum_{e \in T} C(e)$$

The nodes in S are often called multicast *group members*, $|S|$ is the multicast group size. Let $P_T(s, v)$ denote the unique path from s to $v \in S$ on the multicast tree T, if the delays of all source-destination paths on the tree are kept within a given bound Δ, namely:

$$\sum_{e \in P_T(s,v)} D(e) < \Delta \qquad \forall v \in S$$

then tree T is called a *delay-constrained* multicast tree for the given delay constraint Δ. The problem of finding a least cost delay-constrained multicast tree is NP-complete [1]

This paper focuses on the Delay-Constrained Dynamic Multicast routing problem (DCDM), this problem can be formally defined as follows: for a directed network $G=(V,E)$ with non-negative link costs C and non-negative link delays D (note that C and D may change with time), a delay tolerance Δ, the initial delay-constrained multicast tree T_0 with the source node s ($s \in V$) and a series of requests $\{r(t_1), r(t_2), ..., r(t_i), ...\}$ where $r(t_i)$ represents a request of adding a new group member or deleting an existing group member at time t_i($t_1 \leq t_2 \leq t_3 \leq ...$), find a series of delay-constrained multicast trees $\{T_1, T_2, ..., T_i, ...\}$ such that the members of T_i are those of T_0 modified by the requests $r(t_1), r(t_2), ..., r(t_i)$ and the cost of T_i is minimum among all possible choices for tree T_i.

A simple solution for DCDM is to reconstruct a new delay-constrained multicast tree using a static delay-constrained multicast heuristic like [2] for each request. Rebuilding a new tree in this way, however, may cause large disruptions for the current multicast sessions. This is especially unacceptable for real-time multimedia applications.

Most existing dynamic multicast routing heuristics consider only one link metric (link cost or link delay) [6] [7] [8] [9] [10]. The only known existing dynamic multicast routing algorithm that considers more than one link metric is the WAVE algorithm [11]. WAVE works as follows: when a new member v_n joins an existing tree, v_n contacts the source node s with a request **Req**. Starting from s, this request is propagated throughout the existing tree. When a node v_i on the tree receives a **Req**, it will send a response **Rsp** $= (v_i, cost, delay)$ to v_n along the shortest delay path from v_i to v_n, where $cost$ and $delay$ in **Rsp** are the cost and delay of the connection from s to v_n via v_i respectively. The node v_n will receive a set of responses from which it selects one $\mathsf{Rsp} = (v_k, cost, delay)$ and then sends a connection request to node v_k that generated this response. The node v_k will extend the existing tree from itself to v_n along the shortest path from v_k to v_n. WAVE has the following drawbacks: (1) the number of responses could be very large [11]; (2) The source node is always contacted whenever a new node is added, such a solution would not scale well and (3) WAVE does not consider explicitly a given delay constraint.

In this paper, we propose a distributed delay-constrained dynamic multicast routing algorithm (DCDMR) which perturbs the existing tree as little as possible. For a delete request, DCDMR functions exactly like the GREEDY algorithm [6], namely the node to be deleted is simply marked as 'deleted' and if this node is a leaf node on the existing tree, the branch of which it is a part is pruned. In the next section, we will concentrate on how an add request is processed by DCDMR.

3 The Proposed Algorithm (DCDMR)

3.1 Routing information

We first discuss the routing information which needs to be available for the proposed algorithm to do the route computation. Each node in the network should know the delays of all outgoing links and must maintain a delay vector and a cost vector.

The delay vector at node $v_i \in V$ consists of $|V| - 1$ entries, one entry for each other network node. The entry for node $v_j \in V$ $(v_j \neq v_i)$ contains the following items:

- The destination node identity: v_j
- The end-to-end delay of the least delay path $P_{ld}(v_i, v_j)$ from v_i to v_j: $D(P_{ld}(v_i, v_j))$
- The cost of the least delay path $P_{ld}(v_i, v_j)$ from v_i to v_j: $C(P_{ld}(v_i, v_j))$
- The next hop node on the least delay path $P_{ld}(v_i, v_j)$ from v_i to v_j: $id(P_{ld}(v_i, v_j))$

The cost vector at node $v_i \in V$ also consists of $|V| - 1$ entries, one entry for each other network node. The entry for node $v_j \in V$ $(v_j \neq v_i)$ contains the following items:

- The destination node identity: v_j
- The end-to-end delay of the least cost path $P_{lc}(v_i, v_j)$ from v_i to v_j: $D(P_{lc}(v_i, v_j))$
- The cost of the least cost path $P_{lc}(v_i, v_j)$ from v_i to v_j: $C(P_{lc}(v_i, v_j))$
- The next hop node on the least cost path $P_{lc}(v_i, v_j)$ from v_i to v_j: $id(P_{lc}(v_i, v_j))$

The above delay vectors and cost vectors are both similar to the distance vectors of the existing distance vector based routing protocol [12]. The same procedures for maintaining the distance vectors[1] can be used (and slightly ex-

---

[1] These procedures are very simple: the contents of the vector at each node are required to be periodically transmitted to direct neighbors of that node only. As one of the reviewers has pointed out, the critical point for our case is how frequently the cost vector should be transmitted.

tended) for maintaining the delay vectors and the cost vectors. Because our focus is on the routing algorithm, we will not discuss these procedures in this paper. In the following, we assume that the contents of the delay vectors and the cost vectors do not change during the execution of the routing algorithm.

3.2   The Delay-Constrained Unicast Routing Algorithm (DCR)

DCDMR is based on a distributed delay-constrained unicast routing algorithm (denoted by DCR) recently proposed by us [5]. For a source node s, a destination node d and a delay constraint $\triangle$, DCR constructs a path from s to d satisfying the given delay bound $\triangle$. In the following, we give a short description of DCR (for more information please see [5]).

DCR constructs the path one node at a time, from the source node s to the destination node d. If $D(P_{lc}(s,d)) < \triangle$, s reads the next hop node on the least cost path towards d, namely $id(P_{lc}(s,d))$, from its cost vector and sets:
- $next\_node = id(P_{lc}(s,d))$
- $path\_direction = LC$

otherwise, s reads the next hop node on the least delay path towards d, namely $id(P_{ld}(s,d))$, from its delay vector and sets:
- $next\_node = id(P_{ld}(s,d))$
- $path\_direction = LD$
- $delay\_so\_far = D(s, id(P_{ld}(s,d)))$

where $D(s, id(P_{ld}(s,d)))$ is the delay of the link $(s, id(P_{ld}(s,d)))$.
The source node s then constructs a *Path_construction* message and sends it to *next_node*. The *Path_construction* message contains the destination node d and the *path_direction*. If *path_direction* = LD, the *Path_Construction* message contains additionally the given end-to-end delay constraint $\triangle$ and *delay_so_far*, which represents the delay of the already constructed path from s to *next_node*.

When a node v ($v \neq d$) receives a *Path_construction* message, if *path_direction* in the received message is LC or if $delay\_so\_far + D(P_{lc}(v,d)) < \triangle$, node v reads the next hop node on the least cost path towards d, namely $id(P_{lc}(v,d))$, from its cost vector and sets:
- $next\_node = id(P_{lc}(v,d))$
- $path\_direction = LC$

otherwise, v sets
- $next\_node = id(P_{ld}(v,d))$
- $path\_direction = LD$
- $delay\_so\_far = delay\_so\_far + D(v, id(P_{ld}(v,d)))$

Node v then constructs a *Path_construction* message and sends it to *next_node*. Similarly, the *Path_construction* contains destination node d and *path_direction*. If *path_direction* = LD, the *Path_Construction* message contains also the delay constraint $\triangle$ and *delay_so_far*.

If the destination node d receives the *Path_Construction*, it means that the delay-constrained path has been successfully constructed.

The shortest path algorithm in the existing distance-vector routing protocol [12] guarantees to find a loop-free path if the contents of the distance vectors at all nodes are up-to-date and when the network is stable. Under similar conditions, namely if the contents of the delay vectors and the cost vectors at all nodes are up-to-date and do not change during the path construction process, we have also [5]:

Theorem 1: The paths constructed by DCR are loop-free.

Theorem 2: DCR can always find a path from a source node s to a destination node d satisfying a given delay constraint if such a path exists.

Theorem 3: The worst case message complexity of DCR is $O(|V|)$, where $|V|$ is the number of network nodes.

Simulation results in [5] have also shown that DCR has a very good performance regarding the costs of the computed paths.

3.3   The Dynamic Multicast Routing Algorithm (DCDMR)

In the following, we let $D(P_T(s, v_i))$ denote the delay of the unique path from s to v_i on the existing tree T where s is the root (source node) of T and v_i is any other node on T. $D(P_T(s, v_i))$ should be known to node v_i and $D(P_T(s, s)) = 0$.

In DCDMR, when a new member v_n joins an existing tree, v_n contacts node v_p on the existing tree with a request $\mathsf{Req} = (v_n, mode)$ along the shortest delay path where v_p can be any node of T (favourably a node of T in the vicinity of v_n) and *mode* can take one of the two values: FAST and SLOW. DCDMR can work in two different modes, namely the FAST mode or the SLOW mode.

If the *mode* value in Req is FAST, DCDMR takes the following steps:

(F1) Set $v_i \leftarrow v_p$.

(F2) If $v_i = s$ and $D(P_{ld}(s, v_n)) \geq \Delta$, there exists no delay-constrained path from s to node v_n and DCDMR terminates (s may send a "No path found" message to v_n along the shortest delay path); otherwise go to next step.

(F3) If $D(P_T(s, v_i)) + D(P_{ld}(v_i, v_n)) < \Delta$, node v_i computes a path from itself to the node v_n satisfying the delay constraint of $\Delta - D(P_T(s, v_i))$ using DCR (see section 3.2); otherwise go to next step.

(F4) v_i sends the received Req to v_i's parent node v_q on the existing tree (note that according to the following Lemma 1 we should not send Req to v_i's child node(s) on the existing tree). Set $v_i \leftarrow v_q$ and go to $(F2)$.

If the *mode* value in Req is SLOW, DCDMR takes the following steps:

(S1) Set $v_i \leftarrow v_p$.

(S2) If $v_i = s$ and $D(P_{ld}(s, v_n)) \geq \Delta$, there exists no delay-constrained path from s to node v_n and DCDMR terminates (s may send a "No path found" message to v_n along the shortest delay path); otherwise go to next step.

(S3) If $D(P_T(s, v_i)) + D(P_{ld}(v_i, v_n)) \geq \Delta$, node v_i sends the received Req to its parent node v_q on the existing tree (note that according to the following Lemma 1 we should not send Req to v_i's child node(s) on the existing tree), set $v_i \leftarrow v_q$ and then go to $(S2)$; otherwise go to next step.

(S4) Node v_i constructs a response $\mathsf{Rsp} = (v_n, mode, node, cost)$, where $mode$=SLOW, $node$=v_i and $cost$ is determined as follows: if $D(P_T(s, v_i)) + D(P_{lc}(v_i, v_n)) < \Delta$, $cost$ is set to be $C(P_{lc}(v_i, v_n))$; else $cost$ is set to be $C(P_{ld}(v_i, v_n))$. The Rsp is then sent to v_i's parent node and all v_i's child nodes on the existing tree (if $v_i = s$, Rsp is sent to s's child nodes only). When a node v_j on the tree receives the Rsp, it checks

$$D(P_T(s, v_j)) + D(P_{ld}(v_j, v_n)) < \Delta \tag{1}$$

If (1) is not satisfied, according to the following Lemma 1, the Rsp must come from v_j's parent node on T, and Rsp is sent to v_n along the shortest delay path. If (1) is satisfied, v_j determines the cost $C(v_j, v_n)$ of the path from itself to node v_n in the following way: if $D(P_T(s, v_j)) + D(P_{lc}(v_j, v_n)) < \Delta$, $C(v_j, v_n)$ is set to be $C(P_{lc}(v_j, v_n))$; otherwise $C(v_j, v_n)$ is set to be $C(P_{ld}(v_j, v_n))$. The originally received Rsp=$(v_n, mode, node, cost)$ will be modified if $C(v_j, v_n) < cost$. In this case, $node$ is set to be v_j and $cost$ is set to be $C(v_j, v_n)$. Rsp is then sent to all v_j's adjacent nodes (parent/child nodes) on the existing tree except the node from which v_j received the original Rsp.

Node v_n will receive a set of responses from which it selects the one with the least $cost$ value. Let this selected Rsp be $(v_n, mode, v_k, C_k)$. Node v_n then sends a connection request to node v_k that generated this response. Node v_k will compute a path from itself to v_n that satisfies the delay constraint of $\Delta - D(P_T(s, v_k))$ using DCR.

Lemma 1: Let s denote the root of the existing tree T, v_i any non-leaf node on T, v_j any one of v_i's child nodes on T and v_n the new node to be added. If $D(P_T(s, v_i)) + D(P_{ld}(v_i, v_n)) \geq \Delta$, then $D(P_T(s, v_j)) + D(P_{ld}(v_j, v_n)) \geq \Delta$.

Lemma 2: When DCDMR works in the SLOW mode, the largest possible number of responses received by the new node to be added is $L + 1$, where L is the number of leaf nodes on the existing tree.

The proofs of Lemma 1 and Lemma 2 are omitted here for save of space. The above distributed algorithm computes a path from the existing tree to the new node to be added. The worst case message complexity for the SLOW mode is $O(|V|^2)$ and the message complexity for the FAST mode is $O(|V|)$. The computed path and the existing tree can be easily combined to give a new delay-constrained multicast tree. The message complexity of this combination process is $O(|V|)$.

Compared with WAVE, our DCDMR algorithm working in the SLOW mode has the following advantages: (a) the number of responses sent to the new node could be much smaller; (b) In most cases, the source node will send Rsp to its child node(s) instead of the new node to be added, therefore less computation at the source node is needed in DCDMR than in WAVE. Furthermore, some nodes on the tree can stop propagating the Rsp to other tree nodes (for example the source node) and simply send the Rsp directly to the new node and therefore relieve other tree nodes from doing the route computation. (c) DCDMR is specifically targeted to consider a given delay constraint. If DCDMR works in the FAST mode, even less computation at tree nodes is needed and the route computation is very fast.

The DCDMR algorithm is very flexible in the route computation. For example, if a node finds that its multicast capability is depleted (e.g. it can provide only a small degree of multicast) or if not enough resources at the node are available, it can simply send the received Req or Rsp forward to other tree nodes immediately.

4 Performance Evaluation

We have used simulation experiments to evaluate the cost performance of the proposed algorithm and have run our simulations on the multicast routing simulator MCRSIM [4] developed at North Carolina State University. MCRSIM simulates actual ATM networks with fixed cell sizes. The simulator implements a modified version of the random network generator described in [6]. In MCR-SIM, each network node represented a non-blocking ATM switch. Each link was assumed to have a small output buffer. The propagation speed through the links was taken to be two thirds the speed of light. The propagation delay was dominant under these conditions, and the queueing component was neglected when calculating the link delay $D(e)$. In the simulation, the link cost $C(e)$ was set to be the total currently reserved bandwidth on the link. Because the NAIVE algorithm [9] is the only known dynamic multicast algorithm that can compute trees satisfying a given delay constraint (note that NAIVE uses only one link metric) and the bounded shortest multicast algorithm (BSMA) [2] has the best cost performance among all the proposed delay-constrained static multicast heuristics [4], both NAIVE and BSMA were included in the simulation for comparison. In the simulation, the original request Req of the node to be added was sent to the nearest node on the existing tree.

For each run of the experiment we generated a random network with homogeneous link capacities of 155 Mbps and an average node degree of 4. Random background traffic for each link of the network was also generated. The equivalent bandwidth of each link's background traffic was a random variable uniformly distributed between 10 Mbps and 120 Mbps. We then generated an initial multicast tree T_0, which consists of only a randomly generated source node s, and a random series of requests $\{r_1, r_2, ..., r_m\}$. Because there are situations for some applications in which members rarely leave the added group until the end of the application, we first consider the case where every r_i ($i=1,2,...,m$) is an add request (note that $r_i \neq r_j$ if $i \neq j$ for $i,j=1,2,...,m$). The algorithm DCDMR or NAIVE was applied to generate a new delay-constrained multicast tree T_1 for the request r_1 joining T_0, T_2 for the request r_2 joining T_1, ..., T_m for the request r_m joining T_{m-1}. The BSMA algorithm was used directly to generate the tree T_m for the source node s and all the m members to be added without considering the request sequence of the nodes to be added. The experiment was run repeatedly until confidence interval of less than 5% using 95% confidence level was achieved for the cost of the tree T_m. Fig.1(a) shows the percentage excess cost over BSMA versus number of network nodes for the number of add requests of 15 and the delay bound of 50 ms. Fig.1(b) shows the percentage excess cost over BSMA versus number of add requests for the number of network nodes of 60 and the delay bound of 50 ms. From these figures, we see that the DCDMR in the SLOW mode has a cost performance that is always within 10% worse than BSMA, while the DCDMR in the FAST mode and the NAIVE are up to 40 % and 70% worse than BSMA respectively.

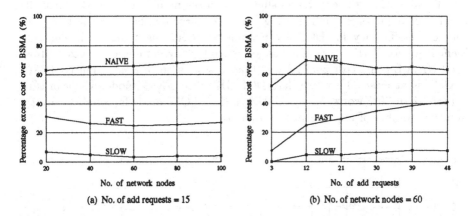

(a) No. of add requests = 15   (b) No. of network nodes = 60

Fig. 1. (a) percentage excess cost over BSMA versus number of network nodes and (b) percentage excess cost over BSMA versus number of add requests, for delay bound=50 ms

Next, we consider the situation where the request may be *add* or *delete*. To determine whether the next request is *add* or *delete*, we compute the following function (similar like that introduced in [6]):

$$P_a(k) = \frac{\alpha(n-k)}{\alpha(n-k)+(1-\alpha)k}$$

where n is the number of network nodes, k is the number of *group* nodes on the current tree and α is the fraction of group nodes on the tree at equilibrium. We then compute a random number r $(0 < r < 1)$. If $P_a(k) < r$, the new request is *delete* and randomly one of the group members on the current tree is determined as the node to be deleted; if $P_a(k) \geq r$, the next request is *add* and a non-group member is randomly selected as the node to be added.

In the simulation, we set α=0.15. Similarly, for each run of the experiment we generated a random network with an average node degree of 4 and random background traffic for each network link. The initial delay-constrained multicast tree T_0 was the resulting tree after processing a random series of 15 add requests for an initial empty tree consisting of only one randomly generated source node (for BSMA T_0 was generated without considering the request sequence). We then generated a series of requests $\{r_1, r_2, ..., r_m\}$ using the method described above. We recorded the cost of the resulting tree after processing the requests $r_1, r_2, ..., r_m$ for DCDMR and NAIVE and the cost of the tree computed using BSMA directly for those group nodes after the request r_m. The experiment was run repeatedly until confidence interval of less than 5% using 95% confidence level was achieved for the recorded cost. Fig.2(a) and Fig.2(b) show the comparison of the cost of the resulting tree obtained by BSMA and that obtained by DCDMR and NAIVE (note that the tree for DCDMR and NAIVE was dynamically evolving with each request). Fig.2(a) shows the results when the number of requests is small and Fig.2(b) shows the case for large number of requests. From both figures it is easy to see that the cost inefficiency for DCDMR in both FAST and SLOW mode increases slightly with the number of requests when the number of requests is small (see Fig.2(a)). For large number of requests (≥ 50), the cost inefficiency for DCDMR remains at about the same level. The cost inefficiency for NAIVE has not changed much for any number of requests. However, NAIVE has a cost performance that is always much worse than DCDMR. NAIVE is at least 20% worse than DCDMR in the FAST mode, which in turn is more than 20% worse than DCDMR in the SLOW mode for small to several hundred number of requests in our simulation.

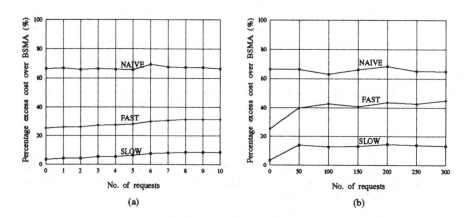

Fig. 2. Percentage excess cost over BSMA versus number of *add/delete* requests, for number of network nodes=60 and delay bound=50 *ms*

5 Concluding Remarks

In this paper, we proposed a distributed delay-constrained multicast routing algorithm (DCDMR) supporting dynamic multicast groups. DCDMR perturbs the existing multicast tree as little as possible when group membership changes. DCDMR scales well because the source node may not be involved in the route computation or needs only small computation. DCDMR has also very good cost performance, simulation results show that DCDMR in the FAST mode is at least 20% better than NAIVE, and DCDMR in the SLOW mode is at least 20% better than DCDMR in the FAST mode. When DCDMR works in the FAST mode, the route computation is very fast; when it works in the SLOW mode, very low cost trees can be computed.

There are a few areas for further research: (a) mechanisms that enable the proposed algorithm to cope with dynamic network environment, (b) the interaction with the routing protocols and resource reservation components, (c) how to deal with the situation where network nodes enter and leave the multicast group simultaneously, etc.

References

1. Kompella, V.P., et al., Multicast Routing for Multimedia Communication, *IEEE/ACM Transactions on Networking*, 1(3),pp.286-292, 1993
2. Zhu, Q., et al, A Source-Based Algorithm for Near-Optimum Delay-Constrained Multicasting, *IEEE INFOCOM'95*, pp.377-385, 1995
3. Wi, S. and Choi, Y., A Delay Constrained Distributed Multicast Routing Algorithm, *12th International Conference on Computer Communication, ICCC'95*, pp.833-838, 1995
4. Salama, H.F., et al, Evaluation of Multicast Routing Algorithms for Real-Time Communication on High-Speed Networks, in *High Performance Networking VI, IFIP 6th International Conference on High Performance Networking*, pp.27-42, 1995 (the multicast simulator MCRSIM can be downloaded from ftp://ftp.csc.ncsu.edu/pub/rtcomm/)
5. Sun, Q. and Langendörfer, H., A New Distributed Routing Algorithm for Supporting Delay-Sensitive Applications, Internal Report, Institute of Operating Systems and Computer Networks, TU Braunschweig, Germany, March 1997 (available from: http://www.cs.tu-bs.de/~sun/dunicast.ps)
6. Waxman, B.M., Routing of Multipoint Connections, *IEEE Journal on Selected Areas in Communications*, 6(9),pp.1617-1622, 1988
7. Kadirire, J. and Knight G., Comparison of Dynamic Multicast Routing Algorithms for Wide-Area Packet Switched (Asynchronous Transfer Mode) Networks, *IEEE INFOCOM'95*, pp.212-219, 1995
8. Bauer F. and Varma, A., ARIES: A Rearrangeable Inexpensive Edge-based On-line Steiner Algorithm, in *proceedings of GLOBECOM'95*, 1995
9. Doar, M. and Leslie, I., How Bad is Naive Multicast Routing, *IEEE INFOCOM'93*, pp.82-89, 1993
10. Effelsberg, W. and Mueller-Menrad, E., Dynamic Join and Leave for Real-Time Multicast, Tenet Report TR-93-056, Computer Science Division, University of California at Berkeley, Oct. 1993
11. Biersack, E. and Nonnenmacher J., WAVE: A New Multicast Routing Algorithm for Static and Dynamic Multicast Groups, in *proceedings of 5th Workshop on Network and Operating System Support for Digital Audio and Video*, 1995
12. Hedrick, C, Routing Information Protocol, Internet RFC 1058, June 1988

An Efficient Software Implementation
of a Forward Error Correcting Code*

Jean-Charles Henrion

Institut d'électricité Montefiore B.28, Université de Liège, B-4000 Liège, Belgium
Email: henrion@montefiore.ulg.ac.be

Abstract. Today, Forward Error-Correcting (FEC) codes are mainly implemented in hardware, and many believe that their complexity prohibit their software implementation. This paper presents in detail how the performances of a software implementantion can be significantly improved. Different levels of optimization which are independent of the working environment are presented and discussed. The coding throughput of 100 Mbps on an UltraSparc 1 shows that FEC codes can be easily added to multimedia applications without requiring dedicated hardware support. As a case study, we use FEC codes to protect AAL5-PDUs from cell losses in ATM networks.

1 Introduction

Communication systems usually rely on two types of error-correction mechanisms. The ARQ (Automatic Repeat Request) techniques, which require data retransmission, are used in many data applications and protocols. The FEC (Forward Error Correction) techniques rely on redundancy in the transmitted messages to recover errors at the receiving side. Three types of applications use FEC rather than ARQ techniques : applications that are sensitive to the delay (e.g. real time systems), Multicast applications (e.g. mutimedia) and applications which cannot retain information for a long time (e.g. satellite). The FEC codes are used to protect information messages which could be altered, or even partially deleted. Each message is segmented into a set of k independent data blocks which are transmitted as n encoded data blocks ($k \leq n$). The receiver will be able to restore the information message if it has received at least k of the n data blocks without errors. Although our coder (decoder) is specific to ATM Networks, the optimized coding and decoding procedures could be easily re-used in another context.

2 A FEC code for ATM Networks

Compared with packet-based networks such as Internet, one of the major characteristics of the ATM networks [1] is that the unit of information transfer is

* This work was partially supported by the European Commission within the ACTS AC051 OKAPI project.

a 53-byte ATM cell. As most users prefer to be able to send variable length packets, the ITU-T and the ATM forum have defined AAL5 and AAL3/4 [2]. Today, most ATM adapters provide efficient support for AAL5, and AAL3/4 is becoming less and less used. For this reason, we choose to design our FEC code above AAL5. The CS and SAR sublayers of AAL Type 5 are summarized in figure 1.

2.1 AAL Type 5

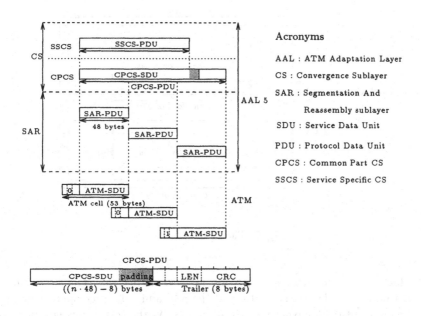

Acronyms

AAL : ATM Adaptation Layer

CS : Convergence Sublayer

SAR : Segmentation And
        Reassembly sublayer

SDU : Service Data Unit

PDU : Protocol Data Unit

CPCS : Common Part CS

SSCS : Service Specific CS

Fig. 1. The AAL Type 5

In ATM networks, cell losses may occur for several reasons. First, bit errors in the ATM cell header may force the discarding of the whole cell if the Header Error Control (HEC) field is unable to correct these errors. Another source of cell losses is the possible congestion in the ATM switches. When cell losses occur with AAL5, we have to distinguish between two different cases. If the last cell of CPCS-PDU is not lost, the receiver will detect an error because the number of received cells does not agree with the LEN field (length of the CPCS-SDU). If the last cell of a CPCS-PDU is lost, the SAR sublayer of the receiver will deliver the received cells concatened with the cells of the next CPCS-PDU(s) upon reception of the last cell of the next CPCS-PDU(s). We also consider the case where ATM cells are misinserted in the cell flow. This rare situation [4] happens when the HEC field is unable to detect an error in the ATM cell header. We assume that the cell loss rate is relatively low and burst losses are sufficiently limited. This corresponds to the CBR and VBR categories defined by ATM forum [3].

2.2 FEC codes

To protect ATM cells, the coder is integrated in the SSCS of AAL5. As it is easier to correct a one-byte erasure in m coded messages than to correct a m-byte erasure in one coded message, we have chosen a FEC matrix similar to the one used by AAL 1 [2]. In this structure, the FEC codes are computed "vertically", while the transmission order is "horizontal".

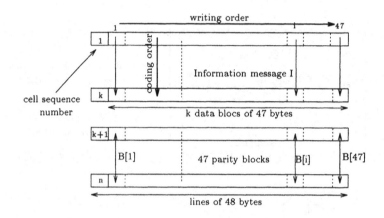

Fig. 2. The FEC matrix

We focused on the RS(255,251) code for several reasons. The first one is the fact that this code is systematic and the add and substract operations are strictly equivalent to the exclusive-OR operation. As the code is systematic, the coded message is constructed by attaching a parity block B to the information message I (figure 2). Each B[i] ($i \in \{1 \cdots 47\}$) is computed with the byte i of the k 47-byte lines. Our implementation of the RS(255,251) code is able to correct, either the erasure of four ATM cells or two erroneous bytes in each of the 47 coded vectors (figure 2). It should be noted that the RS(255,251) code is also theoretically able to correct two deleted cells with one erroneous byte in each of the 47 coded vectors. We could modify the corrective power of our coder by using other codes (RS(255,254), RS(255,253), RS(255,247)), or by adapting the size of the data blocks to be protected.

From an implementation point of view, the RS(255,247) code would certainly be interesting for workstations equipped with 64-bit CPU.

2.3 Using FEC codes with AAL5

The format of the FEC-PDU (figure 3) is a particularization of the FEC matrix shown in figure 2. The main elements of the FEC-PDU are the information message to be transmitted, the parity blocks, the cell sequence number and a trailer. In our implementation, we assume that the user data are passed to the FEC layer as a single contiguous buffer. Instead of copying the SDU into a number of 47-byte lines, we choose to divide it in two parts (BUF_1 and BUF_2). The bytes contained in BUF_2 correspond to the bytes of the original buffer that will be replaced by the cell sequence number. By only moving the contents of BUF_2 instead of the contents of BUF_1 and BUF_2, we avoid costly memory copy operations on the receiver side when the received FEC-PDU is correct. Our FEC sublayer introduces some padding after BUF_1 et BUF_2 so that they both end on cell boundaries. The FEC-PDU contains 47 parity blocks which are needed to protect the data (BUF_1 and BUF_2) as well as the FEC-PDU trailer. The main element of the FEC-PDU trailer is the 64-bit WSC. This error-correcting code is used by the decoder, when it has recoverted some errors, to validate the correction. The UL field which appears in the first line of the FEC-PDU contains the length of the FEC-PDU (in cells). This field is used by the decoder, when the last cell of the CPCS-PDU is lost, to estimate the LEN field.

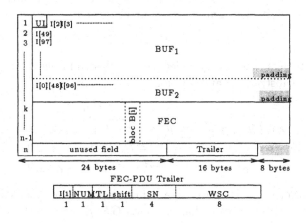

Fig. 3. FEC-PDU (SSCS-PDU)

2.4 The decoding in AAL5

If the CRC and the LEN fields of the received CPCS-PDU are correct, the decoding only involves the displacement of the contents of BUF_2 back in the information message. If an error is signalled by the CPCS to the decoder, it must first detect the cells which are deleted or inserted. The decoder must also stop its execution if it realizes that the corrective power of the code is not sufficient. To estimate the number of generated ATM cells, the decoder uses the LEN field if there exists an integer n such that $LEN = (n \cdot 48) - 8$ bytes. If this is not the case, the decoder will consult the UL field and the sequence number of the last emitted cell. Then, the decoder compares the number of received cells with the number of emitted cells. If equal, it assumes that one or more errors happened in the data information field of the ATM cells and it calls the recovery procedure for erroneous bytes. If there are too few cells, the recovery procedure for cell losses is used. If there are too many cells, it assumes that two FEC-PDUs have been concatened because the last ATM cell(s) of the first PDU has been lost. The decoder will analyze the second PDU because it must move this PDU inside the allocated buffer to take the lost cells of the first PDU into account.

3 Coding Theory

3.1 Basic principles

An information message can be splitted in an ordered list of information elements. The mathematical coding theory [5] notes that the codeword can only take a small set of values (n-uples) in relation with the cartesian product of value finite sets of information elements. We will use the following notation, $GF(q)$ to denote the value finite set of an information element :

$$GF(q) = \{0, 1, \alpha, \alpha^2, ..., \alpha^{q-2}\} \tag{1}$$

For value finite sets, the theory redefines the classical operations of addition, multiplication, substraction and division.

$$x \in GF(q = p^m) \Rightarrow p \cdot x = \overbrace{x + x + \cdots + x}^{p \text{ times}} = 0 \tag{2}$$

$$x \in GF(q), x \neq 0 \Rightarrow \exists \, x^{-1} \in GF(q) \mid x^{-1} \cdot x = 1 \tag{3}$$

$$\alpha^i \cdot \alpha^j = \alpha^{(i+j) mod(q-1)}, \; \alpha^{q-1} = \alpha^0 = 1 \tag{4}$$

In our C implementation, we have defined a set of arrays giving access to the binary representation of Galois elements $GF(q)$, and to the mult and div operations. These arrays are:

$$\forall i \in \{0 \cdots q - 2\} \; galois[i + 1] = \alpha^i, \; galois[0] = 0 \tag{5}$$

$$\forall i \in \{0 \cdots q - 2\} \; conv[i] = \alpha^i, invert[\alpha^i] = i \tag{6}$$

$$\forall x, y \in GF(q) \; mult[x][y] = x \cdot y, \; \forall y, x \neq 0 \in GF(q) \; div[x][y] = y/x \tag{7}$$

3.2 Short codes

We know [5] that the Reed-Solomon RS$(q - 1, q - \delta)$ code is defined over GF(q). If we do not distinguish between the vectorial and polynomial representations of an element list, the RS code can be computed from a generator polynomial G(x) such that $\forall O(x), \exists I(x) \Rightarrow O(x) = G(x) \cdot I(x)$.

Our RS(255,251) code is defined over GF(256) and its minimum distance δ is equal to five. We have also selected the following generator polynomial:

$$G(x) = (x - \alpha)(x - \alpha^2)(x - \alpha^3)(x - \alpha^4) \tag{8}$$

However, we wish encode variable-length message. Thus, the coding procedure uses short codes (n, k) $(n = k + 4, 1 \leq k \leq 251)$ derived from the RS(255,251) code whose minimum distance is $(\delta = 5)$ and whose generator polynomial is $G(x)$. We consider that $(255-n)$ well-defined elements of RS(255,251) codewords are equal to zero. Note that the Galois field GF(256) limits the maximum size of the information message in the FEC-PDU to 11712 bytes (244 cells).

4 Optimization of the coder

We had to reduce the symbolic expressions inherent in the computation of the parity blocks. The main optimization is the definition of a two-dimensional array (*coder*) which greatly reduces the processing time of the coder.

4.1 Symbolic analysis

The reduction method of symbolic expressions is presented below for a Reed-Solomon code $(q-1, q-\delta)$, defined over GF($q = 2^m$), with the minimum distance δ and whose generator polynomial $G(x)$ is:

$$G(x) = (x - \alpha^b)(x - \alpha^{b+1}) \cdots (x - \alpha^{b+\delta-2}) \tag{9}$$

If $I(x)$ is the message to code, there exists $Q(x)$ and $B(x)$ such that $B(x)$ is the parity block of $I(x)$.

$$I(x) = I_{k-1} \cdot x^{k-1} + \cdots + I_1 \cdot x + I_0 \tag{10}$$

$$Q(x) = Q_{k-1} \cdot x^{k-1} + \cdots + Q_1 \cdot x + Q_0 \tag{11}$$

$$B(x) = B_{r-1} \cdot x^{r-1} + \cdots + B_1 \cdot x + B_0 \quad (r = \delta - 1) \tag{12}$$

$$I(x) \cdot x^r = G(x) \cdot Q(x) + B(x) \tag{13}$$

$$\Rightarrow O(x) = I(x) \cdot x^r + B(x) \tag{14}$$

Since $O(x) \ mod \ G(x) = 0$, $O(x)$ belongs to the code. Now, we can factorize Q_i and B_i according to I_i to obtain coefficients $Q_{i,j}$ and $B_{i,j}$ which are independent of $I(x)$.

$$Q_i = Q_{i,k-1} \cdot I_{k-1} + \cdots + Q_{i,0} \cdot I_0 \quad i \in \{0 \cdots k - 1\} \tag{15}$$

$$B_i = B_{i,k-1} \cdot I_{k-1} + \cdots + B_{i,0} \cdot I_0 \quad i \in \{0 \cdots r - 1\} \tag{16}$$

4.2 Implementation of the coder

We can now define a two-dimensional array *coder* allowing the simultaneous update, for each I_j ($j \in \{0 \cdots k - 1\}$), of the four-byte (32-bit internal registers) parity block $B = (B_3 \ B_2 \ B_1 \ B_0)$. $\forall i \in \{0 \cdots 250\} \ \forall j \in \{0 \cdots 255\}$,

$$coder[i][j] = \begin{pmatrix} mult[B_{3,i}][j] \\ mult[B_{2,i}][j] \\ mult[B_{1,i}][j] \\ mult[B_{0,i}][j] \end{pmatrix} \Rightarrow B = \sum_{j=0}^{k-1} coder[j][I_j] \tag{17}$$

where $mult[i][j]$ is defined by equation 7.

Note that the array *coder* can be stored in RAM memory because its size is limited to 256KB, and it only depends on the RS(255,251) code. For each short code, the coder uses a subarray of *coder* ($i \in \{0 \cdots k - 1\}$, $j \in \{0 \cdots 255\}$).

We have also optimized our C implementation by maximizing the use of vectors (one-dimensional arrays) because a two-dimensional array is an array of pointers towards vectors. If the array addresses are stored in the CPU's internal registers, the number of instructions decreases when the software uses vectors (add, load) rather two-dimensional arrays (add, load, add, load). When we update the 47 parity blocks of our FEC-PDU (figure 4) in order to take account of the j line ($j \in \{0 \cdots k - 1\}$), we can define a vector tab_j such that:

$$tab_j = coder[j] \;\Rightarrow\; \forall w \in \{1 \cdots 47\}\; B[w] = \sum_{j=0}^{k-1} tab_j[I_j] \qquad (18)$$

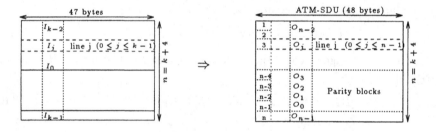

Fig. 4. $I(x)$ and $O(x) = I(x) \cdot x^r + B[w]$

The last optimization consists in the reduction of the memory references by following the ILP principles [8]. For example, the computation of the WSC and the parity blocks are integrated in a single loop.

5 Optimization of the decoder

We have reduced the symbolic expressions inherent in the computation of the syndrome polynomials and in the error-recovery procedures. The result is the definition of two two-dimensional arrays (*decoder* and *table_zero*) which greatly reduces the computation load of the decoder.

5.1 Symbolic analysis

In general, we consider a Reed-Solomon $(2^m - 1, 2^m - \delta)$ code, defined over $GF(2^m)$, whose minimum distance is δ and whose generator polynomial is defined by:

$$G(x) = (x - \alpha^b) \cdot (x - \alpha^{b+1}) \cdots (x - \alpha^{b+\delta-2}) \qquad (19)$$

When a message $O'(x)$ is received, the decoder can determine whether the message is an element of the code by computing the syndrome polynomial. If the syndrome polynomial is equal to zero, the decoder can recover the original message $O'(x) = O(x) + E(x)$ provided that the corrective power of the code is large enough. If the error locations in the received message are known (i.e the errors are erasures), the decoder will be able to reconstruct the original message provided that there are less than δ erasures. In particular, if the number of erasures is equal to $(\delta - 1)$, the decoder has to solve the following system of equations $S = H \cdot O' = H \cdot E$ where H is the parity matrix.

$$H = \begin{pmatrix} \alpha^{(q-2)\cdot(b+\delta-2)} & \cdot \alpha^{2\cdot(b+\delta-2)} & \alpha^{b+\delta-2} & 1 \\ \cdot & \cdot & \cdot & \cdot \\ \alpha^{(q-2)\cdot(b+1)} & \cdot \alpha^{2\cdot(b+1)} & \alpha^{b+1} & 1 \\ \alpha^{(q-2)\cdot b} & \cdot \alpha^{2\cdot b} & \alpha^{b} & 1 \end{pmatrix}$$

If the error locations are not known, the decoder can restore $O(x)$ provided that the number of erroneous bytes is lower than or equal to t, where t is the largest integer such that $2t + 1 \leq \delta$. The Euclidian algorithm [5] is used to compute the error-locator polynomial $\sigma(x)$ and the error-evaluator polynomial $\eta(x)$.

5.2 Implementation of the decoder

First, the recovery procedures must compute the value of the 47 syndrome polynomials $S(x)$ from the parity matrix H of our code.

$$S(x) = H \cdot O(x) = S_3 \cdot x^3 + S_2 \cdot x^2 + S_1 \cdot x + S_0$$

The optimization of the computation of $S(x)$ relies on the two-dimensional array decoder. $\forall\, i \in \{0 \cdots 254\}\, \forall\, j \in \{0 \cdots 255\}$,

$$decoder[i][j] = \begin{pmatrix} mult[j][conv[(4 \cdot i)\%255]] \\ mult[j][conv[(3 \cdot i)\%255]] \\ mult[j][conv[(2 \cdot i)\%255]] \\ mult[j][conv[i]] \end{pmatrix} \Rightarrow S = \sum_{j=0}^{n-1} decoder[j][O_j] \quad (20)$$

where $conv[i]$ is the binary representation of α^i (equation 6).

This array only depends on the RS(255,251) code, it can be precomputed and stored in memory. For each short code (n,k), the decoder uses a subarray of decoder ($i \in \{0 \cdots n-1\}$, $j \in \{0 \cdots 255\}$).

In section 2.4, we have explained in which cases the decoder invokes either the error-recovery procedure for erroneous byte, or the error-recovery procedure for cell losses. These procedures and their optimizations are described below.

Error-recovery procedure for erroneous bytes. This procedure searches the location of one or two erroneous bytes in one corrupted vector to decode O' from the syndrome polynomial $S(x)$. If we particularize the Euclidian algorithm to our RS(255,251) code, we can simplify the computation of the error-locator polynomial $\sigma(x)$ and of the error-evaluator polynomial $\eta(x)$. We obtain :

$$T_1 = S_2^2 + S_1 \cdot S_3 \tag{21}$$

1. If two bytes are in error, $T_1 \neq 0$:

$$T_2 = S_0 \cdot S_3 + S_1 \cdot S_2, \ T_3 = S_1^2 + S_0 \cdot S_2, \ T_4 = S_1^3 + S_0^2 \cdot S_3 \tag{22}$$

$$\sigma(x) = x^2 + (T_2/T_1) \cdot x + T_3/T_1 = (x - \alpha_i) \cdot (x - \alpha_j) \tag{23}$$

$$\eta(x) = (T_4/T_1) \cdot x + (T_3/T_1) \cdot S_0 \tag{24}$$

$$E_i = \eta(\alpha_i)/(\alpha_i + \alpha_j), \ E_j = \eta(\alpha_j)/(\alpha_i + \alpha_j) \tag{25}$$

The positions i and j of the two erroneous bytes is deduced from the value of the polynomial's roots (equation 23). Instead of solving this equation, the decoder uses another two-dimensional array *table_zero* which provides the solutions of equation 26.

$$x^2 + A \cdot x + B = (x - x_1) \cdot (x - x_2) \tag{26}$$

$$table\_zero[A][B] = (x_1, x_2) \tag{27}$$

2. If a single byte is in error, $T_1 = 0$:

$$T_2 = S_2/S_3, \ E_i = S_1 \cdot T_2^2 \tag{28}$$

$$\sigma(x) = x - T_2 = (x - \alpha_i) \tag{29}$$

The position i of the erroneous byte and E_i are directly obtained.

To recover the information message $O(x)$ from $O'(x)$, the decoder computes $O = (O' - E)$ where E is the error vector. This procedure has to be used every time one of the 47 syndrome polynomials is not equal to zero.

Error-recovery procedure for cell losses. This procedure is aware of the position of deleted bytes in each vector O' to decode. It computes the errors from the syndrome $S(x)$ of each O'.

$$E(x) = \sum_{i=0}^{n-1} E_i \cdot x^i \Rightarrow S_j = \sum_{i=0}^{n-1} E_i \cdot \alpha^{(j+1) \cdot i} \quad (0 \leq j \leq 3) \tag{30}$$

For example, the equations relative to the erasure of two bytes (i and j) are given by:

$$E_i = (\alpha^j \cdot S_0 + S_1)/T_2 \tag{31}$$

$$E_j = (\alpha^i \cdot S_0 + S_1)/T_3 \tag{32}$$

$$T_1 = \alpha^i + \alpha^j, \ T_2 = \alpha^i \cdot T_1, \ T_3 = \alpha^j \cdot T_1$$

We can obtain similar expressions for three or four deleted bytes [7]. With this procedure, the computation of T_i is realized once for the 47 vectors.

Optimization of the error-recovery procedures. We have also optimized the recovery procedures by using one-dimensional arrays. Let us for example consider the case where two cells are deleted. Equations (31) and (32) show that we can define four vectors:

$$tab_1 = mult[conv[j]], \ tab_2 = div[T_2] \tag{33}$$

$$tab_3 = mult[conv[i]], \ tab_4 = div[T_3] \tag{34}$$

where the arrays *mult*, *conv* and *div* are defined by equations 6 and 7.

If we divide each syndrome polynomial S[w] ($w \in \{1 \cdots 47\}$), we compute the 94 erroneous bytes $E_i[w]$ et $E_j[w]$ in the following way.

$$S[w] = S_3[w] \cdot x^3 + S_2[w] \cdot x^2 + S_1[w] \cdot x + S_0[w] \tag{35}$$

$$(31) \Rightarrow E_i[w] = tab_2[tab_1[S_0[w]] + S_1[w]] \tag{36}$$

$$(32) \Rightarrow E_j[w] = tab_4[tab_3[S_0[w]] + S_1[w]] \tag{37}$$

6 Performance Measurements

To evaluate the performances of our coder and decoder, we have measured the net coding and decoding throughputs on a 167 MHz Ultrasparc 1. The measurements presented in this section correspond to memory to memory transfer on a single workstation in order to avoid interferences with the ATM adapters and drivers.

The coding test procedure consists in encoding a sufficiently large number of data blocks so that the operation takes several seconds. We use a large number of different data blocks in order to avoid too optimistic results due to cache effects. The array *coder* can go in the CPU cache and this phenomenon will be more and more important when the data block size becomes smaller. In this case indeed, the coder calls a small number of vectors (equation 18).

The decoding test procedure goes on two steps. First, we estimate the needed time (t_1) to code a large number of data blocks and generate randomly a particular error. Then, we estimate the time (t_2) needed to code, generate randomly a particular error and decode each data block. The decoding throughput is computed from ($t_2 - t_1$).

Figure 5 shows the influence of the message size on the throughput. We note that for messages larger than 2KB, the coding throughput exceeds 100 Mbps. By comparison for a 4KB message, we obtain a coding throughput of 20 Mbps with a SparcStation 5 and 45 Mbps with a 133 MHz Pentium PC. If a systematic error occurs (deleted ATM cells, corrupted ATM-SDUs), the recovery procedure limits the data processing throughput to less than 50 Mbps. If the decoder receives less than 50 % erroneous packets of more than 2KB, it is able to sustain the coding throughput. Indeed, the decoding process works at high throughput (> 500 Mbps) if the CRC or the LEN field are correct, because the decoder activity only consists of putting the data bytes of BUF_2 back in the information message (figure 3).

In table 1, we have compared the performances of our code with error-detecting codes developed by D.Feldmeier [6] on our UltraSparc 1. We have

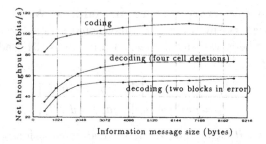

Fig. 5. Memory-to-Memory Throughputs Measurements with 167 MHz UltraSparc 1

Our RS coder (4KB)	108 Mbps
32-bit CRC(array)	162 Mbps
32-bit CRC(shift & add)	269 Mbps
64-bit WSC(recursive)	701 Mbps
read & write	705 Mbps

Table 1. Comparative Performance Measurements with 167 MHz UltraSparc 1

noted that the coding throughput is 30% lower than the throughput (162 Mbps) of a naïve CRC implementation. These throughputs are equivalent because this CRC version (our coder) uses a one-dimensional array for updating the CRC (the parity blocks). By comparison, the best CRC implementation runs at 269 Mbps and the best 64-bit WSC(Recursive) runs at 701 Mbps. We have also noted that the WSC computation and the read & write process (705 Mbps) runs approximatly at the same speed because the CPU activity seems limited by the memory access speed(RAM) on this machine.

Today, most software implementations offer the possibility to tune the dimensions of the FEC code. So, a well-optimized software [12] proposes coding throughputs of 2 to 10 Mbps with a 133 MHz Pentium PC. Our throughput (45 Mbps) with this type of machine shows the usefulness of an optimized implementation which is specific to a particular code.

7 Conclusions

Our coding throughput of 100 Mbps on a UltraSparc workstation shows that an optimized implementation of a FEC code can offer sufficient performances. The coding and decoding procedures have been optimized as follows. First of all, we have particularized the coding and decoding mechanisms to our RS(255,251) code, and we have reduced the polynomial computation by analyzing their symbolic expressions. The result is the definition of two-dimensional arrays which significantly simplify the update of the parity blocks, the update of the WSC, the update of the syndrome polymomials and the work of the error-recovery procedures. A second step of the optimization was the use of vectors defined from these arrays. Finally, we have reduced the number of memory references by following the ILP principles [8].

The software implementation of FEC codes could be integrated in communication protocols of multimedia applications to protect well-defined group of data blocks, eventually with different priority levels [14]. The discussed optimizations can be applied to similar RS codes and are not specific to our particular environment. For example, it is clear that IP datagrams could be protected in a similar scheme.

References

1. M. de Prycker : "Asynchronous Transfer Mode, Solution for Broadband ISDN", Ellis Horwood Series in Computer Communications and Networking, 1993.
2. ITU-T Recommandation I.363 : "Integrated Services Digital Network (ISDN), Overall Networks Aspects and Functions, B-ISDN ATM Adaptation Layer(AAL) specification", 1993.
3. ATM Forum : "Traffic Management 4.0", ATM Forum Specification af-tm-0056.001, 1996.
4. J-R. Louvion, B. Piller : "Performance Measurements and Traffic Characterization on the ATM Pilot Network", European Transactions on Telecommunications, Vol. 7, N° 5, Sept. 1996.
5. H. Imai : "Essentials of Error-Control Coding techniques", Academic Press, Inc., Harcourt Brace Jovanovitch, Publishers, 1990.
6. D. Feldmeier : "Fast Software Implementation of Detection Codes", IEEE/ACM Transactions on Networking, Vol. 1, N° 6, December 1995.
7. J-C. Henrion : "Evaluation du débit maximum d'algorithmes FEC dans un environnement ATM", University of Liège, Institut Montefiore, June 1996.
8. D.D. Clarck, D.L. Tennenhouse : "Architectural considerations for a new generation of protocols", SIGCOMM'90 Proceedings, Philadelphia,Sept. 24-27, 1990.
9. J. Nonnenmacher, E. Biersack : "Reliable Multicast: Where to use FEC", Fifth International Workshop on Protocols for High-Speed Networks, W. Dabbous & C. Diot Eds, Chapman & Hall, 1997.
10. C. Huitema : "The case for packet level FEC", Fifth Workshop on Protocols for High-Speed Networks, W. Dabbous & C. Diot Eds, Chapman & Hall, 1997.
11. A. McAuley : "Reliable Broadband Communication using a burst Erasure Correcting Code", Computer Communication Research Group, Presented at ATM SIG-COMM '90, Philadelphia, 1990.
12. L. Rizzo : "Effective erasure codes for reliable computer communication protocols", Computer Communication review, Vol. 27, N°2, April 1997.
13. P. Karn : URL "http://www.qualcomm.com/people/pkarn/ham.html", "Amateur Audio Digital Communications", May 1997.
14. A. Albanese, M. Luby : "PET - Priority Encoding Transmission", High-Speed Networking for Multimedia Applications, Kluwer Academic Publishers, 1996.

Disk Scheduling for Variable-Rate Data Streams

Jan Korst, Verus Pronk and Pascal Coumans

Philips Research Laboratories
Prof. Holstlaan 4, 5656 AA Eindhoven, The Netherlands
{korst,pronk,coumans}@natlab.research.philips.com

Abstract

We describe three disk scheduling algorithms that can be used in a multimedia server for sustaining a number of heterogeneous variable-rate data streams. A data stream is supported by repeatedly fetching a block of data from the storage device and storing it in a corresponding buffer. For each of the disk scheduling algorithms we give necessary and sufficient conditions for avoiding under- and overflow of the buffers. In addition, the algorithms are compared with respect to buffer requirements as well as average response times.

key words: disk scheduling, continuous media, video server, multimedia, variable rate, buffer requirement, response time.

1 Introduction

Multimedia applications can be characterized by their extensive use of audio-visual material in an interactive way. The presentation of this material requires a continuous data stream from the storage device on which the material is stored to the user.

The magnetic disk is considered most appropriate as secondary storage medium in a multimedia server. It offers a large storage capacity and small random access times at a reasonable cost [Reddy & Wyllie, 1994]. To ensure that magnetic disks are used cost-effectively, several users have to be serviced simultaneously by a single disk or a disk array. This is realized by repeatedly fetching a data block for each of the users and storing it in a corresponding buffer, which has room for a number of data blocks. The buffers are implemented in random access memory (RAM), which is relatively expensive.

Here, we assume that a user consumes data from his buffer, via some communication network, at a rate that may vary between zero and a maximum consumption rate. Users may have different maximum consumption rates. So far, the problem of scheduling variable-rate data streams has received little attention in the literature. Most papers assume constant-rate data streams; see, e.g., Yu, Chen & Kandlur [1993], Özden, Rastogi & Silberschatz [1995], and Korst, Pronk, Aarts & Lamerikx [1995]. Important advantages of considering variable rates are that *(i)* the server can handle variable-bit-rate-encoded data streams and *(ii)* no extra provisions have to be made for handling slow-motion and pause/continue requests from the user. Variable-bit-rate-encoded data streams are defined in, for example, the MPEG-2 standard.

A disk scheduling algorithm determines on-line when and how much data must be fetched for each of the users. It must guarantee that buffers do not become empty or overflow. For the sake of convenience, we speak of buffer underflow if a buffer becomes empty. In addition, a disk scheduling algorithm must also respond promptly to user requests. Reading small data blocks from disk results in relatively small buffer sizes and small response times, but also in many disk accesses per unit of time, i.e., in a less effective use of the disk. Therefore, there is generally a trade-off between, on the one hand, buffer sizes and corresponding response times and, on the other, the effectiveness of disk usage, which determines how many users can be serviced simultaneously.

The aim of this paper is to describe three disk scheduling algorithms for sustaining a number of variable-rate data streams. The algorithms are compared with respect to buffer requirements and response times.

The remainder of this paper is organized as follows. In Section 2, we give a more precise statement of the disk scheduling problem of our interest. Related work is briefly discussed in Section 3. In Section 4, we first consider the simpler case in which the consumption rates are constant. There, we discuss the well-known double buffering or SCAN algorithm. We introduce three disk scheduling algorithms, that are based on the double buffering algorithm, for sustaining variable-rate data streams in Section 5. In Section 6, we compare the buffer requirements and, in Section 7, the average response times of these three scheduling algorithms. Finally, Section 8 contains some concluding remarks.

2　Problem Definition

Figure 1 shows a schematic of a video server. Before giving a more precise statement of the disk scheduling problem, we briefly discuss the three basic components, namely the disks, the users, and the buffers.

Disks.　The disk array consists of one or more magnetic disks, using RAID technology [Patterson, Gibson & Katz, 1988]. The data is striped across all disks in the array, such that a request for a data block results in a disk access on each of the disks in the array. In this way, load balancing problems are avoided and a large composite data transfer rate can be guaranteed. As such, the disk array can be regarded as a single virtual disk. We will therefore often regard the array as a single disk in the following sections. The data on a disk is stored on concentric circles, called tracks. Each track consists of an integer number of sectors, the tracks near the outer edge usually containing more sectors than the tracks near the inner edge of the disk. A disk rotates at a constant angular velocity, so that reading near the outer edge results in a higher data transfer rate than reading near the inner edge.

The time required for accessing data from a single disk generally consists of *seek time*, i.e., the time required to move the reading head to the required track, *rotational delay*, i.e., the time that passes before the required data moves under the reading head once the required track has been reached, and *read time*, i.e., the time required to actually read the data. The sum of the seek time and the rotational delay is called the *switch time*.

The read time depends on the amount of data that is to be read and on the radial position of the track(s) on which the data is stored. The rotational delay per access takes at most one revolution of the disk. The seek time per access is maximal if the reading head has to be moved from the inner edge to the outer edge of the disk, or vice versa. To avoid that we have to take into account such a maximum seek time for each access, disk accesses are generally handled in batches. As the head moves from the inner edge of the disk to the outer edge, or vice versa, the required data blocks are read in the order in which they are encountered by the head. Carrying out such a batch is called a *sweep*. The worst-case total seek time that is required to execute a sweep with n disk accesses has been analysed by Oyang [1995].

For our problem, we assume that the disk array is characterized by a guaranteed data transfer rate r and a switch time function $s(l,m)$. The data transfer rate r gives the minimum guaranteed rate at which data can be read. The switch time function $s(l,m)$ gives the time that is maximally spent on switching when l data blocks have to be fetched in m sweeps, where the l data blocks may be assigned arbitrarily to the m sweeps. The worst-case switch time is defined in this way to allow easy analysis of the disk scheduling algorithms presented in Section 5.

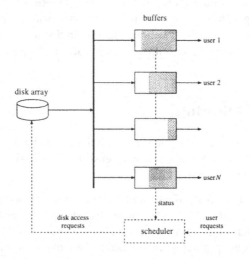

Figure 1: The basic components of a video server.

Users. The users of a server, once admitted service, can be in one of two states: waiting or consuming. Initially, a user is waiting. When sufficient data has been fetched from disk and stored in his buffer, the user can start to consume. In this state, a user i is allowed to consume data from his buffer for an indefinite period of time, at a rate that may vary between zero and a maximum consumption rate $c_i^{\max}$.

The maximum rate that must be allocated for the playback of a variable-bit-rate-encoded MPEG sequence can be determined in an off-line analysis of this sequence, as explained by Dengler, Bernhardt & Biersack [1996]. If we choose $c_i^{\max}$ on the basis of the peak bit rate, which is determined by the size of the largest frame in the sequence, then this generally leads to a $c_i^{\max}$ that is considerably larger than the mean bit rate. If the rate at which data is consumed is averaged for each sequence of, say, N successive frames, then the resulting value for $c_i^{\max}$ can be chosen as the maximum average value, which will be considerably smaller.

We restrict ourselves in this paper to disk scheduling algorithms that offer deterministic guarantees. While a user is consuming, additional data must be repeatedly fetched, in such a way that his buffer neither under- nor overflows. A disk scheduling algorithm is called *safe* if it guarantees that the buffers of consuming users never under- or overflow.

For reasons of simplicity, we assume a continuous model, also called fluid-flow model, with respect to the consumption of data by the users. In practice, data is of course consumed in discrete units. Assuming a continuous model, however, considerably simplifies the analysis of disk scheduling algorithms, while the resulting differences are negligible.

If a consuming user requests other data, he temporarily becomes waiting again. First, his buffer has to be filled with a sufficient amount of new data. The time between the moment that a user request arrives at the server and the moment the user can start consuming the corresponding new data is called the *response time* of this request. Note that delays caused by the communication network are not incorporated in the response time defined above. Disk scheduling algorithms can be compared with respect to their worst-case as well as their average-case response times. Usually average-case times are more important, because the probability of a worst-case situation occurring is usually very low.

Buffers. As already mentioned, buffers are implemented in RAM, which is relatively expensive. Since a consuming user may also cease to consume data for an indefinite period of time, a data block can only be fetched in a given sweep, if at the start of this sweep there is already enough room in the buffer to store it. Otherwise, buffer overflow may occur, causing data which has not yet been read from the buffer to be overwritten.

If a data block is fetched in a given sweep, then it may arrive in the buffer at any moment during the sweep, immediately at the start or only at the end. In the analysis of a possible occurrence of buffer underflow, we assume that a data block arrives in the buffer at the end of the sweep.

Given the above assumptions we can now define the disk scheduling problem as follows. Given a disk with a data transfer rate r and a switch time function $s(l, m)$, find a safe disk scheduling algorithm that can simultaneously service a set U of n users, minimizing a combination of buffer requirements and average-case response times.

Whether more emphasis should be put on minimizing the buffer sizes or minimizing the response times depends on the specific multimedia application at hand. For video-on-demand applications, response times are probably not that important, as long as they are not too large. For highly interactive applications, such as games, response times are very important.

3 Related Work

In this section, we discuss related work on the retrieval of variable-rate data streams from disk arrays. Existing papers propose solutions specifically developed either for handling variable-bit-rate MPEG video or for implementing VCR-like functions such as fast forward at variable speed.

Vin, Goyal, Goyal & Goyal [1994] present a disk scheduling algorithm in which the variability in seek times and frame sizes of MPEG-encoded video is exploited to provide only statistical guarantees. A similar statistical approach is pursued by Rautenberg & Rzehak [1996].

Chang & Zakhor [1994] present two general approaches for handling variable-bit-rate MPEG video, called constant time length (CTL) and constant data length (CDL). In the CTL approach, data blocks that correspond to a constant playback duration are periodically retrieved from disk. The successive data blocks will usually vary in size. In the CDL approach, data blocks of constant size are repeatedly fetched, with the play-back time of a data block varying from data block to data block. Based on the CTL approach, Chang & Zakhor propose an admission control algorithm that offers statistical guarantees. In addition, the authors consider the use of scalable compression to achieve graceful degradation in overload situations.

Chen, Kandlur & Yu [1995] discuss a number of ways to handle variable-rate play-back, e.g., fast forward at different rates, taking into account the interframe dependencies of MPEG video data. They subdivide an MPEG sequence into independently encoded segments and assign these segments to the disks in the disk array in a specific way, to allow the skipping of a variable number of segments between two consecutively retrieved segments. They do not consider the specific problems relating to variable-bit-rate MPEG video.

Dengler, Bernhardt & Biersack [1996] present an interesting retrieval algorithm for variable-bit-rate MPEG video using the CTL approach, which offers deterministic guarantees. The authors determine the maximum size of the data blocks that have to be periodically fetched from disk, based on an off-line analysis of the video data. By averaging the consumption rate over a fixed period, in which a data block is fetched for each user, this maximum size can be chosen considerably smaller than the data block size that can be derived from the peak bit rate of the MPEG sequence. The maximum block size can be further reduced by averaging over more than one period, allowing the admission of more users. However, this increases the buffer sizes as well as the response times. The scheduling algorithm proposed by Dengler, Bernhardt & Biersack presupposes constant-frame-rate consumption and retrieves data blocks in periods of fixed length.

The disk scheduling algorithms that we present in this paper differ from the algorithm of Dengler, Bernhardt & Biersack in the sense that we use periods of variable length, which leads to considerably smaller average-case response times.

4 Scheduling for Constant-Rate Data Streams

Before focusing on variable-rate data streams, let us first consider the simpler case of constant-rate data streams. Hence, we assume that user i consumes data at a constant rate of c_i. For ease of reference, the data stream that is being consumed by user i will often be referred to as data stream i below.

For this special case, we discuss the *double buffering* algorithm (DB), which is based on the well-known SCAN algorithm, originally introduced by Denning [1967]. The SCAN algorithm has been adapted by a number of authors for handling continuous data streams; see, e.g., Gemmell [1993], Kandlur, Chen & Shae [1991], Kenchammana-Hose-kote & Srivastava [1994], and Rangan, Vin & Ramanathan [1992].

All n data streams are serviced by the double buffering algorithm as follows. Let the time axis be divided into periods of length P. During each period, a data block is fetched for each data stream in a single sweep of the disk reading head. To minimize the buffer requirements, P should be chosen as small as possible. The size B_i of a data block for data stream i equals Pc_i, which is the amount of data user i consumes during one period. In this way, for each user, exactly one data block is fetched and one data block is consumed in every period.

The number of data streams that can be guaranteed to be serviced simultaneously is of course bounded by the disk's data transfer rate r. Let $C = \sum_{j \in U} c_j$. Then the users can only be serviced simultaneously if $C < r$.

The buffer for a data stream has room for exactly two data blocks. A data block that is fetched from disk in a given period is consumed in the next period. If a user i has just been admitted service or has requested another data stream, then the consumption of this new data stream can start at the end of the period in which the first data block was fetched for this stream. This time is denoted as T_i^{start}.

We first derive a necessary and sufficient condition for the safety of DB. The worst-case time T_s necessary for fetching a data block for each of the data streams in one sweep is given by

$$T_s = \frac{\sum_{j \in U} B_j}{r} + s(n, 1).$$

Now, the period length P must be at least T_s, or

$$P \geq \frac{\sum_{j \in U} B_j}{r} + s(n, 1). \tag{1}$$

Since $B_i = Pc_i$ for each $i \in U$, we can rewrite $\sum_{j \in U} B_j$ as

$$\sum_{j \in U} B_j = P \sum_{j \in U} c_j = PC. \tag{2}$$

By combining (1) and (2), we can prove the following result.

Theorem 1. *Let U be a given set of n users, with $C = \sum_{j \in U} c_j$. Let $C < r$ and let the buffer for each $i \in U$ have room for two data blocks of size B_i. Furthermore, let $B_i/c_i = B_j/c_j$ for each pair $i, j \in U$. Then, the double buffering algorithm is safe if and only if for each $i \in U$*

$$B_i \geq \frac{r\, s(n, 1)}{r - C} c_i. \tag{3}$$

Proof. The sufficiency of (3) can be demonstrated as follows. If each sweep is guaranteed to be completed within P time units, i.e., if $T_s \leq P$, then neither buffer under- nor overflow will ever occur. This can be shown as follows. At time T_i^{start}, the buffer for stream i contains exactly one data block. Now, if $T_s \leq P$, then in each following period one data block will be consumed, and one data block will be fetched from disk for this stream. Consequently, at the start of each following period, the buffer again will contain exactly one data block. Since at the start of a period the buffer contains sufficient data for consumption during that period, buffer underflow will not occur. Since there is already room for one data block at the start of a period, buffer overflow will not occur either. We next prove that $T_s \leq P$ is equivalent to (3). By combining (1) and (2), it is easily shown that $T_s \leq P$ is equivalent to $P \geq \frac{r \, s(n,1)}{r-C}$. Since $B_i = P c_i$ for each $i \in U$, elimination of P yields the required result.

The necessity of (3) can be demonstrated as follows. If (3) does not hold, then there is an $i \in U$ such that $B_i / c_i < \frac{r \, s(n,1)}{r-C}$. This implies that $P < \frac{r \, s(n,1)}{r-C}$, which is equivalent to $T_s > P$, i.e., the resulting period P is not large enough to accommodate a worst-case sweep. Thus, buffer underflow may occur. □

The worst-case response time of DB is $2P$. This occurs if a request is issued just after the start of a period. The consumption of the new data can then start only at the end of the next period. The average-case response time is $\frac{3}{2}P$, because on average we have to wait $\frac{1}{2}P$ before a request can be handled in the following period, which takes another P.

We end this section with the observation that DB requires the consumption from the buffers to be strictly synchronized with the data retrieval from disk. At the start of each period, there must be exactly one data block in the buffer of each consuming user. Even for constant-rate data streams this strict synchronization may be difficult to realize in practice.

5 Scheduling for Variable-Rate Data Streams

If the consumption rates of the users vary over time, then the double buffering algorithm cannot prevent buffer under- and overflow by just using $c_i^{\max}$ instead of c_i in Equation (3). This can be inferred as follows. Since the consumption rate may be equal to zero, one can only fetch a data block in a given period, if there is room for this block at the start of this period. Otherwise, buffer overflow may occur. However, if at the start of some period, there is an amount of $B_i + \varepsilon$ in the buffer for user i, for some ε with $0 < \varepsilon < B_i$, and the buffer can store exactly $2B_i$ of data, then the next data block can be fetched only in the next period. This block may arrive at the end of this period, in which case user i may have consumed $2B_i$ of data so that buffer underflow may have occurred.

Since the execution of sweeps can no longer be synchronized with the consumption of data from the buffers, it is of no use to allocate a worst-case time for each sweep. A new sweep is started immediately upon completion of the previous one, instead of every P time units. This can improve the average-case response times considerably, as will be shown in Section 7. A user can (re)start consuming at the end of the sweep in which the first data block has been fetched. As in Section 4, this time is denoted by T_i^{start}. Furthermore, we assume that successive sweeps are consecutively numbered.

Next, we discuss three disk scheduling algorithms for sustaining variable-rate data streams, each using this variable-period approach.

5.1 Triple Buffering Algorithm

A straightforward way of generalizing the double buffering algorithm, such that it can handle variable consumption rates, is obtained by using $c_i^{\max}$ instead of c_i in Equation (3) and by extending the buffers such that they can store three data blocks. By analogy, this algorithm is called the *triple buffering* algorithm (TB).

It works as follows. A data block of constant size is fetched for data stream i in a given sweep, only if there is room for another data block at the start of the sweep. Again, a data block must contain enough data to survive at least a worst-case sweep. With respect to the safety of TB, we derive the following result.

Theorem 2. *Let U be a given set of n users, with $C = \sum_{j \in U} c_j^{\max}$. If $C < r$ and the buffer for each $i \in U$ has room for three data blocks of size B_i, then the triple buffering algorithm is safe if and only if for each $i \in U$*

$$\frac{B_i}{c_i^{\max}} \geq \frac{\sum_{j \in U} B_j}{r} + s(n, 1). \tag{4}$$

Proof. The sufficiency of (4) can be demonstrated as follows. Assuming that (4) holds, it is easy to see that, for each user $i \in U$, at most B_i data will be consumed in a single sweep. Buffer overflow will never occur, because a data block will be fetched in a sweep only if there is already enough room for this data block at the start of the sweep. The proof that buffer underflow cannot occur is by contradiction. Let j be the sweep that is started at time T_i^{start}, and let j', $j' \geq j$, be the first sweep in which buffer underflow can occur for user i. Clearly, $j' > j$, since at the start of sweep j there is exactly B_i data in the buffer, so buffer underflow cannot occur in sweep j. Let $b_i(k)$ be defined as the amount of data in the buffer for user i at the start of sweep k. Now, by definition of j', $b_i(j') < B_i$. Since at most B_i data is consumed in each sweep, we obtain that $b_i(j' - 1) < 2B_i$. Clearly, either $b_i(j' - 1) < B_i$ or $B_i \leq b_i(j' - 1) < 2B_i$. The first case contradicts the assumption that j' is the first sweep in which buffer underflow can occur. The second case contradicts that $b_i(j') < B_i$, since during sweep $j' - 1$ a data block will be fetched from disk for user i, as $b_i(j' - 1) < 2B_i$, while at most B_i data will be consumed in this sweep. Consequently, in both cases we derive a contradiction.

The necessity of (4) can be shown as follows. If (4) does not hold, then buffer underflow can occur for some user i, for which (4) does not hold, in the sweep starting at T_i^{start}. At T_i^{start}, the buffer for user i contains exactly B_i data. If this sweep is of worst-case duration, user i consumes at maximum rate, and user i is serviced at the end of the sweep, then buffer underflow will occur. □

It is remarked that letting the user wait another two sweeps before he can start consuming does not solve the problem of buffer underflow in the second part of the proof of Theorem 2.

That indeed a buffer must be large enough to store three data blocks and not less than three, can be inferred from the following. Suppose that we reserve buffers of size $(2+x)B_i$, $0 \leq x < 1$. Now, if at the start of a given sweep the buffer contains $(1+x+\varepsilon)B_i$ data, with $\varepsilon > 0$ and $x + \varepsilon < 1$, then no data can be fetched for user i during this sweep. The next data block may only arrive at the end of the next sweep. In the meantime, $2B_i$ may have been consumed, which would result in buffer underflow.

We now derive an expression for the minimum buffer requirements as follows. Since we have to store three data blocks for each data stream, we have to buffer a total of $3\sum_{j \in U} B_j$ data. By adding all B_i's, using Equation (4), we derive a lower bound on

$\sum_{j \in U} B_j$, which is given by

$$\sum_{j \in U} B_j \geq \frac{r \, s(n, 1)}{r - C} \, C, \tag{5}$$

where $C = \sum_{j \in U} c_j^{\max}$. This lower bound can be attained, if we assume that $B_i / c_i^{\max} = B_j / c_j^{\max}$ for each pair $i, j \in U$, from which we derive that $\sum_{j \in U} B_j = B_i / c_i^{\max} C$. Using this expression in Equation (5), we derive that

$$\frac{B_i}{c_i^{\max}} \geq \frac{r \, s(n, 1)}{r - C}.$$

Now, by choosing

$$B_i = \frac{r \, s(n, 1)}{r - C} \, c_i^{\max}, \tag{6}$$

this lower bound is indeed attained. Since each sweep is started immediately upon completion of the previous one, a data block will not be fetched for each user in each sweep. As a result, the number of data blocks that have to be fetched in a single sweep can be considerably smaller than n. We give an example in Section 7.

5.2   Variable-Block Double Buffering Algorithm

An alternative approach for generalizing the double buffering algorithm is obtained by fetching data blocks of variable size. This algorithm is called *variable-block double buffering* algorithm (VDB).

Instead of fetching a constant-size data block whenever there is room for one in the buffer, one could fetch, in a given sweep, an amount of data that is guaranteed to fit in the buffer, i.e., the size of a data block is chosen identical to the amount of room that is available at the start of the sweep, with a maximum of B_i as given by (6). The buffer for data stream i must have room for two data blocks of size B_i.

In that case, it is easily seen that neither buffer underflow nor overflow will occur. Let $b_i(j)$ denote the amount of data in the buffer for user i at the start of sweep j. Buffer overflow will not occur, since a data block will never exceed the room that is available at the start of the sweep in which it is fetched from disk. Buffer underflow will not occur either. At time T_i^{start}, there is exactly B_i data in the buffer, i.e., enough to survive the first sweep. Let this sweep be denoted by j. Furthermore, for each sweep $j' \geq j$, it holds that if underflow does not occur in sweep j', it will not occur in sweep $j' + 1$, as is shown as follows. By assumption, $b_i(j') \geq B_i$ and the amount of data fetched for user i in this sweep equals $2B_i - b_i(j')$. Since user i consumes at most an amount B_i of data, $b_i(j' + 1) \geq b_i(j') + 2B_i - b_i(j') - B_i = B_i$, i.e., user i survives sweep $j' + 1$.

Unless a user does not consume from his buffer, a data block will be fetched in each sweep.

Note that reading data blocks of variable size may impose additional constraints on the layout of the data on disk and the granularity of striping, since we must guarantee that a data block can be fetched by a single disk access.

5.3   Dual Sweep Algorithm

Finally, we propose the *dual sweep* algorithm (DS), which operates as follows. In each sweep, a constant-size data block is fetched for data stream i whenever there is room for one in the buffer at the start of the sweep, *unless* a data block for this stream has already been fetched in the previous sweep. Hence, for each pair of successive sweeps at most one data block is fetched for each data stream.

Let B'_i be the size of the data blocks that are repeatedly fetched from disk for user i by the dual sweep algorithm. B'_i equals the maximum amount of data consumed in two successive sweeps. Since a data block of size B'_i is sufficient to survive two successive sweeps, a buffer must only have room for two such data blocks.

Theorem 3. *Let U be a given set of n users, with $C = \sum_{j \in U} c_j^{\max}$. If $C < r$ and the buffer for each $i \in U$ has room for two data blocks of size B'_i, then the dual sweep algorithm is safe if and only if for each $i \in U$*

$$\frac{B'_i}{c_i^{\max}} \geq \frac{\sum_{j \in U} B'_j}{r} + s(n, 2). \qquad (7)$$

Proof.   The sufficiency of (7) can be shown as follows. It is easy to see that buffer overflow will never occur, since a data block will be fetched in a sweep only if there is already enough room for this data block at the start of the sweep.

With respect to fetching data blocks for stream i, let the set of sweeps be divided into yes-sweeps and no-sweeps. In a yes-sweep, a data block is fetched for stream i; in a no-sweep, no data block is fetched for stream i. Between two successive yes-sweeps there is at least one no-sweep. At time T_i^{start}, at the start of sweep j, there is exactly B'_i data in the buffer.

With respect to buffer underflow, we first show for each $i \in U$ that if at the end of a yes-sweep there is at least B'_i data in the buffer, then there is also at least B'_i data in the buffer at the end of the next yes-sweep. We consider two cases. If between these two yes-sweeps there is exactly one no-sweep, then in the time between the completion times of the two yes-sweeps a block of size B'_i is fetched, while at most B'_i data is consumed. On the other hand, if there are more than one no-sweeps between the two yes-sweeps, then at the start of the last no-sweep there must be more than B'_i data in the buffer, since otherwise it would not have been a no-sweep. Again, between the start of this last no-sweep and the completion of the succeeding yes-sweep a block of size B'_i is fetched and at most B'_i is consumed.

During any such no-yes subsequence the buffer will not underflow, since at the start at least B'_i data is available while at most B'_i data is consumed. Furthermore, during the remaining no-sweeps underflow will never occur.

The necessity of (7) can be shown analogously as the necessity of (4) in Theorem 2. $\qquad \square$

By analogy with the minimum buffer requirements for TB, we can derive that B'_i is minimal if

$$B'_i = \frac{r \, s(n, 2)}{r - C} c_i^{\max}. \qquad (8)$$

6 Comparing Buffer Requirements

Using Equations (6) and (8) we can compare the minimum buffer requirements of the three disk scheduling algorithms. The relation between the block sizes B_i and B_i' can be rewritten as $B_i'/B_i = s(n,2)/s(n,1)$. To give some quantitative results, we use the characteristics of the Seagate Elite 9 disk. We assume that this disk can offer a guaranteed data rate of 44 Mbit/s. For $n = 5, 10, 15, 20$, and 25, the resulting ratio B_i'/B_i equals 1.185, 1.102, 1.070, 1.053, and 1.045, respectively. It is noted that, for $n > 10$, B_i' is at most 10% larger than B_i. To give an impression of the buffer requirements, Table 1 presents values of B_i and B_i' for different values of n and $c_i^{\max}$, assuming that all users have an identical maximum consumption rate and that one Seagate Elite 9 disk is used. The buffer sizes for TB and VDB are given by $3B_i$ and $2B_i$, respectively. The buffer

	B_i						B_i'				
	$c_i^{\max}$ (Mbit/s)						$c_i^{\max}$ (Mbit/s)				
n	0.5	1.0	1.5	2.0	2.5	n	0.5	1.0	1.5	2.0	2.5
5	0.055	0.117	0.188	0.269	0.363	5	0.065	0.139	0.223	0.319	0.430
10	0.107	0.244	0.430	0.693	1.094	10	0.118	0.269	0.474	0.764	1.206
15	0.165	0.415	0.840	1.721	4.633	15	0.177	0.444	0.899	1.841	4.957
20	0.232	0.658	1.691	7.890	–	20	0.244	0.693	1.781	8.308	–
25	0.309	1.025	4.494	–	–	25	0.323	1.071	4.696	–	–

Table 1: Values of B_i and B_i', in Mbit, for different values of n and $c_i^{\max}$.

sizes for DS are given by $2B_i'$. From (6), it is easily seen that $\lim_{C \to r} B_i = \infty$. If, for example, we have 29 users, with $c_i^{\max} = 1.5$ Mbit/s for each user i, and $r = 44$ Mbit/s, then $B_i = 66.121$ Mbit.

7 Comparing Average Response Times

To compare the average response times of the algorithms, we have carried out simulations for both contant and variable consumption rates. In this paper, we only discuss the constant rate results. The results for variable consumption rates are comparable with the results presented here. In the simulations, we restrict ourselves to the case in which user requests can be considered independent. This is realised by creating enough time between two successive requests.

In order to compare the fixed-period double buffering algorithm with the three variable-period algorithms presented in this paper, we assume that all streams consume at the same constant rate of 1.5 Mbit/s. For the disk parameters, we used the characteristics of the Seagate Elite 9 disk. This disk has a rate that varies from 44 Mbit/s for the inner tracks to 65 Mbit/s for the outer tracks. During each simulation the number n of users is fixed. The size of the data blocks that are repeatedly fetched for a user depends on the disk scheduling algorithm that is being used and $n_{\max}$, the maximum number of users that is admitted service simultaneously. Response times are measured by repeatedly generating a request for other data by one of the users. The time between two successive requests is chosen uniformly from the interval $[13, 28]$. Each request is issued by a randomly chosen user. Each simulation is based on a total of 50,000 requests.

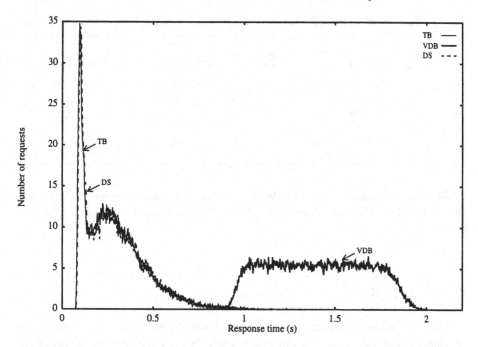

Figure 2: Frequency diagram of the response times observed for TB, VDB, and DS, in case of constant-rate simulations where $n_{max} = 25$ and $n = 25$.

Table 2 gives the average observed response times and corresponding 99% quantiles, for TB, VDB, and DS. The 99% quantile is defined as the smallest value for which 99% of the values observed are smaller than this value.

For the three algorithms, Figure 2 gives the frequency diagrams of the observed response times when $n_{max} = 25$ and $n = 25$. We observe that all three algorithms have average response times that are considerably smaller than those of DB, as discussed in Section 4. For example, if $n_{max} = 25$, then a period equals 3 seconds, resulting in an average response time of 4.5 seconds for DB. Hence, starting sweeps immediately upon completion of the previous one can indeed improve the average response times considerably. In case of TB and DS, users have to wait only a small time before their request is taken into account: either the disk is idle at the moment the request is issued or a sweep with only a few disk accesses is being carried out. This results from the differences in the average and worst-case data transfer rate, rotational latency, and seek time. In case of VDB, a user has to wait, on average, half a sweep in which for each of the n users a small data block is retrieved, before the request is taken into account. In the following sweep one large data block for the user that issued the request and $n - 1$ small data blocks for the remaining users have to be retrieved. This results in considerably larger average response times than those of DS and TB.

n	TB					VDB					DS				
	n_{max} 5	10	15	20	25	n_{max} 5	10	15	20	25	n_{max} 5	10	15	20	25
5	44.5	34.9	39.9	55.4	110.2	125.5	129.9	137.4	152.8	204.7	40.7	35.2	40.9	57.2	113.8
	97.3	*86.1*	*97.5*	*131.3*	*243.8*	*180.2*	*184.2*	*192.3*	*208.6*	*265.1*	*102.3*	*91.1*	*100.0*	*133.4*	*239.0*
10		65.2	54.5	68.1	126.9		265.6	273.6	288.8	340.4		62.8	55.7	70.1	131.1
		146.9	*141.2*	*181.0*	*326.7*		*368.8*	*377.5*	*393.4*	*447.4*		*161.1*	*151.9*	*191.6*	*336.3*
15			89.9	89.9	154.0			465.7	481.6	532.7			89.9	92.8	158.5
			213.8	*242.8*	*424.3*			*637.3*	*654.1*	*707.2*			*234.8*	*262.8*	*446.6*
20				134.1	199.8				785.9	836.6				139.3	205.7
				335.7	*571.5*				*1065.9*	*1119.3*				*378.6*	*604.6*
25					280.6					1404.8					289.7
					743.9					*1888.2*					*825.0*

Table 2: Average response times and 99% quantiles (in milliseconds) for TB, VDB, and DS in case of constant-rate simulations. The former are printed in roman, the latter in italics.

8 Conclusions

In this paper we have considered three safe disk scheduling algorithms for sustaining multiple heterogeneous variable-rate data streams. We can draw the following conclusions.

Using a disk scheduling algorithm that starts a new sweep immediately upon completion of the previous one, instead of executing sweeps strictly periodically, leads to much better average response times. This already holds for constant-rate data streams, as we have seen in Section 7, and can be shown to be even more interesting for variable-rate data streams.

With respect to the buffer requirements, the triple buffering algorithm uses 50% more buffer space than the variable-block double buffering algorithm. The dual sweep algorithm uses at most 10% more buffer space than the variable-block double buffering algorithm if the number of users is larger than 10, assuming that the disk parameters of the Seagate Elite 9 disk are used. For larger numbers of users, the buffering overhead is even smaller.

With respect to the average response times, the results for the triple buffer and the dual sweep algorithms are more or less the same. In this respect, these algorithms perform considerably better than the double buffer algorithm with variable blocks.

To conclude, DS combines the small buffer requirements of VDB with the favourable response times of TB.

References

Chang, E., and A. Zakhor [1994], *Proc. of the 1st Int. Workshop on Community Networking Integrated Multimedia Services to the Home*, San Francisco, July 13-14, 127-137.

Chen, M.-S., D.D. Kandlur, and P.S. Yu [1995], Storage and retrieval methods to support fully interactive playout in a disk-array-based video server, *Multimedia Systems* **3**, 126-135.

Dengler, J., Ch. Bernhardt, and E. Biersack, Deterministic admission control strategies in video servers with variable bit rate streams, in: B. Butscher, E. Moeller, and H. Pusch (Eds.), *Proc. Workshop Interactive Distributed Multimedia Systems and Services*, Berlin, March 4-6, 245-264.

Denning, P.J. [1967], Effects of scheduling file memory operations, *Proc. of the 1967 AFIPS SJCC* **30**, 9-21.

Dey, J.K., C.-S. Shih, and M. Kumar [1994], Storage server for high-speed network environments, *Proc. of the SPIE* **2188**, 200-211.

Gemmell, D.J. [1993], Multimedia network file servers: Multi-channel delay sensitive data retrieval, *ACM Multimedia* **6**, 243-249.

Kandlur, D.D., M.-S. Chen, and Z.-Y. Shae [1991], Design of a multimedia storage server, *IBM research report*, June 1991.

Kenchammana-Hosekote, D.R., and J. Srivastava [1994], Scheduling continuous media in a video-on-demand server, *Proc. of the Int. Conf. on Multimedia Computing and Systems* **5**, 19-28.

Korst, J., V. Pronk, E. Aarts, and F. Lamerikx [1995], Periodic scheduling in a multimedia server, *Proc. of the 1995 INRIA/IEEE Symposium on Emerging Technologies and Factory Automation, ETFA'95*, Paris, October 10-13, 205-216.

Özden, B., R. Rastogi, and A. Silberschatz [1995], A framework for the storage and retrieval of continuous media data, *Proc. of the Int. Conf. on Multimedia Computing and Systems*, Washington, May 15-18, 2-13.

Oyang, Y. [1995], A tight upper bound of the lumped disk seek time for the scan disk scheduling policy, *Information Processing Letters* **54**, 355-358.

Paek, S., and S.-F. Chang [1996], Video server retrieval scheduling for variable bit rate scalable video, *Proc. of the IEEE Int. Conf. on Multimedia Computing and Systems*, Hiroshima, June 17-23, 108-112.

Patterson, D.A., G.A. Gibson, and R.H. Katz [1988], A case for redundant arrays of inexpensive disks (RAID), *Proc. of the ACM Conf. on Management of Data*, 109-116.

Rangan, P.V., H.M. Vin, and S. Ramanathan [1992], Designing an on-demand multimedia service, *IEEE Communications Magazine* **7**, 56-64.

Rautenberg, M., and H. Rzehak [1996], A control for an interactive video on demand server handling variable data rates, in: B. Butscher, E. Moeller, and H. Pusch (Eds.), *Proc. Workshop Interactive Distributed Multimedia Systems and Services*, Berlin, March 4-6, 265-276.

Reddy, A.L.N., and J.C. Wyllie [1994], I/O issues in a multimedia system, *Computer* **27**, 69-74.

Vin, H.M., P. Goyal, A. Goyal, and A. Goyal [1994], A statistical admission control algorithm for multimedia servers, *Proc. ACM Multimedia*, San Francisco, 33-40.

Yu, P.S., M. Chen, and D.D. Kandlur [1992], Design and analysis of a grouped sweeping scheme for multimedia storage management, in P.V. Rangan (Ed.), Proc. of the 3rd Int. Workshop on Network and Operating Systems Support for Digital Audio and Video, La Jolla, CA, *Lecture Notes in Computer Science* **712**, 44-55.

A Novel Data Placement Scheme on Optical Discs for Near-VOD Servers

Shiao-Li Tsao, Yueh-Min Huang, Chia-Chin Lin, Shiang-Chun Liou, and Chien-Wen Huang*
Department of Engineering Science, National Cheng-Kung University,
Tainan, Taiwan, R.O.C.
Institute for Information Industry*, Taipei, Taiwan, R.O.C.
E-mail: sltsao@system.es.ncku.edu.tw

Abstract

Recent advances on computing and communication technologies have made it possible to provide digital video services through network. One of the economic and practical applications is near video-on-demand systems. In this paper, a novel data placement scheme on optical discs is proposed for providing near video-on-demand services. By applying the scheme, the seek time of an optical disc drive is almost eliminated, and the bandwidth utilization can be optimized. Then, the maximal number of supported streams from an optical disc drive with the minimal buffer requirement can be easily achieved. The scheme can be applied not only to CD-ROMs storing MPEG-1, but also to Digital Versatile Disks (DVDs) storing MPEG-2 in the near future.

1. Introduction

By advances on network and storage technologies, the development of video servers is growing rapidly. However, it seems not practical to provide on-demand video services for a large number of users accessing a huge amount of video content due to the high cost currently. For example, there should be about 1.5Gbps of storage and network bandwidths as well as 1000GB storage space required in order to provide 1000 MPEG-1 movies for serving 1000 users   simultaneously. One of the economic and practical applications is near video-on-demand(NVOD) system. NVOD system broadcasts a video on several channels at different starting times with a fixed interval. Users can either choose one of the channels to watch, review, or skip a segment of the video by changing the channel[1][2]. In the manner, the resource reservation of a NVOD server is made per channel, which is different from that made per user of a True-VOD server.

Compared with True-VOD (TVOD) servers, NVOD servers require a lower storage bandwidth. Storing video data on optical discs seems a low-cost and easy-maintenance approach for NVOD systems. However, an optical disc drive usually has a   longer seek time than that of direct access devices. To overcome this problem, [5] analyzed   head scheduling, caching strategies, and data placement   for a CD-ROM-based multimedia storage system. In [3], they proposed a 4 pick-ups CD-ROM drive to support 4 independent streams from a single CD-ROM drive for NVOD servers or library systems. Recently, the maximal bandwidth from an on-the-shelf optical disc drive is much larger than the bandwidth requirement for the playback of the stored video, e.g., 10Mbps CD-ROM drive for MPEG-1 video with 1.5Mbps average consumption rate. However, the previous studies did not consider the data placement issues in maximizing the number of supported streams from an optical disc drive. In this study, we propose a novel data placement scheme on optical discs to exploit the property of periodical playback of a NVOD server to reduce the seek time overhead. By applying the proposed method, the number of supported streams from an optical disc drive can be optimized.

The rest of the paper is organized as follows. In Section 2, we describe the storage architecture for a NVOD server. In Section 3, we elaborate a novel data placement scheme and a buffer management strategy for a NVOD server. In Section 4, an example by applying the proposed scheme is presented and discussed. Finally, we conclude the study in Section 5.

2. System Architecture

For a NVOD server, video streams are retrieved from the storage device and delivered to the network. These streams start at different times by a regular fixed interval. To transfer the data to the network, one approach is that the requests for the same channel are grouped together and the stream can be transmitted by using multicast mechanism. An alternative approach is that the video can be broadcasted on several channels on a cable TV network such as Hybrid Fiber Coax (HFC) network. Due to the regular fixed interval between starting times on the channels for the same video, the maximal initial delay for new requests is upper bounded. For example, if we allocate a 60-minute movie to 6 channels, the initial delay for any new request is at most 10 minutes.

We propose optical disc drives as the storage subsystem for NVOD servers for the sake of low cost and easy maintenance. However, optical disc drives usually have longer seek time than direct access devices, and result in lower bandwidth utilization. To overcome this problem, we exploit the periodical playback property of a NVOD server on designing the data placement scheme to reduce the seek time overhead. As the result, an optical disc drive can output the maximal number of streams simultaneously. A NVOD server may have several optical disc drives containing video discs. A video may be stored on one or more discs. Figure 1 depicts the basic concept of the proposed storage architecture. Optical disc drives retrieve data and dispatch them to the individual buffer of the channels.

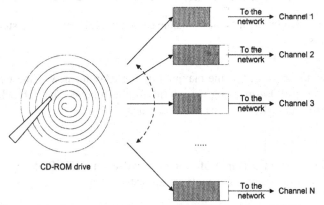

Figure 1 Basic concept of the proposed architecture for a NVOD server

3. Data Placement Scheme

3.1 Data placement on a single optical disc

To describe the proposed data placement scheme, we adopt CD-ROMs and MPEG-1 video data as the optical discs and the video format. The scheme is also feasible for MPEG-2 on Digital Versatile Disks (DVDs). Suppose a MPEG-1 file is divided into M blocks. According to ISO9660 standard[4], the block size is recommended to be larger than 2 Kbytes. We assume that the duration of the video is T seconds. The block size, B, of the video file is defined as the sector size on a CD-ROM. Then, we have the relation : $T \cdot C_r = B \cdot M$, where C_r is the consumption rate of a video stream. Suppose N channels are allocated for a video, the length of the video is padded with dummy video to make $\frac{M}{N}$ become an integer for accommodating our design. The dummy video can be some advertisement or preview. The length of the padded dummy video is at most N blocks, which is insignificant compared with the length of a video. Furthermore, the consumption rate of a video stream is assumed to be a constant in the current study. The meanings of the notations used in this paper are summarized on Table 1.

Table 1 The meanings of the notations

Notation	Meaning
M	The number of blocks in a video file
N	The number of channels allocated for a video
B	The size of a block (KB)
T	The duration of a video (sec)
C_r	The consumption rate of a video stream(KB/sec)
I	The starting interval between two neighbor channels of the same video (sec)
D	The ideal bandwidth of an optical disc drive (KB/sec)
S_{full}	The time for a full stroke seek (sec)
K	The number of optical disc drives allocated for a video

Since the video broadcasts on N channels, the interval between the starting times on N channels is :

$$I = \frac{T}{N},$$

where I can be also regarded as the maximal initial delay for new requests. M blocks are placed on a CD-ROM by using the interleaving technique. The ith block is placed on the jth sector of a CD-ROM. Thus, we define that :

$$j = N \cdot (i \bmod \frac{M}{N}) + \left\lfloor \frac{i}{M/N} \right\rfloor$$

Figure 2 shows an example of placing a video on a CD-ROM, on which the video contains 12 blocks and 3 channels are allocated for it.

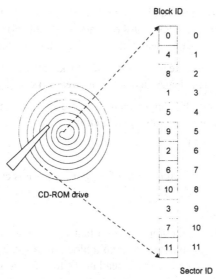

Block ID

0	0
4	1
8	2
1	3
5	4
9	5
2	6
6	7
10	8
3	9
7	10
11	11

CD-ROM drive

Sector ID

Figure 2 An example of the proposed data placement scheme on a single optical disc

In order to guarantee the continuous playback for each stream, the block size must be large enough for each client's consumption during a service round. Here, a service round is defined as the process that the server retrieves the required blocks for all steams once. Moreover, we assume that the videos replay on the channels. We take above example for describing our proposed scheme. Since the streams of the same video start at a fixed interval to each other, the required blocks for the streams are separated by several blocks, e.g. the required blocks are separated by $\frac{M}{N} = \frac{12}{3} = 4$ blocks in above case. In other words, if the first stream requires the 0th block, and then the 4th block and the 8th block will be required on behalf of the second and the third streams. After retrieving the first 3 blocks, the server enters a new service round and retrieves the 1st, 5th, and 9th blocks on behalf of three streams. Similarly, the server retrieves the 3rd, 7th, and 11th blocks in the fourth service round, respectively. After this service round, the third stream ends and then replays, meanwhile, the next required blocks for the first and second streams are the 4th and 8th blocks. The server seeks back to the beginning sector of the CD-ROM, and retrieves the 0th, 4th, and 8th blocks again. Unlike the first service round, the delivery sequence has been changed to the 3rd, 1st, and 2nd streams, respectively. Although the delivery sequence has been changed, the required blocks for all streams can be still retrieved within a service round. The continuity of video playback is also held by applying dual-buffer scheme. The basic concept of dual-buffer scheme is that when the server is retrieving blocks in the ith service round, the data retrieved in the $i\text{-}1^{th}$ service round is being transferred to the network. In that way, the retrieval sequence of blocks can be reordered in a service round, which is independent on the delivery sequence. According to the underlying strategy, we retrieve the data from the first to the last sectors sequentially, and then go back to the first sector. Thus, we can eliminate the seek time between blocks except of the last full stroke seek, and the bandwidth utilization from an optical disc drive can be optimized.

Then, we can derive the maximal number of supported streams by an optical disc drive. The total time to retrieve all the data on an optical disc including the final seek is $\frac{B \cdot M}{D} + S_{full}$, where D is the ideal bandwidth from the optical disc drive, and S_{full} is the time of the full stroke seek. In fact, M blocks stored on an optical disc are retrieved and $\frac{M}{N}$ blocks are dispatched to each stream, i.e. the time to consume $\frac{M}{N}$ blocks by any stream must be longer than the time of retrieving all blocks on the optical disc. Thus, we have the following inequality :

$$\frac{B \cdot M}{D} + S_{full} \le \frac{M}{N} \cdot \frac{B}{C_r} \ \ ......(A).$$

From inequality (A), the maximal number of supported streams can be derived as :

$$N = \left\lfloor \frac{D \cdot M \cdot B}{C_r \cdot (M \cdot B + S_{full} \cdot D)} \right\rfloor.$$

From the above floor function, we learn that few bandwidth of the optical disc drive is wasted since the consumption time of a block for a stream may not be just the time period of a service round. The drive head needs to wait for a while and enters the next service round to ensure that all the buffer are consumed. This situation is illustrated in Figure 3. However, if the head idles and misses the next consecutive block, there will be at least one rotational latency time to locate the block again. In most cases, the rotational latency of an optical disc drive is longer than the waiting time of entering the next service round. In order to solve this problem, a buffer management is needed to determine when to idle for a period. The buffer management strategy will be elaborated in Section 3.3.

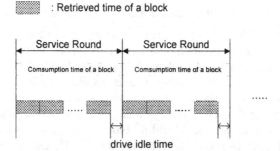

Figure 3 The consumption time of a block may not be the same with the time period of a service round

3.2 Data placement on multiple optical discs

Occasionally, the length of a video is too long to place on a single optical disc (or we may want to allocate more channels for the video). Thus, more than one optical disc drive is needed for the video. Indeed, our proposed data placement scheme is also feasible for multiple optical discs. Suppose K disc drives are allocated for the video. We evenly divide the video into K segments whose sizes are all smaller than the maximal stored space of one optical disc. Here, the length of a video is also padded with dummy video to ensure that $\frac{M}{K \cdot N}$ is an integer. Each optical disc drive serves $\frac{N}{K}$

streams simultaneously (We assume $\frac{N}{K}$ is also an integer). The padded video is at most $K \cdot N$ blocks. For each disc, a segment of the video is placed by applying the same scheme described in Section 3.1. The formal description for the location of each block can be defined as:

$$l = \frac{i}{M/K},$$

$i' = i \bmod M/K$, and

$$j = N \cdot (i' \bmod \frac{M}{N}) + \left\lfloor \frac{i'}{M/N} \right\rfloor,$$

where l is the identification of the drives, and j is the identification of the sector storing the ith block.

An example is illustrated in Figure 4, in which 2 discs and 6 channels are allocated for a 24-block video. By definition, the interval of required blocks between two neighbor streams of the same video is : $\frac{24}{6} = 4$. Then, the first stream requests the 0th block, and the 2nd, 3rd, 4th, 5th, and 6th streams request the 4th, 8th, 12th, 16th, and 20th blocks, respectively. After two service rounds, the heads move back to the first sectors and retrieve the 0th, 4th, 12th, 16th and 20th blocks again for the 6th, 1st, 2nd, 3rd, 4th, and 5th streams, respectively. Although the serving drives for streams are changed, the required blocks for all streams can guarantee to be retrieved within a service round. In this manner, the seeks except of the full stroke one for both drives are all eliminated.

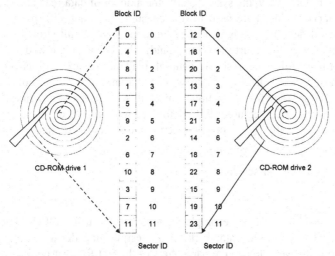

Figure 4 An example of the proposed data placement scheme on two optical discs

3.3 Buffer management

As mentioned in Section 3.1, the consumption time of a block may be longer than the time period of a service round. If the drive does not wait for a while and enter the new service round, some residual data must be occupied in the buffer. The size of the residual buffer occupied in the memory can be denoted as $C_r \cdot (\frac{B}{B_r} - N \cdot \frac{B}{D})$. For

the ideal case, when the drive head retrieves the data from the innermost to the outermost sectors, i.e. each stream has been served for $\frac{M}{N}$ service rounds, the total amount of data read ahead for a stream can just compensate for the consumption during the final full stroke seek. In other words, the minimal required buffer of a stream for reading ahead data is $\frac{M}{N} \cdot C_r \cdot (\frac{B}{C_r} - N \cdot \frac{B}{D}) = C_r \cdot S_{full}$. In most cases, the

amount of the data read ahead is larger than that is consumed when the full stroke seek occurs. Thus, the drive should idle for a while to avoid the buffer overflow. However, the idle time for a drive must be one or more periods of a rotational latency since if the drive misses the consecutive block, the time to locate the block again must be at least a revolution latency. In order to guarantee the accumulated data in the buffer which is enough for compensating for the full stroke seek, we read ahead data at the first several service rounds until the amount of data reaches $C_r \cdot S_{full}$. After the first

several service rounds, the drive head is allowed to idle for a revolution if the accumulated data is enough for the consumption of a full stroke seek and a revolution. If B_a denotes the accumulated data and R_i represents the time for one revolution while the head locating at the ith sector, the head is allowed to idle for one revolution when $B_a \geq (S_{full} + R_i) \cdot C_r$. Figure 5 shows an example to describe the situation of the

idle for a revolution. When the system starts, the amount of data read ahead in the first and second service rounds are used for the consumption for a full stroke seek. Until the accumulated data are enough for the consumption of a full stroke seek and one rotational latency(in the fourth service round), the drive is allowed to idle for a revolution. Meanwhile, the residual data of read ahead are still enough for the consumption of a full stroke seek. Thus, we can guarantee that the size of accumulated buffer data is larger than the consumption of a full stroke seek at any time. Therefore, we have the maximal required buffer for reading ahead of each stream as : $(S_{full} + R_{max}) \cdot C_r$, where R_{max} denotes the maximal rotational latency of the optical disc drive. The total amount of required buffer is $(S_{full} + R_{max}) \cdot C_r + 2 \cdot B$, where $2 \cdot B$ is required for dual-buffer scheme.

4. Design Example

In this section, a design example is described by using NEC CDR-1410A 8 times CD-ROM drive as the framework for the testing. In order to get more accurate simulations results, we obtained the parameters of the CD-ROM drive by experiments. Figure 6 shows the seek times for crossing various distances expressed in terms of sector numbers. The rotational delay can be shown as a saw-tooth riding on the curve in Figure 6. The sustaining data transfer rate of NEC CDR-1410A by the experimental result is 1240 KB/s. Meanwhile, we suppose the encoding scheme of video data is MPEG-1 which is 1.5Mbps constant consumption rate.

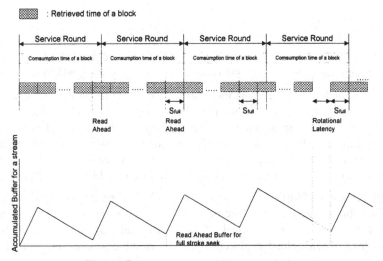

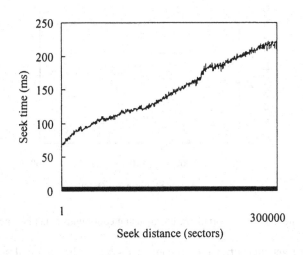

Figure 5 The situation for idling a revolution

Figure 6 The relation between seek time and seek distance of NEC CDR-1410A

We consider a 120-minute video and 12 channels are allocated for it, i.e. 10 minutes starting latency between channels. Due to that the maximal storing space of a CD-ROM (assumed the sector size is 2324 Bytes [6]) is the length of a 70-minute MPEG-1 video[6], at least two CD-ROMs are necessary to store the video. In Figure 7, we simulate the maximal number of supported streams under the original continuous data placement on CD-ROMs with dual-buffer strategy. We find that the number of supported streams from a CD-ROM drive is at most 4 and 400KB buffer per stream is required. On the other hand, the maximal supported streams from the CD-ROM drive by applying the proposed data placement scheme can be calculated by the following equation :

$$N = \left\lfloor \frac{D \cdot M \cdot B}{C_r \cdot (M \cdot B + S_{full} \cdot D)} \right\rfloor.$$

After calculating, we derive that 2 CD-ROM drives are needed, i.e. $K = 2$, and the size of data stored on a CD-ROM is $\frac{120}{2} \cdot 60 \cdot \frac{1.5Mbps}{8} = 675MB$ (594836 sectors). We have the maximal number of supported streams of a CD-ROM driver: 6, which is larger than that of the original continuous data placement. Therefore, the supported streams by two drives can accommodate for the number of allocated channels. If the number of allocated channels are more than the number of steams that the drives can support, more drives are needed for the video's allocation. Since $\frac{594836 \text{ blocks}}{2 \text{ discs} \cdot 6 \text{ channels}} = 49569.66...$, the video is padded with 4 dummy blocks to ensure that the blocks placed on a CD-ROM is complete. According to the underlying strategy, the first 297420 blocks are placed on the first CD-ROM, on which the blocks are placed by interleaving 6 sectors. The residual 297416 blocks and 4 dummy blocks are placed on the other CD-ROM. The data placement for the example is shown in Figure 8.

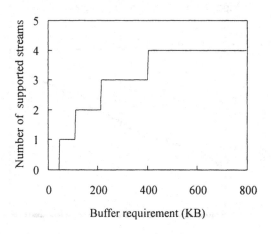

Figure 7 The number of supported streams by using continuous data placement scheme

In order to get the buffer requirement per stream, the maximal seek time and rotational latency must be first evaluated. We have $S_{full} = 234ms$ and $R_{max} = 11ms$ according to experiments if the amount of stored data on the CD-ROM equals 675MB. Thus, we can derive the buffer requirement per steam as:

$$(S_{full} + R_{max}) \cdot C_r + 2 \cdot B = (0.244 + 0.011) \cdot \frac{1.5Mbps}{8} + 4648 \text{ bytes} \approx 50KB$$

From the design example, we find that the required buffer size for each stream is very small since the seek times are almost eliminated.

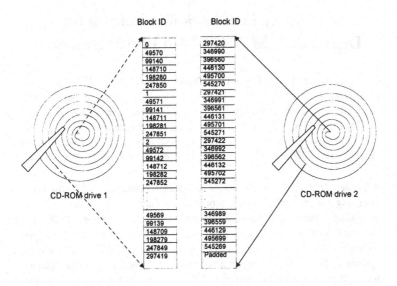

Figure 8 Design example by applying the proposed data placement scheme

5. Conclusion

In this paper, we proposed a novel data placement scheme on optical discs for supporting near video-on-demand services. The method fully exploits the property of periodical playback of a NVOD system to optimize the utilization of optical disc drive bandwidth. Thus, the maximal number of supported streams with the minimal buffer requirement can be easily achieved. The method can be applied not only to MPEG-1 on CD-ROMs, but also to MPEG-2 on DVDs in the near future.

References

[1] Thomas D. C. Little, Dinesh Venkatesh, " Prospects for Interactive Video-on-Demand", IEEE Multimedia Magazine, Fall 1994.

[2] W. T. Wong, L. Zhang and K. K. Pang, "Video on Demand Service Policies", Proc. of IEEE Singapore Intl. Conf. on Networks, pp560-564, July 1995.

[3] K. Shioda, "Digital Optical Disc Technology Applied in Broadcasting", IEE Intl. Broadcasting Convention, pp37-42, Sep. 1994.

[4] William Frederick Jolitz, Lynne Greer Jolitz, "Inside the ISO-9660 File system Format", Dr. Dobb's Journal, pp80-89, Dec. 1992.

[5] Vijnan Shastri, et. al., "Design Issues and Caching Strategy for CD-ROM based Multimedia Storage", Proc. of Multimedia Computing and Networking, pp30-47, Jan. 1996.

[6] Philips Consumer Electronics, "Video CD Specification Version 2.0", July, 1994.

A Priority Feedback Mechanism for Distributed MPEG Video Player Systems

Kam-yiu Lam[1], Chris C.H. Ngan[1] and Joseph K.Y. Ng[2]

Department of Computer Science[1]
City University of Hong Kong
83 Tat Chee Avenue, Kowloon
HONG KONG
Email: cskylam@cityu.edu.hk

Computing Studies Department[2]
Hong Kong Baptist University
224 Waterloo Road, Kowloon
HONG KONG
email:jng@comp.hkbu.edu.hk

Abstract. In this paper, a priority feedback mechanism is proposed for adjusting the priorities of the processes dynamically in a video server according to the current status of the clients – the status can be reflected by the number of frames being dropped in a period of time. The priority feedback mechanism is implemented in a distributed MPEG video player system. Different priority mapping functions are suggested for the priority feedback mechanism to cater for different service requirements of the clients, especially when the clients are demanding different quality of services. Experiment results have shown that when the video server is serving several clients where every client may demand a play speed of their own, the use of the priority feedback mechanism can effectively improve the whole system performance.

1 Introduction

In recent years, a lot of work has been devoted to the design and development of distributed MPEG video player systems in which the video server may connect to a number of clients and each client may demand the same or different video streams at different play speeds. To support a high quality of service (QoS), the transmission rate of the video frames from the video server must be high and the transmission has to be highly predictable in every step from the video server to the clients until they are displayed. If some of the frames cannot be displayed on time, they are of no use to the clients and have to be dropped. This can seriously affect the QoS of the video display.

To provide a high QoS to the clients, sophisticated policies have to be adopted on managing the resources such as the CPU in the video server, the network, the buffers and the CPU in the clients. The simplest way is to use some static mechanisms for resource allocation. However, these mechanisms have their problems since they require priori knowledge on the system environment and the requirements of each process. These assumptions are not valid and the required information is not available in most distributed MPEG video player systems in which the system environment can be highly dynamic and not controllable. For example, in a broadcasting news on the Internet system, the clients can be anybody who has connected to the news server. The clients may request videos with different sizes (resolution) and to play video at different speeds. It may even change the play speed of a video while it is being displayed. On the other hand, the transmission delay of the frames in the network is uncontrollable. It is difficult to predict the latency in the Internet due to the differences in bandwidth among the different links in the network and the dynamic workload in the network. The only more controllable component in the system is the video server.

To achieve a better performance, a more practical way is to dynamically adjust the server itself to the changing environment so that the required QoS can be delivered to the clients.

To make the system more adaptive, one recent suggestion is to use the *software feedback mechanisms* [1]. Although they have been shown to be effective in a single client MPEG video player system – only one client is connected to the server at each time, they are not suitable for multiple-client systems in which the video servers are connected to a number of clients. The reason is that the effectiveness of a software feedback mechanism is highly dependent on the length of the feedback cycle which is the periodic time interval between the time a client sends out the feedback signal and the time when the video server generates the response. In a conventional Unix environment (a time-sharing environment), this period is dependent on how many clients are connecting to the server and how the operating system in the video server serves the processes. To make the feedback effective, the length of the feedback cycle should depend on the current status of the clients. The client, which is in a "worse" status, should be served first so that the length of its feedback cycle will be shortened[1].

In this paper, we have designed and developed a *priority feedback mechanism* for the scheduling of the server processes, which are serving the clients, in a multiple-client distributed MPEG video player system. The priorities of the server processes are adjusted dynamically according to the observed status of their clients. In the design of priority feedback mechanism, one important consideration is that the additional overhead for priority scheduling must be low [4]. Although different priority scheduling algorithms have been proposed, most of them are very complicated and require the assumptions on the priori knowledge of the processes [7]. Thus, they are not suitable for distributed MPEG video player systems . Instead, a simple priority mapping function is more preferable.

The implementation of the priority feedback mechanisms requires real-time supports from the underlying operating system. Priority scheduling in the video server is supported by using the real-time extensions provided by a general purpose operating system – Solaris 2.5. The reason of choosing a general purpose Unix environment is that it is a well-known environment. Although some real-time operating systems may provide better supports for scheduling of the processes, these systems are not common to our regular audience.

2   System Architecture

Figure 1 shows the basic architecture of a distributed MPEG video player system. It consists of a video server which is connected to a number of clients through a network, e.g., the Internet. The video server runs on a host as a daemon process. It accepts requests from the clients and forks video server sessions (server processes). Each server process is responsible for servicing a particular client. In our proposed mechanism, the priorities of the server processes will be adjusted dynamically.

Each client consists of a set of collaborating processes: the controller (CTR), the video buffer and the decoder. They share the same memory containing buffers and data structures for input parameters, system status and performance statistics.

---

[1] The reason will be explained in the later sections.

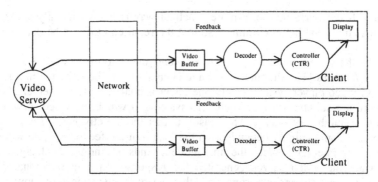

Fig. 1. The basic structure of a distributed MPEG video player

During the playback, compressed video frames are sent from the video servers to the clients according to the specified play speed through the network. The play speed is measured in the number of frames per second (fps). The video buffer is a temporary buffer space for storing the received frames. The decoder then retrieves frames from the video buffer and decompresses the frames. It sends the decompressed frames to the controller which provides an interface and supports operations such as fast forward and backward, variable speed play and random positioning.

3 Software Feedback in Distributed MPEG Video Player Systems

Figure 2 shows the basic structure of a software feedback mechanism [2]. It consists of two main components: the feedback filter and the control algorithm. Feedback signals are sent from the system under control to the feedback filter to eliminate transient noise. The algorithm compares the output of the feedback filter against a goal specification. A response is then generated for adjusting the system under control to keep its status within the defined performance objective.

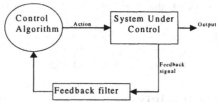

Fig. 2. Software feedback mechanism

A variety of software feedback mechanisms have been designed for monitoring the status of the clients in a distributed MPEG player systems [1]. For example, in applying the software feedback mechanism for buffer management, the buffer level in the client will be monitored. The observed buffer level is used as the feedback signal and inputs into a control algorithm which generates a signal to the server based on the specified threshold buffer level. According to the received signal, the server will make different responses. For example, if the buffer of a client is lower than certain threshold value, the server will send the video frames to the client faster so that to restore to the threshold level. If the buffer level is high enough, the transmission rate of the video frames will be reduced. By ensuring sufficient number of frames in the buffer, it is expected that the display of the video should be smooth.

Using the software feedback mechanism alone may not be able to ensure that the video server can generate timely response for their clients. For example, at time t1, a feedback signal is generated at which the buffer level of the client is at 80% of the threshold buffer level. The server receives the feedback at time t2 and generates a response at time t3. The feedback cycle length is thus equal to t3 − t1. In a time-sharing environment, each server process will receive the same amount of CPU service time in each CPU service cycle. The feedback cycle length is then dependent on the number of server processes (number of clients) in the system. If the length is very long, e.g., t3 >> t1, the buffer level may become even lower at the time when the response is generated. Therefore, it will take a longer time to restore the level to the threshold value and a lot of frames will be dropped during this period.

Another problem with this approach is that the status of the clients may not be the same. The amongst of time spent on a client is independent of its current status. This may serious degrade the performance of a client which is in a poorer status. Consider a distributed MPEG video player system with a video server connected to n clients: C1, C2, ..., Cn. If the play speeds requested by the clients, C1, C2, ..., Cn, are R1, R2, ..., Rn, respectively and R1 > R2 > ... > Rn. Under the impact of a transient network overload, the performance of the clients are affected. Since the play speed of C1 is much higher than the play speed of Cn, it is likely that C1 will be more affected than the other clients. Thus, the status of C1 is worse than the status of Cn. For example, at some instance, the buffer level of C1 is at 0.1 of the threshold level while Cn is at 0.7 of the threshold level. Since the same amount of time is allocated to each server process in every CPU service cycle, the time required for the buffer level in C1 to restore to the threshold value will be much longer than the time required in Cn. During the "restoring" period, a lot of frames will be dropped in C1 and its performance will be more affected than other clients.

4 Priority Feedback Mechanism and Priority Mapping Functions

4.1 Priority Feedback Mechanism

Treating all the clients equally is not be a good approach to monitor the status of the clients. A better approach is to use the *priority feedback mechanism* in which a higher priority is assigned to the server process which is serving a client in a poorer status. By adjusting the priorities of the server processes dynamically, the CPU can be made more "responsive" to the changing status of the clients [3].

In the design of priority feedback mechanism, we have to determine what are the level of control and the feedback signals. A distributed MPEG video player system can be divided into different components with different levels. Therefore, different levels of feedback mechanisms can be defined in the system for the control of different components. However, the use of multi-feedback mechanisms in a system will incur a heavy overhead. Thus, we choose to apply the feedback mechanism in the highest level, between the server processes and the clients. The feedback signal is the QoS provided to the clients. In our implementation, we choose the number of frames being dropped within a fixed period of time[2] as the feedback signal. If the number is larger, the status of the client is poorer and it should receive "more services" from the video server than the clients which are dropping lesser frames. The number of frames being dropped in a period of time x is input into a *priority mapping function* to calculate the priority of the server process, Priority(P):

Priority(P) = priority_mapping_function(# of frames being dropped in period x)

---

[2] Other QoS can be used.

The calculation of the number of frames being dropped in a period is based on the feedback streams sent from the clients. In a client, its current status is monitored continuously. It sends a feedback stream to the server aperiodically (whenever its status is deviated from the normal) to tell the server what its current status is. The stream contains information on the frames to be dropped in the next n frames. For example, if the pattern of the feedback stream received from the client is "100111100011" (a zero indicates that the frame has to be dropped). The second, the third, and the eighth to the tenth frames need to be dropped by the server. After the server process has sent out the first frame, it notices that two frames – the 2nd and the 3rd frame – have to be dropped. Then, the number "two" will be input into the priority mapping function to calculate the priority of the server process for the period of sending the forth frame. After sending the forth frame, the priority of the process will be restored to normal, e.g. zero, as no frame has to be dropped before sending the next frame, the 5th frame.

4.2 Priority Mapping Function

We have designed three priority mapping functions to cater for different service requirements of the clients. They are:
(1) *Linear* function;
(2) *Increasing* function; and
(3) *Decreasing* function.

In the *Linear* function, the priority mapping of a process is directly proportional to the number of frames being dropped:

Priority(P) = x + number of frames to be dropped in the period
    where x is the default priority of the process (e.g., the priority of a process when it is at the normal status)

The problem of the *Linear* function is that it treats all the clients the same way according to their number of frames being dropped. If the play speed of the clients in a system is not the same, linear priority mapping will favour the higher play speed clients as they drop more frames as compared with the low play speed clients.

In the *Increasing* function, the rate of increase in priority increases with the number of frames being dropped:

Priority(P) = (number of frames to be dropped in the period)^2 / k
    where k is a constant.

It has the effect of making the priority feedback mechanism less sensitive to the number of frames being dropped if it is small. It is suitable for the clients with high play speeds as they always have to drop frames (dropping a small number of frames is a consider to be normal status to them.)

The *Decreasing* function is the reverse of the increasing function in which the rate of increase in priority decreases with the number of frames being dropped.

Priority(P) = $\sqrt{k \times \text{(number of frames to be dropped in the period)}}$

It has the effect of making the priority feedback mechanism highly sensitive to dropping frames if the number of frames being dropped is small. It is suitable for the clients with low play speed such that they seldom have to drop frames. Dropping even one frame means that their status is very poor, and the server processes have to be raised up to higher priorities.

By using different values of k, we can obtain different distributions for the mapping functions and control the intersection point of the curves. In order to cater for the different requirements of the clients, a mixed priority mapping method is used in which each client has its own priority mapping function, which may be increasing or decreasing. If the play speed required by the client is high, an increasing mapping function will be used for the client. On the other hand, if the required play speed is low, a decreasing mapping function will be used so as to make the performance of the client more sensitive toward dropping frames. The definition of the threshold level can be based on the nature of the video such as its size and resolution. If the video is a large with high resolution, a lower value may be used. By using the mixed priority mapping method and different values for the threshold level, different requirements of the clients can be served separately.

5 Implementation

Our distributed MPEG player system is modified from the distributed MPEG video player system developed by the Oregon Graduate Institute (OGI) [1]. The main difference is in the video server in which we have implemented our priority feedback mechanism. Real-time scheduling of the server processes is supported in the video server by the underlying operation system, Solaris 2.5.

The distributed MPEG video player system is running on top of the Solaris 2.5 operating system which supports some real-time features such as priority scheduling, memory lock and asynchronous I/O [5]. In Solaris 2.5, 60 levels of priorities can be defined for the real-time processes. All the server processes are defined as real-time processes. The priority feedback mechanism assigns a priority to the server process within this 60 priority levels. When a real-time process is running, it will only be pre-empted by a higher priority real-time process.

6 Experiments

We have compared the performance of our distributed MPEG player system with the distributed MPEG video player developed from OGI (without priority feedback mechanism) which has been shown to perform well in various aspects [1].

6.1 Experiment Setup

The basic configuration of the system is as follows:

Video Server: Pentium PC (90 MHz)
Clients: 8 Pentium PC (90 MHz) – (all clients are on the same network)
Network: Ethernet (10Mbps). Server and clients are on the different sub-net
Video stream: A racing car clip with frame size of 320x240.
The total number of frames is 3596 at 30fps (about 2 minutes)
File size is 25,462,128 bytes
Average frame size ratio (I : P : B) is 4.8 : 3.3 : 1
Bit rate variations (bps) is from 5,760 to 1,134,720
The average frame size is 7080bytes
Picture group pattern IBBPBBPBBPBB

6.2 Experiment Measurement

In the experiments, we measure the percentage of frames displayed (PD) by a client and the relative improvement (RI). They are defined as:

$$PD = \frac{\sum_{\text{all clients}} \text{\# of displayed frames}}{\text{number of clients} \times \text{total number of frames in the video file}}$$

$$RI = \frac{PD \text{ with priority feedback} - PD \text{ without priority feedback}}{PD \text{ without priority feedback}}$$

Using the number of displayed frames alone is not sufficient for measuring the QoS of a system [6]. The performance is also dependent on how a client drops the frames. In most cases, if the client drops the frames more evenly, its performance should be better. Thus, another more important QoS indicator is the smoothness of the playback. One smoothness measure is the deviation of presentation jitter from the desired value of zero [1,6]. Consider a video frame sequence of $f_0, f_1,..., f_n$ and a playback sequence of these frames are $f_{j0}, f_{j1},..., f_{jm}$. At each logical display time j (where $j > 0$ and $j < n$), the logical time error is calculated as $e_j = j - j_k$, between the expected frame f_j and the actually displayed frame f_{jk} where $j_k < j$ and $j_{k+1} > j$. In [1], based on error sequence, E, $(e_0, e_1,..., e_n)$, smoothness, S, of a playback is defined as:

$$S = \sqrt{\frac{\sum_{e \in E} e^2}{n}}$$

With this definition of smoothness, S, a lower value of S indicates a smoother playback. We define its relative improvement SRI as:

$$SRI = \frac{S \text{ without priority feedback} - S \text{ with priority feedback}}{S \text{ with priority feedback}}$$

6.3 Experiment Results

6.3.1 Impact of Number of Clients

In this set of experiments, the play speeds of the clients are set to be different. One client requests a play speed of 30 fps and the others request a play speed of 12 fps. Table 1 and Table 2 summarise the results when the number of clients for low play speed (12 fps) is increased from 3 to 7.

	PD				RI	
Play speed	30 fps		12 fps		30 fps	12 fps
# of clients	With PB	Without PB	With PB	Without PB	---	---
4	29.86%	27.86%	96.25%	94.51%	7.18%	1.84%
5	30.03%	28.19%	92.69%	90.68%	6.53%	2.22%
6	30.03%	28.19%	92.69%	90.68%	6.53%	2.22%
7	30.17%	29.19%	87.87%	85.64%	3.35%	2.6%
8	28.64%	25.56%	91.83%	78.97%	12.05%	16.28%

Table 1. The number of displayed frames and relative improvement by varying the number of clients with low play speed.

	Smoothness (S)				SRI	
Play Speed	30 fps		12 fps		30 fps	12 fps
# of clients	With PFB	Without PFB	With PFB	Without PFB	---	---
4	1.574	1.670	0.528	0.662	6.10%	25.37%
5	1.669	1.831	0.435	0.531	9.71%	22.01%

6	1.648	2.324	0.670	0.875	41.02%	30.60%
7	2.002	2.469	0.998	1.109	23.33%	11.12%
8	2.34	3.180	0.683	1.350	35.90%	49.40%

Table 2. The smoothness and relative improvement by
varying the number of clients with low play speed.

Although the results are fluctuated due to the unpredictable network traffic during the experiments, it can be seen from Table 1 that the number of displayed frames in the client at high play speed (30 fps) with priority feedback is consistently larger than the one without priority feedback. Although the improvement in PD is small, the relative improvement is much larger and is in the range of 3% to 12%. The small improvement in PD is due to the high play speed. At a higher play speed, the number of dropped frames will be unavoidably large due to the higher demand on the system resources. The improvement with priority feedback is more significant if we look at the smoothness of the playback which is shown in Table 2. A relative improvement of up to 40% is observed in SRI with priority feedback. With the used of the priority feedback mechanism, the status of the high play speed client can be closely monitored. Whenever they are deviated from their normal status, responses will be generated and served immediately so that their status can be restored to normal in a short period.

As depicted in Table 1, the number of displayed frames (PD) in the clients with low play speed (12 fps) is also larger with priority feedback than without priority feedback. The relative improvement (RI) is also consistent although is marginal in most cases. From the results, we can see that the clients with low play speed also benefit from the priority feedback mechanism as well. Assigning a higher priority to the high play speed client, which drops more frames, does not affect the performance of the low play speed clients. If we look at Table 2, we can see that the improvement in smoothness is generally greater in the low play speed clients than in the high play speed client. Since their play speed is low, under most of the time, the frames can be displayed in time and do not need to be dropped. A small drop in number of frames can seriously affect their performance. With the priority feedback mechanism, the deviation from normal status can be detected and rectified immediately.

6.3.2 Impact on Various Play Speed

In our first set of experiments, we emphasis on whether giving more attention to a high play speed player will affect the performance of the lower play speed clients. Hence, we have only one high play speed client and a maximum of 7 low play speed clients. In this set of experiments, the system has eight clients with play speed of 6, 12, 18, and 30 fps. For each play speed, we have two clients. The reason to have this set up is to investigate the problem on how the priority feedback mechanism will affect the clients at various play speeds. The experiment results are presented in Table 3.

Play Speed (fps)	PD		Smoothness (S)		RI	SRI
	With PFB	Without PFB	With PFB	Without PFB	---	---
6	96.94	95.56	0.64	0.77	1.44%	20.31%
12	88.89	76.28	1.03	1.47	16.53%	42.72%
18	48.68	47.96	1.84	1.95	1.50%	5.98%
30	28.46	26.10	2.67	3.02	9.04%	13.11%

Table 3 shows the results with various play speed of eight clients.

As can be seen in Table 3, consistently with the results from the first set of experiments, the results with priority feedback are consistently better than without priority feedback for all the clients in terms of both number of frames displayed and smoothness of playback. The improvement in smoothness is up to 40%. This further confirms the believe that the priority feedback mechanism can improve the overall system performance and make the system more adaptive to the changing environment.

6.4 Comparing the Performance of Different Priority Mapping Function

We repeat the above two sets of experiments with different priority mapping functions and the results are shown in the Figures 3 to 5. Figure 3 and Figure 4 show the smoothness in the high and low play speed clients respectively. Consistent with the results in Section 6.3, the performance with priority feedback (linear, increasing, decreasing and mixed) is always better than the system without priority feedback. Among the four mapping functions, the mixed priority mapping method produced the best results most of the time as the clients with different service requirements are served with the priority mapping functions specially designed for them.

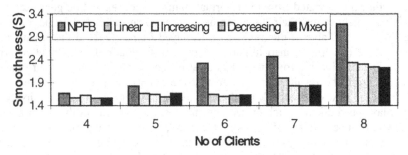

Fig. 3. Smoothness of High Playing Speed Client

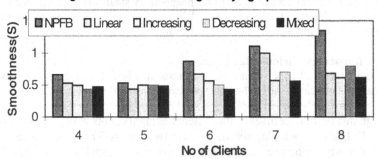

Fig. 4. Smoothness of Low Playing Speed Clients

In Figure 5, it is obvious that the higher the play speed, the poorer the performance. That is not avoidable and it shows at Figure 5. When we compare the performance within the same group (at the same speed), the system without the priority feedback (NPFB) always give a poorer performance, larger smoothness values. Among the different mapping functions, the mixed priority mapping method again produces the best results most of the time. It performs especially well at play speed between 12 – 30 fps.

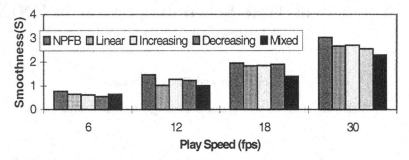

Fig. 5. Smoothness of Various Playing Speed Clients

7 Conclusions

In this paper, we have proposed a priority feedback mechanism to improve the performance of a distributed MPEG video player system in which the video server is connected to a number of clients and each client may request different videos at different play speeds. In the priority feedback mechanism, the priorities of the server processes are adjusted dynamically in a video server according to the observed status of the clients so that the client, which is in a poorer status, can receive more attention from the server. It is expected that by spending more time with the client, its status can be restored to the normal more quicker. Different priority mapping functions have been suggested for the priority feedback mechanism to cater for different service requirements of the clients.

Experiments have been performed to investigate the improvement of using the priority feedback mechanism and the performance of different priority mapping functions. Although the results are fluctuated due to unpredictable network traffic, significant improvement in the whole system performance are obtained with the priority feedback mechanism in terms of number of displayed frames and smoothness of playback. Amongst the proposed priority mapping functions, the mixed priority mapping method gives the best overall performance as the requirements of different clients are satisfied by using different priority mapping functions.

References

[1] Shanwei Cen, Calton Pu and Richard Staehli, "A Distributed Real-time MPEG Video Audio Player", In Proceedings of the 5th International Workshop on Network and Operating System Support of Digital Audio and Vudeo (NOSSDAV'95), April 18-21, 1995.

[2] Calton Pu and R. Fuhere, "Feedback-Based Scheduling: a Toolbox Approach", In Proceedings of 4th Workshop on Workstation Operating Systems, October 14-15, 1993.

[3] Veronica Baiceanu, Crispin Cowan, Dylan McNamme, Calton Pu and Jonathan Walpole, "Multimedia Applications Require Adaptive CPU Scheduling", In Proceedings of Workshop in Multimedia Resource Management, December 1-2, 1996.

[4] C.L. Liu and J.W. Layland, "Scheduling Algorithms for Multiprogramming in a Hard-Real-Time Environment", Journal of ACM, volume 20, number 1, pp. 46-61, 1973.

[5] Bill O. Gallmeister. Programming for Real World POSIX.4. O'Reilly & Associates, Inc., 1995.

[6] A. Vogel, Brigitte Kerherve, Gregor von Bochmann and Jan Gecsei, "Distributed Multimedia and QOS: A Survey", IEEE Multimedia, volume 2, number 1, pp. 10-18, 1995.

[7] Shuichi Oikawa and R. Rajkumar, "A Resource-Centric Approach to Multimedia Operating Systems", In Proceedings of Workshop in Multimedia Resource Management, Dec. 1-2, 1996.

A Soft Real Time Scheduling Server in UNIX Operating System

Hao-hua Chu and Klara Nahrstedt*

University of Illinois at Urbana Champaign, Urbana IL 61801, USA

Abstract. We present a soft real-time CPU server for Continuous Media processing in the UNIX environment. The server is a daemon process from which applications can request and acquire soft real-time QoS (Quality of Service) Guarantees. Our server architecture addresses in addition to other multimedia CPU scheduling extensions properties such as fairness, QoS brokerage and enforcement, and security. Overall it provides (1) protection among real-time(RT) processes, (2) fairness among RT and non-RT processes, (3) rate monotonic scheduling, and (4) a fix to the UNIX security problem. We have implemented our soft real-time CPU server in the SUN Solaris 2.5 Operating System, and we have shown through experiments that our soft RT server provides predictable QoS for continuous media applications.

1 Introduction

Continuous media such as Video and Audio are becoming widely-used in computer applications nowadays. To preserve their time-sensitive (temporal) behavior, multimedia applications require that the underlying systems provide soft real time Quality of Service (QoS) guarantees.[2] However, in the current UNIX multi-user, multi-process, and time-sharing (TS) environment, these applications do not perform very well when they are scheduled concurrently with the traditional non-RT applications such as text editors, compilers, web browsers, or computation-intensive tasks. These soft real-time (RT) applications also do not perform very well when other RT applications are scheduled concurrently. Part of the problem is due to untimely scheduling of processes rather than insufficient CPU capacity. This paper addresses this problem and presents an extension to the current UNIX scheduler - a soft real time server - to schedule soft RT applications in the more predictable fashion.

One possible solution to serve RT application in UNIX environment is to dedicate the entire system to serve only one RT application. This involves blocking services to all other RT or non-RT applications and users. This solution avoids the potential scheduling problem, and it also defeats the UNIX environment goals of supporting multi-user, multi-process and time-sharing properties. Therefore, this solution is not feasible in the UNIX environment.

---

* This work is supported by National Science Foundation Career Grant, under contract: NSFCCR96-23867 and CISE Research Infrastructure: A Shared Distributed Facility for Multimedia Signal Processing and Visualization with Applications to Human Computer Intelligent Interaction.

[2] Soft real time guarantee means that there could exist a time when guarantees are violated, e.g. due to page faults, but this violation is bounded.

Table 1. UNIX/POSIX.4 RT/TS priority class with the rate monotonic emulation.

Priority	Class	Process
higher ↑	fixed priority	RT process with smallest period
		..
		RT process with largest period
	dynamic priority	Any TS process
		..

Another more feasible solution is the current RT extension to UNIX. The UNIX/POSIX.4 real time extension provides fixed priorities to real time applications. The priority scheduling rule dictates that higher priority processes are scheduled before the lower ones in a preemptive fashion. RT processes are assigned higher fixed priorities, whereas non-RT processes are assigned lower dynamic priorities. As a result, the RT processes are served before non-RT processes, and higher priority RT processes are served before lower priority RT processes. This fixed priority mechanism provides a convenient way to implement the rate monotonic (RM) algorithm because the ordering of priorities between the RT processes depends on the ordering of the process rates (the length of their periods) as shown in the Table 1. Under this RM schedule[3], the RT processes with smaller periods are executed first, followed by RT processes with larger periods, and then non-RT processes. Prior to the start of any RT process, the schedulability test is performed by checking its total CPU demand so that including this new process, the CPU resource allocation will not exceed the CPU capacity. But this schedule has many problems described below.

– Priority should represent the importance of a process rather than whether this process is RT or non-RT. For example, is user A playing a song (a RT application) more important than user B writing a term paper (a non-RT application)? RT and non-RT applications must share the CPU time fairly. It is also unreasonable to assign priority for RT applications based on the length of their periods. For example, is a video player running at 30 frames per second more important than another video player running at 20 frames per second? This is called the *fairness* problem.
– It offers no protection between applications. There isn't any mechanism to enforce deadline and to guard against overrun and monopoly of CPU by a faulty or greedy RT process at high priority. Frequent overruns from a high priority process can cause starvation to other processes at lower priority. A run-away RT process at very high priority can even block most of the system processes and lock up the entire system. This is called the *enforcement* problem.
– It requires root privilege to use the fixed priority. From a UNIX security viewpoint, it is impossible to give every user root privilege to run RT applications. This is called the *security* problem.

In this paper, we describe a soft RT scheduling server in UNIX OS environment that will address the *fairness, enforcement, and security* problems mentioned above without compromising the flexibility and efficiency of the UNIX Operating System. The approach of our RT server is general, it requires no modification to the kernel and it can be easily ported to almost any derivations of UNIX systems. Our RT server is intended for scheduling of soft RT applications only, such as multimedia applications that can tolerate some variations in temporal quality of service.

The paper is organized as follows: Section 2 describes the related work; Section 3 explains the scheduling server architecture; Section 4 describes the implementation on SUN Solaris; Section 5 shows the experimental results; Section 6 presents the concluding remarks.

2   Related Work

The area of accommodating scheduling of soft RT applications on the current UNIX platforms was addressed by several groups. Goyal, Guo, and Vin [2] implemented the *Hierarchical CPU Scheduler* in the SUN Solaris 2.4. The CPU resource is partitioned into hierarchical classes, such as Real-time and Best-Effort classes, in a tree-like structure. A class can further partition its resource into subclasses. Each class can designate a suitable scheduler to meet the objective of its processes(leaf nodes). Protection between classes is achieved by the *Start-time Fair Queuing* (SFQ) algorithm which is a modification of the Weighted Fair Queuing Algorithm. SFQ is a fair scheduler that schedules all the intermediate classes and subclasses according to their partitioned resources. The major disadvantage of this approach is that their implementation requires modifications to the Solaris kernel scheduler. Fair sharing also does not translate directly into applications QoS guarantees that require a specific amount of CPU allocation and a constant periodicity.

Mercer, Savage, and Tokuda [4] implemented the *Processor Capacity Reserves* abstraction for the RT-threads in the RT Mach Operating System. A recent version [1] supports dynamic Quality adjustment policy. A new thread must first request its CPU QoS in the form of *period*, and *requested CPU usage in percentage* during the reservation phase. Once it is accepted, a *reserve* of CPU processing time is setup and it is bound to this new thread. Any computation time done on behalf of this process, including all service time from various system threads, is charged to this reserve. The reserve is replenished periodically by its requested CPU usage time. This concept is similar to the *Token Bucket*. The accurate accounting of system service time is a superior but costly feature. It requires non-trivial modifications and computation overhead inside the UNIX kernel to support this abstraction, such as keeping track of the reserves database, and passing the client process's reserve to and between system threads.

Yau and Lam [8] implemented the Adaptive Rate-Controlled Scheduling, which is a modification of the *Virtual Clock* Algorithm. Each process specifies a reserve rate and a period for its admission control phase. During its execution phase, the reserve rate is adjusted upward or downward to match its actual usage rate in a gradual fashion. It is called *rate adaptation*.

Kamada, Yuhara, and Ono [6] implemented the *User-level RT Scheduler* (URsched) in the SUN Solaris 2.4. The URsched approach is based on the POSIX.4 fixed priority extension and its priority scheduling rule. The scheduler runs at the highest

Table 2. URsched priority structure

	Priority	Process
RT class	highest	URsched scheduler
	2nd highest	Running RT process
	..	Not used
TS class	any	Any TS processes
RT Class	lowest	Waiting RT processes

possible fixed-priority, the RT process waits its scheduling turn at the lowest possible fixed-priority (called the waiting priority), and the active RT process runs at the 2nd highest fixed-priority (called running priority). The priority structure is shown in Table 2. The scheduler wakes up periodically to dispatch the RT processes by moving them between the waiting and the running priority; during the other time, it just sleeps. When scheduler sleeps, the RT process at the running priority executes. When no RT processes exists, the non-RT processes with dynamic priorities execute using the fair time sharing scheduler of UNIX. This provides a simple mechanism to do RT scheduling in UNIX. It also has many desirable properties:

- It requires no modification to the existing UNIX/POSIX.4 kernels. The scheduling process can be implemented as an user-level application.
- It provides the flexibility to implement any scheduling algorithms in the scheduler, e.g. rate monotonic, earliest deadline, or even the Hierarchical CPU Scheduler.

In this paper, we will apply lessons learned from the previous approaches and extend the existing solutions to build a higher level soft RT scheduling server. In addition to the properties of the URsched, we provide the following functionalities:

- New Design of the *Soft RT Server Architecture*.
- Implementation of *Rate Monotonic Scheduling*.
- Provision of *Protection* among real-time processes.
- Fixing the *Security problem* in UNIX Operating System.

3 Server Architecture Design

Our server architecture contains three major components–the *broker*, the *dispatcher*, and the *dispatch table* as shown in Figure 1. Each component is described in details in the following sections.

3.1 Broker

The broker receives requests from client RT processes. It performs the admission control test to determine whether the new client process can be scheduled. If it is schedulable, the broker will put the RT process into the waiting RT process pool by

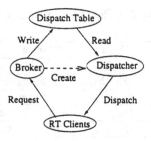

Fig. 1. The Soft RT server architecture

changing it to the waiting priority. The broker also computes a new schedule based on a desirable scheduling algorithm, the new schedule is written to the *dispatch table*.

The broker process is a root daemon process running at a normal dynamic priority. It can be started at the system boot time, like any other network and file system daemons. It will wake up when the new client request arrives. The broker needs to be a root process so that it can change processes into the fixed priority. The broker process does not perform the actual dispatching of the RT processes, instead it will fork a separate real-time *dispatcher* process. The reason is that the admission and schedulability test in the broker may have variable computation time, hence it may affect the timing of dispatching. The admission and schedulability test do not need to be done in real time; as a result, the broker runs at a dynamic priority. The separation of RT dispatching in the *dispatcher* process and the non-RT *schedulability test* in the *broker* process is an essential feature that allows both *dispatcher* and *broker* to do on-line computation without compromising the precision of RT processes dispatching.

The client RT processes must start their processing at the non-RT dynamic priority level. The broker and dispatcher will change them into the fixed RT priority when they are accepted and dispatched. This is an improvement over the current UNIX environment, our scheme allows any user to run processes at the fixed priority in a fair and secure manner.

3.2 Dispatch Table

The *dispatch table* is a shared memory object where the *broker* writes the computed schedule to and the *dispatcher* reads from in order to know how to dispatch RT processes. It is locked inside memory for efficient read and write. The dispatch table contains a repeatable *frame* of slots, each slot corresponds to a time slice of CPU time. Each slot can be assigned to a RT process pid, a group of cooperating RT process pids, or be free which means yielding the control to the UNIX TS scheduler to schedule any non-RT processes. Consider the example in Table 3. The repeatable *frame* is 80*ms*, and it contains 8 time slots of 10*ms* each. The sample dispatch table is a result of a RM schedule with the pid 721 at *period=20ms, execution time=10ms,* and pid 773/774/775 at *period*=40*ms*, *execution time*=10*ms*. There are 2 free slots, which means 20*ms* out of every 80*ms* of CPU is allocated to the non-RT processes.

Table 3. Sample Dispatch Table, Real Time

Slot Number	Time	Process PID
0	$0 - 10ms$	721
1	$10 - 20ms$	773 774 775
2	$20 - 30ms$	721
3	$30 - 40ms$	free
4	$40 - 50ms$	721
5	$50 - 60ms$	773 774 775
6	$60 - 70ms$	721
7	$70 - 80ms$	free

The minimum number of free slots is maintained by the broker to provide a fair share of CPU time to the non-RT processes. In the Table 3, 25% (20*ms* out of 80*ms*) of the CPU is guaranteed to the non-RT processes. The site administrator can adjust the non-RT percentage value to be what is considered fair. For example, if the computer is used heavily for RT applications, the non-RT percentage can be set to a small number, and vice versa.

3.3 Dispatcher

The *dispatcher* is a process running at the highest possible fixed priority. The dispatcher process is created by the broker and it is killed when there are no RT processes to be scheduled in the system. When there are only non-RT processes running, the system has no processing overhead associated with the RT server.

The dispatcher contains the shared memory dispatch table and a pointer to the next dispatching slot. At the beginning of the next dispatch slot, a periodic RT timer signals the dispatcher to schedule the next RT process. The length of time to switch from the end of one slot to the start of the next one is called the *dispatch latency*. The dispatch latency is the scheduling overhead which should be kept at a minimal value.

The dispatching process is similar to the URsched approach. Consider the dispatch table in Table 3 at time 50*ms*. The dispatcher is moving from slot 4 to slot 5, the following steps are taken.

- The periodic RT timer wakes up the dispatcher process, and the process 721 is preempted (1 context switch).
- The dispatcher changes the process 721 to the waiting priority and processes 773/774/775 to the running RT priority (4 system calls to set priority).
- The dispatcher puts itself to sleep, and one of the processes 773/774/775 are scheduled (1 context switch).

The program code segment that corresponds to the above steps is executed repeatedly, and is locked into memory to avoid costly page faults. The dispatch latency can be bounded by the time to do 2 context switches and (the maximum number of processes in any 2 adjacent slots) set-priority system calls. The dispatcher follows the schedule in the dispatch table and it will not allow processes to overrun. The execution time of the RT processes execution time is protected from each other.

3.4 RT Client

Our client's system QoS request uses the simple RT Mach's form of QoS specification: *period=p, CPU usage in percentage=u*. For example, the specification ($p = 10ms, u = 20\%$) means that $2ms$ out of every $10ms$ are reserved for this RT process. There are restrictions to the size of the period. The hardware restricts the size to be greater than the software clock resolution. Smaller period also leads to smaller time slice, which results in higher number of context switchings and inefficient CPU utilization.

The users can compute the client's required period (e.g. 40 frames per second video player has a period of $25ms$), but the computation of the CPU usage percentage is very difficult. The UNIX command *rusage* will not give an accurate system time because some of the system threads execution times may not be accounted, as described in the Process Reserves Model in the RT Mach [4]. The user can get a good estimation by a few trial-and-error runs of the client programs.

The execution time also depends on the state of memory contention and the resulting number of page faults. We are currently designing a *memory broker* where the RT process can reserve memory prior to their execution.

The client RT process may block while making a system call. During this blocking time, the client is not executing. But since the kernel is serving the system call for the client, the blocking time should be considered as a part of client's execution time (as in the case of Process Reserves Model).

4 Implementation

We have implemented our server architecture on a single processor Sun Sparc 10 running Solaris 2.5 OS. The Solaris OS has a default global priority range (0-159), 0 the least importance. There are 3 priority classes:RT class, System class, and TS class. The RT class contains fixed priority range (0-59), which maps to the global priority range (100-159). The dispatcher's priority is 59, the running priority is 58, and the waiting priority is 0. The waiting priority 0 needs to be mapped to the lowest global priority 0, and it must be lower than any TS priorities. This can be done by compiling a new RT priority table *RT_DPTBL* inside the kernel.

The changing priority is done by using the *priocntl()* system call. Its average cost is measured as $175\mu s$. The average dispatch latency (2 context switch + 2 *priocntl()* is measured as $1ms$. The interval timer is implemented using *setitimer()* with the maximum resolution of $10ms$ and it is the smallest time slot size. The overhead is 10%, which is acceptable.

5 Experimental Result

We have tested our soft RT server with two experiments which run multimedia application(s) and non-RT applications concurrently. The metric to evaluate the performance of our CPU server is the *jitter* of the continuous media in multimedia application(s).

5.1 Experiment 1

The first experiment consists of the mixture of the following 4 popular applications running concurrently:

- The Berkeley mpeg-play (version 2.3) plays the TV cartoon Simpsons mpeg file at 10 frames per second.
- The gcc compiler compiles the Berkeley mpeg-play code.
- Our program computes the *sin* and *cos* table using the infinite series formula.
- The Latex program formats this paper.

The graphs in Figure 2 show the measurement of *frame jitter* on the mpeg-play under the above specified load. The left graph shows the result of the normal TS UNIX scheduler without our server. The right graph shows the result of our server with 60% CPU reserved every 50*ms* to the mpeg-play. Using the UNIX TS scheduling, noticeable jitter over 100*ms* (equivalent to 1 frame time) occurs frequently—61 times out of the 650 frames (65 seconds). The largest jitter is about 350*ms* (over 3 frames time), which is clearly unacceptable. Using our server, noticeable jitter over 100*ms* occurs only once. The Table 4 shows the elapsed time of the other three applications—compute, compile, and Latex. Using our server, their execution time increases by no more than 20%, which is acceptable for non-RT applications.

Table 4. Elapsed Time of the compute, compile and Latex applications for Experiment 1 and 2.

	Experiment 1			Experiment 2		
	Compute	Compile	Latex	Compute	Compile	Latex
TS UNIX	75.11s	141.36s	28.93s	102.84s	283.10s	33.70s
Server	90.90s	160.33s	33.24s	132.86s	293.04s	69.53s

5.2 Experiment 2

The second experiment adds a second mpeg-play that plays the same mpeg file at 5 frames per second. The graphs in Figure 3 show the measurement of jitter on two mpeg-play applications. The top graphs show the results of the 10 frames per second mpeg-play and the bottom graphs show the results for the 5 frames per second. The

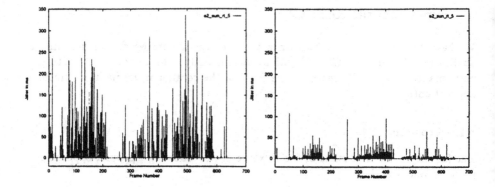

Fig. 2. Frame Jitter Measurement for mpeg_play at 10 frames per second on the SUN Sparc 10 Solaris 2.5 platform. The left graph shows the result for the TS UNIX Scheduler. The right graph shows the result of our RT scheduling server with 60% CPU reserved every 50ms.

left graphs show the results of the normal TS UNIX without our server. The right graphs show the results of our server with 60% CPU reserved every 50ms to the first mpeg_play, and 30% CPU reserved every 50ms to the second mpeg_play. Using the UNIX TS scheduling, noticeable jitter over 100ms occurs frequently–153 times out of 650 frames (65 seconds) for the first mpeg_play, and 10 times out of 325 frames for the second mpeg_play programs. The largest jitter is about 400ms (4 frames time), which is unacceptable. Using our server, noticeable jitter over 100ms occurs 0 and 1 time for the first and second mpeg_play programs. The Table 4 shows the elapsed time of the other 3 applications. Using our server, their average execution time increases by 50%, which is still acceptable for non-RT applications.

6 Conclusion

We have shown through experiments that our soft RT server provides predictable QoS for Continuous Media applications in the UNIX environment. It addresses *fairness* by maintaining a minimum amount of CPU time for the non-RT processes, it offers *protection* by disallowing RT process to overrun their time slots, and it provides *security* through the broker interface.

References

1. Chen Lee, Ragunathan Rajkumar, and Cliff Mercer. "Experience with Processor Reservation and Dynamic QOS in Real-Time Mach". *Multimedia Japan*, 1996.
2. Pawan Goyal, Xingang Guo, and Harrick Vin. "A Hierarchical CPU Scheduler for Multimedia Operating System". *The proceedings of Second Usenix Symposium on Operating System Design and Implementation.*

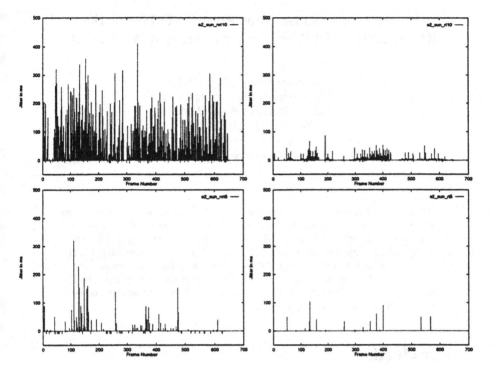

Fig. 3. Jitter Measurement for the two mpeg_plays at 10fps (top) and 5fps (bottom). The left graphs shows TS UNIX. The right graphs shows our server with 10fps (top) 60% CPU reserved every 50ms, and 5fps (bottom) 30% CPU reserved every 50ms.

3. Bill O. Gallmeister. "Programming for the Real World: POSIX.4". O'Reilly & Associates, INC. 1995.
4. Clifford W. Mercer, Stefan Savage, and Hideyuki Tokuda. "Processor Capacity Reserves: Operating System Support for Multimedia Applications". *IEEE International Conference on Multimedia Computing and Systems.* May 1994.
5. Jason Nieh, James G. Hanko, J. Duane Northcutt, and Gerard A. Wall. "SVR4 UNIX Scheduler Unacceptable for Multimedia Applications". *Fourth International Workshop on Network and Operating System Support for Digital Audio and Video.* Nov 1993.
6. Jun Kamada, Masanobu Yuhara, Etsuo Ono. "User-level Realtime Scheduler Exploiting Kernel-level Fixed Priority Scheduler". *Multimedia Japan,* March 1996.
7. Sandeep Khana, Michael Sebree, and John Zolnowsky. "Realtime Scheduling in SunOS 5.0". *USENIX Winter 1992 Technical Conference.*
8. David K.Y. Yau and Simon S. Lam. "Adaptive Rate-Controlled Scheduling for Multimedia Applications". ACM Multimedia Conference '96, Boston, MA, Nov 1996.

Stored Video Transmission Across ATM Networks: RVBR versus RCBR Service for Interactive Applications

Stefan Gumbrich, Jürgen Heller
IBM European Networking Center
Vangerowstr. 18, D-69115 Heidelberg
Phone: +49 6221 59-4369
Fax: +49 6221 59-3300
E-mail: gumbrich@heidelbg.ibm.com

Abstract. : The transmission of stored video from a server to a client across an ATM network is considered. In this context the performance of the renegotiated VBR (RVBR) and renegotiated CBR (RCBR) services is investigated for interactive applications with bounded reaction times. We present two basic results. First, for delay sensitive applications smoothing the video stream in advance is much more efficient, than adding buffer to the network switches. Second, RVBR service leads to higher network utilization than RCBR or CBR service, independent of the call admission control scheme.

1 Introduction

A broad range of applications requires the playback of stored video over high speed networks. Examples include digital libraries, interactive virtual environments, shopping and entertainment services. To allow for near constant quality video delivery, variable-bit-rate (**VBR**) compression of the video is necessary. The varying quality of constant-bit-rate (**CBR**) compressed video is not acceptable, as viewers are accustomed to constant quality service since the beginning of the motion picture era.

To allow for continuous playback at the client, strict end to end quality of service (QoS) must be guaranteed by the network. ATM's real time-VBR (rt-VBR) [TM4.0] and CBR service can guarantee the necessary QoS (delay, jitter, cell loss). Two new services (or service enhancements) that are under discussion for ATM are renegotiated CBR service (**RCBR**) and renegotiated VBR service (**RVBR**). These services allow renegotiation of the reserved bandwidth (Source Traffic Descriptor) throughout the duration of a connection. This is called dynamic bandwidth allocation. The RCBR and RVBR services are considered necessary for video transmission, since VBR compressed video exhibits significant rate variability, often spanning time scales of several minutes [GaWi94], [GrKe95] due to scene changes.

There are in general two classes of real time applications that require transmission of stored video: First, applications with user interaction and second applications without or with limited user interaction. Examples for the first class are: video on demand (**VoD**) with full VCR functions (pause, jump forward, rewind, etc.), news on demand, interactive movies and video games. Examples for the second class are digital video broadcast and VoD with very restricted rewind and jump forward functionality. The question that arises is: Which ATM service is most efficient for a given class of video applications.

For **applications with limited user interaction** schemes based on CBR and RCBR services are proposed by [McRo96], [Sale96], [LiChi96].

- CBR service is a good option for digital video broadcast (TV). It is possible to multiplex a large amount of videos [LiChi96] at the source and to reserve a constant bandwidth for the cumulated bandwidth requirements on the path to a client. This makes use of the statistical multiplexing gain of a large group of for example 50 TV channels. On the contrary, for interactive applications statistical multiplexing is only feasible at the network switches, since a client typically requests only one, or a very small number of video streams.

- RCBR service is used in the optimized schemes of [McRo96], [Sale96] which are based on workahead smoothing. In this approach video data is transmitted ahead of schedule and stored in a large buffer with known size at the client. This allows to send at piecewise constant rate (RCBR service). These schemes are not applicable for the class of applications with user interactions, since a user interaction (e.g. jump forward) can lead to a large amount of unnecessarily sent data, and a long time to fill the client buffer before constant playback is possible again. A time-laps of several seconds is not acceptable for most interactive applications. The additional waiting time for a response to a jump forward is up to 30 seconds for the Starwars video with the proposed 2MByte client buffer in the scheme of [McRo96].

It is still open which ATM service is most suitable for the class of **interactive applications**. In this paper the services RVBR and RCBR are compared in the context of interactive video applications to find an answer. Our metric is the number of real video streams that can be accepted by a switch in the network, while guaranteeing a given QoS. This is motivated by the possible reduction of the costs for an individual stream if a higher number of connections (increased network utilization) is possible. For given link speed, video streams, segmentation algorithm and buffer sizes the number of acceptable connections is mainly determined by the chosen ATM service and the connection admission control (**CAC**) scheme implemented in the ATM switch. The CAC decides in each switch, if an additional stream with given bandwidth demands and QoS requirement can be accepted.

For CBR and RCBR the deterministic CAC and for VBR services the statistical CAC is mandatory. For RVBR service statistical or deterministic CAC can be used. RVBR service with deterministic CAC is compared to CBR and RCBR service in chapter 3 and 4. In chapter 5 statistical CAC for RVBR service is investigated and compared to the performance of RVBR service with deterministic CAC.

2 Parameters for Real Time Services

A traffic contract is negotiated for each ATM layer connection at the UNI. It consists of a source traffic descriptor (**TD**), a set of QoS parameters and the conformance definition based on the Generic Cell Rate Algorithm (GCRA)[UNI3.1]. The TD can be changed without setting up a new connection for RCBR and RVBR services [Cho-Li95],[ReiRa95],[LakYa96]. In this section it is described how the parameters for the different ATM services are determined.

For VBR and RVBR service the source traffic descriptor consists of the elements Peak Cell Rate (**PCR**), Sustainable Cell rate (**SCR**), Maximum Burst Size (**MBS**) and Cell Delay Variation Tolerance (**CDVT**). For CBR and RCBR service it consists only of PCR and CDVT. The CDVT is independent of the source traffic and is therefore provided by the network adapter. It is the users task to determine the parameters PCR, SCR and MBS. An algorithm, based on the empirical envelope [Cruz91], [Cruz95], [KnWr95], [WreLi96], is used for stored VBR video to select those parameters in such a way, that no cells are tagged or discarded by the GCRA policing algorithm. In [GuEm97] it is described how the TD is determined for a given video stream.

Additional parameters for RVBR and RCBR service are the positions for renegotiating the bandwidth requirements. An optimized algorithm to determine the renegotiation points is presented in [GuEm97]. It is shown that this solution is superior to other well known RCBR and RVBR schemes.

In the case of dynamic bandwidth allocation it may happen, that a renegotiation request for higher bandwidth is denied. In case of a denied renegotiation request the bandwidth requirements of the video are adapted to the allowed bandwidth by reducing the picture quality as little as possible. The allowed bandwidth is at least equal to the bandwidth that was reserved in the last segment. Due to the possible reduction in perceived quality in the case of adaptation an important QoS requirement for renegotiated services is the probability of a failed renegotiation request. If we assume a mean renegotiation rate of 10 seconds and a length of 90 minutes for the video session then it is acceptable for a user to watch one segment of 10 seconds length with reduced quality. This results in a QoS parameter of 10^{-3} for the renegotiation failure probability. Therefore a service based on dynamic bandwidth allocation provides a statistical service on the level of segments since it is possible for a renegotiation request for more bandwidth to fail. The advantage compared to traditional statistical services is the higher level of control for the user due to bandwidth adaptation. The adaptation during overload situations avoids uncontrollable packet drop and leads to higher subjective quality for the user [GuSa95],[GaPu97].

3 RVBR Service with Deterministic CAC

A deterministic connection admission control (CAC) will ensure that no packets are dropped or delayed beyond their negotiated delay bound, if the sending end system conforms to the accepted TD. The CAC relies on the deterministic ATM source traffic descriptor. The traffic constraint function b(t) described by ATM's VBR TD is:

$$b(t) = min(PCR \cdot t, SCR \cdot t + BT \cdot SCR) \qquad (1)$$

with $BT = (MBS - 1) \times (\frac{1}{SCR} - \frac{1}{PCR})$

The time invariant deterministic bound b(t) bounds the source over every interval of length t. Denoting A[s1,s2] as a connections arrivals in the interval [t1,t2] the traffic constraint function guarantees that $A[s, s+t] \le b(t)$ for all s, t > 0.

Deterministic CAC relies on the delay analysis techniques of [Cruz91][ZhFe94]. Their work shows that the delay bound provided to a connection varies with the service disciplines. If we assume for simplicity a FCFS scheduler then the deterministic CAC checks wether equation (2) holds.

$$\sum_{j=1}^{N} b_j(t) \le L \cdot t + L \cdot d_k \qquad \text{for all t} \qquad (2)$$

In equation (2) L denotes the link speed, serving N connections with upper delay bound d_k. Each connection is bounded by its constraint function $b_j(t)$, j=1,...,N. In [KnZh95], [KnZh96a] this form of CAC was investigated for different videos and values of delay bound d_k with the traffic constraint function based on Knightly and Zhang's DBIND model. Note that $d_k > 0$ means that cells are buffered in each switch for a time up to d_k. This accounts for a delay of $\sum d_k$ at the end to end delay.

In [GuEm97] it was shown, with smoothing based on frame period averaging (cells within one frame period are equally spaced [Graf94]), that RVBR service with deterministic CAC based on equation (2), reduces the bandwidth requirements significantly. Here, the impact of the smoothing delay d_s versus delay bound d_k, on the number of admissible streams per link is compared. We restrict ourselves to small smoothing delays of 0 to 300 ms to account for the interactive applications with bounded times on the reaction to a users interaction. The smoothing technique used, is workahead smoothing as this is most suitable for stored video. Workahead smoothing allows to send cells ahead of schedule and to buffer them at the receiver. For applications without interaction this kind of smoothing can avoid any additional delay [Sale96]. For applications that allow arbitrary access to different points of a video (most interactive applications) additional delay is introduced after an interaction. This delay is equal to the time needed to fill the receiver buffer with enough data to allow for continuous playback after an interaction. The delay depends on the amount of workahead data expected at the actual point of the video and the bandwidth that is reserved for the actual segment. It is possible to avoid this start-up delay by calculating a new renegotiation schedule after each interaction. Unfortunately this is not feasible due to the complexity of the algorithms for finding an optimized renegotiation schedule [GuEm97], [Sale96], [McRo96], [KnZh96a], [GrKe95].

We restrict the start-up delays for interactive applications by the smoothing bound d_s. A traffic smoothing delay of d_s prohibits to send video data more than d_s ahead of schedule.

3.1 Simulation Model

The number of connections that are accepted by the CAC while guaranteeing a desired QoS level are determined by simulation. For the simulations we used VBR MPEG coded video traces, kindly provided by the University of Würzburg (ftp-info3.informatik.uni-wuerzburg.de). The TD for each video stream was determined such that no cells have to be tagged or discarded at the UNI by the policing algorithm (GCRA).

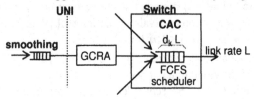

Fig. 1. network model

In the model presented in Figure 1, N video streams are multiplexed onto one output link of a switch. Each stream is transmitted starting with an independent random starting point. The starting points are distributed uniformly between 0 and the length of the trace. A stream that reaches the end of the trace wraps around to the beginning. For each simulation a time frame of 22 minutes where all traces are active is evaluated. The video of each source is segmented according to the optimized algorithm in [GuEm97]. The renegotiation requests are signaled to the network together with the TD for each new segment. The average length of a segment is about 10 seconds. Moreover there is at least one second between two renegotiation requests of a connection to avoid overloading of the signalling system of the switches. The number of denied renegotiation requests for all N video streams is measured and the renegotiation failure probability of the actual set of starting points is calculated. We ran 5000 simulations with independent starting points for a given value of N. From those 5000 simulations the 99% percentile was determined, to make sure that the chosen renegotiation failure probability of 10^{-3} holds for at least 99% of the starting point combinations. This reflects the cautious behavior of a CAC algorithm for renegotiated services that has to give statistical guarantees for the renegotiation failure probability, not knowing the future behavior of the sources. Such an algorithm for the DBIND model is presented in [KnZh96a].

For simplicity a first come first serve (FCFS) scheduler is assumed which leads to the deterministic CAC described by equation (2). Using more advanced scheduling disciplines [KnWr95],[Cruz95] allows different delay bounds on a link, but leads to the same conclusions.

Our delay smoothing scheme reduces the peaks of a video trace as much as possible, while guaranteeing an upper bound d_s on the introduced delay. For workahead smoothing this is achieved by sending cells no more than d_s ahead of schedule. Note that this is necessary for interactive applications to restrict the response time after an interaction.

Furthermore it is assumed that the TD for each connection is preserved by a rate controller [ZhFe93] in each switch. This avoids that buffer requirements for a connection increase with the length of the path. Due to the rate controller each switch on the path of a connection can treat the traffic descriptor identically. Therefore it is sufficient to investigate only one switch on the path.

For a trunk size of 155 Mbit/s the usable bit rate excluding the specific frame structure of an ATM SDH Physical Layer is 149.76 Mbit/s. After accounting for an overhead of 1.2% for the MPEG Elementary Stream, 2.6% for MPEG Transport Stream, 2.2% for ATM AAL 5 and 11% for ATM layer the available link rate is 128 Mbit/s for the video data

3.2 Evaluation

The effectiveness of the deterministic CAC with VBR and RVBR schemes is evaluated in this section. This is done in the context of interactive applications with strong requirements on the reaction time. The reaction time is the time a user has to wait after an interaction until the desired response gets visible. This is for example the waiting time after an action in an interactive video game. The propagation-, buffering-, switching-, smoothing-delay and the processing time in the end systems contribute to this reaction time. Reaction times in the range of 0 to 0.5 seconds are reasonable to allow for interactive behavior. Since most delays are fixed the smoothing delay and the buffering delay are investigated. Therefore two cases are compared:

1. The scheme of [KnZh96a] where additional buffer is introduced at the switch to allow for smoothing the bursts when multiplexing is achieved at a link. This buffering delay is bounded by the value d_k in each switch. The overall buffering delay is the sum of the delays in each switch on the path from server to client.Smoothing is based on frame period averaging [Graf94].

2. Our delay smoothing scheme with bounded delay d_s. No additional buffering is performed in the switches ($d_k=0$).

In Figure 2 and Figure 3 the number of admissible connections with deterministic CAC for different values of d_k and d_s is presented for the videos Mtv2 and Movie (length 22 minutes). The number of connections for the RVBR service is determined for the 99% percentile of a renegotiation failure probability of 10^{-3}. To demonstrate the performance gain of renegotiated service the admissible number of connections for a CBR scheme (identical to a VBR service based on deterministic CAC) is also shown.

It is obvious that the number of admissible connections is considerably higher for the delay smoothing scheme (d_s) than for the scheme based on buffering in the switch (d_k), even though only one switch was considered on the path from the server to the client. If a more realistic scenario with at least 4 switches on the path from client to server is considered, the number of connections for a given value d_s has to be compared to the number of connections for $d_k=d_s/4$. In this case the delay smoothing scheme (d_s) outperforms the scheme based on buffering in the switch (d_k) by at least a factor of 2. From these results we conclude that the delay smoothing scheme is the better choice in the case of interactive video applications based on stored video. The reasons are:

1. higher number of admissible connections, which leads to lower costs

2. reduced complexity of the CAC algorithm since the buffer calculation is omitted

Comparing the renegotiated service (RVBR) with the CBR service (Figure 2, Figure 3) shows obvious performance gains for the RVBR service. The large difference in admissible number of connections for bounded smoothing delays clearly states that renegotiation should be used in the context of constant quality video transmission for interactive applications.

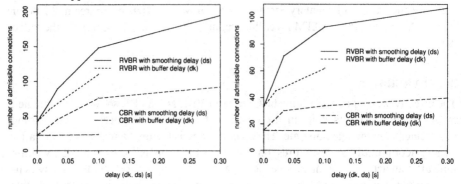

Fig. 2. deterministic CAC for video **Movie**   **Fig. 3.** deterministic CAC for video **Mtv2**

4 RCBR versus RVBR Service with Deterministic CAC

With no extra buffer (d_k=0) the deterministic CAC allows as many connections for the RVBR as for the RCBR service. This puts up the question about the advantage of RVBR over RCBR service with deterministic CAC.

The main reason to use RVBR service is that there is a considerable amount of *unreserved bandwidth* left for services without real time requirements that rely on reservation. In ATM [Sig4.0], [TM4.0] the non real time-VBR (nrt-VBR) and available bit rate (ABR) service are defined for data applications that need QoS guarantees on cell loss, but no explicit guarantee on cell delay and jitter. Therefore the ATM services ABR and nrt-VBR rely on reservation of bandwidth. In this section we investigate the amount of unreserved bandwidth that can be used by these services. Therefore a buffer is introduced into the ATM switches to allow for buffering of non real time cells. The amount of buffering is described in terms of delay since this value is of most interest for the user. The investigations are based on the model described in section 3.1. For the examples in Figure 4 the maximum amount of Asterix, Dino or Mtv2 video streams is accepted by the deterministic CAC as explained in section 3.2. For the RVBR service additionally to this maximum amount of real time streams some nrt-VBR or ABR streams can be accepted. The mean bandwidth in Figure 4 describes the average amount of bandwidth that is available for the non real time service. This adds up to 40-60 Mbit/s on the 155 Mbit/s link investigated if a bounded delay of maximum 1 second per switch is allowed. This has to be compared to less than 10 Mbit for the RCBR service. We further determined the minimum bandwidth available within the duration of the connections. This reflects the maximum size of a nrt-VBR stream that can be accepted by the CAC, next to a worst case combination of rt-VBR streams for the duration of the investigation time (22 minutes). For a bounded delay of 1 second per switch this adds up to 10-20 Mbit/s. This has to be compared to a value of less than 5 Mbit/s for the RCBR service.

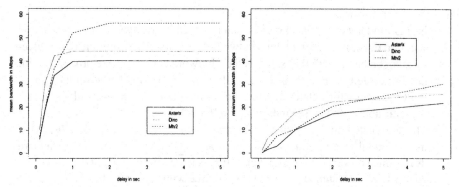

Fig. 4. nrt-VBR bandwidth available next to maximum number of admissable VBR streams

Under the assumptions that

- only small smoothing delays are applied to the rt-VBR videos (valid for interactive applications)

- a large delay of at least 1 second can be applied to the nrt-VBR streams per switch (about 600KByte buffer per link)

we conclude, that for the deterministic CAC scheme the RVBR service is superior to the RCBR service. This is due to the large amount of additional non real time (nrt-VBR or ABR) streams that can be accepted by the CAC.

5 RVBR Service with Statistical CAC

In this section we consider RVBR service based on traditional statistical CAC schemes and compare this to the RCBR service with deterministic CAC (section 3).

Typically statistical CAC schemes [ZKST97], [Guer92] are based on statistical models described by statistical parameters. The problem concerning ATM networks is, that ATM relies on the deterministic parameters PCR, SCR and MBS [UNI3.1] for policing and signaling. These deterministic parameters specify an upper bound on the mean rate and burst size. On the contrary the statistical CAC schemes usually rely on a mean rate, mean burst size and sometimes additional parameters to describe the tail of the distribution. It is shown that statistical multiplexing with this kind of parameters can achieve a high multiplexing gain given that enough sources are multiplexed to a link [ZKST97].

To use statistical CAC for ATM networks we have to derive statistical parameters from the given deterministic parameters. We chose mean rate m = SCR and mean burst size b = MBS for each connection. This is obviously an overestimation of the required bandwidth, but the best choice to be compliant with the ATM standards. We investigated the Gaussian approximation [Guer91], [Onvural] and the equivalent capacity [Guer92], [Kelly91], [Vakil93].

•Gaussian approximation

In this model each connection i is characterized by its average bit rate m_i and standard deviation σ_i. The total bandwidth C required by n multiplexed connections is the bandwidth that satisfies the following condition: the probability for the aggregated bit rate to exceed C is less than ε. This is based on the central limit theorem which states that the sum of n random variables results in a random variable with Gaussian distribution [Guer92].

The aggregated rate is defined as $C = m + \alpha \sigma$ with $\alpha = \sqrt{2 \cdot ln(\varepsilon^{-1}) - ln(2\pi)}$.
m is the total mean bit rate and equal to $m = \sum m_i$, and σ is the total standard deviation. The variance σ_i of a connection is given by $\sigma_i^2 = m_i \cdot (PCR_i - m_i)$, if connection i is assumed to be a On-Off source with stationary bit rate.

•equivalent capacity

As a general definition the equivalent capacity is the virtual portion of bandwidth required by a source in order to obtain a given grade of service from the network [RaCa95]. We use the formula of [Guer91], [Guer92] for the calculation of equivalent capacity since it is the most complete.

Equivalent capacity assumes On-Off sources with exponentially distributed burst and idle period as input. If the source has nonexponential burst and idle periods it has to be mapped by an approximation technique called moment matching [Guer91] or by an indirect measurement based approach [Guer92]. We use the moment matching technique for a source with deterministic On-period. The minimum of PCR and equivalent capacity is used to allocate the bandwidth for a connection.

•HBIND

The HBIND approach [KnZh96a] defines a statistical CAC scheme based on the deterministic DBIND model. Therefore it is more straightforward to map this approach to ATM networks. The approach is based on the principle described by equation (2). The main difference is the traffic constraint function which is replaced by a random variable of Gaussian distribution. This random variable is built by the rules described for the Gaussian approximation. The main difference is the determination of σ_i from the deterministic TD.

The result of section 3 (use of smoothing is more advantageous for interactive applications than buffering at the switch) also holds in the context of statistical CAC. With delay bound $d_k=0$ the HBIND scheme is reduced to a scheme very similar to the Gaussian approximation. Because of the higher complexity of HBIND and the very similar results of HBIND and Gaussian approximation in our investigations (for $d_k=0$ and with ATM's TD), HBIND is not further discussed in this paper.

For interactive applications renegotiation has to be used together with statistical CAC schemes. This is due to the fact that the traffic characteristics depend on user interactions that are not predictable at connection setup time. Therefore a TD chosen at connection setup time has to be very conservative or cells will be tagged or discarded by the policing algorithm at the UNI. This would lead to a bad performance (low number of admissible connections) or a unacceptable decrease in picture quality due to arbitrary cell loss. Therefore VBR service without renegotiation is not applicable for interactive applications.

5.1 Results

The different CAC schemes are compared by simulation. The simulations are based on the model described in section 3.1. For each CAC scheme the maximum number of admissible videos that can be transmitted simultaneously on a link while guaranteeing a renegotiation failure probability of 10^{-3} for 99% of the start point combinations is determined.

In Figure 5 to Figure 8 the number of admissible connections depending on the smoothing delay are shown for the CAC schemes: Gaussian approximation, equivalent capacity and deterministic CAC. Representative for our investigations the videos Mtv2 and Movie are shown for link rates of 45Mbit/s and 155Mbit/s.

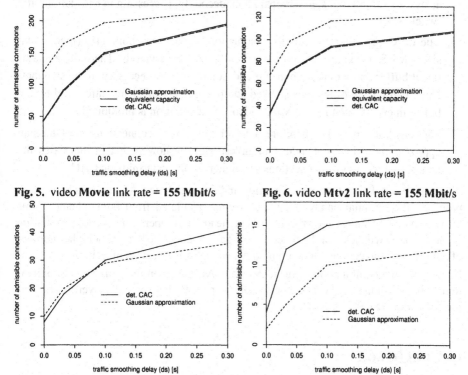

Fig. 5. video **Movie** link rate = 155 **Mbit/s**   **Fig. 6.** video **Mtv2** link rate = 155 **Mbit/s**

Fig. 7. video **Movie** link rate = 45 **Mbit/s**   **Fig. 8.** video **Mtv2** link rate = 45 **Mbit/s**

In Figure 5 and Figure 6 it gets obvious that for a link rate of 155Mbit/s the *equivalent capacity* leads to nearly the same number of admissible connections as the deterministic CAC. For a link rate of 45Mbit/s the admissible number of connections for equivalent capacity and deterministic CAC are identical. The reason is, that the equivalent capacity is an approximation, based on the assumption of short burst periods and low statistical Multiplexing capabilities. This results in allocation of PCR for nearly all segments of the videos. Therefore the additional overhead of calculating the equivalent capacity is not justified independent of the link rate.

The best scheme in terms of admissible number of connections depends on the ratio of traffic parameter size to link rate. The *Gaussian approximation* leads to much better results than the deterministic CAC for a link rate of 155Mbit/s (Figure 5, Figure 6). Nevertheless the gain of Gaussian approximation over deterministic CAC decreases with increasing traffic smoothing delay d_s (decreasing ratio of PCR/SCR). For a link rate of 45 Mbit/s (Figure 7, Figure 8) the Gaussian approximation does not hold and the deterministic CAC leads to the best or at least comparable results for most streams.

From these investigations it can be concluded that a combination of statistical CAC based on Gaussian approximation and deterministic CAC is the preferred choice for RVBR service. The reasons are as follows:

1. For many interactive applications smoothing is allowed only in the range of less than 100 ms. This is due to the acceptable reaction times for interactive perception which is in the range of 250 ms [HeSa89]. The delay of 250ms is the sum of the delays for smoothing, signal propagation, decoding, processing in end stations and switches, etc.

2. Due to renegotiation and the overestimation of the mean burst size (mean burst size = MBS) the Gaussian approximation works conservativly. Therefore no additional buffer that would account for additional delay is necessary in the switches. Note, that our investigations showed additional delay of 21 ms introduced by the buffer in each switch for the Mtv2 video if renegotiation is prohibited.

3. For low link rates and therefore a lower number of connections the Gaussian approximation does not hold. Our investigations showed that in this cases the admissible number of connections are higher for the deterministic CAC.

A combination of Gaussian approximation and deterministic CAC reserves bandwidth based on the minimal bandwidth requirements calculated from both schemes. This leads to slightly increased processing overhead of 4 operations per renegotiation request in the switches. On the other hand a gain of about 30% to 60% higher number of admissible connections for smoothing delays up to 100ms is achievable.

In conclusion a combination of the statistical CAC based on the gaussian approximation and deterministic CAC is the best choice for the RVBR service. Nevertheless simple deterministic CAC is also feasible.

6 Conclusion

Our goal was to determine the appropriate ATM service for interactive applications with bounded reaction time, that are based on real time transport of stored video. The ATM services CBR, VBR, RCBR and RVBR were investigated for this purpose. The main assumption on interactive applications is, that for this class of applications the time for the reaction to a user interaction is bounded. The investigations led to the following results:

- The VBR service is not useful since the traffic characteristics of a video stream are changed by user interactions. Therefore it is not possible to determine the proper

TD in advance. This can lead to unacceptable picture quality or low network utilization.

- The CBR service can accept only a small number of streams, which leads to very low network utilization compared to RCBR or RVBR service.
- The RCBR service is defeated by the RVBR service even for the same deterministic CAC scheme. The RVBR scheme allows for a considerable amount of additional non real time streams (nrt-VBR, ABR) next to the maximum amount of acceptable real time-VBR streams.
- The RVBR service with statistical CAC based on a combination of Gaussian approximation and deterministic CAC leads by far to the highest network utilization (highest amount of acceptable video streams). This is especially true for video applications with small reaction times to user interactions.

From this we conclude that the RVBR service is the proper choice for interactive applications.

7 Acknowledgments

The authors thank Achim Schneider for excellent technical support.

8 References

[ChoLi95]   S. Chong, S. Li, J. Ghosh: Predictive Dynamic Bandwidth allocation for Efficient Transport of Real Time VBR Video over ATM, IEEE Journal on Selected Areas in Communication, Vol 13, No1, Januar1995

[Cruz91]   R.L. Cruz: A Calculus for Network Delay, Part I and II, IEEE Transactions on Information Theory, Vol.37, No.1, January 1991

[Cruz95]   R.L. Cruz: Quality of Service Guarantees in Virtual Circuit Switched Networks, IEEE JSAC Vol.13, No.6, August 1995

[GaPu97]   S. Gara, G. Pujolle: Study of MPEG VBR sequence quality transmitted with rt-VBR and ABT/DT contracts, IEEE ATM'97 Workshop, May 1997, Lisboa, Portugal

[Graf94]   M. Graf: Traffic Shaping for VBR Video, Proceeding 4th Open Workshop on High Speed Networks, Telecom Bretagne, Brest France 1994

[GrKe95]   M. Grossglauser, S. Keshav, D. Tse: RCBR: A Simple and Efficient Service for Multiple Time-Scale Traffic, ACM Sigcomm 95, August 1995

[Guer91]   R. Guerin: Equivalent Capacity and Its Application to Bandwidth Allocation in High-Speed Networks, IEEE JSAC, Vol. 9, No7, Sept. 1991

[Guer92]   R. Guerin, L. Guen: A Unified Approach to Bandwidth Allocation and Access Control in Fast Packet-Switched Networks, INFOCOM'92

[GuEm97]   S. Gumbrich, H. Emgrunt, T. Braun: Dynamic Bandwidth Allocation for Stored VBR Video in ATM Endsystems, Seventh IFIP Conference on High Performance Networking (HPN'97), White Plains, NY, May 1997

[GuSa95]   S. Gumbrich, J. Sandvoss: Comparison of Media Scaling Techniques for Real Time Packet Video Transport, IEEE HPCS, Aug 1995, Mystic, Connecticut

[HeSa89]	D. Hehmann, M. Salmony, H. Stüttgen: Transport Services for Multimedia Applications, IFIP WG6.1/WG6.4, Workshop on Protocols for High Speed Networks North Holland, 1989
[Kelly91]	F. Kelly: Effective Bandwidth at Multiclass Queues, Queuing Systems, No9, 1991
[KnWr95]	E. Knightly, D. Wrege, J. Liebeherr, H. Zhang: Fundamental Limits and Trade-offs of providing deterministic guarantees to VBR Video traffic, CM Sigmetrics '95, Ottawa, Canada, May 1995
[KniZh95]	E. Knightly, H. Zhang: A New Approach to Support Delay-Sensitive VBR Video in Packet-Switched Networks, NOSSDAV 1995, Durham, New Hampshire, 18-21 April 1995
[KnZh96a]	E. Knightly, H. Zhang: Connection Admission Control for RED-VBR, a Renegotiation-Based Service, IFIP IWQoS '96, Paris, France, 1996
[LakYa96]	K. Lakshman, R. Yavatkar: An Empirical Evaluation of Adaptive QoS Renegotiation in an ATM Network, NOSSDAV 1996, Zushi, Japan
[LiChi96]	S. Liew, C. Tse: Video Aggregation: Adapting Video Traffic for Transport Over Broadband Networks by Integrating Data Compression and Statistical Multiplexing, IEEE JSAC, Vol.14, No.6, August 1996
[McRo96]	J. McManus, K. Ross: Video on Demand over ATM: Constant-Rate Transmission and Transport, IEEE JSAC, Vol.14, August 1996
[Onvural]	R. Onvural: ATM Networks: Performance Issues, Artech house, 1994
[RaCa95]	C. Raffaelli, F. Callegati, M. Casoni: Evaluation of Call Blocking in a Multiuser ATM Environment Based on Definitions of Equivalent Capacity, EFOC&N, Brighton, June 1995
[ReiRa95]	D. Reininger, G. Ramamurthy, D. Raychaudhuri: VBR MPEG Video Coding with Dynamic Bandwidth Renegotiation, IEEE International Conference on Communications 1995, NJ, USA, p1773-1777
[Sale96]	J. Salehi, Z. Zhang, J. Kurose, D. Towsley: Supporting Stored Video: Reducing Rate variability and End to End Resource Requirements through Optimal Smoothing, ACM Sigmetrics, May 1996
[Sig4.0]	ATM UNI Signalling Specification Version 4.0, ATM Forum/95-1434, April 1996
[TM4.0]	ATM Forum Traffic Management Specification Version 4.0, ATM Forum/95-0013, February 1996
[UNI3.1]	ATM User-Network Interface (UNI) Specification Version 3.1, TM Forum, 1995
[Vakil93]	F. Vakil, A Capacity Allocation Rule for ATM Networks, IEEE Globecom 1993
[WreLi96]	D. Wrege, J. Liebherr: Video Traffic Characterization for Multimedia Networks with Deterministic Service, in Proc. IEEE Infocom 1996
[ZhFe93]	H. Zhang, D. Ferrari: Rate Controlled Static Priority Queuing, IEEE INFOCOM 1993
[ZhFe94]	H. Zhang, D. Ferrari: Improving utilization for deterministic service in multimedia communication, International Conference on Multimedia Computing and Systems, Boston, May 1994
[ZKST97]	Z. Zhang, J. Kurose, J. Salehi, D. Towsley: Smoothing, Statistical Multiplexing and Call Admission Control for Stored Video, To appear in IEEE JSAC Special Issue on Real-Time Video Services in Multimedia Networks

Improving Clock Synchronization for MPEG-2 Services over ATM Networks

R. Noro and J.P. Hubaux

Telecommunications Services Group, TCOM Laboratory
Swiss Federal Institute of Technology, 1015 Lausanne, Switzerland
e-mail:{noro, hubaux}@tcom.epfl.ch
URL: http://tcomwww.epfl.ch/

Abstract. A stable clock indication is needed for the processing and playout of MPEG-2 streams. ATM networks can support the transfer of such streams but contribute significantly (as well as other system operations) to delay and delay variation. The MPEG-2 standard specifies strong temporal constraints for processing and presentation. MPEG-2 is also based on the assumption of a common time base shared among sender and receiver. Clock synchronization covers this requirement and is traditionally performed using PLLs. Unfortunately, their linear nature cannot simultaneously guarantee rapidity and good clock reconstruction. The non-linear technique we propose in this paper is able to offer such guarantees and also can handle higher amounts of delay variations. Such technique is not restricted to the scope of MPEG-2 clock synchronization and can also succeed in the reconstruction of time bases for generic distributed applications. It does not require any specific device, like very precise oscillators or filters.

1 Introduction

The MPEG-2 standard is becoming one of the most popular standards for the storage and transmission of high-quality audiovisual material. The standard specifies an audio and a video coding layer as well as a system layer, which defines the syntax and operations for transferring MPEG-2 streams over B-ISDN networks.

MPEG-2 allows several rate control methods for video: the more common are the *Constant Bit Rate* (CBR) control and the *Variable Bit Rate* (VBR) control. The latter is gaining more and more attention from the Multimedia Network community and has several advantages over CBR: reduced end-to-end transfer delay, better quality of video and simpler buffer management in the encoder/decoder. Nevertheless VBR traffic suffers a higher *Cell Loss Ratio* (CLR) and *Cell Delay Variation* (CDV).

MPEG-2 integrates a timing model that assures the synchronized presentation of audio/video streams and prevents buffer underflows/overflows, providing that sender and receiver share a common time base. This is obtained through a clock synchronization mechanism that locks the receiver's clock (slave) on the sender's one (master): the clock synchronization can fail if significant delay variations are introduced by the network and/or protocol's layers. In this paper we propose a non-linear clock synchronization method that is able to handle strong delay variations and does not require particular hardware devices at the receiver side.

The paper is organized as follows. In Section 2 we give a brief overview of MPEG-2 and its timing model, in Section 3 we describe how MPEG-2 streams map onto ATM networks and in Section 4 we explain the traditional clock synchronization approach. In Section 5 our non-linear technique for clock synchronization is discussed and a comparison with classical techniques is proposed. Finally, in Section 6 we derive some conclusions and further developments for the proposed method.

2 MPEG-2 timing model

The MPEG-2 standard is subdivided into three main parts: audio [1], video [2] and system [3]. Each audio/video stream is independently coded and an *Elementary Stream* (ES) is produced for each medium. Private data can also be streamed with independent syntax.

At the higher level, the *Logical Units* [4] of information are defined as *Presentation Units* (PUs) and correspond to *Access Units* (AUs) in the compressed domain. Due to the predictive nature of MPEG-2, equally sized PUs may generate varying size and differently ordered AUs. The ES are then organized in a sequence of *Packetized Elementary Stream* (PES) packets. Each of these packets is composed by an *header* and a *payload*, filled by contiguous bytes of an ES. A PES packet contains one or more complete AUs and its length is usually variable.

For the transmission over error-prone supports, such as ATM networks, the PES packets are further split into 188-byte, fixed length *Transport Stream* (TS) packets. These packets are suitable for the multiplexing of several media streams into a single multiplexed stream. The TS multiplexer labels the packets belonging to the same elementary stream with a unique *Program Identifier* (PID) field. The density of the packets with a given PID in the multiplexed stream is medium-dependent: for example, video packets can be six times more dense than corresponding audio packets. Fig. 1 illustrates the logical architecture of MPEG-2.

The timing in MPEG-2 standard is designed to support both *inter-* and *intra-stream synchronization* [5].

Intra-stream synchronization is based on two types of *Time Stamps* (TSs). Indeed, in the header of a PES packet we can include the *Decoding Time Stamp* and *Presentation Time Stamp* (DTS and PTS) of the first AU contained in that packet. They indicate the time at which the AU must enter the decoder and the time at which it must be delivered to the presentation device. DTS and PTS can be different because display order can differ from decoding order (e.g., P frames of video streams). The AUs not labeled with DTS and PTS are displayed according to the known frame/sample rate and the organization of the ES: anyway, DTS/PTS fields must be encoded in the single stream every 700 ms (at least). Intra-stream synchronization is then possible if a common clock is shared among the sender, the receiver and the multiplexed streams.

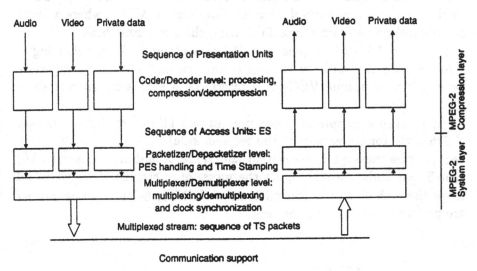

Fig. 1. MPEG-2 compression and system layers.

Inter-stream synchronization covers this task. The sender's clock is periodically indicated to the receiver by means of the *Program Clock Reference* (PCR) field (at least one every 100 ms, if the TS structure is used). The receiver must lock its local clock on the incoming PCR in order to supply a "clean" time base for decoding and presentation purposes. The MPEG-2 standard specifies the following features for the system clock:

- *System Clock Frequency*: 27 MHz.
- *Frequency Drift*: 540 Hz.
- *Rate of Change* in the *System Clock Frequency*: 75×10^{-3} Hz/s.

Locking of clock is usually achieved through *Phase Locked Loops* (PLLs) mechanisms. In this paper we address the problem of clock synchronization comparing traditional PLLs with a new, non-linear approach that seems more appropriate for use on top of jittered connections, such as ATM.

3 Mapping of MPEG-2 streams over ATM

We consider in this section the approach taken by the ATM Forum in mapping MPEG-2 streams onto ATM cells for audiovisual services. The ATM Forum specifies that a *Single Program Transport Stream* (SPTS) is carried by one single *Virtual Channel Connection* (VCC), where a SPTS is a multiplexed stream of media sharing the same time base.

Between the TS level (sequence of 188-bytes long packets, belonging to a SPTS) and the ATM cells level (sequence of 5+48-bytes long cells, belonging to a unique VCC) two interfaces are present, as plotted in Fig. 2.

First, a *Network Adaptation* layer merges two TS packets into a message of 376 bytes (or conversely at the receiver side).

Second, this message is passed to the *ATM Adaptation Layer* (AAL) type 5 that adds a trailer, splits the message in chunks of 48 bytes and maps these chunks onto the payload of ATM cells. Reverse operations are performed at the receiver side.

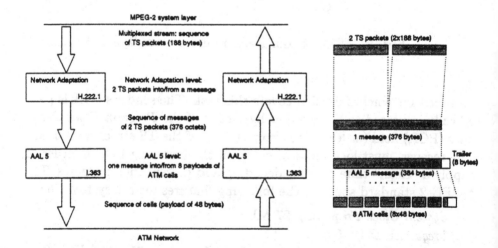

Fig. 2. ATM Forum mapping of audiovisual services onto ATM.

The specification of ATM Forum is designed for CBR MPEG-2 [6]: nevertheless, it is expected that a close approach will be chosen for the next VBR MPEG-2 specification. Moreover, it is noticeable that the presented scheme delegates dejittering and synchronization tasks to the application and error correction issues to the physical layer.

4 Clock synchronization by means of PLLs

Mapping of MPEG-2 Transport Streams onto ATM networks introduces delay: the processing and transfer operations contribute to this factors of de-synchronization.

The introduced delay has two components. First, a *Fixed Delay* that can be predicted, evaluated and/or bounded at run-time: this delay affects in the same manner all the elements of the stream (i.e., video, audio, DTS, PTS and PCR indications). It turns out that it can be easily compensated by delaying all the DTS/PTS indications of the same, suitable value. The system presents a latency that usually is less than a second. Second, a *Statistical Delay* or *Jitter* with an unpredictable distribution around his long-term average value of zero.

The jitter influences the clock reconstruction scheme at the receiver: indeed, if not equalized, early arrivals of packets with PCR indications cause buffers to be wasted at higher rate than required, PUs to be discarded due to the incoherent PTS indication and so on. Late arrivals, conversely, cause overflows of buffers and reduction of presentation rate. The PLL mechanisms allow clock synchronization despite the presence of noisy PCR.

PLLs was originally designed for phase-locking purposes of modulated signal carriers in the early thirties [7]. In the classical implementation (Fig. 3) the receiver employs a local oscillator to reproduce the incoming signal.

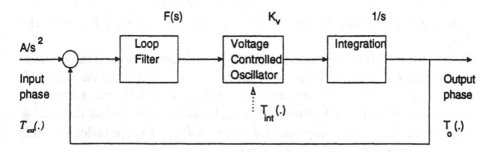

Fig. 3. Classical PLL.

For coherent phase-locking, the local frequency should be "close enough" to that of the incoming signal. The frequency shift is driven by the loop filter via a voltage applied to the *Voltage Controlled Oscillator* (VCO). The phase error fed a low-pass filter that modulates the frequency to minimize the phase error. The phase error is averaged over a given period of time to minimize the effects of noise in the incoming signal. Only a little information is needed by the local oscillator to lock on a stable input signal.

4.1 Mathematical model of PLLs

Mathematically, the analog PLL is modeled by the following *closed-loop transfer function* in the Laplace domain (refer to Fig. 3):

$$H(s) = \frac{T_o(s)}{T_{ext}(s)} = \frac{K_v F(s)}{s + K_v F(s)} . \tag{1}$$

By means of this transfer function, it can be demonstrated that an input ramp $T_{ext}(s) = \frac{A}{s^2}$ is perfectly tracked if the loop filter $F(s)$ is characterized by (at least) one pole located in the origin. Therefore, the simplest way to do this is to set a loop filter with the following transfer function:

$$F(s) = K_p + \frac{K_i}{s} , \tag{2}$$

with a *proportional* (K_p) and an *integrative* (K_i) action. The PLL is then described by a *second order loop*:

$$H(s) = \frac{2s\xi\omega_n + \omega_n^2}{s^2 + 2s\xi\omega_n + \omega_n^2} , \tag{3}$$

where:

$$\omega_n = \sqrt[2]{K_i K_v} , \ \xi = \frac{1}{2}\sqrt[2]{\frac{K_v}{K_i}} K_p , \tag{4}$$

are, respectively, the *natural frequency* and the *damping factor* of the loop.

When using PLL in the case of MPEG-2 clock synchronization, we move in the *discrete domain*. Samples of the input ramp enter the discrete PLL at given arrival times and are affected by delay and delay variations (i.e., the noise factors). With some algebra, it can be shown that the analog PLL described above maps onto a discrete PLL. Computationally, this can implemented in the following way:

$$T_o(k) = T_o(k-1) + V(k-1)(T_{int}(k) - T_{int}(k-1)) \, ,$$
$$V(k) = K_v[prop(e(k)) + int(e(k))] \, ,$$
$$prop(e(k)) = K_p e(k) \, ,$$
$$int(e(k)) = int(e(k-1)) + K_i e(k) \, ,$$

where $T_o(\cdot)$ is the locked clock, $T_{int}(\cdot)$ is the free internal clock value and $e(\cdot) = T_{ext}(\cdot) - T_o(\cdot)$ is the error, assuming that PCRs are carried as $T_{ext}(\cdot)$. The index k indicates the arrival of the $k-th$ PCR.

4.2 Simulation results of PLLs

We notice that the dynamical behavior of this PLL is driven by the pair K_p, K_i (alternatively ω_n, ξ): they determine the amount of residual jitter after locking and the locking speed of the PLL. For different cases of K_p, K_i, we simulate the transmission of PCR indications (at 50 ms intervals) affected by a fixed delay of 4 ms and a normally distributed jitter of average zero and 1 ms variance. The initial difference between frequencies is 2000 *parts per million* (ppm). In Fig. 4 we plot the residual frequency error (in ppm) for three cases of K_p, K_i (corresponding values of ω_n and ξ are derived).

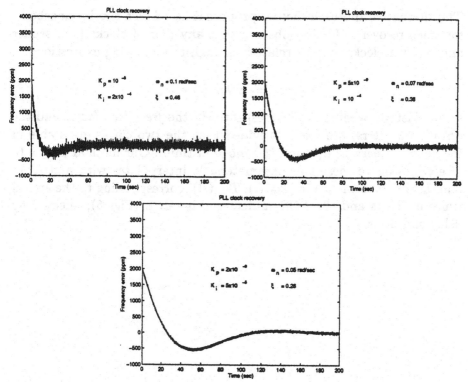

Fig. 4. Dynamic behavior of classical PLLs.

These PLLs reveal a symmetrical behavior: either we have a short transient phase and a poor clock reconstruction or transient is longer and clock is better locked. This is due to the linearity of PLLs: short transient is obtained through a large filter band, hence noise enters the system, while good phase locking implies a narrow band, hence long transient.

5   The non-linear technique for clock synchronization

The linear characteristics of classical PLLs do not fullfill both the requirements of fast and robust locking. Moreover, PLLs are not optimized for clock locking and/or recovery (i.e., when an input ramp enters the system), because they were originally designed to perform phase locking of periodic signals. The method itself and the needed circuitry can be bypassed by the proposed technique.

5.1   Mathematical approach to the non-linear technique

The alternative approach that we propose is based on a non-linear scheme for clock recovery. We first observe that any pair of clocks (i.e., sender and receiver clocks) can be related through the following expression:

$$T_{ext}(\cdot) \;=\; a\, T_{int}(\cdot) + b\,, \tag{5}$$

at any instant, where a and b (respectively the *frequency factor* and the *offset* parameters) are fixed, provided that the two clocks are stable in frequency (otherwise $a(\cdot)$ and $b(\cdot)$ move in time). Our method is able to directly estimate these two parameters. It interpolates with a straight line the points in the plane $(T_{int}(\cdot), T_{ext}(\cdot))$, corresponding to the arrival times of PCRs and the PCR indications it-selves (Fig. 5), affected by delay and jitter.

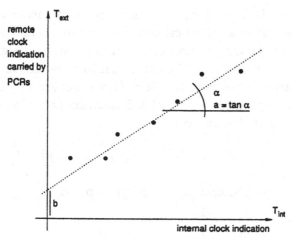

Fig. 5. Interpolation of PCR indications.

It is non-linear in the sense that we use a min-square interpolation. Slope and offset of the straight line are the tracked parameters a and b and are evaluated as follows:

$$(\bar{x}, \bar{y}) = \left(\frac{\sum_{i=0}^{n} x_i}{n}, \frac{\sum_{i=0}^{n} y_i}{n}\right) \quad \text{is the baricentric point,}$$

$$cotan(2\alpha) = \frac{\sum_{i=0}^{n}[(x_i - \bar{x})^2 - (y_i - \bar{y})^2]}{2\sum_{i=0}^{n}[(x_i - \bar{x})(y_i - \bar{y})]},$$

$$a = tan(\alpha), \quad b = \bar{y} - a\bar{x}.$$

Where, for the sake of simplicity, we have indicated with x and y the values along, respectively, the axes $T_{int}(\cdot)$ and $T_{ext}(\cdot)$.

5.2 Implementation of the algorithm

In order to be implemented, the algorithm requires to be recursive and computationally light. It can be tailored to follow the closed loop approach, then resulting in two interconnected blocks as in Fig. 6.

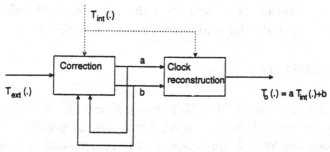

Fig. 6. Separated functionalities of the non-linear technique.

Reconstruction block recovers the remote clock by means of the internal clock and the pair a, b. Correction block updates a and b according to the incoming PCRs and the previous values: it can be designed to handle any length of the PCR trace (1 to ∞), thanks to recursion. If we assume that clocks have close frequencies ($a \simeq 1$), we can simplify the evaluation of α. On the receipt of the $n-th$ PCR indication ($x(n)$, $y(n)$), we update the system in the following way:

$$(\bar{x}^{(n)}, \bar{y}^{(n)}) = (\frac{(n-1)\bar{x}^{(n-1)} + x(n)}{n}, \frac{(n-1)\bar{y}^{(n-1)} + y(n)}{n}),$$

the updated baricentric point,

$$\alpha^{(n)} = \frac{\pi}{4} + \beta^{(n)},$$

$$\beta^{(n)} \simeq -\frac{\sum_{i=0}^{n} x(i)^2 - \sum_{i=0}^{n} y(i)^2 - n(\bar{x}^{(n)^2} - \bar{y}^{(n)^2})}{4\sum_{i=0}^{n} x(i)y(i) - 4n\bar{x}^{(n)}\bar{y}^{(n)}},$$

referring to n as the current PCR, to $n-1$, $n-2$,..., 0 as those previously received.

The performance of the present technique is compared to the classical PLL in the same test conditions: PCR are sent every 50 ms, affected by 4 ms of fixed delay and a normally distributed jitter with zero average an 1 ms of variance. In Fig. 7 are plotted the residuals frequency errors for three different cases of window length (i.e., the length of the PCR trace used to estimate a and b). Here we guess an initial frequency equivalent to the local one and we adapt it by filling the PCR trace with the incoming indications.

The non-linear clock recovery scheme allows a fast-locking of the receiver clock. Moreover, the locked clock is really more jitter-free and tracks very close the sender's clock.

A comparative analysis of PLL and non-linear technique is plotted in Fig. 8. We consider the transient duration and the residual frequency error as the features of the two clock recovery schemes. It is worth to notice that, with an equal residual frequency error, we can easily reduce the transient duration by a factor ten: this represents, therefore, a more efficient way to satisfy the timing requirements of MPEG-2.

6 Conclusion

Clock synchronization of MPEG-2 streams with jitter (where jitter is introduced by network and network protocols) is possible with PLLs: they can lock the local frequency onto the remote one or speed up the transient period associated with locking, but not both simultaneously.

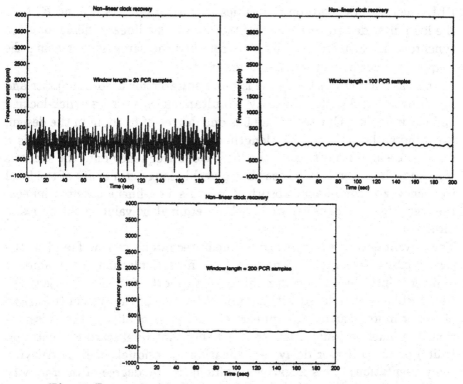

Fig. 7. Dynamical behavior of non-linear clock recovery.

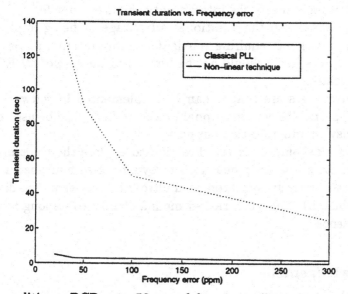

Test conditions: PCR rate 50 ms, delay 4 ms, jitter variance 1 ms

Fig. 8. Compared performances of PLLs and non-linear technique.

PLL's behavior depends on the values of the parameters K_p and K_i (i.e., the loop filter designed to absorb jitter). PLL is a linear scheme for clock synchronization: increasing jitter is roughly the same as increasing the frequency error in a proportional manner.

In this case, a non-linear scheme is more suitable for absorbing jitter and synchronize clocks: non-linearity simultaneously satisfies quick-locking and clock-fidelity. Our technique has one degree of freedom in the design, the window length (i.e., the length of the PCR trace), and embeds a min-square algorithm to map the local clock frequency onto the sender's frequency. It is worth to notice that possible long-term drifts of the local frequency are completely absorbed by this non-linear scheme: indeed, the algorithm compensates $a(\cdot)$ with an equivalent value in the opposite direction.

The advantages of using this non-linear technique can be found in the case of interactive applications (i.e., [8]), in particular if a strong amount of jitter is introduced by communication support. If we plan to use traditional PLLs, we are faced with an initial latency of the system (transient of one or more seconds can appear) or to a poor quality of the presentation (too much residual jitter, so it is likely that we discard Presentation Units) or to an higher delay (equalization of residual jitter by delaying the presentation). This negatively influences the degree of interactivity of the application itself. Instead, using the proposed non-linear technique, latency is reduced by a factor ten, the clock has less jitter and the application more meets the interactivity requirements [9].

Moreover, thanks to the additional advantage of being implementable in software without requiring particular computational power, the efficiency of such scheme should not be limited to the case of MPEG-2 clock synchronization.

Other advantages are that it can be implemented in software without requiring particular computational power and can also be used on top of very "noisy" communication supports.

Further work should go in the direction to explore the benefits in terms of presentation quality, when used to synchronize multimedia streams, but also to precisely synchronize multiple end systems for distributed application with time constrained media, thanks to its long term jitter-independence.

Acknowledgement

The authors would like to thank S. Koppenhœfer for his technical comments and for the improvement of the editorial quality of this manuscript.

References

1. ISO/IEC, *Information technology - Generic coding of moving pictures and associated audio information - Part 2: Audio*, Oct. 1994. DIS 13818-2.
2. ISO/IEC, *Information technology - Generic coding of moving pictures and associated audio information - Part 3: Video*, Oct. 1994. DIS 13818-3.
3. ISO/IEC, *Information technology - Generic coding of moving pictures and associated audio information - Part 1: Systems*, Oct. 1994. DIS 13818-1.
4. G. Blakowski and R. Steinmetz, "A Media Synchronization Survey: Reference Model, Specification, and Case Studies," *IEEE JSAC*, vol. 14, pp. 5–35, January 1996.
5. R. Steinmetz and K. Nahrstedt, *Fundamentals in Multimedia Computing and Communications*. Englewood Cliffs, NJ, USA: Prentice - Hall, 1995.
6. ATM Forum, *Video on Demand Specification*, 1.1 ed., Oct. 1996.
7. F. M. Gardner, *Phaselock Techniques*. New York, USA: John Wiley & Sons, Inc., 1979.
8. A. Basso, O. Verscheure, and R. Noro et al., "A Multimedia Architecture for Medical Tele-Imaging over ATM," in *CAR '96*, Paris, France, June 1996.
9. R. Noro, "Un mécanisme distribué de synchronisation appliqué à un cas de téléconsultation," in *JDIR '96*, Paris, France, September 1996.

VoR: A Network System Framework for VBR over Reserved Bandwidth

Yasunori Matsui[1]* and Hideyuki Tokuda[2]

[1] NTT corporation
[2] Keio University, Japan***

Abstract. Distributed multimedia systems must handle multiple continuous media streams at once. The network subsystem must thus be able to transmit aggregation of VBR streams effectively. Data smoothing and shaping is necessary for high quality transmission. Flexibility is also required to handle different types of data formats and contents. In this paper, we propose the VoR system, a network subsystem framework for VBR continuous media transmission. It controls stream multiplexing by the scheduler in the network driver along with the QoS coordinator middleware. The system provides a way for application programs to interact with the underlying network. The system is designed on the principle of policy and mechanism separation, which enables to flexibly adapt to application dependent inter- and intra-stream QoS tradeoffs. Our preliminary result using MPEG2 streams indicates that while simple multiplexing gains some bandwidth utilization, better multiplexing can be achieved by providing more elaborate data dependent policies.

1 Introduction

Computer network of the next generation must be able to transmit high-quality continuous media with low cost. ATM and other high bandwidth, reservation capable network technologies are expected to provide us a way of handling such data.

The tradeoff between cost and service quality nevertheless exists. Not every application requires the highest quality all the time, many can be compromised if available resources become scarce. This tradeoff is heavily dependent on applications and its use. For example, a few bit errors in a X-ray picture may cause erroneous diagnosis, while pictures of teleconference can be continued if the voice session is not interfered. The underlying network subsystem must have flexibilities to support these situations.

In this paper we propose VoR, a sender-side network architecture aimed to handle multiple VBR streams over a reserved bandwidth. Efficiency is achieved by gaining bandwidth utilization through priority-based stream multiplexing.

* Work done while the author was in Keio University.
*** This work has been supported in part by Keio-MKng project sponsored by Information-technology Promotion Agency (IPA), Japan.

QoS request and translation mechanisms are developed for both user to driver downcall and driver to user upcall. These calls also initiate media scaling. The system is designed to be flexible in that different stream coordination and media scaling policies can be implemented as modules, independently of common mechanisms supporting it.

The rest of this paper is organized as follows. In Section 2, related works are discussed. Background of our work as well as conceptual framework is described in Section 3. Section 4 presents the design of VoR system in detail. In Section 5 the result of preliminary experiment using several MPEG2 streams is shown. Section 6 concludes the paper.

2 Related Works

Statistical analysis and modeling of VBR streams and its theoretical application to network engineering has been a hot research topic in recent years. Heyman and Lakshman [6] discussed error rate versus queue length at the ATM switches when transmitting VBR data with LRD (Long Range Dependence) characteristics. This work assumes no traffic shaping at the sender

A comprehensive survey of QoS support for distributed multimedia system is found in Aurrecoechea et al. [1]. Käppner and Wolf [7] presented the object model for media scaling of multiple streams. In HeiTP [4], media scaling based network system was introduced. It splits a data stream into multiple sub-streams and transmit them by both guaranteed and best- effort paths. Media scaling is performed based on data formats at the transport level. Our framework extends this notion further to coordinate multiple streams through reflecting their semantics.

A design of QoS controlled ATM system was described as a part of integrated multimedia system built on the Chorus operating system [3]. The network system has EDF-based cell level scheduler whose parameters are set at the connection establishment time. However, dynamic QoS control and support for media scaling is not included in the work. The design of our system is based on the policy and mechanism separation and open implementation [8].

3 Network Control for VBR Streams

The forthcoming generation of multimedia applications will handle multiple continuous media streams at once. As an example, consider a telemedicine application. Doctors sitting at distant places are discussing over video streams from microscopes during an operation. There are audio and visual streams for each doctor, streams from microscopes, and perhaps real-time data of patient's heart beat and blood pressure. Our work is aimed to provide an infrastructure for such application.

In this section, the characteristics of VBR continuous media stream (hereafter referred to as VBR-CM) and resource reservation capable network is first presented. How multiple VBR-CM can be transmitted over the network is then discussed.

3.1 Transmitting VBR Continuous Media over Network

High quality continuous media data consumes large bandwidth. Raw pictures used for pathological diagnosis requires 6 Gbps[4][5]. Data compression reduces the data size substantially, the result of which forms VBR streams.

Delay sensitiveness of continuous media stream demands elaborate network control. It is widely conceived that resource reservation or priority based packet processing at the network is necessary. A connection-oriented network model fits well to the concept of reservation, by allocating necessary resource over established virtual circuit. Virtual circuit also acts as a firewall against ill-behaved streams which tries to monopolize all the available resource in the network. Policing in the network nodes also enforces established bandwidth. If the data going through a circuit violates network parameters, switches react by dropping or delaying data.

When transmitting VBR-CM streams over network, reserving a single virtual circuit for each stream is not efficient. The discrepancy between the peak and average rate of VBR-CM leads to the resource underutilization. There are two ways of improving bandwidth utilization. One is to use intelligent encoder to produce CBR streams. This is achieved by changing the compression parameters, such as changing the quantization step size of DCT coefficients. The other scheme we propose is to multiplex VBR streams onto a single virtual circuit. Here, peaks and valleys of the streams are flattened by counterbalancing.

We consider VBR multiplexing has advantages over CBR compression. CBR compression sacrifices data quality when the scenes which are difficult to compress are encountered. With VBR compression, the quality of streams are kept constant when multiplexing is going well. However multiplexing VBR streams may result in overflow or burst. There are long-term overflow (order of minutes) by the scenes with larger entropy, or short-term overflow attributed to the abrupt scene changes. Unless the network system responds to overflow by reducing the data rate, its presentation quality degrades.

In fact, some works on VBR over high-speed networks implicitly or explicitly assume that all the data is pushed into the network regardless of the network parameters. From a practical point of view, they seem to have the following deficiencies. First, because drops are uncontrollable from users, it is not always clear how drops affect the presentation quality. Compressed data is more vulnerable to data drops than original data. Different frame types are affected differently by the data drops. In the case of MPEG, because I frames are much larger than B frames, I frame data is more likely to be dropped than B frames. However, I frames form a basis of subsequent P and B frames, its data loss will affect the frames in the same GOP (Group of Pictures).

---

[4] This bandwidth was calculated on a picture for SHD (Super High Definition TV, 2048 x 2048 pixels) display.

Secondly, queues at the switches to absorb and shape overflow sometimes cause too long delay for live real-time streams. There are many factors of end-to-end delay other than queuing. For instance, the speed of light is dominant in WAN environment. The measured RTT from Tokyo to the east coast of the U.S was 170msec through trans-pacific ATM line[5]. Delay in codec is an another factor. Because MPEG uses backward prediction algorithm, buffering delay for several frames is inherent in encoding. Because people will perceive delays of over 200msecs, there are often few margins left for queuing for smoothing.

Controlling the transmitting node to obey the established network parameters is therefore necessary. This will reduce unnecessary data loss by the policing in the network. Queuing at the sender is different from queuing at the network nodes in that sender host can perform discard operation by considering data format. When data is packed from the most important to the least important, the effect of discarding latter end is less affective. This is achieved by hierarchical coding such as defined in the MPEG standard [9].

The transmitting node can determine whether or not to transmit overflow data, depending on the delay to be absorbed by other elements of the transmission path.

3.2 Media Scaling Request from the Network

Overflow of one to several frames is handled by the method described in the previous section. In addition, there are longer resource shortage. The simplest example is adding new streams, which is common in teleconference. Moreover, VBR streams have a burst period caused by scene changes. The quality of some streams must be then compromised for the sake of others. This is highly dependent on the contents and the use of streams. Discarding voice data disturbs communication more than dropping some data in video. When doctors are discussing the quality of the speaker's voice should be higher than others, or on the window system the picture in the foremost window needs to have a higher priority. Different frame types also require different handling.

As can be seen from the above examples, intra- and inter-stream tradeoffs exist in the QoS control. These data types and contents dependent control are unable to be handled by a generic, uniform algorithm. It is therefore a flexible framework for conducting media scaling is necessary. It must be able to plug-in versatile policies and modules for determining and performing media scaling.

4 VoR System

This section describes the VoR system, a sender-side network system framework for VBR continuous media. The system is composed of three elements, namely Network Driver, Stream Coordinator and Coordinator Interface Objects.

---

[5] This RTT was measured by ping program from NTT lab in Tokyo to AT&T lab in New Jersey.

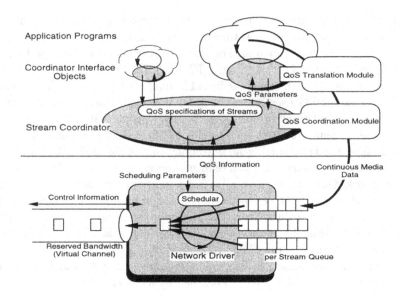

Fig. 1. Overview of VoR System

These components translate QoS, multiplex streams into a virtual circuit, performs traffic shaping and controlled data discard, and provide a flexible framework for media scaling. Mechanisms for determining and establishing bandwidth reservation is not the focus of this paper. We rather focus on making the best use of given bandwidth through network control. It is therefore "admission control" in general sense is not discussed here, only admission control within the virtual circuit is considered.

Fig. 1 shows the overview of VoR system. There are one Network Driver, one Stream Coordinator and Coordinator Interface Objects for each streams. Application programs generate VBR-CM streams and pass it to the driver. The Network Driver's scheduler multiplex and pushes data into the virtual circuit. Scheduler in the Network Driver and Stream Coordinator is provided only for better transmission of VBR streams. It does not guarantee the performance.

For each stream, there is a Coordinator Interface Object. It acts as an interface between applications and Stream Coordinator for QoS translation, media scaling and admission test. Stream Coordinator coordinates QoS requests for different streams, performs admission test, and at the time of congestion it chooses a victim stream and requests media scaling. Media scaling can be issued from both sender and receiver applications. Coordination calculation is performed by Stream Coordinator. When the media scaling request is issued, Stream Coordinator does inter-stream QoS mediation. Each applications are responsible for choosing QoS parameters for intra-stream media scaling and actually perform media scaling.

The following subsections describe the roles of each component in detail.

4.1 Network Driver

The Network Driver sends and receives VBR streams. It has additional features beside the basic functionalities of the traditional network device driver. Here we focus our discussion on new features of the sender side. Receiver side is designed the same as the traditional drivers.

There are two major added functionalities: driver's scheduler and per stream queues. When an application issues write(), the data written is first enqueued into the stream queue (Fig. 2). Each write() calls passes a whole unit of data of each timing constraints (eg. picture frame) to the Network Driver. Network Driver then determines when to send data to the network board. For each transfer of data from the host to the network board, the scheduler calculates total bytes to send for each stream. Transmission cycle interval is determined by the DMA buffer size and the bandwidth. When one cycle of transfer is finished, the scheduler is waken up by the network board to prepare for the next transmission. If the discard flag for the per-stream queue is true, deadline missed data is discarded and its counter is incremented. If the flag is false, all the remaining data in the per-stream queue is pushed out in FIFO. Collected statistics are used for determining long-term network congestion. The bandwidth allocation algorithm in the scheduler is a priority based, modified weighted fair scheduling. This algorithm guarantees minimum allocation of bandwidth for each stream, at the same time filling up all the available bandwidth.

Along with the data interface such as open(), read(), write(), it has a control interface to the Stream Coordinator. Stream Coordinator interacts with the Network Driver through this interface, such as obtaining network information and setting scheduling parameters. This is described in the next subsection.

4.2 Stream Coordinator

Stream Coordinator is implemented as a user level middleware. It is a component for dynamic QoS control which coordinates requests from application programs and sets scheduling parameters in the Network Driver. For flexible QoS admission test and QoS coordination calculation, Stream Coordinator provides APIs for plug-in modules (Fig. 3). Programmers using the VoR system must provide coordinating modules through these interfaces. Coordinating modules can be a calculation algorithm or a GUI based parameter adapter.

Stream coordinator accepts QoS specification requests of streams through Coordinator Interface Objects. QoS specification from the Coordinator Interface Objects are given by the stream independent QoS parameters, which is shown in scqos_t (Fig. 3). Here, delay_jitter and lossrate are reflected to the discard flags at the per-stream queue (Fig. 2). Admission test and dynamic QoS adaptation is done by calc_netqos() function. If the specified QoS cannot be satisfied, result is set to the affordable QoS parameters. Application programs can then consider sending the stream after performing media scaling. When higher priority stream upgrades its QoS, calling calc_netqos() may cause lower streams to degrade. This notification is described later.

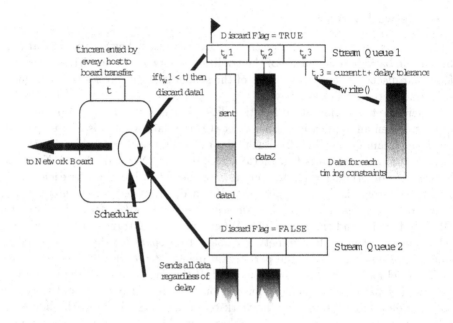

Fig. 2. Data Scheduling in the Network Driver

```
/*    Stream Independent QoS specification  */
struct {
        int priority;      /* Stream's priority */
        int min_bw;        /* bandwidth, bytes/sec */
        int mean_bw;
        int max_bw;
        int period;        /* Timing constraints of data */
        int delay;         /* maximum allowable queue delay */
        int delay_jitter;  /* maximum allowable queue jitter */
        int lossrate;      /* Data loss by network policing */
} scqos_t;
/*
 * APIs for Stream Coordinator
 */
/* Called from Coordinator Interface Objects */
int calc_netqos(scqos_t* spec, scqos_t* result);
/* APIs for Plug-in Modules */
void init_calc();
void calc_schtbl(scqostbl_t* sq, schtbl_t* stp);
```

Fig. 3. Stream Independent QoS Specification and APIs for Stream Coordinator

When a new set of QoS parameters is set for the streams, Stream Coordinator calculates scheduling table by `calc_schtbl()`. User specified QoS is passed by `scqostbl_t`.

4.3 Coordinator Interface Object

Streams are generated by application programs. In other words, it is the application programs that know the characteristics of its streams best. They are responsible for data modification and media scaling. Coordinator Interface Objects forwards QoS change notification from Stream Coordinator to the application programs. Applications then react by performing media scaling.

Different streams have different media formats with different QoS parameters. As for Stream Coordinator, a uniform set of QoS parameters is necessary for coordination calculation. Coordinator Interface Object provides interface for QoS translation module which acts as a bridge between application and the Stream Coordinator. The APIs for both application and QoS translation modules are given in Fig. 4. The usage of these interfaces is discussed in the next subsection.

```
/*
 * API functions
 */
boolean_t isNetworkQoSChanged;
void cio_init(str_t strID);
void cio_specqos(void* sqos);
void cio_getnetqos(void* sqos);
/*
 * Interface for QoS Translation Modules
 */
void SQoStoNQoS(qos_t* nq, void* sq);
void NQoStoSQoS(void* sq, qos_t* nq);
```

Fig. 4. API and Interface for QoS Translation Modules of Coordinator Interface Object

4.4 Application Programs

Application programs generate streams. Data API is a traditional `open()`, `write()`, `close()` paradigm, except for the system unique `strID` is provided for identifying different streams. QoS ignorant programs will not use QoS control APIs. They push data at the best-effort basis. QoS aware programs open control path by calling function `cio_initcio()` of the Coordinator Interface Object (Fig. 4). This binds Coordinator Interface Object to the stream with `strID`. Through the Coordinator Interface Object, application programs specify QoS in terms of stream dependent QoS parameters. It is then translated into stream independent network QoS by SQoStoNQoS(), which is a plugged-in module for Coordinator Interface Objects.

When an application program supports media scaling, it must poll coordinator interface object's variable isNetworkQoSChanged. Media scaling is requested to the application only synchronously to stream generation. This is because handling media scaling request is not urgent in VBR transmission, and the code of the application programs will be complexed if interrupt handling must be considered. When the stream is chosen as a victim or more network resource becomes available, it is notified through this variable. cio_getnetqos() must then be called to get the new QoS parameters specified by the network. cio_getnetqos() calls NQoStoSQoS() to translate network QoS into stream dependent QoS.

NQoStoSQoS is also a plugged-in module. Application programs perform media scaling of data and sets new QoS parameters as described before.

5 Preliminary Evaluation

The implementation of the VoR system is underway on RT-Mach operating system [10] running on the off-the-shelf PC and ATM board. We are conducting preliminary experiments with policy modules for MPEG streams on a simulator. In the experiment, several MPEG2 streams are multiplexed on a virtual circuit and their deadline miss rate is evaluated. The stream we used have different frame patterns for each GOP, as IBBPBBPBBPBB, IBBPBBBPBBPBBB etc. In order to keep the bandwidth utilization rate constant, the net bandwidth is increased by the degree of multiplexing.

Table. 1 shows the change of deadline miss rate of a sample MPEG2 video stream along with the different degree of multiplexing. It shows that the deadline miss rate decreases by multiplexing, but when too many streams are present it increases again. When a few streams are multiplexed, bandwidth margins gained is dominant. When many streams are present, however, the probability of I frame collision increases. By providing format dependent MPEG2 policies such as delaying I frames of lower priority streams, better multiplexing is expected. Detailed experiment is in progress.

Degree of Multiplex	Deadline Miss Rate
1	3 %
2	0.1 %
4	0.04 %
8	0.2 %

Table 1. Deadline Miss Rate vs. Degree of Multiplex, Network Utilization 32 %

6 Conclusion

This paper proposed the VoR network system for transmitting multiple VBR streams over a reserved virtual circuit. The system aims to maximize the utilization of bandwidth, at the same time coordinate streams in order to improve presentation quality. To achieve this, streams are multiplexed, shaped and media scaled within a disciplined framework. Detailed evaluations of policies for specific types of data formats is in progress using VoR framework.

References

1. Aurrecoechea, C., Campbell, A., Hauw, L., "A Survey of QoS Architectures", *Multimedia Systems Journal, Special Issue on QoS Architecture (to appear)*, 1997
2. Clark, D.D., Tennenhouse, D.L., "Architectural Considerations for a New Generation of Protocols", In *Proceedings of ACM SIGCOMM 1990* .
3. Coulson, G., Campbell, A., Robin, P., Blair, G., Papathomas, M., Hutchison, D., "The Design of a QoS Controlled ATM Based Communications System in Chorus", *IEEE Journal on Selected Areas in Communications, Special Issue on ATM LANs, Vol13, No.4, May 1995.*
4. Delgrossi, L., Halstrick, C., Hehmann, D., Herrtwich, R., Krone, O., Sandvoss, J., Vogt, C., "Media Scaling for Audiovisual Communication with the Heidelberg Transport System", In *Proceedings of ACM Multimedia 1993.*
5. Fujii, T., Ishimaru, K., Sawabe, T., Suzuki, J., Ono, S., "Transmission Characteristics of Super High Definition Images with MPEG2", *HTDV '95*, 1995.
6. Heyman, D.P., Lakshman, T.V., "What Are the Implications of Long-Range Dependence for VBR-Video Traffic Engineering?" *IEEE/ACM Transactions on Networking, Vol.4, No.3, June 1996.*
7. Käppner, T., Wolf, L.C., "Media Scaling in Distributed Multimedia Object Services ", In *Proceedings of 2nd IWACA*, 1994.
8. Kiczales, G., "Towards a New Model of Abstraction in Software Engineering", In *Proceedings of the IMSA '92 Workshop on Reflection and Meta-level Architecture*, 1992.
9. "Information Technology-Coding of Moving Pictures and Associated Audio", ISO/IEC 13818-1/2/3
10. Tokuda, H., Nakajima, T., Rao, P., "Real-Time Mach: Towards a Predictable Real-Time System", In *Proceedings of USENIX First Mach Symposium*, 1990.

EPK-fix: Methods and Tools for Engineering Electronic Product Catalogues*

A. Knapp, N. Koch, M. Wirsing (LMU München)
J. Duckeck, R. Lutze (mediatec GmbH, Nürnberg)
H. Fritzsche, D. Timm (TU Dresden)
P. Closhen, M. Frisch, H.-J. Hoffmann (TH Darmstadt)
B. Gaede, J. Schneeberger, H. Stoyan, A. Turk (FORWISS Erlangen)

Abstract An electronic product catalogue (EPC) is a computer controlled information system with multimedia product presentations and navigation facilities. The paper presents the results of the EPK-fix project for the systematic construction of EPCs. These include a software engineering process model, a high level specification language for EPCs, and an integrated set of tools supporting the entire EPC development process. The EPK-fix process model supports the classical development phases: requirements analysis, specification and design, implementation, and test. In each of the phases emphasis is put on the particular multimedia aspects, human machine interaction, production of prototypes, and the quality of the produced documents.

Introduction

Electronic Product Catalogues (EPCs)[1] are computer controlled information systems with important multimedia (especially visual) product presentations and navigation facilities. They are almost always equipped with a shopping bag administration feature.

EPCs are an inexpensive alternative to paper catalogues, but a high quality design is still related to elevated costs, because there are no appropriate production tools available. There are *catalogue providers* introducing a multimedia presentation to the market, *catalogue developers* designing and producing EPCs with the assistance of software and multimedia experts, *testers*, and *users* or *end-users*.

State of the art technologies to produce EPCs are still far from being easy to use and efficient and they show serious weaknesses. New methodologies and specific tools for EPCs are needed. They must support the complete *life cycle* of EPCs starting with the analysis of the catalogue providers requirements, continuing with the catalogue design up to the functional tests. These tools have

---

* This work was supported by the BMBF project EPK-fix (Förderkennzeichen 01 IS 250). Corresponding author: N. Koch, Institut für Informatik, Ludwig–Maximilians–Universität München, Oettingenstraße 67, D–80538 München, kochn@informatik. uni-muenchen.de

[1] German: EPKe (Elektronische Produktkataloge)

to be easy to use, reduce the amount of EPC development time, and permit a low-cost production of catalogues. These are neccessary prerequisites for the acceptance of a new system, especially in small and medium size organizations.

We present the results of the EPK-fix project for the systematic construction of EPCs. Methods have been developed and a collection of integrated tools (RASSI, SASSI, GASSI, TASSI) has been designed for efficient specification, production, and validation of EPCs. The requirement analysis is carried out with the help of structured interviews which are guided and recorded using the first tool. The result of this analysis is the base for an EPC design written in the specification language EPKML. EPKML is a high level HTML-like language that is particularly designed to support the description of EPCs. The second tool supports the specification construction with a set of specialized editors. The third tool automatically generates different catalogue versions in Java from the EPKML specification. Finally, the test assistant helps to ensure the quality of the produced catalogues. It provides static and dynamic test strategies as well as strong support for manual testing of media objects.

The first section of this paper outlines the state of the art of EPCs, the development process, and the resulting architecture of the EPK-fix system. Sections two to six describe the specification language and the tools respectively. In the last section some conclusions are delineated.

1 Developing EPCs with EPK-fix

The analysis of about 40 EPCs [6] has demonstrated that they go far beyond paper catalogues with cross references. They offer services like search features, demos to show how to use the catalogue, games, query language, enquiries through telephone communication and fax, or on-line ordering. The constituents we identified in each EPC are: *structure, layout, direction, database,* and *services.*

- The *structure* is the skeleton of the catalogue; it comprises a graph or hierarchy of themes and pages.
- The *layout* is the static description of pages and their contents.
- The *direction* describes the dynamic facilities for pages and themes that allow for user interaction and the navigation through the catalogue.
- The *database* component provides all the information about offers, in such a way that it can easily be searched, exchanged, and maintained.
- The *services* add comfort to the EPC allowing i.e. administration of orders, user registration, access to help functions, online communications.

We observed that working with these EPCs can mainly be divided into installation, presentation, search, selection, and order steps. Depending on their relative importance , we distinguish between the following catalogue types: *Presentation, Search,* and *Order* catalogues; see [6].

The Development Model. The development process for EPCs requires an informal preanalysis followed by the phases: *requirements analysis, design, implementation, and test.* For each theme a tool has been developed considering the above mentioned aspects *structure, layout, direction, database,* and *services.* Using a *formal specification* based on a specification language it is possible to take into account the application requirements and to avoid inconsistencies among the different tools. Characteristics of the EPK-fix development process are the emphasis put on the multimedia aspects, the human-machine interaction, the production of prototypes, and the quality of the produced documents. The resulting process is a kind of spiral model [2] that leads to the final EPC through successive revisions and refinements.

Requirements Analysis. Human dialogue is central to the analysis process in engineering. The fundamental analysis activity is to carry out *structured interviews.* This interviewing process is divided into three major steps [8]: preparation, interview, and edition. During *preparation* a questionnaire is assembled considering all aspects (see Section 1) of an EPC. The selection of the questions is guided by predefined generic checklists. The *interview* is a complete (audio-)recording of the conversation between the EPC developer and the catalogue provider. Arbitrary multimedia information are attached to the corresponding questions and answers of the interview. Finally, the *edition* constructs the resulting analysis document from written (transcribed) notes and acoustic data.

Design. The informal catalogue description recorded in the analysis document provides the basis for the design of an EPC. The *catalogue developer* carries out his work with the help of appropiate *editors* which generate automatically a formal specification of the catalogue. Work starts with the analysis document specifying all the aspects and details of the intended catalogue. The developer supplies the structure and the layout of example EPC pages (or templates).

Integrated Software Generation. A general advantage of our approach is the speedup given to the otherwise time consuming and error-prone implementation phase. A generation assistant produces a prototypical EPC version with additional interfaces to communicate with the other development tools. These interfaces provide *capture* and *replay* facilities that can, for instance, be used to present a catalogue prototype to the developer in exactly that state which is refered to by a given analysis document. For testing purpose, catalogue elements can be addressed by their unique identifier (provided by the design tool). Event handling of the user interactions is done by a separate module allowing to treat simulated user input generated by the test assistant in exactly the same way as real end-user input.

Testing. Quality assurance is a central point of our development process. We aim at complete not only sample testing. In the generated EPCs the large number of media-objects are tested automatically via rules. Dynamic tests are used to ensure correctness of time-dependent multimedia processes. Manual validation of the layout is supported by test agents. The test reports are fed back to the other development steps.

The Architecture of the EPK-fix System. The EPK-fix system components rely on the formal description language EPKML. It integrates the following four tools: RASSI, SASSI, GASSI, and TASSI.

- EPKML is a specification language that makes the description of the static and dynamic aspects of EPCs possible.
- The Requirements analysis ASSIstant (RASSI) supports the informal recording of information (text, sound, images, video) that results at the requirements analysis stage based on structured interviews.
- The Specification ASSIstant (SASSI) is responsible for EPC design based on the results of the RASSI tool and generates an EPKML specification. The catalogue developer is assisted by efficient and powerful editors.
- The Generation ASSIstant (GASSI) translates the EPKML description specifying the EPC into a general programming language (e.g. Java).
- The Testing ASSIstant (TASSI) performs static tests on the catalogue description in EPKML and a dynamic validation on the EPC generated by GASSI, using test data especially prepared for that purpose.

2   The Specification Language

The design of the EPKML language was guided by basic aims like easiness of learning and extensibility as well as more EPC-specific considerations like the integration of database features or navigational support. EPCs on CD-ROM and on the World Wide Web (WWW) create an alternative channel to traditional paper catalogues for product sale and service offer. Since these catalogues may coexist it was our goal to develop a declarative language that allows an easy common specification.

EPKML is HTML-like. It is defined as an instance of SGML (ISO-standard 8879, [4]). The Standard Query Language (SQL) is integrated into the language for easy access to relational databases. EPKML has primitives for navigation flow and allows connection to external languages via applets like in HTML. The language integrates a conceptual view of EPCs, this means a structured, annotated, multimedia, and highly automatic front-end to a database (see Section 1). We restrict ourselves in this paper to only a few aspects of the syntax and pragmatics of the language EPKML that distinguish it from other mark-up languages. A detailed description including a formal structural operational semantics is presented in [5]. Appendix A shows one page of the tourist catalogue example and its EPKML specification. Below we describe some characteristics of the language and reference the lines of the specification.

Structure. Products are organized in hierarchies, so-called themes (<theme>, 01). The developer defines the products belonging to these themes (<extension>, 02) and the presentation of these products (<template>, 07). These templates are to be filled with actual product data obtained from the database. Whenever a special layout is desired, it is possible to define <exceptions> for those products with their own <template>. A hierarchical structure is achieved via the definition of sub-themes for a <theme>. It is also used to generate automatic navigation facilities supported by the command tags <next>, <previous>, <up>, <down>, and <back>, which branch to the next or previous theme in a given hierarchy, to the first one below or above, or back in the history of visited themes, respectively.

`<question-form>`, `<shopping-bag>`, `<shopping-list>`, `<order>`. In the example a `<registration-form>` (44) is mentioned, that permits the personalization of the catalogue. A `<shopping-bag>` is a template for a list of products that have been selected as "products to buy". For more details see [5].

Audio-Visual Elements. The visual (layout) features of EPKML are a superset of those of HTML. Introducing e.g. `<window>` (08) we make EPKML window oriented instead of screen oriented. It is worthwhile to mention `<flowbox>` that distributes layout elements surrounding e.g. an image or text. Multimedia is integrated by adding the time-dependent elements `<video>` (15), `<slide-show>`, and `<audio>`. All these elements may be customized in advance using `<style>`s (08) by defining `<stylesheet>` for them.

Interaction. A large part of EPKML is concerned with interactive elements, like `<button>` (28, 41) or `<pulldown-menu>`. Interactive elements are provided with a specifiable method, e.g. `<on-click>` (30, 43) for `<button>` or `<on-option>` for `<browser>`. For the navigation through the theme structure, there are precustomized elements with standard behaviour like `<previous-button>` (38), and `<next-button>` (40). Of course, this behaviour can be changed or extended.

Database. Database access is achieved via the `<sql>` tag (03, 20). Text within the scope of this tag must be written in standard SQL.

3 The Requirements Analysis Assistant

The *Requirements Analysis Assistant* (RASSI) is a software tool that supports the task of interviewing. It contains five integrated submodules, that serve to manipulate and convert the basic object types *checklists*, *questionnaires*, *interviews*, *documents*, and *protocols*. Checklists enumerate all important aspects that have to be addressed during EPC design and development. Interviews with the catalogue provider are performed on the basis of quetionnaires resulting in protocols. The following are the RASSI submodules: The *checklist editor* manipulates new or predefined generic checklists. The *questionnaire editor* is used to assemble a questionnaire from a checklist by formulating questions and adding explanatory documents. The *interview assistant* performs a complete audio recording of the interview, allows for synchronously written notes, and links arbitrary multimedia material. The *protocol editor* supports revisiting and combining mutiple interviews. Finally, the *presentation assistant* generates an analysis document. RASSI is well-integrated into the EPK-fix toolbox: the chosen format of the analysis document (HTML) ensures immediate document exchange via the WWW enabling distributed workgroups to interact efficiently.

The *Tourist Catalogue Example* (Figure A) illustrates the development steps of EPCs throughout this paper. For information elicitation in the *system analysis* phase, a questionnaire has been created starting from a predefined checklist. As shown in Figure 1 the questionnaire contains the topic "Screen elements for city pages" which is part of the main topic "Layout and micro-direction". The initial checklist mentioned the topic "product pages" which is now replaced by "city pages". The topic "Grouping, placement, and attachment" was discussed

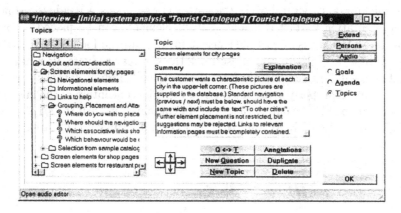

Figure 1. Main window of the interview assistant in RASSI

with the catalogue provider asking four specific questions. Explanations and multimedia annotations can be attached without any restriction. After the interview, the essential information from lower-level topics is summarized. The result of the final analysis document (HTML) can be viewed with any standard WWW browser.

4 The Specification Assistant

The Specification Assistant supports the transformation of an informal description of an EPC to a formal specification. A qualified developer uses SASSI's set of powerful editors to compose an EPC according to the information gained by the requirements analysis. SASSI is based on an intuitive graphical user interface, it is implemented in Smalltalk, and its architecture is based upon the design evolution presented in the DIADES user interface builder [3].

Main features of the SASSI component are to support a complete specification of the analysis information, to protocol which analysis objects led to which specification objects, to make sensible assumptions where analysis information is incomplete, to provide version information in order to make recovery possible, and to generate syntactically and static-semantically correct EPKML output. These features are supported by an object-oriented class library of specification object classes which reflects the EPKML elements. When specifying an EPC the user simply composes a hierarchy of specification objects similar to an abstract syntax tree which can then be used to generate EPKML output.

SASSI can handle all catalogue aspects (described in Section 1) with the following editors: *structure editor*, *layout editor*, also responsible for specifying direction and integrating services, and *database editor*, where database access can be formulated using SQL statements. The SASSI application starts with the structure editor window showing a view of the predefined initializing catalogue structure, which is subsequently refined according to the example. On double-clicking a structural node the layout editor opens a window displaying the representation part of the node. Layout elements can be placed according to the RASSI analysis of customer needs, making up the look of the catalogue pages. Entries, e.g., information describing the various products, images or even animations, will be taken mainly from the customer product database and have to be referenced properly using the SQL editor. The conception of the layout editor is similar to DTP layout applications. In combination with the structure and the SQL editor the EPK-fix approach goes beyond ordinary DTP applications.

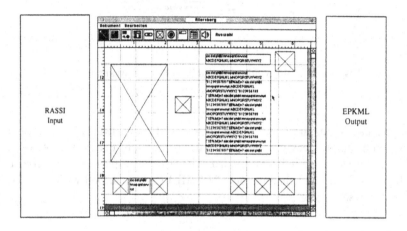

Figure 2. The SASSI user interface

To reference the requirements analysis input there exists an additional window in which analysis units coming from RASSI are displayed. Combined with a third window transcribing the generated EPKML specification, this results in a three-window approach (see Figure 2) similar to [9] and the split-screen approach used in Oberon [7]. Working in either of the two *editor* windows, the user simply has to highlight the paragraph in the RASSI window corresponding to the work he actually performs, and all objects created will reference the identificator of this unique analysis information unit.

5   The Generation Assistant

The purpose of GASSI is to generate catalogues based on their formal specification. Targetted output platforms include a paper version (i.e. LaTeX), an HTML version and an EPC implemented in Java. Within the scope of this paper, we will concentrate on the presentation of the latter's generation.

The generation of EPCs relies on a library of extensible generic classes (that are reusable, and reliable due to their automated testing by TASSI) providing EPC-specific components ranging from simple layout elements up to modules that perform *services* (e.g. online-help ordering-facilities). The generation of software implementing a formally specified multimedia system is done by analogy with compiler phases [1] as follows:

Parsing comprises lexical and syntactical analysis of a given specification. An EPKML conform document defines a tree structure of EPKML elements that can be parsed by a standard SGML parser yielding a more easily processable text output. GASSI calls the parser *nsgmls* and uses its output for further processing.

An *Intermediate Representation* of syntactically correct specifications allows for semantic analysis and optimization. The internal representation of a specification consists of *specification objects*, that are nodes of a tree reflecting the specifications structure. Each node contains an SGML element allowing to access the attributes of this element which had been set in the specification.

The objects are specific to the EPKML element they represent by providing specialized methods to generate code implementing the desired layout and behaviour for this element. These methods encapsulate implementation details like different classes of the library an EPKML element is mapped to depending on its content. Besides, methods for restructuring the tree are available.

Sourcecode Generation is done according to the syntax of the output format and the results of the specific optimization. The basis of the code generation is a class library realizing common features of EPCs which are used for a specific implementation either by direct instantiation or by subclassing and instantiation of the new subclass. *Specification objects* have access to a mapping table defining which library class has to be used for the elements implementation. For these classes miscellaneous Java-syntax-conform expressions can be retrieved or composed, e.g. unique Java identifiers; constructor calls; variable declarations or templates for subclasses.

The structure of the specification that implies a default structure for the EPC by the nesting of <themes> and therein contained instances for <templates> is used to build a module realizing the navigational primitives (e.g. <next> or <up>). For that purpose, a method returns the concatenation of Java instructions that sequentially build a corresponding tree. In the example specification (Appendix A), there is only one <theme> (01) visible. The <template> (07) defines a common layout for the instances to create from the database entries, which are returned by the SQL statement (04–05) within this theme's <extension> (02). The default navigation enables the user to switch from one instance to another using either the <next-button> (40) or the <previous-button> (38).

Compilation is finally accomplished by existing tools that are integrated in the GASSI system via facilities for their invocation on all generated files in the target directory and input/output redirection.

6   The Test Assistant

TASSI serves the analytical quality assurance within the EPK-fix development process. With TASSI all those quality features are examined, which can not constructively be ensured by the other EPK-fix tools. TASSI especially inspects the features that are important for an EPC user, i.e. functionality, robustness and usability of the catalogue.

TASSI should enable a tester to systematically and mostly automatically scrutinize *static* and *dynamic* aspects of an EPC without the help of other tools. A strong support for *manual* tests is given. TASSI includes the following features: fully graphical user interface, declarative specification of general and catalogue-specific automatic tests via rules, automatic execution of static tests of all media-objects, support of dynamic tests by a test agent (automatic navigation of the EPC to untested objects, automatic execution of static tests of dynamic objects), error classification via browser, context-sensitive specification and requirements presentation, capture and replay of manual test input, complete test state administration, and automatic test document generation.

In more detail, firstly the EPKML instance is fed into an automatic *static analysis* which aims at a rule-based static test of each EPKML element and each used database object. To determine most used database objects and at least one used set of attributes and elements of most EPKML elements a dynamic model of the EPKML instance is automatically generated. This model only excludes outcomes of applets and text inputs.

In the *Tourist Catalogue Example* all database objects of the shown page, i.e. one video reference (15), one image reference (12), eight texts (11, 29), and seven link (31) destinations, are determined. The type of a database object is obtainable from the surrounding EPKML element at its usage point. Type-specific tests are carried out on all atomic EPKML elements, e.g. if "To other cities" (39) is spelled correctly or if $cities.video$ (15) references an AVI video. Top-level grouping EPKML elements, i.e. pages and windows (08), are examined for group errors each time the container contents changes, i.e. in the example anytime a new city page is entered. Group errors are those errors which only occur if some elements are used simultaneously. In the example for instance it could be verified, that the text colours are in adequate contrast to the background colours or that not too many different fonts are used on the page. Each error found is recorded. Each tested EPKML element and database object is marked.

In a second phase the actual EPC is *dynamically tested*. The aim is to test the execution of each branch of the EPKML instance, to inspect manually all media-objects and to validate automatically media-objects and elements that have not been recognized in the static analysis. The EPC is navigated to untested parts by a test agent. A tester essentially only acknowledges the end of manual testing of a set of elements, enters errors into a form and requests the navigation to the next element set. In the case of text input and applet calls, the tester is responsible for choosing sensible input for systematic testing.

The test agent tries to carry out all tests of an element set before navigating to the next one and also attempts to follow intuitive navigation paths. In the example, first the city page (07) is shown. After the execution of the video (15) the tester is asked to confirm the end of manual testing. Then clicking one of the buttons which lead to untested parts is simulated. If there was no such material, the quit button would be chosen. The tester can always steer the test agent to places of her choice.

7 Conclusions and Further Steps

We support the *whole life cycle of the development process of* EPCs by developing an integrated package of tools for the production of electronic product catalogues. These tools aim at the efficient production of low cost electronic product catalogues. Through the features and services observed in current EPCs [6], we identified the components and characteristics of the language EPKML and the features of each assistant. The *catalogue developer* is assisted from the beginning of the first interview by special editors and a presentation assistant. During the catalogue design and generation phase the assistance is realized with other editors and class libraries. Tests accompany the entire development process including the final EPC.

In our approach, an EPC is the result of cooperating experts working at various places on different aspects of the catalogue. In order to keep track of the correct versions of the documents and programs produced so far, version management is required. We are currently implementing a WWW based project server, which allows to define and manage users and projects with appropriate access capabilities. Furthermore, the server manages revisions of documents and identifies official releases of software modules.

References

1. A. V. Aho, R. Sethi, and J. D. Ullman. *Compilers—Principles, Techniques, and Tools*. World Student Series. Addison-Wesley, 12 edition, 1995.
2. B. Boehm. A spiral model of software development and enhancement. *IEEE Computer*, 21(5):61–72, May 1988.
3. I. Dilli and H.-J. Hoffmann. DIADES II: A Multi-Agent User Interface Design Approach with an Integrated Assessment Component. SIGCHI bulletin, ACM Press, April 1991.
4. C. Goldfarb. *The SGML Handbook*. Clarendon Press, Oxford, 1994.
5. A. Knapp, N. Koch, and L. Mandel. The Language EPKML. Technical report 9605, LMU München, November 1996.
6. N. Koch and L. Mandel. Catalogues on CD-ROM: The State of the Art. Technical report 9610, Ludwig–Maximilians–Universität München, December 1996.
7. M. Reiser. *The Oberon System—User Guide and Programmer's Manual*. Addison Wesley, 1991.
8. A. Turk and H. Stoyan. Erfassung, Verarbeitung und Dokumentation natürlichsprachlicher Äußerungen in der Anforderungsanalyse. In E. Ortner, B. Schienmann, and H. Thoma, editors, *Natürlichsprachlicher Entwurf von Informationssystemen*, pages 32–46. Universitätsverlag Konstanz, May 1996.
9. A. Wasserman and P. Pircher. A Graphical, Extensible Integrated Environment for Software Development. *ACM SIGPLAN Notices 22:1*, 1987.

A   Example

In this section we present the EPKML specification of the tourist catalogue page.
Note, that some detailed layout information is omitted.

```
01 <theme name="general">
02  <extension result="cities">
03   <sql>
04    select name,vid,img
05    from tourist
06   </sql>
07   <template name="city-page">
08    <window name="city-window"
              style="bavaria">
09     <set name=city value=
         "$cities.name$">
10     <frame name="header">
11      <p>$city$
12      <img src="$cities.img$">
13     </frame>
14     <frame name="left-col">
15      <video src=
          "$cities.vid$">
16       <stop-button>
17      </video>
18     </frame>
19     <frame name="right-col">
20      <sql result="out">
21       select issue subtheme
22       from tourist
23       where theme='$city$'
24      </sql>
25      <itemize>
26       <make-items from="out">
27        <frame>
28         <button>
29          $out.issue$
30          <on-click>
31           <open name=
              "$out.subtheme$">
32          </on-click>
33         </button>
34        </frame>
35       </make-items>
36      </itemize>
37     </frame>
38     <previous-button>
39     <p>To other cities
40     <next-button>
41     <button name="to-reg"
              style="city">
42      <img src=
          "icons/reg.jpg">
43      <on-click>
44       <open name="reg-form">
45      </on-click>
46     </button>
      ...
59    </window>
60   </template>
61  </extension>
62 </theme>
63 <main>
64  <open name="general">
65 </main>
```

CHEOPS:
Adaptive Hypermedia on World Wide Web

Salvatore Ferrandino[1], Alberto Negro[2], and Vittorio Scarano[2]

[1] XCom Wide Communication
Via E.De Filippis 107/A –P.co Luciano
84013 Cava De' Tirreni (SA) – Italy
[2] Dipartimento di Informatica ed Applicazioni "R.M. Capocelli"
Università di Salerno
84081 Baronissi (SA) – Italy

Abstract. Because of stateless characteristics of HTTP, it is not possible for HTTP servers to adapt responses on the basis of previous interactions with the same user. Although efficient, this is a limiting aspect of the protocol, particularly for Web documents that are directed toward a broad audience, including experts and novices and an adaptive response would be helpful in providing information with the right *style*.
We present here the key concepts of CHEOPS, a system that is being developed as a tool that can be used by hyperdocument designers to provide their document with adaptivity and navigational aids. Version 0.9 of CHEOPS, actually available, is implemented by several CGI-BIN scripts that enables an HTTP server to interact with a user and follow her on the "Path to Knowledge". Solution is transparent to the user and to the client, and has a moderate impact on server performances. Further directions of the work toward version 1.0 is also presented.

1 Introduction

The World Wide Web (WWW) is an information retrieval system on the Internet that can be considered the first real global hypermedia network. It was conceived as an environment where information from any source can be accessed in a consistent and simple way from any place in the network.

The World Wide Web was developed in 1989 at CERN as a means of sharing information through the organization. Since then, WWW's growth has been exponentially fast, by reaching tens of millions of users through the whole Internet.

Although originally developed to facilitate information sharing (especially in academic environments), the World Wide Web has large potentials as an educational support, especially for distance learning. In fact, several educational systems based on WWW have been developed in the recent past [1,7].

It is our opinion that in order to fully exploit WWW potentiality in the educational field, server response must be aware of user's behaviour so that it can take into account the level of knowledge and, as a consequence, provide the user with documents that she can read, given her background/capacity.

This ability is well beyond the usual interaction with students through exercises and automated response for exercises or tests [1]. In fact, what we propose and implement here is to have an "opaque" mechanism, that, based on the history of user's requests, can provide a different document that is likely to be understood and lightly challenging while not being intimidating.

Such a mechanism can be also helpful to avoid the risk that users can "*have trouble in finding the information they need*" [10] because of the large amount of information available. As an example, many WWW servers of official institutions (like Universities, Colleges, Research Labs and so on) include a "Presentation" of the institution and its size can be as small as few Kilobytes of text or as large as hundreds of Kilobytes including text, images, audio, history and so on. On the increasingly overcrowded network, downloading the "Presentation" can take up to few minutes, when to make a "hit" few lines would be usefully delivered to a "passer-by" and could induce her to continue the visit of the site. An adaptive response from server could slowly increase the amount of information given to the user and, therefore, avoid the risk of an "information overloading" that could "scare her off" the site.

HTTP and Adaptivity.

HyperText Transfer Protocol (HTTP) is a very simple, fast and efficient Internet protocol, for searching and retrieving information from a server: the client makes a TCP-IP connection to the host, sends a request document and the server sends back to the client its answer, i.e., the document required or an error message. Because of efficiency requirements, HTTP 1.0 was designed as a stateless protocol, that is, the server does not keep any state on behalf of the client. In this way, HTTP servers can handle a large number of requests at the same time.

Although very efficient, this design requirement is limiting many possible activities that the server can perform on request. Examples are the ability of analyze the efficacy of cross references within the same Web site, build statistical profiles of users and to adapt its behaviour on the basis of previous interactions.

Solutions to this limitations fall in two categories: those proposing an extension to the HTTP 1.0 protocol and those devoted to desing a new protocol. In the first category, we see, for example, an extension to the HTTP 1.0 protocol [8] which is a refinement of the "Cookie" solution proposed by Netscape [9]. On a different direction lies the effort to overcome this limitation in the next future by designing HTTP 1.1 [4], where it is allowed to use the same TCP/IP connection to perform multiple operations.

Both kinds of solutions offer advantages but also suffer from some limitations. New HTTP 1.1 is still being designed and so are the browsers that can take advantage of the multiple operations capability. Multiple operations should be performed in HTTP 1.1 by keeping the connection open between client and server (*Persistent* connections). The mechanism looks ill-suited for mantaining a session in an educational hyperdocument. In fact, typical usage would be to browse through a hyperdocument for a non-negligible amount of time, which can be burdensome on the server-side, being a child process of the HTTP server devoted exclusively to the goal of serving the user requests.

On the other side, the "cookie" extensions proposed suffer from the drawback that they can be used on specific (also if widespread) browsers like Netscape 2.0 and following versions [9]. Also, widespread complaints over similar mechanisms that can be used to mantain surveillance of users accessing a Web server are reported over the Internet and taken into account by several authors [3].

CHEOPS System.

CHEOPS is more oriented toward a "server-side", application-specific solution to the adaptivity requirement, with particular care in avoiding overloading the server with many tasks. Its architecture is designed to limit concerns on users' privacy: the session that CHEOPS keeps track of is strictly limited to the "current" hyperdocument and is not extended on user's behaviour on the whole Web site. Moreover, CHEOPS, although specific to educational application, can be also generalized to other similar situations when presenting information at the right rate (for the user) is crucial.

Potentialities of WWW as an educational tool were immediately recognized and many results are available on the educational use of the Web for distance learning [1,5–7]. In particular, in [6] the need for a "WWW work session" (using our terminology) in an educational setting was recognized and an application based on CGI-BIN programs was developed that adds a similar capability to the Web server.

Our solution differs from [6] in two major aspects: first, CHEOPS is a design system that should make easy for a designer to add adaptivity to a hyperdocument in a modular way. Especially with the extensions planned (and currently under development) for version 1.0, this goal looks quite reachable. The second one is an architectural difference: in [6], each time a session was started there was a sleeping process on the server which each successive request would interact with. That would pose two problems: the first is that one needs (as also pointed out by the author in [6]) an additional mechanism to eliminate "hanging" unwanted sleeping process ("zombie" processes that do not correspond to an active session anymore). Then one should consider carefully the impact on server performances: if many sessions are to be taken care of concurrently, there would be a proliferation of spawned processes with a consequent, possible, performances degradation on heavily loaded servers.

It should be noted that, from a design point of view, one does not need a permanent "sleeping process" to interact with during the whole session, but only a process that loads the server only when needed, that is, *at the moment of the request* and, if correctly instructed, can recognize the request as belonging to a specific session.

Our solution, therefore, adopts a different strategy: each time a session is started, the server creates a user's profile, "distributing" (virtually) a "session card" *embedded within the hypertext* file that is sent back to the browser. Each successive requests to the server will be done with this session card and, therefore, recognized as belonging to a specific session.

CHEOPS System is non-obtrusive, widely usable, (one does not need specific browsers to use our system), has moderate impact on server load (putting all the "memory" of the session where it belongs, i.e. on the disk) and is modular enough to be easily installed on a Web server for a specific hyperdocument.

CHEOPS is implemented through several CGI-BIN scripts, used to build "on-the-fly" documents that depend on user's previous inquiries at the same server *within the same session*. CGI-BIN scripts are program run on the server and triggered by input from a browser.

Structure of the paper. In the next section, the system design and architecture of CHEOPS is presented with a brief description of requirements, the Session Protocol, the Knowledge Model we propose and few implementation details. Then, in section 3, we conclude by showing a case study that led to the development of CHEOPS and giving the guidelines of future work that is currently planned.

2 System Architecture

CHEOPS, actually being developed at the Dipartimento di Informatica ed Applicazioni of the University of Salerno in Italy is a "server-side" implementation of a session-based interaction model based on several specifications required by an educational, WWW-based, software.

CHEOPS version 0.9 consists of several Unix Shell scripts that provide the designer of a hyperdocument with an automatic mechanism that satisfies user's requests according to previous inquiries.

We begin the section by defining several requirements, identified during a previous project briefly described in section 3, that led us to the design of CHEOPS. Then, we show our Session Protocol and illustrate the Knowledge Model which CHEOPS is based on. Finally we provide a description of the implementation, with few key concepts of CHEOPS design.

Analysis of requirements. Being focussed on developing a design tool for hyperdocument to be employed in the educational field, several requirements were identified during a preliminary project:

- *Lecture-type Interaction:* The user should be able to interact with a WWW hyperdocument that makes her feel like having a teacher/instructor/field-expert *"on the other side of the screen"*. To reach such a goal, it is necessary to have tools able to distinguish between requests of different users. It is also recommended the usage of context-sensitive help, where by context it is not only meant the part of hyperdocument currently visited but also the user knowledge and previous interactions with the help system.

- *Categories:* It is often the case that there is a "natural" subdivision of the hyperdocument in categories that are well-suited to the argument. It is important that user is able to check her "knowledge level" for each category and also able to choose to further explore a category that, willingly or not, was ignored at a previous step.

- *History Mechanism:* User should be able to check the path followed during her interaction with the hyperdocument: being able to trace back previous interactions (within the same session) is helpful in developing and reviewing the actions taken.

- *Avoid the "Too much" syndrome:* Overloading the user with too much information, too early is a bad policy well-known to lecturers.

- *Tours:* User should be able to learn according to her own character: some users would like to "explore" on their own the hyperdocument while others could be more at their ease by being provided a guided tour through a part of the document.

It was clear, during the design of CHEOPS, that a WWW hyperdocument that develops the full potentialities of above mentioned requirements needed adaptivity in the server response. Furthermore, the project was to be realized by using "off the shelf", state-of-art tools that were not to make heavier server load.

Given the requirements above specified, we are ready to present, in the next subsection, our model of the interaction between user and server.

The Session Protocol. The session protocol that has been implemented is based on the following sequence of actions:

1 - Introduction. A new user is introducing herself to the server through the choice of an "introductory" link.

2 - Beginning of the Session. The server responds by sending the user a *Session Card*, that is a unique, opaque, identifier that will be used in the following interactions by the client thus identifying itself as "known" for the system.

3 - Session. An (unlimited) number of interactions user-server where each action is accounted for and used by the server in order to adapt its future responses to the same user.

4 - End Session. The Session Card is destroyed and no longer valid.

2.1 Knowledge Model

When a hyperdocument is designed for a broad audience, it is customary to face several problems in trying to make its *style* acceptable for novices, amateurs and experts at the same time. They all have different, some time opposite needs: a novice may need information that can be boring for an expert since it makes her loose time in finding the more advanced part of the hyperdocument. On the other side, a novice can be scared and confused by an excessive usage of technicality and leave the site before she could have the chance to access information that are suitable for her experience in the field. Then, great care must be taken to fulfill this important, we dare say paramount, requirement if one is asked to use at its best such an efficient communication media like the World Wide Web.

Another orthogonal[1] need is to help the user (whichever level and experience they have) in navigating through the (sometimes huge) amount of information. This is often achieved by subdividing data in a limited number of *categories* that can be seen as providing an hypertext of a small degree of traditional linearity within the hypertext itself.

---

[1] Use of the word is not casual, as the patient reader will soon discover.

Putting all together, the view we propose as a model for dealing with the "shapeless" amount of information is a geometric one: let us build a *Knowledge Pyramid*, having as bases two regular polygons of the same shape but of different size, the smallest on the top. Each vertex of the polygon represents one of the categories which the designers subdivided the whole amount in; the edges of the pyramid represents the experience level in each category.

The experience acquired by a user is defined as the surface obtained interpolating the intersection of the current experience levels on each category. The learning process can be seen as trying to get the surface as low as possible.

A possible variation of the model can be to have fewer category on the top than on the bottom. This corresponds to the natural mechanism to avoid mentioning some categories if the user may not be ready and actually may be disturbed by that.

THE KNOWLEDGE PYRAMID

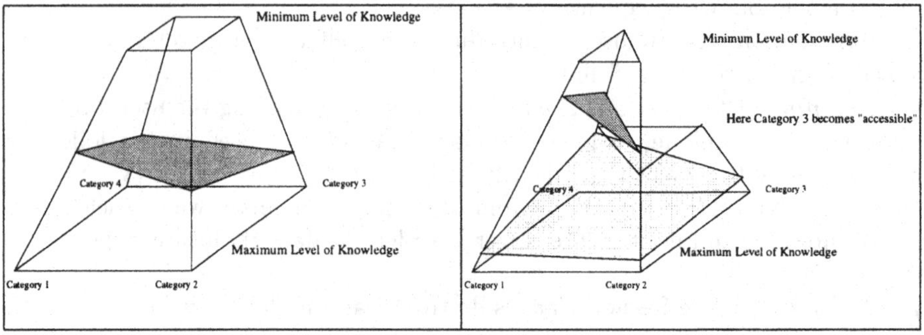

Fig. 1. Left. The Knowledge Pyramid: corners represent categories and the surface is the "current level" of knowledge of the subject. **Right.** The Model when Category 3 is accessible only given a certain knowledge of the subject.

CHEOPS uses the Knowledge Pyramid as an important navigational aid to the user: at each step, the user is able to see what her confidence for each category is (or is supposed to be by CHEOPS) and act as a consequence. She may choose to explore a part of the hyperdocument she left behind, or she may want to go deeper in a category or she may go on with her hypertextual visit of the document.

2.2 The Implementation

CHEOPS 0.9 provides the hyperdocument with a mechanism based on the Knowledge Model that makes explicit for the user the access to different categories further differentiated by knowledge level. The mechanism is non-obtrusive both for designers and users. In the first case, in fact, hyperdocument is built as usual, the only difference being the way "internal" anchors are defined (that is as links to CGI-BIN scripts with arguments) and the fact that for each HTML document some information (like category and level) has to be inserted in a configuration file. For the user, nothing is lost of the well-known hypertextual browsing of the document, only few additional capabilities are added and placed on two toolbars, non-obtrusively placed at top and bottom of the page, available at user's will.

The main mechanism is implemented through a CGI-BIN script that takes as input the document required by the user and her session card and builds "on-the-fly" a document that has all the links changed according to the current session card and introduces, if required, two toolbars as header and footer to the page.

We describe, now, the files used by CHEOPS and the CGI-BIN scripts that make part of version 0.9 and, finally, present a sort of "designer guide".

CHEOPS **Support files.** There are three files involved in the system behaviour of CHEOPS 0.9. The first is a *configuration file* located in the starting directory of the hyperdocument and called **general.dat**. In this file, the designer is asked to introduce (in this version through a text editor) information about the hyperdocument in general and about each HTML file.

The general information about the hyperdocument includes the name and location of the scripts, a prefix used to identify the *user profile* and *history* files (more details follow immediately), location of a temporary directory (under the Web root) where temporary images can be stored, location of a directory where history files are archived, the starting page of the hyperdocument, the starting page of the hyperdocument ("home page") which the user can return to at the end, and the names of the Categories.

For each HTML file, the designer is asked to include information about the category and the level the file belongs to, if one wants the header and bottom toolbar to appear (more details follow) and the context-sensitive help file.

The *user profile* file is generated by one of the scripts at browsing time and contains information about the current levels of expertise the user has reached during her inquiries in each category. User profile is located in the /tmp directory and is prefixed as indicated by the designer in the configuration file and suffixed by the session card of the session they refer to. Prefixes are unique and identify CHEOPS data on a specific hyperdocument: CHEOPS could run on the same server for different hyperdocument. Profile files are not supposed to be altered by hand and are deleted at the end of the session. A recurrent procedure that cleans up profile files should be executed at a fixed amount of time (each day or week) in order to delete information of users who choose to exit the browsing without the explicit "End Session" link.

The *history* file is also script–generated, and holds information, in HTML format, on user requests within the same session. The history file is also kept in the /tmp directory and, as well as the user profile file, is prefixed as indicated in the configuration file and suffixed by the session card of the session and is not supposed to be altered by hand. The designer can choose whether history files are deleted or not at the end of the session by setting in the configuration file an empty target directory. We think that history files provide a useful mean to the designer to check efficacy of links and the possibility of elaborate statistics on accesses to the hyperdocument and, therefore, suggest that, at least during a test phase, history files should be saved and checked.

CHEOPS CGI-BIN **Scripts.** CHEOPS 0.9 consists of several Unix shell scripts (CSH); the main component is the script named **builder**. This script takes two input parameters: the first one is what we call the *session card*, while the second is the HTML file actually required by the user. Its output consists of the required HTML file, where (if required by the configuration file) two toolbars are introduced. Three are the main tasks that **builder** is to accomplish: *(1)* insert a *header toolbar*; *(2)* change the reference to the hyperlinks within the same hyperdocument in such a way to include the current session card; *(3)* insert a *bottom toolbar*.

The *header* toolbar offers hyperlinks to the four Category Summaries with the actual confidence level (according to the session card that is a unique identifier in the system) plus hyperlink for the context-sensitive help and a hyperlink to another script that gives the user the chance of explicitly changing her confidence levels (more details follow later in this section). Moreover, the name of the "current category" (that is the one which the document required belongs to) is shown in capital letters and a GIF image is placed on the right part, which represents the current view that user has of the Knowledge Pyramid.

Image of Knowledge Pyramid is built "on-the-fly" through a C program called **gifcreate** that uses GD 1.2, a graphic library for GIF creation [2]. The image is prefixed and suffixed as profile and history files and is located in the directory indicated in the configuration file.

At this point, after inserting (if so required by the configuration file) the header toolbar, the script provides as output the HTML file required, taking care of instantiating each occurence of the pattern SESSION_CARD with the value of the session card passed as first parameter. The designer takes care, in fact, to insert as first parameter of the internal links (i.e. links to the **builder** script) the pattern SESSION_CARD. That allows CHEOPS recognizing further choices made within the same session.

The *bottom toolbar* gives hyperlinks to the beginning of the whole document, to another script that let the user see the history of her browsing of the hyperdocument, to the "End Session" part of the session protocol previously described and to a general help file.

The second script involved in CHEOPS is the **level-modification** script. It is important to give the user the chance to modify current confidence levels in such a way to access, if desired, more complete information on a subject or if the user realizes that "she has gone too far" and is confused by the technicalities presented at that level.

This script proposes the user with a form where she can modify confidence levels, where the current value is shown as a default choice. The **action** associated with the form is the **builder** script itself. This gave an additional degree of difficulty to the **builder** script that has to recognize whether it has been called from an "ordinary" link (that is with two parameters) or it has been called from the **level-modification**. A *Back* hyperlink is also provided to give the user the possibility to go back to the previous document without changing levels.

The third script is the **view-history** script that allows the user to see the history file that is step-by-step updated by **builder** each time the user takes an action. Each entry is available as an hyperlink so that the user can access files that she previously visited. Each manual levels modification (i.e. through the **level-modification** script) is also shown.

The Session Protocol. The Session Protocol described at the beginning of this section is implemented through different ways of setting the first parameter (i.e. the session card) of **builder** script. At the beginning, the designer provides an entry point to the hyperdocument by calling the **builder** script with first parameter set to **START**. In this case **builder** will have (transparently to the user) a different behaviour, taking care of getting a new session card and initializing the history file before doing the ordinary actions previously described.

The "Close Session" hyperlink provided in the bottom toolbar is a link to **builder** with first parameter set to **END** and, in this case, the script will take care to provide the user with a conclusive page that has the home page link as indicated in the configuration file. The script will also delete the user profile from the **/tmp** directory and move the history file to some other directory so that there is no possible interference with successive sessions.

A designer guide. CHEOPS 0.9 is designed taking into account the needs of designers: not much additional work is required in order to make a hyperdocument adaptive to user's responses. Here is a brief check-list for the designer:

- Design the WWW hyperdocument as usual: HTML files with hyperlinks pointing to other files within the whole hyperdocument.

- Categorize the files in 4 categories with 3 knowledge levels. Version 1.0, actually under development, will remove any constraint on the number of categories and confidence levels.

- Edit the default configuration file provided with the distribution. Insert all the information relevant to the hyperdocument being designed as its directory, home page, the first file to be shown, names of the categories and their *short names* and so on. Short names of categories are strings with no spaces, used by **builder** to get the name of the files containing Category Summaries, indexed by the confidence level.

- Build the Category summaries; often as the experience suggests, summaries are already been developed during the usual design of the hyperdocument.

- Change the internal links so that they call the **builder** script with first parameter **SESSION_CARD** and second parameter the name of the file which the link points to. This is a step that will be removed by version 1.0 that will change automatically internal links through the **builder** script.

- Provide an entry point to the document as a call to **builder** with first parameter START.

3 Conclusions and Future Developments

The design of CHEOPS has been highly influenced by a project (in collaboration with Echos Studio Media, a Musicology association) of an hypertextual document about Giovan Battista Pergolesi, a Neapolitan composer of the XVII century. In fact, hyperdocuments on Theory of Music represent a useful paradygm since the argument can be accessed both by a novice and by an expert that can be interested in accessing (for example) Pergolesi's handwriting manuscript in order to check and ascribe to him some newly discovered manuscript.

CHEOPS 0.9 is actually undergoing further modifications and improvements toward version 1.0 that should be available soon. Information on the updated version of CHEOPS can be retrieved at http://stromboli.dia.unisa.it/CHEOPS/ where a demo of its capabilities is also available.

We would like, in the next versions of CHEOPS, to make the designer able to provide in the configuration file the number of categories (i.e. not limited to 4 as it is in version 0.9) and the the number of available confidence levels (currently 3 levels are allowed). Moreover we plan to include an automatic mechanism to present "long" HTML files in several pieces linked together through **Next** and **Previous Page** links. We also want to make life easier for the designer so that the designer can build her hyperdocument as usual, test links correctness and then let CHEOPS add the adaptivity to it.

Future versions of CHEOPS (probably developed in a more portable language like PERL or C) will also include multi-faceted knowledge pyramid (fewer categories on the top and more on the bottom) and possibilities to insert a more exotic mechanisms for the adaptivity. New versions of CHEOPS should give the designer, at least, a choice among several adaptivity response algorithms. A significative improvement would be a mechanism to allow the "enrollment" of users so that they can re-use hyperdocuments starting where they left. The idea is to enclose the session-based protocol within a layer that supports user-based access (autenthication, user management etc.)

References

1. D. Dwyer, K. Barbieri, H.M. Doerr. *"Creating a Virtual Classroom for Interactive Education on the Web"*. Proc. of WWW 95, Third Int. Conf. on World Wide Web.
2. T.Boutell, "GD 1.2, *A graphics Library for fast GIF creation"*. Quest Protein Database Center, Cold Spring Harbor Labs (US).
3. P.M. Hallam-Baker, D.Connolly. *"Session Identification URI"*. W3C Working Draft WD-session-id-960221.
4. R.Fielding, H. Frystyck, T. Berners-Lee. *"Hypertext Transfer Protocol, HTTP 1.1"*. HTTP Working Group Internet Draft.
5. N. Hammond, L. Allison. *"Extending Hypertext for learning: An investigation of access and guidance tools"*. In Sutcliffe,A. and Macaulay, L. (Eds.): *People and Computers*, Cambridge University Press, 1989.
6. B. Ibrahim, *"World-Wide Algorithm Animation"*. Computer networks and ISDN Systems, Vol. 27 No.2, Nov. 1994, Special Issue of the First World Wide Web Conference, pp.255-265.
7. B. Ibrahim, S.D. Franklin. *"Advanced Educational Uses of the World Wide Web"*. Proc. of WWW95, 3rd International Conference on World Wide Web.
8. D.K. Kristol. *"Proposed HTTP State Management Mechanism"*. HTTP Working Group Internet Draft (2/22/96).
9. Netscape Comm. Corp. *"Persistent Client State HTTP Cookies"*. 1995.
10. J. Nielsen. *"Hypertext and Hypermedia"*. Academic Press Ltd, 1990.

Application Output Recording for Instant Authoring in a Distributed Multimedia Annotation Environment

Hartmut Benz, Steffen Fischer, Rolf Mecklenburg, Andreas Wenger

Institute of Parallel and Distributed High-Performance Computing (IPVR),
University of Stuttgart, Breitwiesenstrasse 20-22, D-70565 Stuttgart,
{benzht, fischest, mecklerf, aswenger}@informatik.uni-stuttgart.de

Abstract. Application Output Recording (called AOR from now on) is the task of recording the output of an application so that the reproduction of the recording is sufficiently similar to the original output performance. The paper proposes a taxonomy to structure the task and describes a generic architecture covering the complete range of possible complexities in a recorder. Variables determining the usability potential, technical aspects and design decisions are defined. The usage of AOR is described in the context of the European Union funded ACTS-project DIANE (Design, Implementation and Operation of a Distributed Annotation Environment), which implements a generic distributed multimedia annotation environment running on broadband systems. A usage scenario clarifies the utilisation of AOR in DIANE. DIANE is evaluated in a telemedicine and a tutorial environment. Special requirements can be derived from these and have implications to the process of recording generic applications. A short overview on the first implementation of AOR in DIANE is given. The taxonomy is applied to solutions from related work as well in research as in the commercial sector.

1 Introduction

Application Output Recording is the task of recording the output of applications. The output is recorded in such a way that the behaviour of the application and the interaction of the user with it can be replayed at a later time. Since there are several approaches to AOR, we will propose a taxonomy, which reflects all influences on this task. A generic architecture for AOR will be introduced.

What is the reason for making recordings of the behaviour of and the interaction with an application? The most straightforward answer is: to achieve something like a playable demo of the application. In a more sophisticated environment, the recorded output may be included in a multimedia document with the possibility to add other media to the document. In this way the application output could be, for instance, commented by voice and telemarkers.

Application output as part of a multimedia document gives the user the opportunity to talk about applications in a new way. With AOR one can ask an expert questions about applications, while at the same time showing him what went wrong. The expert can use the original document to attach his comments to. In this way it is possible to start a news-group like discussion regarding the output of any applications.

The process of "attaching comments" may be referred to as "making an annotation to a document". A system incorporating AOR is able to include virtually everything performed by a computer into documents.

The EU-funded ACTS-project DIANE is developing such an annotation environment. DIANE is conceived as a service allowing users to create, exchange and consume multimedia data easily. The basic concept to be supported by DIANE is that of a multimedia annotated document consisting of two distinct parts: recorded application output and annotations given by a user in various media. By providing techniques to record output of arbitrary applications, DIANE defines an annotation recorder which can be used in a generic manner with one user interface to arbitrary applications.

This paper is structured as follows: Section 2 proposes a taxonomy for application output recording. The taxonomy focuses on the users point of view, but also considers technical aspects like costs and complexity. A generic architecture derived from the taxonomy is presented in Section 3. To show the relevance of the taxonomy, Section 4 deals with the EU project DIANE and how AOR is realised for this environment. Other approaches to AOR will be shortly described and classified according to the taxonomy in Section 5.

2   Application Output Recording

The aim of AOR is to record the output of a given application in such a way that the reproduction of this recording is sufficiently similar to the original output performance. This approach to AOR is entirely user oriented since it focuses on factors a user will immediately be concerned with and perceives technical and implementational aspects as dependent on them.

Our analysis of usage scenarios for AOR revealed that the process can be characterised by two independent variables determining the potential usability: *replay controllability* and *processing capabilities*. Dependent on these are the technical variables *requirement for recording awareness* of the application, *recording data format* and *timing of recording*, the latter determining the *point of data interception*. Independent of these is the *selection of the recording source*. The general variable *security* affects all other variables. Fig. 1 shows these variables and their most important dependencies. Restrictions to the possible variable values are propagated forward and backward along the lines. User requirements influence technical variables; technical requirements and limited resources pose restrictions to the usability potential. The following sections discuss these variables and their dependencies in detail.

2.1   Variables Determining Usability Potential

This section describes the variables that will primarily determine the usability potential a user may draw from AOR. The focus of these variables is task-

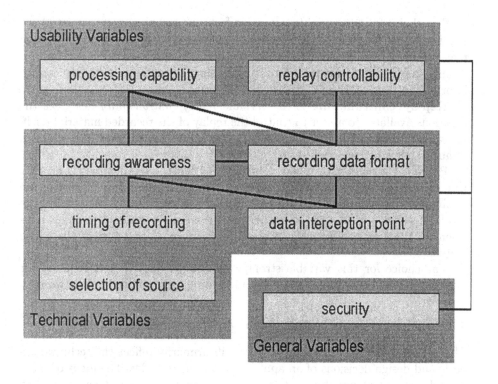

Fig. 1. Variables and dependencies for AOR classification.

oriented and can be more easily communicated to a user in form of use-cases or similar methods than the more technical aspects.

Processing Capabilities: The variable *processing capabilities* describes how the recorded data may be reused in addition to their basic presentation. It basically determines the structural and semantical content of the recorded data. The value of this variable is a description of intended post replay usages, e.g. copying of text from the presentation, identifying the application the recording was taken from, or even starting this application with or without accessing the same data as at the time of recording.

A basic processing capability only requires a surface representation of the media. To acquire this it would suffice to record the analog signal to the display to capture the screen output of an application. The recording would be faithful to the original with perhaps a minor loss of image quality and smoothness. Nevertheless, there would be no information available to distinguish the windows of the target application from any other window on the screen (recording these other windows may even violate security requirements). A very high processing capability would be available if the original application would be used as the player to present the recording [ManPra95-1]. In this case every processing capability of the original application would be available to the user.

The choice for this variable strongly affects the technical requirement of the application to be aware of the recording since only a cooperative application may provide sufficient data to satisfy the given requirements. Furthermore, it strongly affects the data format for recording since the data must contain enough information to provide for the chosen capabilities.

Replay controllability: The variable *replay controllability* describes the functions available to a user to control the replay of the recorded material itself. Its value is a list of VCR like functions, for example play forward, fast forward, jump to index, or play backward.

Providing a rich set of replay controls with short reaction times is an easy way to increase user satisfaction ([Shn92]). Failing to provide even one of the minimal set of functions deemed essential by the users can easily render the whole tool unacceptable. On the other hand even one additional, perhaps non-essential, function can severely increase the complexity and cost of the whole system.

The choice for this variable strongly affects the data format required for recording since, for example, not every data format allows playing backward.

2.2 Variables Determining Technical Consequences

This section describes the variables that will primarily reflect the technical aspects and design decisions of an application output recorder. The focus of these variables is system oriented and they are used to represent restrictions due to limited resources (e.g. storage-capacity). Restrictions to these variables will most commonly originate in the controlling department of the customer and the design team of the AOR-supplier.

Application Awareness for Recording: The technical variable *requirement for recording awareness* describes whether the recorded application is aware of the recording or not. Requiring applications to be aware of being recorded may severely simplify AOR since such an application is likely to provide data much more suitable for recording and replay than others. Thus, this choice would also severely restrict the range of possible applications to record. A very simple example for an aware application could be a (slightly modified) text editor Emacs. Through an API it could make its internal state (e.g. edited files, cursor position, etc.) available to the recorder. During replay these states would be fed back to the editor to recreate a presentation similar to the original session.

Recording Data Format: The technical variable *recording data format* describes the format of the raw data that is recorded. It represents all recorded application relevant information. Its value is the specification of a data structure. Almost all other aspects of the system affect the data format.

Point of Data Interception: The technical variable *point of data interception* describes where the data to be recorded is acquired (Fig. 2, left side). Its value describes for each media recorded where its respective media data will be accessed and how this access is achieved. Example values for this variable are "screen output recorded from analog signal to the monitor", "screen output recorded by intercepting the X-protocol using a virtual X-device", or "audio output recorded by intercepting audio data stream to the virtual audio device".

One effect of the selection of the point of interception regards platform dependency. Only when acquiring the data of the application itself (and nowhere else), platform independency of the recording tool can be achieved. All other interception points impose hard- and software-specific approaches on the design of the tool.

Selection of Recording Source: Another important issue of AOR is the selection of what is to be recorded, e.g. which area or application. To record application output the user has to select an application or a part of an application. This can be achieved by either providing a dedicated recording area where everything to record is dragged into or a direct selection mechanism, for example selecting the application with the mouse or identifying it by name.

Both methods focus on the visual output to the screen where deictic input devices are common (mouse, light pen, etc.) and feedback of selection is easy (e.g. modify window borders, background colour, etc.). If AOR encompasses additional media like audio, robot control, or network communication, the selection, acquisition of data, and feedback get more difficult.

The decision which method to use in an AOR is primarily affected by human factors (usability) and ease of implementation; it is independent of the general capabilities of AOR discussed above.

Timing of Recording: The technical variable *timing of recording* describes how the recording of individual media "frames" is initiated. The term "frame" corresponds to the same term used for a video, which is replayed with a certain rate of frames per second. Possible values for this variable are continuous with fixed frame rate, automatic when output changes, and manual on user-demand. The most important considerations in choosing this variable are the amount of data produced and the quality (smoothness) of the replay.

2.3 General Variable: Security

The need for security in AOR is most obvious and affects every other aspect of the AOR system. Generally, the security requirements for the recorded output are at least as high as they were for the original material and applications. They may be higher due to the fact that the recorded material, in essence, contains an interpretation of the security relevant material, e.g. the personal view of the recording user. Additionally, nothing should be recorded which the user is not aware of. Therefore, the important usability aspect is to ensure that the user is, at all times, aware which applications or part of applications are recorded. Failing to provide this would pose a severe security risk since the user may accidentally record confidential material without noticing.

3   Generic Architecture for AOR

This section presents a generic architecture for an application output recorder. Fig. 2 shows the data-flow diagram of the architecture of the recorder; replay is not shown.

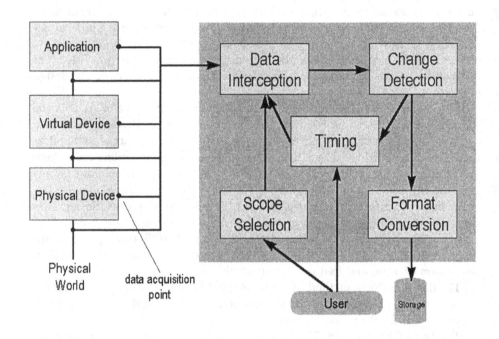

Fig. 2. Generic Architecture for AOR.

The application to be recorded with its runtime environment (virtual and physical devices) is shown on the left. The Data Acquisition Point determines the location where the data to be recorded can be acquired. Depending on the point of acquisition, the structuring of the data ranges from semantically very high level intercepted directly from the recorded application to semantically very low intercepted after physical devices (e.g. point a camera to the screen and put a microphone in front of the speaker). The component Data Interception gathers all the raw data from one or more Data Acquisition points according to the currently selected scope. The scope is managed by the component Scope Selection which maintains a list of objects that have to be recorded. This list may dynamically change due to the dynamic structure of the recorded application (i.e. windows may be opened or closed) or due to interactive scope modifications by the user. The component Change Detection may optionally reduce the amount of recorded data. Mainly, dependencies between recorded frames in the selected scope are being processed. The component Timing tells Data Interception when it's time to acquire data from the recorded application. It may be running in continuous mode with a determined frame rate, or it may be interactively controlled by a user (i.e. pressing a key to record). Finally, the component Format Conversion provides for transformation facilities to generate portable representations of the recorded data (e.g. MPEG).

4 Application Output Recording in DIANE

The EU-funded ACTS-Project DIANE is developing a multimedia annotation environment, which is making use of an AOR facility. In DIANE the values of the variables for AOR had to be defined according to predefined requirements.

In the DIANE project, several requirements have been defined for AOR. The requirements directly result from the scenarios DIANE will be evaluated in (telemedicine and tutorials). In a telemedicine environment very strong security requirements exist: No confidential information shall be recorded. The user has to have full control of what is recorded. Everything is forbidden, except the user explicitly allows it to be recorded.

DIANE has to support a broad range of applications (e.g. wordprocessors, imaging utilities, etc.). A goal for AOR in DIANE, therefore, has to be that applications are completely unaware of being recorded. Modifications of the recorded applications are impossible, since this implies that generic interfaces to the internal states of recorded applications would be needed. Such interfaces would definitely be a rather nice thing, but they are not available at the time being.

Annotations in DIANE are recorded on top of running presentations where one main requirement is that the running presentation can be manipulated in a flexible way. That is, the user has full control with a set of VCR commands like rewind, pause, fast-forward, and jump over the context of the running presentation.

Checking these requirements on application recording in DIANE against the usability and technical variables described in Section 2, the project decided to go for a low structured output recording (LOR) approach combined with parts of abstract script recording for change detection and data reduction. LOR provides for an easy to use, secure, flexible and portable way to record applications. The technique will be used for visible and audible media of applications.

According to the strong security requirements, a "recording area" approach is implemented. The user has full control over what is recorded ("what you see is what you record") by dragging everything to record into the recording area.

4.1 Technical approach

This section will describe how the recording facilities in DIANE have been built with technical details and obstacles we experienced on Sun Solaris.

Grabbing of visual content In the X Window environment, low structured output grabbing of images is performed with repeatedly calling the Xlib method

XGetImage. There is no generic way to get raw pixel data because the access to the framebuffer is hardware-dependent. Additionally, the colourmap has to be recorded and during playback it has to be installed to ensure correct representation.

The X Window System is maintaining a window hierarchy where windows are parts of the screen with certain attributes like colourdepth, colourmap, position, etc. Every application owns at least one top level window, called Application Top-level Window (ATL). Each ATL is a direct sibling of the root window. Buttons, textfields, canvases are all subwindows of this ATL and hence can have different attributes than it's parent.

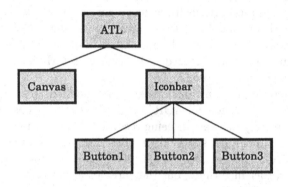

Fig. 3. Application Window Hierarchy.

When XGetImage copies the contents of a window from the framebuffer to a user accessible buffer, it copies everything from the area in the framebuffer according to the dimensions of the window. If any part of the window is obscured by another window, the content of this window will be overlaid over the original window because that is the only data which is actually stored in the framebuffer. The content is only valid if the depth of the obscuring window is the same as for the underlying window. Therefore every subwindow with different attributes has to be grabbed in a separate run and put in a so called application window hierarchy (AWH) tree.

An example for an AWH with different attributes is a video player where the video is replayed in 24bit colourdepth and the VCR controls are made up of 8bit icons. What is actually done in DIANE, is that all the AWHs in the recording area are grabbed and combined to a 24bit colour image of JPEG format. To reduce the data load in continuous mode, a simple change detection is implemented. The X-Protocol stream is intercepted, and only in case of drawing requests the recording area is grabbed. We use a Parallax graphics board with support for hardware JPEG-compression. Tables with compression ratios and compression times can be found in the appendix Section8.

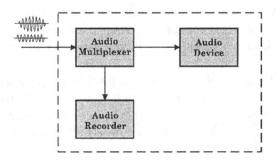

Fig. 4. Generic Audio Recording.

Grabbing of audible content In DIANE, the technical approach to record audio output of applications is to write a pseudo device-driver that intercepts audio input and output streams and multiplexes them to the recorder and to the original device-driver. All the interfaces of the original device-driver therefore have to be shadowed in the pseudo device-driver.

A drawback of the implemented solution is that all audio output and input of the system is recorded. No distinction between different applications can be made out of the stream of audible content passed into the recorder.

5 Related Work

Several approaches to AOR exist as well in research as in the commercial sector. These approaches will be classified according to the proposed taxonomy and compared to our realisation of AOR in DIANE.

Section 5.1 focuses on a paradigm developed at the University of Michigan. Section 5.2 gives links to three commercial products on application output recording, which cannot be described in this paper.

5.1 The Session Capture and Replay Paradigm

This AOR paradigm centers on the concept "WYSNIWIST" (what you see now is what I saw then), using the recorded application as a player. According to [ManPra95-1] the paradigm has to be classified as follows:

Processing capabilities: Recreation of user session: Using the recorded data, the recorded application is run, trying "[...] to recreate the look and feel of the original session [...].", ([ManPra95-1], S. 2). Editing of user session: This is an open issue, but aimed for.

Replay controllability: {play, stop, fast forward, fast replay, jump to index}: The realisation of these functions are described in the paper of Manohar and Prakash ([ManPra95-1]).

Application awareness for recording: Applications are required to be able to record and reset their state, that is, to be aware of the recording process.

Recording data format: The application output is captured in so called session objects. The session object contains information about user interaction with the application, audio annotations, application states and resource references (files, fonts, etc.).

Point of data interception: Application itself and window system. There are no information concerning the point of interception to audio data.

Selection of recording source: There is no explicit mentioning of this point. Nevertheless, it is implied that the scope of recording is always the application and additionally the audio device.

Timing of recording: Since it is streams, that are recorded, timing of recording is continuous.

Security: There is no reference of security features.

5.2 Commercial Products

There are some commercial products available on the market, which cannot be discussed here due to constraints on the length of the paper. Nevertheless we want to give the reader the possibility to inform himself regarding these products and to decide, how the variables are met.

Lotus ScreenCam: http://www.lotus.com/screencam/
Microsoft Camcorder: http://www.microsoft.com/office/office97/camcorder,
HyperCam: http://www.hyperionics.com/www/hypercam.html/

6 Conclusion

In this paper we described a taxonomy for application output recording, which is useful in the design and implementation of a generic tool for the recording of output of any application. Several parameters have to be weighed by the designer of such a tool to find a balance between usability and complexity.

We also described, how AOR is realised in the EU-funded project DIANE, which utilises an AOR facility in the creation process of multimedia annotations. Due to the use of AOR, DIANE is able to include the output of any legacy application into a multimedia document, which then can be annotated by other documents or be an annotation itself.

In the project DIANE several requirements are imposed by the users, which directed our choice of the values for the taxonomy variables. This choice, which we call LOR (low structured output recording), enables DIANE users to create presentations of visible and audible application behaviour with full replay controllability and also full control of the data, which are recorded, thus providing maximum security.

References

[BeFiMe97-1] Hartmut Benz, Steffen Fischer, Rolf Mecklenburg: Architecture and Implementation of a Distributed Multimedia Annotation Environment: Practical Experiences using JAVA, to appear in the Proceedings of the DAIS97, Cottbus.

[BeBeFiMe97-2] Sandford Bessler, Hartmut Benz, Steffen Fischer, Rolf Mecklenburg: DIANE: A Multimedia Annotation System, in Serge Fdida, Michel Morganti (Eds.): Multimedia Applications, Services and Techniques, Lecture Notes in Computer Science, Proceedings of the ECMAST'97 - Second European Conference, Springer Verlag, 1997.

[ManPra95-1] Nelson R. Manohar, Atul Prakash: The Session Capture and Replay Paradigm for Asynchronous Collaboration, in H. Marmolin, Y. Sundblad, K. Schmidt (eds.): Proceedings of the 4th European Conference on Computer-Supported Cooperative Work, ECSCW'95, Kluwer Academic Publishers, 1995, pages 149-164.

[ManPra95-2] Nelson R. Manohar, Atul Prakash: Dealing with Synchronization and Timing Variability in the Playback of Session Recordings, in Proceedings of the Third ACM Multimedia Conference, 1995, pages 45-56,

[Nye89] Adrian Nye: X-Protocol Reference Manual Volume Two, 5/89, O'Reilly & Associates Inc.

[Nye90] Adrian Nye: X-Protocol Reference Manual Volume Zero, 2/90, O'Reilly&Associates Inc.

[Nye93] Adrian Nye: X-Protocol Reference Manual Volume One, 3/93, O'Reilly & Associates Inc.

[Shn92] Ben Shneiderman: Designing the User Interface: Strategies for Effective Human-Computer Interaction, Addison-Wesley, Reading, Mass, 2nd edition, 1992

[Sun94] SunSoft: XIL Programmers Guide, 1994, Sun Microsystems.

[ZimGil94] Martha Zimet, Stephen Gildea: Record Extension Protocol Specification, 1994, Version 1.13, X-Consortium Standard.

Video Data Management in *Media Controller:* A Distributed Multimedia Surveillance System

Akihiro Tsukada, Toshihiko Hata, Fumio Matsuda,
Kazuya Sato and Minoru Ozaki

Mitsubishi Electric Corporation,
Industrial Electronics & Systems Laboratory,
8-1-1 Tsukaguchi-Honmachi, Amagasaki, Hyogo 661, Japan

Abstract. This paper describes a multimedia surveillance system, the Media Controller. The Media Controller is designed for industrial surveillance and monitoring applications, and it provides the realtime recording and retrieval of digital video data. Considering the characteristics of industrial surveillance applications, the distributed recording of video data is shown to be necessary for surveillance systems. A method of managing video data in Media Controller is also described. It can effectively manage surveillance video data which is continuous (realtime), distributed, temporal, and partial. We have implemented a prototype, which confirmed the effectiveness of the Media Controller system.

1 Introduction

In the field of industrial surveillance systems such as security systems or plant monitoring/control systems, there is an increasing requirement for multimedia surveillance using video cameras.

Digital video data enables fast, random searches of data on a disk drive, and is easy to link with traditional surveillance systems, for example to be displayed with the sensor data of the same time period. These features of digital video surveillance systems can help operators easily recognize the condition of the system under surveillance, and take quick and appropriate action. Focusing on the above points, we are developing a multimedia surveillance system named the Media Controller[1, 2]. In this paper, software design and audio/video data management in the Media Controller system are described, based on the characteristics of industrial surveillance applications.

There are some related works which introduce digital video data into surveillance/control systems. *Mercuri* [3] is a distributed, realtime process-control system which handles digital video data and provides image processing functions, aiming for automatic control. When some problems are detected in a video data and an alarm is issued, this system automatically transfers and stores the video data segment in a node called the "history site". It is one of the few examples of multimedia surveillance systems that take accounts of hard realtime (process control) systems, but it has some drawbacks.

First, video data triggered by false alarms wastes the resources of the network and disk areas. Secondly, the centralized recordings at the history site can cause bottlenecks in the CPU, disk access or the network resources, when lots of alarms occur in a short period. Lastly, only video data related to an alarm is recorded, and data not related to any alarm is discarded. The essential source of these problems is in the centralized recording.

In the security field, there are some works on digital closed circuit TV (CCTV) systems [4, 5]. These systems continuously gather distributed video data through networks to a certain site and record the data there, even when there is nothing unusual happening and thus no need to store video data. This results in load concentrations in the network, disk arrays and the CPU of the storage server. So these systems are not suitable for applications which require lots of high quality video cameras.

Video on demand (VOD) systems like Star Works[6] have the same problem when they are applied to surveillance systems. In order to record a large amount of surveillance video data, both the network and the storage server site are heavily loaded, because these systems record video data at the single site.

Media Controller is a platform for multimedia surveillance systems, which aims to provide a solution to above problems of digital video surveillance systems at a reasonable cost and support cooperation between surveillance video data and other systems like process control systems. It stores video data at distributed camera system sites, each of which can capture, store and transmit digital video data of surveillance video images.

This paper is organized as follows. In Sect.2, the architecture requirements that industrial surveillance systems should meet are discussed, and the software architecture of Media Controller system is presented. Sect.3 describes video data management in the Media Controller system, which efficiently manages and organizes distributed continuous video data. Sect.4 presents the current prototype of Media Controller and its implementation. The paper concludes with Sect.5.

2 Distributed Surveillance System

2.1 Distributed Recording and On-Demand Transmission

Listed below are the characteristics of industrial surveillance applications and their requirements for surveillance systems.

- Large numbers of video cameras
  Because industrial surveillance systems can have a lot of video cameras (over 100), it is impossible for operators to watch all the cameras all the time, and as long as nothing unusual is happening in the system, it is unnecessary to keep watching these cameras. Therefore, an effective method of surveillance is to store the video data and afterwards selectively watch only necessary or important video data.
- High quality video data
  Industrial surveillance systems require high quality video data, (for example, size of 640x480, 30 frame per seconds (fps), VCR quality), which means lots of resources are required per camera.

– Multiple access to the same video camera
  In surveillance systems, it is required that a video camera does not stop recording, even while retrieving recorded data from the same camera, for example to check the past situation in the case of an accident. Thus multiple and simultaneous access to video data is required.

In order to meet the above requirements, we can say that video surveillance systems should have a distributed architecture, where each node should digitize and record video data from only one or a few video cameras.

If a system has an centralized recording architecture (Fig. 1(a)), unnecessary data is also transmitted through the network from each camera to the node (storage server), wasting the network's bandwidth. And in order to accommodate large amounts of high quality video data, a high performance network is required and the storage server must have a high speed CPU and disk drives, or it might need a disk array (RAID) system to allow lots of simultaneous disk access. Therefore, centralized recording architectures tend to be expensive and do not have scalability according to the number of video cameras.

On the other hand, if a system has a distributed recording architecture (Fig. 1(b)), video data is transmitted through a network only when it is requested, allowing efficient use of the network bandwidth. Besides, since each node handles only a small number of cameras and discards old, unnecessary data, it doesn't need expensive hardware like a RAID system.

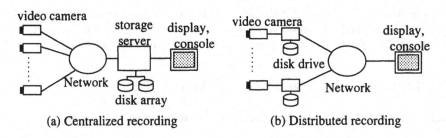

(a) Centralized recording   (b) Distributed recording

Fig. 1. Recording architectures of distributed surveillance systems

2.2 Video Data Management

There is one problem with distributed recording. Since video data is recorded and managed separately on distributed nodes, it is not easy for an application to search for and access video data. Therefore, a global management mechanism is required which manages all the video data distributed over a system.

Also, it is important for an intelligent surveillance system to link surveillance video data with other data, for example, sensor data from a plant control system, a 3D-model of a plant. That is, integration of different types of data is required.

However, some of the small surveillance systems do not have such requirements, and in that case, it is desirable to be able to exclude these functions from the system, in order to reduce overheads and cost.

2.3 Media Controller System

The design of the Media Controller system is based on the requirements of the surveillance applications mentioned above. Figure 2 shows an overview of the system. Video data is digitized, compressed and recorded onto hard disk drives on each distributed node, called "MC-L" (Media Controller local node), and only requested video data is transmitted in realtime from an MC-L through a digital network to an output node called "MC-C" (Media Controller center node). The system has the following characteristics.

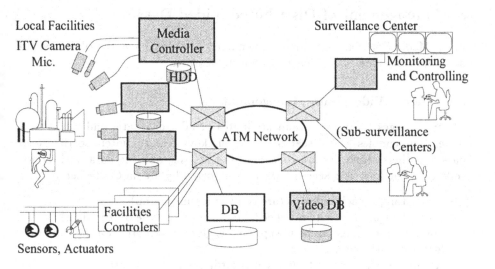

Fig. 2. Media Controller system

— Distributed, realtime capturing and recordings of high quality video data
— Simultaneous multiple access to the recorded data in a hard disk drive
— Efficient recording policy (in addition to the *Live* video data):
  • *Continuous*: high quality recordings of the latest period.
  • *Event*: automatic recording for sometime before and after an alarm
  • *Time-lapse*: low quality, sparse recording of the latest long period
  • *Backup*: permanent copying of data from one of the above 3 types
— Pipelined, on-demand transfer of video data from MC-L to MC-C
— Quality of Service (QoS) control per transfer

Hierarchical Software Structure. The software structure of Media Controller consists of the four functional layers. Below is an overview of the functions of each layer, with the lowest layer fist.

- *Data acquisition*: distributed video input, compression, and recording
- *Data transfer*: on demand, realtime transfer and decompression
- *Global management*: video data and resource information management, including global search for a video data, QoS control, synchronization.
- *Integration*: linking surveillance video data with other systems such as [9].

The Media Controller system can run without the upper two layers, the global management layer and the integration layer, in the case of a small surveillance application which does not require any integration. This hierarchical structure provides functional scalability to the Media Controller system.

3 Management of Distributed Video Data

In the Media Controller system, a node named AVDS manages distributed video data and resources. AVDS has the following features.

3.1 Global Video Data Management

AVDS manages the location and attributes of all video data, and works as a server that enables global search by various keys. Only pointers to the video data and their attributes are collected from MC-Ls, and the data itself is not collected. The following search keys are supported by Media Controller.

- Time-stamp: (each video picture is given a time-stamp)
- Camera identifier: (without specifying location or type of recording)
- Spatial information: each video picture is given information on the angle and location of the camera.
- Alarm: (which triggered the event recording of a video data)

Since AVDS manages the above attributes of each video data item, it can realize the global search of distributed video data, regardless of the recording type or location of the recording node, and the timestamp and camera status attributes are useful when a video data is used together with other systems. In this paper, we focus on the management with respect to the timestamp.

In order to efficiently manage video data, Media Controller uses a simple data structure called a "time-window" for storing video data information.

Data Management using Time-Window. Time window is used to describe information with a recorded portion of video data. Pictures produced from the same camera during a certain period, called video segment below, are managed together by one time-window data. It includes the following elements.

- T_s: time-stamp of the oldest picture in that video segment.

- T_e: time stamp of the latest picture in that video segment.
- r: frame rate during the time from T_s to T_e
- other attributes common to all pictures in the video segment, such as camera identifier, picture size etc.

With the time-window, video data is easily searched for both as a streaming sequence of pictures, and as a picture in a separate set of pictures. When video data is searched for as a video segment or streaming sequence, it is searched for by the time-window unit. When video data is searched for as a picture or a still image (frame), the time-stamp of each frame in a video segment can be calculated from T_s, T_e, and r in the time-window. Other attributes that are common to all frames in the same video segment are checked with the search conditions only once, without checking through all pictures in the video segment one by one.

In order to efficiently manage video segments which are continuously updated into the latest period, like "continuous recording" in MC-Ls, T_s and T_e can be specified using *relative time* to the present time, whose designated time gains automatically as time goes on. Therefore, if both T_s and T_e are specified using relative time, the length specified by T_s and T_e stays unchanged, but the interval defined by T_s and T_e slides on as time goes on. There is no need to update management information.

With these two types of time, relative time and absolute time, both dynamic video segments like those created by continuous recording and static video segments like those created by event recording are managed uniformly and efficiently.

3.2 Resource and QoS Management

In large-scale distributed surveillance systems, not only data but various resources are distributed in many places. By collecting resource information and controlling resource usage, the following advantages can be obtained.

- QoS control of video data retrieval:
  In order to achieve end-to-end QoS control or guarantees of video data transfer, resource reservation within a single node is not enough, and system-wide, cooperative reservation is required. Centralized management of all the resource information in the system is suitable for that purpose.
- Support for the system operations:
  Global search for available resources in the system and network-transparent access to distributed resources are useful for the operation and maintenance of the system.

In the Media Controller system, AVDS collects resource information in the system. We have not developed the above features yet, and we are now studying the method of global QoS control and resource reservations for video data transmission. We also plan to use an ATM network to reserve the bandwidth for each stream, or to prepare middleware which provides guaranteed maximum delay for network transmissions[7, 8].

3.3 Synchronization

It is very effective in surveillance applications to display multiple video streams synchronously, for instance when multiple video cameras are used to watch certain related equipment together in a plant control system. Media Controller provides two types of synchronization, in accordance with the accuracy of synchronization required by applications.

Rough synchronization. AVDS arranges the time for transmission to start for each video data stream, so that pictures with the same timestamp arrive and are displayed at almost the same time. This arrangement is based on the current transmission delays and image decompression delays of each stream. For example, let the transmission delay of stream i be x_i, and the decompression (decoding) delay be d_i, then the timings to issue the instruction to start transmission for m streams are calculated as below. Assuming $(x_1 + d_1) \geq (x_2 + d_2) \geq \ldots \geq (x_m + d_m)$, stream 1 should be started first, and stream 2 should be started $(x_1 + d_1) - (x_2 + d_2)$ later, likewise, stream i should be started $(x_{i-1} + d_{i-1}) - (x_i + d_i)$ after stream $(i - 1)$ is started. Here d_i is the constant parameter of the decoding device, and x_i can be measured or estimated from the running video stream transfer.

This synchronization method is easy to implement, but accuracy of synchronization is relatively low, because transmission delays can fluctuate dynamically, and arrivals of the command to start transmission can be delayed and vary from stream to stream. However, as a small difference in output timings between streams does not matter with the human eyes, this synchronization mechanism works well with low overhead, as long as the system is not heavily-loaded.

Strict Synchronization. There are some cases where exact synchronization is required. For these cases, Media Controller provides another synchronization method which buffers data from each stream and waits for all the pictures which must be synchronized to arrive. This method can synchronize video streams to the accuracy of a frame, but the transmission delay for each stream increases because of buffering.

4 Implementation of Media Controller

4.1 Prototype of Media Controller

We developed a prototype Media Controller system to evaluate its architecture, functions, and performance. Figure 3 shows the hardware components of the prototype system.

It has two MC-L nodes, one MC-C node, one AVDS node, and one user-interface (UI) node that controls the system interactively. We used a 100 Mbit Ethernet network, instead of ATM, as we have not developed the resource reservation features yet, and also due to the high availability of 100Base-T. All nodes are commercial personal computers with 200 MHz Pentium CPUs.

Each MC-L has two video cameras, and 2 GB SCSI hard disk drives are used in the MC-Ls, one per camera. All video data is digitized and compressed in realtime using Motion JPEG compression/decompression cards. The video data outputs from MC-C are overlayed to the console of the UI node by a superimposer device.

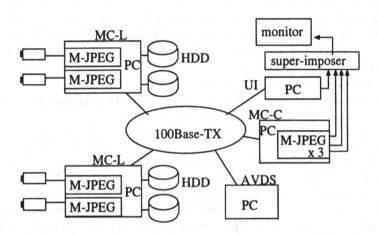

Fig. 3. Prototype of Media Controller

The prototype realized the following functions of Media Controller system.

– Recording and transmission of video data (corresponds to data acquisition layer and data transmission layer), which includes
  • Realtime transfer of live/recorded video data of 9 Mbps Motion JPEG. Data-rate can be configured for each camera as an system parameter, and frame-rate can be reduced for each transfer dynamically.
  • 3 types of recording for each video camera (continuous, event, timelapse)
– Global video data management
  • Global search of video data by the time, camera, and alarm.
  • Rough synchronization of multiple video data streams.

4.2 Design Characteristics and Evaluation

Software Design Characteristics. The software implementation of the prototype system has the following characteristics, which are necessary or effective for industrial surveillance applications.

– Realtime recording and realtime transmission:
  In order to ensure realtime recording and transmission without losing data, we used an original method for efficient realtime disk access [2], which, in

summary, accesses successive frames together at a time to reduce the seek actions of a disk drive.

– Software component blocks for the pipelined stream transfer:
We have implemented two types of software component blocks, the *bucket* and *pump*, for constructing a video stream transfer. The bucket is an abstracted buffer where video data is stored and retrieved, such as memory buffers, encoder devices, decoder devices, certain areas of a disk drive, and network devices. The pump is a process that transfers video data from one bucket to another. These software blocks work concurrently to make a pipeline, and abstract the differences of hardware or types of I/O devices, providing flexibility and extensibility.

– Video data format including extra attributes:
Each picture data is given time-stamp and camera information, in order to be used together with other types of data easily.

Evaluations. Though we have only just implemented the prototype and not started detailed evaluation yet, the prototype works fairly well as we intended at the beginning.

Because of the efficiency of the distributed recording architecture, the prototype smoothly retrieves and plays back 3 streams simultaneously with high quality (9 Mbps). Each MC-L can simultaneously record with two cameras, each one consisting of 3 types of recording (continuous, event, timelapse) without loss of data, and at the same time, it can simultaneously transmit 3 streams of 9 Mbps video data, or 6 streams of 5 Mbps to the MC-C. Notice that these performance are obtained using normal personal computers and hard disk drives (instead of using RAID).

With respect to the rough synchronization of multiple streams, we tested the synchronization of 3 streams and observed a small synchronization gap of approximately within several tens of milliseconds under a slight load, showing its effectiveness as a simple synchronization method.

5 Summary

In this paper, we introduced the Media Controller system and its video data management method. Based on the requirements and characteristics of industrial surveillance applications, Media Controller has a distributed recording architecture and a hierarchical software structure, which provide efficiency and scalability to the system. Media Controller uses a data structure called a "time-window" to manage distributed video data segments. The time-window structure is especially effective for such data as the recording the latest period from a video camera.

The prototype showed that the architecture and functions of Media Controller work well with high quality video streams, using inexpensive hardware. We are now developing the next version of the prototype system, and plan to design and evaluate its functions including QoS guaranteed transfer, linkage with other systems, and spatial search for video data.

Acknowledgments

The authors would like to thank the referees for their helpful suggestions.

References

1. F. Matsuda, et al., "Multimedia Systems for Industry — disk access —", Proc. of the 1996 IEICE General Conf., D-331, pp. 119-120, 1996(in Japanese).
2. K. Sato, et al., "Multimedia Systems for Industrial Surveillance", Proc. of IS&T/SPIE's Electronic Imaging '97: Science and Technology(EI'97), Vol.3020, pp. 182-193, Feb. 1997.
3. A. Guha, et al., "Supporting Real-Time and Multimedia Applications on the Mercuri Testbed", IEEE J. Select. Areas. Commun., Vol. 13, No. 4, pp. 749-763, May 1995.
4. A. McLeod, "Digital Video Storage for Security Applications", Proc. Int. Carnahan Conf. on Security Technology, IEEE, pp. 382-387, 1995.
5. M. J. Cattle, "The use of Digital CCTV in an Airport Car-Park Application", Proc. Int. Carnahan Conf. Security Technology, IEEE, pp. 180-185, 1995.
6. F. A. Tobagi, et al., "Star Works – A Video Applications Server", Dig. Pap. COMPCON, Spring, pp. 4-11, 1993.
7. D. D. Kandlur, et al., "Real-time communication in multihop networks", IEEE Trans. on Parallel and Distributed Systems, Vol. 5, No. 10, pp. 1044-1056, Oct. 1994.
8. I.Mizunuma, et al., "Real-Time Communication in Plant Monitoring/Controlling Systems with ATM Networks", Proc. RTCSA'95, pp. 243-250, Oct. 1995.
9. H. Simakawa, et al., "Acquisition and Service of Temporal Data for Real-Time Plant Monitoring", Proc.Real-Time Systems Symposium, pp. 112-119, Dec. 1993.

Personalised News on Demand: The "HyNoDe" Service

*The HyNoDe consortium**

Abstract

This paper presents the approach of the ESPRIT project HyNoDe (EP 22160 - Hypermedia News on Demand) towards a novel Personalised News-On-Demand (PNoD) Service. The HyNoDe Service Personalised feature refers to customisation of the service to suit individual needs and preferences of the clients based on the information filtering concept. The service addresses all the involved phases in the news publishing industry, from News Capturing up to end-user News Presentation. The paper compares the HyNoDe service with existing ones, and presents the architecture of the service, and the HyNoDe modules and tools related with each of the news publishing phases.

1. Introduction

First we introduce the notion of personalised news-on-demand and give a short overview on existing PNoD systems. The rest of the paper is structured as follows: Section 2 presents the architecture of the HyNoDe system. In Section 3, the News-oriented procedures of the HyNoDe system are described. Section 4 covers aspects related to Quality of Service (QoS) and 5 presents the system/service management and user-oriented procedures. Section 6 contains implementation details and section 7 concludes the paper.

1.1 Personalised News on Demand

The advanced capabilities of the emerging information and telecommunication technologies that trigger the decrease in cost of software, hardware and telecommunication charges, as well as the increasing importance of quality information in modern economies are fundamental drivers of multimedia services [Anal94]. The combination of real time communication with sufficient bandwidth, guaranteed by the evolving network architectures [Ovum95], creates opportunities for a wide range of residential as well as business services and promises a new era in information acquisition and delivery through the provision of on-demand and personalised services. The *on-demand* mode enables clients to schedule the viewing time of programs and manage the information presentation by pausing, resuming and in general controlling the navigation in the information space. The *personalised* feature refers to customisation of the service in order to suit the individual needs and preferences of the clients. Personalising the service to meet the clients interests is

---

* The HyNoDe Consortium partners are: INTRACOM SA, Hellenic Telecommunications and Electronics Industry - IBM Deutschland Informationssysteme GMBH - COSI - ETNOTEAM, Multimedia Systems Institute of Creete (MUSIC) - University of Ottawa - Il Sole 24 Ore RADIOCOR Agenzia Giornalistica Economico Finanziaria - IMERISSIA S.E.A - TH Darmstadt.

based on the information filtering concept, where intelligent mechanisms are used to suppress automatically the delivery of information that does not directly interest the user.

1.2 Related Approaches

A number of news publishers have already provided electronic access to their news using WWW browsers over the Internet. Nevertheless, the majority of the existing applications lack the personalised and on demand features that give added value to the service, the latter mainly due to the limited bandwidth of the current packet WAN networks.

In the area of personalised systems, MIT's FishWrap (running at http://www.sfgate.com) provides the user with textual news stories that are personalised through categories and a fixed keyword list. The selection of stories is made by full text search and the user's personal newspaper is created at logon time. IBM's infoSage (recently cancelled) extended this system by a free keyword search at any node of the category tree and an optional e-mail delivery instead of Web access. This service adds the news notification process to news-on-demand. In the business news domain, the pushing kind of delivery has been established as the rule. PointCast (found at http://www.pointcast.com) concentrates on the delivery approach. It delivers hypertext news, sometimes containing images, and runs on Windows only. at PointCast restricts personalization to channel selection. InfoComm (http://www.infocomm.ca) provides only short textual ticker news, but supports various media from computer through fax to pager. Grayfire (http://www.grayfire.com) delivers categorised articles through e-mail or fax. Farcast (found at http://www.farcast.com) offers e-mail delivery based on categories, but also full-text search operations both in an online and an off-line mode (the agents doing the latter are called „droids"). It is noteworthy that all of the dedicated business news services concentrate on the individual news story and do not offer multimedia features in their news systems. This also holds for the agent based search and filtering approaches of the artificial intelligence community, e.g. [Mitc94], [Kara94]. Brown et al. describe in [Brow95] the integration of information retrieval with multimedia by allowing free text searches on subtitles. They do not use a thesaurus in this case, but apply character stripping to compute relevance. They apply only a retrieval model, just like Hyon et al. in [Hyon96], who support attribute-based searching on documents that can be prepared in various formats. They consider also editing of these documents, but they assume slow off-line processing using a storyboard-based approach. For continuous media support in NoD systems, Georganas proposes a dedicated continuous media server that supports QoS negotiation and synchronization in [Geor97].

To relate the distributed multimedia issues of HyNoDe to other approaches, systems of other domains like video on demand or multimedia mail were considered. The GLASS project implemented the authoring and distribution of multimedia contents [Glas96]. This work heavily used the multimedia standard MHEG and helped to prove and improve MHEG's appropriateness for storing and communicating multimedia data in distributed systems. In [Glas96] a distributed multimedia authoring environment is shown. It defines a proprietary authoring data representation and exports a textual MHEG notation. Within the BERKOM Multimedia Mail project (MMM), [Thim94] describes the dynamic composition of multimedia documents, controlled by the user's preferences in e.g. QoS or topic. Like HyNoDe, MMM has the notion of different user roles, but is oriented towards workgroups and computer experts in contrast to HyNoDe relying only on common users' expertise and end systems. Caloini et al. describe a news distribution system in [Calo96], which provides personalised views on documents by application of alternative styles that can be used with a document. Their approach is also used to handle errors. Martinez et al. use the concept of views into the hypermedia system [Mart96], too, with a support for non-unique addresses of referenced contents and follow-up documents. The CHIMP system [Selc96] introduces personalisation by using the concepts of filters. Filtering considers user access rights and end-system capabilities and provides user-dependent projections of each document.

2. The HyNoDe System

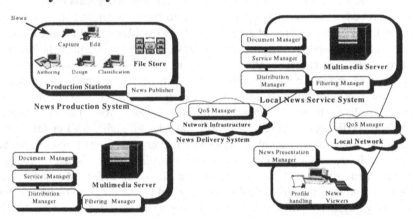

Figure 1. The Architecture of the HyNoDe System

The design goals of HyNoDe were mainly defined according the needs of professional news providers and 'professional news users'. This resulted in various constraints concerning delivery speed, target environments, user interfaces and system partitioning. The wide range of end user network connectivity options, ranging from dial-up modems to ATM leased lines, demands individually scaled news contents and system's knowledge about the users' connections. Despite just having PC type end systems, there is a wide variation in the power of the systems, which also influences the multimedia delivery to individual end users. The domain of news provisioning has time constraints, therefore a NoD system requires near real-time behaviour. Two content providers are among the partners of the HyNoDe consortium who have extensive experience both in the real-time and non-real-time news business. Their experience led to a system design for fast composition of news stories and delivery to the customers, while other parts of the system, which are not directly involved in the delivery of news, are not considered time-critical. The chapter 3 illustrates the HyNoDe system architecture following the path of a single news story Figure 1 depicts the architecture of the HyNoDe system. The basic components of it are closely related to the phases of production, storage, distribution, delivery and presentation of News Stories.

3. News Provision

This section presents the news life-cycle from collection through the News Production System up to delivery and presentation. The intermediate steps include content preparation, news authoring, filtering, news storage and retrieval, and news propagation through a hierarchy of servers.

3.1 Delivery Path

Content Preparation: A news story in the HyNoDe system is written using one of the tools that are common to the infrastructure of the news agency. The News Story is a multimedia document; it contains information in various formats such as text, hypertext, graphics, audio and video and also encapsulates any associations among them (e.g. synchronised play out of audio and video components). The author is required to move the news body (text, video, etc.) to a specific directory to make it available for further processing.

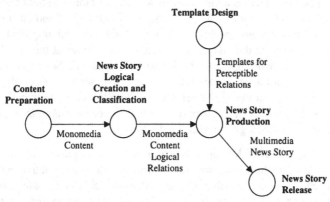

Figure 2: Data Flow Scheme in the News Authoring Process

Classification: Through this, the news story becomes available to the news editor. The editor is responsible for the classification of the story. The main issue of the editor's work, deciding the correctness and relevance of a story, is the same in the HyNoDe system. The editor is also responsible for making the logical connection between the central news story and additional information. In the HyNoDe framework, news stories are independent entities that do not require layout consideration or a positioning of the story with respect to other available stories. Rather, categories and keywords must be assigned to a news story, which are then used to decide whether the story fits to the profile of a consumer or not. To do this, the Media Relations Tool is developed within HyNoDe. At the end of this step, the news story, which is now attributed with category, keywords and additional content data, is passed on to the tool that assigns the layout.

Production: This layout tool is called the News Production (NP) Tool. The NP Tool allows the editor or the presentation designer to decide quickly for one layout from those templates that can present the news story and all the additional contents that have been assigned to the story by the editor. It is important for the content providers that no full-fledged multimedia editor is necessary for this step but that the connection of informative data and multimedia enhancement is made quickly. Although such a connection could also be made automatically on the fly, it was decided for the textual news stories of the HyNoDe system to involve human interaction. This is mainly because enforcing consistency of available templates and the selection made from these templates may not be accepted by the content providers at this stage. After the decision about the connection of template and contents, the NP Tool generates a multipart document for the transmission to the storage component.

Storage: The storage component is the Regional Multimedia Server, which stores all attributes of the news story, now aggregated in a news story descriptor, in a database and forwards this descriptor to the next component, the Filtering Manager. If that component determines that the story is relevant for at least one client, all of the files of the news story are copied into dedicated filesystems. While the Regional Multimedia Server is the only server involved in the current installations, the later installations of the system will be organized as a hierarchy of servers for more efficient distribution of the news stories, because in a real world situation the number of end customers of the HyNoDe system may become prohibitively large, and their geographical location is assumed to be arbitrary. In the hierarchical system, each server at the next lower level in the server hierarchy will appear to the Regional Multimedia Server as a client. Not all uploaded information is forwarded to the next server, but only the story descriptors. If the server at the next level decides from the set of known client profiles (a client may be a customer or another server) that the content is relevant, it requests the transmission from the Regional Multimedia Server.

Distribution: The Filtering Manager determines that a requested upload operation handled by the Multimedia Server was successful. In case that an active end-system is directly attached to the server, the Filtering Manager checks whether one or more customers interested in the news story are logged onto the system. Although the system is basically an unconnected service based on HTTP, so a connection monitoring application is running on each end-system indicating that a customer is currently using the service. This application maintains an open connection to the nearest server, through which notifications of newly arrived stories are forwarded.

Delivery: A notification indicates that the new story has been made available, and it provides a simple means (a hyperlink) to retrieve it. There is no fixed format for the notification, but an installation-defined template is available to construct the notification from the attributes of the incoming news story. This allows the decision about the amount of data to be made at installation time. In contrast to pure push systems, the customer is not only a consumer of the incoming news. News stories that are not immediately consumed are stored and not lost. A HyNoDe user can specify notifications for a subset of all interesting news by specifying multiple profiles.

3.2 Filtering and Searching

News Filtering

The HyNoDe News Filtering Manager automatically selects information based on a user's pre-recorded profile such as preferences of interests, with the main objective to eliminate news stories that do not directly interest the user. The filtering scenario of the HyNoDe system can be described as a set of independent operations which are depicted graphically in Figure 3.

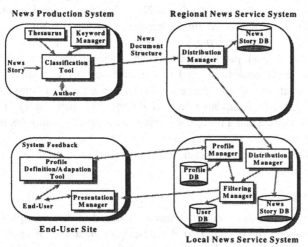

Figure 3. Filtering Process

At the end-user's site, the user formulates interactively a preference vector that reflects the his current interests. He specifies his preferences about news authors, categories, language, content provider and a set of weighted keywords may be specified. This preference vector is then transferred to the Profile Manager at the Local NSS that the end-user is connected to and is stored in the Profile Database in order to be used during the filtering process.

At the news production phase, the Classification Tool generates the essential filtering information, which is included in the News Story Descriptor and in general is considered as the identification vector of the news story. Specifically, documents are characterised by a set of attributes provided by the news author which include category, language and content provider. In addition, a document is described with a set of keywords which is also provided by the news story author. For each keyword, the number of times the keyword appears in the document, as well as the number of times the keyword appears in all the documents of its collection, are used to determine the importance (weight) of the keyword in the collection (category or sub-category).

The above suggest that a dictionary for keywords (handled by a Keyword Manager), exist for every information provider, language and sub-category. The system maintains statistics for the number of occurrences of each keyword in the documents of every collection. These statistics are updated as new documents are coming into the system in order to be able to compute similarities. Finally, a thesaurus mechanism is associated with the dictionary. It allows the news author to find synonyms for infrequently appearing keywords, and to find specialisation for very frequently appearing keywords.

This information is then communicated through the Distribution Manager to the corresponding Regional NSS and on a similar way to the Filtering Manager of the Local NSSs.

As soon as a new story arrives at the Local NSS, the filtering mechanism is triggered. For every news story published, the Filtering Manager matches the identification vector associated with it with the preference vectors of the clients and assigns a similarity measure to each piece of information. If the similarity measure estimated is within the correlation bounds defined in the specific profile, then the document is considered to be interesting for the client. In such a case the list of unread news stories of interest of the user is updated to include the new news story. If the user is connected to the system and a preference exists in his profile stating his interest in real time notification of arrival of important news stories, a message is sent to the Presentation Manager of the end-user, triggering a notification to the end-user about this new news story.

News Searching

All information needed to execute the search is stored in the descriptors. Due to the fact that descriptors are copied into the descriptor database of each local server, searching can be done locally. As a search result the user gets a list of matching news, sorted by a ranking value. The matching news are not necessarily available on the local server. This fact is visualised by using a traffic light paradigm. A flag with the colour red, yellow or green at each matching news describes the quality of service which may be achieved when accessing the news. Green means that the nearest copy of the news is located either on the local server or on another server in the upper server hierarchy which supports guaranteed QoS for the connection to the user. A yellow flag indicates that the news may be located on a system where the connection to the user may provide an acceptable throughput to transfer the news but there is no guaranteed QoS for the connection to guarantee a smooth replay of continuos media. A red flag indicates that the news has first to be downloaded onto the local server before it is accessible for the user. This happens when the news is only stored in the archive.

HyNoDe provides two search interfaces. One very simple interface uses HTML forms. The user can fill in some simple search parameters to create a standard query. The receiving CGI script transforms the form parameters into a search query conforming to the HyNoDe search language. More elaborated queries can be created using a search language. The language consists of any number of triples combined by the boolean operators AND and OR. A triple consists of a matching condition, a flag, and a value. The matching condition names one of the attributes, a relation (is equal, is unequal, is smaller, is greater, ...) and a value. If a triple is marked „mandatory“, the condition has to be valid for a news to be selected. If the flag is „optional" this triple is only used for the ranking part of the search query which follows the selection part.

4. Management

4.1 HyNoDe QoS Manager

The HyNoDe project invests special effort to serve the quality of service requirements of the NoD service. The QoS Manager gives emphasis to the *Application Level QoS* since there are constraints to the realisation of a complete *Network Level QoS Manager*. Most constraints are imposed by the capabilities of the underlying communication protocols and the existing network infrastructure.

The HyNoDe *Application Level QoS* is based on the concept of the *alternative media*. A HyNoDe News Story may have for each monomedia component more than one monomedia alternatives at different quality levels. A News Story is delivered to the end-user with the best possible quality based on the available media alternatives in the news story, the end system capabilities, the user preferences, and the network connection capabilities.

For each of the monomedia components, a list of alternatives is kept in the *Descriptor* of the HyNoDe News Story. In addition, the QoS Manager has access to information about the hardware and software capabilities of the client's terminal as well as his/her preferences for the media types he wishes to be presented with. In this way, the QoS Manager decides which of the available alternatives match the user's needs and preferences best and delivers an appropriate news story instance, the *on demand* view of the story for the particular client. In this way, the application can flexibly serve requests for limited delivery delay (interactivity level), duration of the overall presentation and preferred presentation format.

Regarding the Network Level QoS, for each of the media components, the requirements for real-time transfer and sufficient network bandwidth are examined and serviced appropriately. Based on the media types supported, the QoS Manager decides the network interfaces and resources it should allocate to deliver the new story to the client. In this way, continuous data can be downloaded to the client using RTP/UDP whereas discrete data can be delivered using TCP/IP over low-speed reliable channels.

4.2 System/Service Management

Some additional issues are considered to be important for the successful commercialization of the HyNoDe News on Demand application. These issues are the management of the HyNoDe system and the management of the NoD Service offered.

The HyNoDe system management monitors and controls the access and the status of the system. This includes status/performance monitoring, system configuration, network management (connectivity control, and network configuration), fault detection, and error/crash recovery.

On the other hand, the HyNoDe service management monitors and controls all the services offered by the system. This includes: access control, server security, copyright protection, subscription processing, accounting and billing. Electronic payment methods had been considered but postponed until a standard in this area has been settled.

5. Conclusions & Future Plans

Innovative information services with strong need of multimedia content, as the Personalised News-on-Demand Service, prove to be a driving force for the deployment of Advanced Communication Infrastructures. Key components of the Service are the NPS, the NSS and the NDS. Within the NSS, the Filtering manager adds an important new asset, compared to other existing News Services.

ESPRIT Project HyNoDe (Hypermedia News on Demand) aims to design and develop an efficient and market-oriented PNoD service. This will be achieved through the adaptation and integration of existing IT components needed in the phases of news authoring, storage and delivery. Appropriate mechanisms will be developed enabling news filtering according to each end-user profile.

An Initial prototype implementation of the HyNoDe system has been completed and a trial phase with real world actors is expected to begin before the end of June 1997. Consequently, it is expected that the presentation of this work at IDMS97 will also include the evaluation results of the trial phase.

The programming languages used in the implementation are C++ and Java. The database for storing the Descriptors of the News Stories, the user data, the user profiles and the accounting info has been built using the *Relational Database Management System DB2* 2.1.2 from IBM, while all the database accessing functionality has been implemented using the *ODBC* interface. The HyNoDe Server operates on Solaris 2.5.x and AIX 4.1.4 (or later) platforms while the clients need to be capable of running Java-enabled browsers.

The following components will be added in the future: A profile optimiser will support the user in the management of his profiles. This component will register and evaluate the users behaviour to propose modifications of the profiles with the goal to adapt the profiles to the changing interests of the user. Additionally to the simple and the enhanced interface to searching a third one will be implemented. It will provide a graphical means to formulate complex search queries.

References

[Anal94]   Multimedia in Telecoms, Simon Norris, Susan Ablett, Analysys Publications 1994.

[Brow95]   Automatic Content-based Retrieval of Broadcast News, M. G. Brown et al., ACM Multimedia 95, Nov. 5-9, 1995, San Francisco

[Calo96]   Extending Styles to Hypermedia, A. Caloini et al., Proceedings of ICMCS 96, pp. 417-424

[Geor97]   Multimedia Applications Development: Experiences, N. D. Georganas, Journal of Multimedia Tools and Applications, Vol. 4, No. 3, May 1997

[Glas95]   GLASS: A Distributed MHEG-Based Multimedia System, Helmut Cossmann et al., COST 237 Proceedings, 1995

[Glas96]   GLASS-Studio: An Open Authoring Environment for Distributed Multimedia Applications, Brian Heumann et al., IDMS'96 Proceedings, 1996

[Html96]   HTML 3.2 DTD

[Hyon96]   A Web-based Archive and Retrieval System for Multimedia Production, J. Hyon et al., Proceedings of SPIE Multimedia Storage and Archiving Systems Proceedings, 1996, Boston, pp. 23-32

[Kara95]   A Computational Market for Information Filtering in Multi-Dimensional Spaces, G.J. Karakoulas and I.A. Ferguson, in Working Notes of AAAI-95 Fall Symposium Series on AI Applications in Knowledge Navigation and Retrieval, 1195, pp. 78-84

[Mart96]   Generic Information Engine for Use in Information System Servers, J. M. Martinez et al., Proceedings of SPIE Multimedia Storage and Archiving Systems, 1996, Boston, pp. 278-289

[Mheg96]   ISO/IEC 13522-5, ISO/IEC JTC1/SC29/WG12: Multimedia and Hypermedia information coding Expert Group (MHEG), 1996

[Mitc94]   Expirience with A Personal Learning Assistant, T. Mitchell et al., Communications of the ACM, 37(2):80-91, 1994

[Ovum95]   Applications for the Superhighway: Market drivers, John Moroney, John Matthews, Ovum Reports 1995

[Selc96]   CHIMP: A Framework for supporting Distributed Multimedia Document Authoring and Presentation, K. Selcuk et al., ACM Multimedia 96, Nov. 18-22, 1996, Boston

[Thim94]   A Mail-Based Teleservice Architecture for Archiving and Retrieving Dynamically Composable Multimedia Documents, Heiko Thimm et al., COST 237 Proceedings, 1994

[Wolf96]   Resource Management for Distributed Multimedia Systems, Lars C. Wolf, Kluwer, 1996

Broadband Video Conference
Customer Premises Equipment

H. Brandt, C.Tittel, P.Todorova, J.J. Tchouto, M.Welk
GMD-FOKUS, Hardenbergplatz 2, D-10623 Berlin

Abstract: The introduction of advanced multimedia services in ATM based B-ISDN networks requires a new generation of Customer Premises Equipment (CPE), e.g. workstations, for interactive user dialogue in a multimedia service provisioning environment. One of the most attractive services to be offered in the future is Broadband Video Conferencing (B-VC) where audio, video and other media can be exchanged in real time among a group of users in two or more locations in the network.

1. Introduction

This paper proposes a software design approach for a video conference end system [1] serving as Customer Premises Equipment (CPE) for audio and video connections used in a Broadband Video Conference. The video conference application will need a video conference server, which will not be described here, and a video conference end system for every involved user. The end system is under development and will be used as CPE in the ACTS-INSIGNIA project.

The objectives of the ACTS-INSIGNIA project are to define, to implement and to demonstrate an advanced architecture integrating Intelligent Networks (IN) and B-ISDN networks. The project will provide the technological basis for setting up switched connections between a selected set of National Hosts, and to provide advanced communication services such as broadband virtual private networks (B-VPN), interactive multimedia services and combination between these [2].

Today many video conference systems exists. So the first question to decide about the CPE architecture was "What are the properties of the available systems?". A great number of applications are designed on the base of IP like the Multimedia Collaboration (MMC) [3]. These systems provide hardware independence and good quality video and audio. On the other hand it is hard to investigate the benefits of B-ISDN service classes, because service classes are currently not supported by IP over ATM implementations. Other products like the K-Net CellStack Video [4] transfer video and audio directly over B-ISDN networks. They support signalling and traffic management as well, so the benefits of ATM for video conferencing can be evaluated. But it is nearly impossible to adapt the signalling protocol implementations to the project objectives. Also hardware independence can not be achieved. This is true for video conferencing system like FVC PictureTel [5] as well. The video conference application described in this document was designed to meet the INSIGNIA project goals, to make replacement of hardware specific and protocol specific parts as simple as possible, to achieve high quality audio and video and to evaluate the practical advantages of B-ISDN networks.

2. Reference Configuration

The video conference CPE will be used as one part of the service architecture defined in the ACTS-INSIGNIA project. The reference architecture is shown in Fig. 1.

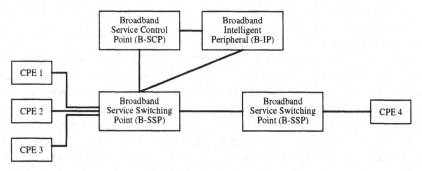

Fig. 1. **INSIGNIA reference configuration**

The video conference CPEs are in charge of video and audio transmission. They must communicate with the service logic located in the network. The B-SSP receives the service request from the CPEs, triggers the service logic and handles routing issues. The logic for video conferencing is placed within the B-SCP and B-IP. Together they handle the validation of user requests, database access and initiate service actions.

Although the CPE is able to transmit audio and video by it's own, the video conference application needs interworking between all mentioned components.

3. System Overview

The Broadband Video Conference end system is a high end SPARC workstation designed for native ATM applications. It runs an operating system with real time capabilities and is equipped with special video and ATM hardware. The system runs a number of processes, which together form the video conference end system application [6]. The general architecture of the video conference application is show in Fig. 2.

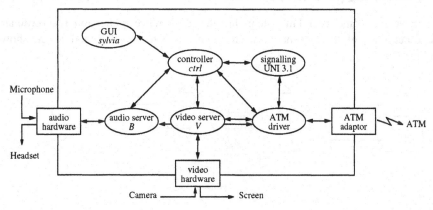

Fig. 2. **Application architecture**

There are six different processes show in Fig. 2:

- the graphical user interface *sylvia*,
- the audio server *B*,
- the video server *V*,
- the controller *ctrl*,
- the signalling stack and
- the ATM driver.

The audio server has two tasks: it takes audio input from the microphone and transmits it to a number of network connections and it takes input from network connections, mixes the input, buffers it to compensate for delay jitters and puts it on the end system speaker or headphones.

The video servers takes video frames from the video hardware and transmits the frames to a number of network connections and it takes video frames from network connections and displays these frames on the workstation monitor. The video and audio server are designed to be reusable for other applications.

The controller is the central part of the video conference application. It interfaces with all other parts through well defined interfaces and controls network connections. The controller is the only part of the application allowed to communicate with the signalling stack of the system.

Although video conferencing may be done with the controlling processes alone (ctrl provides a TCL based interface which may be used to implement a text based user interface), the graphical user interface is used to simplify usage of the video conference software. The user interface also provides additional functionality.

The signalling stack manages ATM connection setup and tear down. It interfaces only with ctrl, which allows easy replacement of the signalling stack [7],[8].

The ATM driver is part of the operating system and provides an AAL5 interface to the ATM network.

All parts of the Broadband Video Conference software communicate through well defined protocols. This allows simple substitution without breaking the other modules.

4. Controlling process (ctrl)

The controller is the central module in the video conference application. It co-ordinates the functioning of all other modules. The connections to the other modules are shown in Fig. 3.

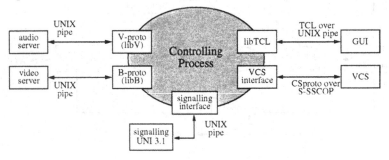

***Fig. 3.* Controller connectivity**

The controller is organised as a single threaded event-driven program. Events are generated by messages arriving on the different connections and by timers.

At start-up ctrl tries to connect the video and audio server. If both are successful connected, video and audio capabilities are requested from the servers and stored internally. Next the signalling entity is contacted and initialised. Three ATM Service Access Points (ASAPs) are created on which connection requests for controlling connections, audio connections and video connections are received.

4.1 Transport Protocol

The controller uses a simplified version of SSCOP [9] as transport protocol. The protocol is stripped down to the bare function of a reliable transport protocol over AAL5. Unsupported features are f.e. resynchronisation, but current estimates have shown, that this simplified protocol is sufficient to transmit control messages.

4.2 Access Protocol

The controlling process can be accessed via Tool Command Language (TCL) [10]. The TCL interface to the controller allows the implementation of different kinds of user interfaces. One is a text based interface, which is directly implemented in TCL. It has low requirements on performance, but is not as intuitive as the graphical interface. The graphical interface is implemented in Java and interfaces with ctrl through the text based interface.

There are two ways of interaction between an application and control functions: by means of commands and events. Commands are issued by the user (or the graphical user interface). Events are created internally by the controller. In the case of an event the controlling process tries to find an event handler (a TCL procedure) with a predefined name. If no such handler exists, the event is ignored. If a handler exists it is called with event specific arguments. Event handlers are always called synchronously, i.e. when the controller is not executing a command.

5. Audio Server

The audio server connects multiple bi-directional network connections to the audio device. It's design is loosely based on [11] and [12], but has some unique properties. The audio servers structure is shown in the following Fig. 4.

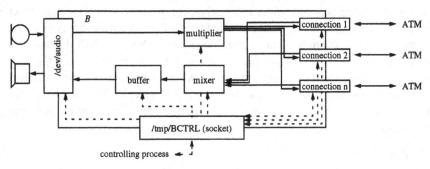

Fig. 4. **Audio server structure**

The Sparc audio hardware consists of an A/D converter, which converts the audio signal, received from one of three possible input ports into digital audio samples and a D/A converter, which converts audio samples into an analog audio signal, which may be played on up to three output ports. This device is driven by the operating system's /dev/audio driver. The hardware supports a number of different audio formats, like 8-bit A-law and μ-law [13] or 16-bit PCM, which can be selected dynamically.

The audio server currently supports only one audio format which must be chosen at server start up and cannot be changed at run-time. It expects all other servers to support exactly the same format. The transport protocol is a greatly simplified version of the real-time transport protocol [14].

Incoming audio samples are first buffered by the audio driver. The size of this buffer may be configured and is currently set to 20 msecs. This is a compromise between performance and delay. Therefore each 20 msecs the audio server receives one buffer with audio samples from the driver. It adds the necessary protocol header and transmits the buffer to each of the network connections. This event is also used to move the same amount of samples from the playout buffer to the audio device output queue. In this way receiving and sending of audio samples becomes synchronous.

The receiving side is much more complicated than the sending side. The central part of the receiver process is the playout buffer. It is a ring buffer, which is able to hold up to 32 kbyte of sample data. The need for such a buffer arises, because many components in the audio chain between the microphone and the loudspeaker induce delay and delay jitter. As experience has shown, most delay jitter is introduced in the paths from the audio server input (reading from /dev/audio) to the output of the AAL5 frames on the network. Measurements have shown, that the maximum interval between two frame may be as large as 400 msecs, values between 0 and 40 are common. This jitter is introduced by the scheduling of applications, streams modules and drivers and can only be prevented by pushing the audio module nearer to the hardware. Therefore the receiver must buffer some amount of samples to compensate for the jitter. This amount is currently a fixed value of 20 msecs. This gives a total delay of 60 msecs not counting the network delay.

Mixing is done directly into the playout buffer. An insertion point is computed based on the estimated time difference between sender and receiver and the playout delay. If this point falls outside of certain ranges, the incoming packet is dropped. Else the samples in the packet are converted to linear encoding and added into the playout buffer. The time difference between sender and receiver is estimated every 2 seconds, or when a series of 3 packets falls outside the allowed time range. The algorithm is adaptive.

The audio server is controlled via a controlling socket. This is a UNIX domain socket, which is posted into a well known place and can be opened by the controlling process (*ctrl*). The controlling process may announce new connections to the server, remove existing connections and control port and audio parameters like loudness and balance. The server informs the controlling process about certain events: changes in the port configuration and closed connections.

6. Video Server

The video server is a real-time process and is able to receive and send multiple video streams at the same time. The server structure is shown in Fig. 5.

The current video server implementation is based on the Parallax video card. A server for the SunVideo card is in preparation. The Parallax video card consists of a frame grabber with X11 extension and a hardware implemented compression algorithm. The Parallax frame grabber supports PAL, NTSC and SECAM video signals and offers a maximum picture size of 768x567 pixels for PAL/SECAM format and 640x603 for NTSC format. The frame rate for this picture size is 25 fps for PAL/SECAM format and 30 fps for NTSC format. The digitized picture can be compressed by using the hardware implemented MJPEG algorithm. The quality of the compressed image can be changed on request from the controlling process. The Parallax XServer stores the compressed image in a shared memory buffer provided by the video server V and returns a header to the image. The header contains width, height, size and quantisation factor of the compressed image. The necessary protocol header is added and the buffer is transmitted to each of the network connections.

On the receiving side the video server gets a whole image - including the header - from the video transport protocol. Because the Parallax video card can only show one image at one time on the screen, it is necessary to make the receiving side of the video server a little more complex than the sending one: A buffering mechanism and a simple state machine is used to allow multiple connections to use the same output device.

The video server is controlled via a controlling socket. This is a UNIX domain socket, which is posted into a well known place and can be opened by the controlling process (ctrl). The controlling process may announce new connections to the server, remove existing ones and control port and video parameters like frame rate, video size and video quality. The server informs the controlling process about certain events: changes in the port configuration and closed connections.

7. Graphical User Interface

The Graphical User Interface (GUI) *sylvia* simplifies the usage of the Broadband Video Conference application. It provides the user with an (almost) intuitive interface, so that the user doesn't need to learn the text based language offered by the controller. Additionally the GUI hides implementation-details and provides enhanced functionality.

To interact with the audio, video and the video conference server, *sylvia* is connected through a TCL interface to the controlling process (*ctrl*). The interface is based on a UNIX-pipe.

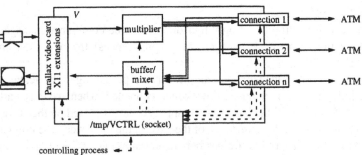

Fig. 5. **Video server structure**

Sylvia is written in Java. The decision for implementing the GUI in Java is based on the following features.

- Java is a modern object-oriented programming language.
- A Java interpreter is available for almost all platforms, so *sylvia* should be portable to PCs and other platforms.
- Java allows multi-threading.

7.1 Message passing

The communication between different parts of the user interface and between the interface is based on the exchange of messages. Messages are passed between objects in a hierarchical way. Every message not destined for the object that originated the message is send to the parent of this object. The parent object informs other objects about the received message. With this mechanism it is possible to run multiple dialogues in parallel.

7.2 Additional functionality

The video conference service provides basic functions, which are not always intuitive for the user. *Sylvia* tries to hide the elementary scenarios within high level scenarios which are easier to use. One example is the creation and establishment of a conference. For this example the video conference application needs to invoke two different scenarios:

- Create Conference to define the conference and
- Establish Conference to start the defined conference.

The graphical user interface combines this two scenarios into one, which allows the user to define a conference and automatically establish it. The basic scenarios are still accessible.

Sylvia also allows the user to store and retrieve the preferred configuration. One example is the invitation handling. For every invitation to a conference *sylvia* provides the user with the choices to accept it automatically or to interact by means of a dialogue.

Audio preferences can define default audio setting like loudness, balance a.s.o for every new connection, video preferences can do the same for video parameters like picture size or frame rate. Layout preferences give the user total control over the desktop by saving default colour, size and position of different windows.

7.3 Dialogues

Sylvia consists currently of only four dialogues (the conference manager, the conference creation, conference management and preferences), the local video and the remote videos.

The Conference-Manager shows all information pertaining to the actual active conference [15]. This is the main window, which is opened, when *sylvia* is started. It allows the user to invoke all the basic and complex scenarios (based on the current context).

The local video displays the video of the user on the local side. The user can edit audio- and video-parameters like frames per second and loudness with a toolbar. This local video is always needed with the Parallax based video server.

There is one remote video window for each conferee (except the locale one) in the active conference. Every video has a toolbar for editing audio- and video-parameter like frames per second and loudness.

On the Conference Creation window the user can create a new conference out of the user-list, the group-list and the permissions. The conference record is built locally and sent to the conference server when the user presses the create-button. If this action is successful, the newly assigned conference identifier is shown in the number field.

With the management-option the user can edit and delete conferences. The dialogue is the same as in Conference Creation except for the number-field which always contains the conference identifier.

The Preference-dialogue allows the user to define default values for audio-, video- and layout-settings, to defines pathnames to ctrl and the servers. It also offers access to the extended functionality (invitation handling etc.).

8. Experimental Results

The experiments shown in the following section contain results of measurements done during the integration phase of the INSIGNIA project. Because the first trial of the IN-SIGNIA project is still ongoing, more results will be available till autumn. The first experiments presented in this paper are concentrated on the quality of the video transmission and on the utilization of the B-ISDN network.

One of the aspects we were interested in was to investigate the use of the provided B-ISDN service classes and the traffic management operations. For this experiment we connected two CPEs through a private network and inserted background traffic in the private network. Then we started the video conference application on both CPEs and used two different methods of transmitting the video frames:

1. For the first experiment an Unspecified Bit Rate (UBR) connection was used to connect both CPEs. This is the service class used by "IP over ATM".
2. For the second experiment a Constant Bit Rate (CBR) connection was used to connect both CPEs. Also the traffic shaping option was used on the CPEs.

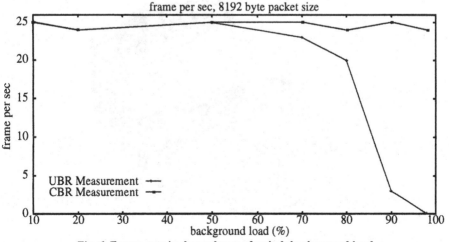

Fig. 6. **Frame rate in dependence of switch background load**

For both connection types, the receiving CPE counted the correct transmitted video frames as well as the interarrival time between two correct frames. Fig. 6 shows the resulting frame rate on the receiving CPE. The experiment shows, that the behaviour of both connection types is very similar as long as the background load on the link stays beyond a certain degree. If the UBR traffic type is used, cell and frame loss cannot be avoided, but the point at which frame lost occurs depends on the quality of the private network equipment. The result of this measurement can be summarized as follows: B-ISDN service classes should be used for transfer of video data if there is a high background load on the involved network or if the properties of the network equipment are not known.

The second experiment compares the received frame rate with the number of active conferees in a conference and the size selected for the received video frames. As shown in Fig. 7, the frame rate drops for each additional active conferee as well as for each larger video sizes. Further investigations are needed on this topic.

9. Conclusions and future work

First experiments with the multimedia CPE described above in a real environment have been done within the INSIGNIA Project trial in spring 1997. The CPE was used as an end system on an integrated IN/B-ISDN architecture providing video conferencing for real users.

The current implementation of the CPE supports a good quality video conference. Depending on the size of the video, more conferees could participate in a conference without losing too much quality. Performance bottlenecks, like the operating systems or introduced delay, were found and solutions will be adopted to the video conference system. The application proved to be efficient to demonstrate the objectives of the INSIGNIA project, it will be used in the second phase of the project as well. Main issues for

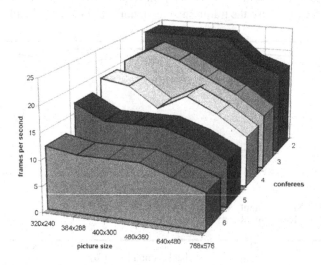

Fig. 7. **Frame rates in dependence of picture size and number of conferees**

future work are performance optimization, signalling enhancements and useability improvements. It is also planned to make this system available for real users in the second phase of the project.

10. References

[1]   H.Brandt, C.Tittel, P.Todorova, J.J.Tchouto,M.Welk, "Broadband Video Conference Workstation", presented as a poster to EuropIA'97, Sixth International Conference on the Application/Implications of Computer Networking in Architecture, Construction, Design, Civil Engineering and Urban Planning, Edinburgh, 2-3 April, 1997

[2]   INSIGNIA Project Description (URL: http://www.fokus.gmd.de/nthp/insignia/entry.html)

[3]   BERKOM-Teleservice Multimedia Collaboration (MMC) (URL: http://www.deteberkom.de/projekte/texte/MMCe.html)

[4]   Product Portofolie K-Net (URL: http://www.k-net.co.uk/products/prodintr.htm)

[5]   First Virtual Corporation (URL: http://www.fvc.com)

[6]   H.Brandt, C.Tittel, P.Todorova, "Broadband Video Conference in an Integrated IN/B-ISDN Architecture", accepted for presentation to NOC'97 European Conference on Network & Optical Communications, Antwerp, Belgium, June 17-20, 1997

[7]   ATM Forum UNI Specification, Version 3.1

[8]   ITU-T Rec. Q.2931 "B-ISDN-DSS2-User Network Interface Layer 3 Specification for Basic Call/Connection Control"

[9]   ITU-T Rec. Q.2110 "B-ISDN - ATM Adaption Layer - Service Specific Connection Oriented Protocol (SSCOP)"

[10]   John K. Ousterhout, "Tcl and the Tk Toolkit", Addision-Wesley, Reading, 1994

[11]   H.Schulzrinne, "Voice communication across the Internet: A network voice terminal", Technical Report TR 92-50, Dept. of Computer Science, University of Massachusetts, Amherst, Massachusetts, July 1992

[12]   T.M. Levergood, A.C. Payne, G.James, G.W. Treese, and L.C.Stewart, "Audiofile: A network-transparent system for distributed audio applications", Technical Report 93/8, Digital Equipment Corporation, Cambridge Research Lab, Cambridge, Massachusetts, June 1993

[13]   ITU Recommendation G.711 "Pulse code modulation (PCM) of voice frequencies", 1988

[14]   RFC 1889, H.Schulzrinne, et al., "RTP", January 1996

[15]   H. Hussmann, F.J Herrera Galvez, R. Pasquali, P.Todorova, V.Venieris, F.Zizza, "A transnational IN/B-ISDN integrated network for the provisioning of multimedia services", submitted to ECMATS'97, 2nd European Conference on Multimedia Applications, Services and Techniques, Milan - Italy, May 1997.

Radio Webs -
Support Architecture for Mobile Web Access

Reiner Ludwig, Norbert Niebert, Raphaël Quinet

Ericsson Eurolab Deutschland GmbH,
Ericsson Allee 1, 52134 Herzogenrath
e-mail: {eedrel,eednni,eedraq}@eed.ericsson.se
Fax: +49 2407 575 400

Abstract: The speed of Internet access over GSM can be improved by optimizing the performance of TCP/IP connections over the air link, and by using smart proxies for HTTP transfers or other high-level protocols. The goal is to improve the performance as perceived by the user, which implies the minimization of the end-to-end latency. This document describes the setup of GSM data connections and points to a number of possibilities to improve its basic performance. It then takes a more detailed look into WWW access and discusses how a double HTTP proxy can improve the web access over a (very) slow link. This double proxy is made of two parts: a small and simple HTTP proxy on the mobile station and a more complex proxy serving several users from a specific Internet Cellular Access Gateway. The users on the mobile station can configure their usual WWW browser to use the proxy on their computer - the mobile proxy - for accessing Web pages and everything will work as before, except that the overall performance should be considerably better.

Keywords: TCP, HTTP, GSM, performance, protocol optimization, WWW, proxies, proxy pair, distillation, caching, pre-fetching.

1. Introduction

There is an increasing demand for accessing information while on the move. This is most easily achieved by using wireless networks, particularly the ubiquitous GSM. On the service side the TCP/IP protocol suite and applications installed on top of it have clearly taken the market and created a common service platform. As the Internet was initially designed and implemented for a fixed network infrastructure, the involved protocols are tailored to an environment where terminals communicate via links with relatively steady characteristics but may experience congestion in the network. When using GSM networks, however, some of these protocols run rather inefficiently and some even fail completely. This is due to various assumptions about network loss rates which do not apply to wireless links.

The single slot GSM bearer provides a relatively low bandwidth to the mobile data user at a relatively high cost. Combined with the poor protocol performance mentioned above, this leads to a usually low perceived quality of Internet connectivity over GSM. This paper discusses the issue of increasing performance of Internet protocols and WWW access in particular over GSM networks. This means, among other things, trying to achieve higher throughput and lower delays for the mobile user.

2. General Background

The current situation for Internet access over GSM is illustrated in Fig. 1.

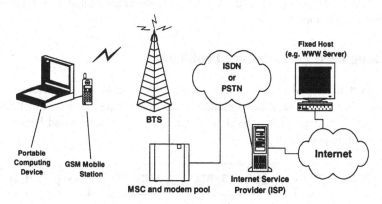

Fig. 1. Internet Access over GSM Today

The GSM subscriber uses the GSM mobile station as a wireless modem to dial into their Internet Service Provider (ISP) through the public telephone network. A link layer protocol such as PPP is used between the portable computing device and the Internet Service Provider. This setup is used today but offers no enhancements to protocols and suffers from call setup delays of 25-30s. The next logical step is therefore to provide direct Internet data access without involving a modem pool thus avoiding slow analog signaling. The general architecture for the direct Internet data access is shown in Fig. 2.

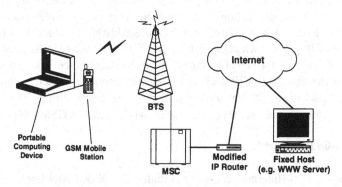

Fig. 2. Basic Architecture for Direct Data Access

Notice that in Fig. 2, three parts of the original picture are missing. Firstly, there is no third-party ISP needed —though still possible— since the mobile user can access Internet hosts directly. Secondly, there is also no need to go through a public telephone network. This improves the user's service, and also adds value to the operator's network: it is now possible for him to provide an Internet access service to GSM subscribers. Thirdly no modem pool is involved and an all-digital connection can be achieved. However, there are still problems to be solved. The most important of these is the poor performance still experienced for the mobile user due to inherent TCP/IP and HTTP protocol problems. These problems are explained in more detail in the following sections.

3. Recognized Performance Problems

It is important to realize why Internet access over GSM is considered to be suboptimal. This section identifies a number of the most important issues that contribute to this fact. Fig. 3 gives an overview of the involved protocols and between which communicating peers they span:

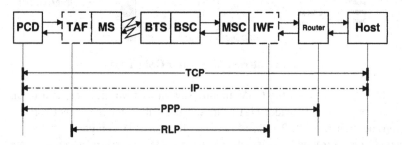

Fig. 3. Protocols and location of their entities

The central goal is to establish a TCP connection between the Portable Computing Device (PCD) and the Internet host and by this to allow the transport of application data in both directions. The underlying IP is just responsible for delivering IP packets from the source to the destination. The Point-to-Point protocol (PPP) provides the necessary negotiation of link options in the link establishment phase as well as the encapsulation of IP packets when the link is open for data transport. As shown in Fig. 3, PPP packets are used for signaling during the link establishment and configuration, as well as for the transport of IP packets and again when the connection shutdown is signaled. The packets are all sent serially over the same physical link. The following paragraphs explain the weaknesses of each layer, when used in a GSM environment.

3.1 Low Radio Bandwidth

Typical fixed network Internet access is via dialup PSTN or ISDN lines, with bandwidths ranging between 28.8 Kbps to 128 Kbps. In contrast, the GSM single slot data bandwidth of 9.6 Kbps is rather low, providing at best a third of the fixed network equivalent.

3.2 TCP Specific Problems

The Transmission Control Protocol (TCP) is perhaps the most important of the Internet protocols, since the majority of Internet applications make use of it. Unfortunately TCP does not adapt well to wireless networks. This is largely because TCP attributes all packet loss to *congestion* in the network, and not to lossy links or links with variable delays caused by link layer retransmission. Thus, should TCP experience packet loss or time-out, it invokes a congestion control mechanism such as slow start or exponential timer backoff. While these mechanisms are suitable in the case of network congestion, they are totally unsuited to dealing with packet loss, where a far better solution is to retransmit the lost packets immediately. Reducing window sizes is really the wrong action in GSM with its long round trip delays.

3.3 WWW (HTTP) Specific Problems

World Wide Web access must be considered one of the primary user applications of today, and hence its associated protocol, HTTP, is one of the most important ones in the Internet. HTTP is a request/response protocol, in which each piece of data is requested individually by the client, and then returned by the server if access if allowed. HTTP/1.0 opens a new TCP connection for every request, which adds unnecessary overhead on all network types and causes particularly poor performance over wireless networks due to the above-mentioned problems with TCP.

In summary, Internet performance over GSM is considered poor for a combination of reasons, ranging from radio-link characteristics to application protocol problems. In most cases, however, some work has already been done, or is in progress, to try to overcome the problems. A survey of some of this work is given in the next section.

4. Current Approaches

4.1 Increasing the GSM Data Bandwidth

Two distinct approaches for increased bandwidth data services are being considered at present by ETSI for GSM Phase 2+. These are the High-Speed Circuit Switched Data service (HSCSD) and the General Packet Radio Service (GPRS). The former is a variant of the single-slot data service currently available in GSM in which multiple TDMA slots may be used between the BTS and a single mobile station, thus achieving data rates of at least 64 Kbps.

However, for variable and bursty data traffic sources, the above is not so suited. Instead, a packet-based service, GPRS, will be provided. This will allow non time-critical data to be transmitted with a "best effort" quality of service — i.e. packets are transmitted only if free slots are available. The maximum throughput supported in GPRS ranges from 14.4 Kbps (single slot) to 115 Kbps (if all eight slots are used).

4.2 Improving the TCP Throughput

End-to-End TCP Modifications/Enhancements.

These approaches include changes of TCP's retransmission policies, optimization of the segment size, use of information from the mobile network to aid handover [7] and extensions of the TCP protocol, such as selective acknowledgments [5] and explicit packet loss notification [6]. The main advantage is that these mechanisms are easy to implement and do not require additional resources in the mobile network. However, for some of the approaches, modifications on the side of the fixed host are necessary, making them useless with existing Internet hosts.

Performing Local Retransmission.

An additional link-layer protocol is employed to retransmit lost data on the wireless link only. The idea behind this approach is to shield the TCP hosts from the peculiarities of the wireless link. The use of the non-transparent GSM data mode as a bearer service for TCP connections can be considered as an example for this approach (with RLP being the link-layer protocol). Other protocols, like the Snoop protocol described in [11], take advantage of some knowledge of the TCP protocol and intercept certain acknowledgments from the mobile host in order to suppress unnecessary retransmission from the fixed network host. This avoids throughput degradation caused by competing retransmission of TCP and the link-layer protocol.

The Split Connection Approach.

This involves a mobile support agent (MSA), located in the mobile network, which handles the TCP connection with the fixed host on behalf of the mobile host. The TCP connection is terminated at the MSA, and a different connection (and possibly a different protocol) is used for communication between the mobile host and the MSA. Examples of this method are the I-TCP [8] and Mowgli [9] protocols. This approach can be quite demanding in terms of resource usage in the mobile network. However, it also promises the best throughput improvements and enables the use of advanced services not possible with the other methods (like e.g. automatic handling of transactions with a fixed-net host by the MSA while the mobile user is disconnected). Another important consideration is the loss of TCP's end-to-end semantics that is introduced by this approach. For example, one host can receive acknowledgments (generated by the MSA) for packets that haven't even arrived at the peer host. Some applications may need additional session-layer protocols to compensate for this.

4.3 Enhancing the HTTP Protocol

As was mentioned, a major problem with HTTP currently is the number of different TCP connections required. The recent RFC for HTTP/1.1 [10] addresses this issue by allowing the use of *persistent* connections. These are TCP virtual circuits, which may be used in a pipelined manner for multiple client requests and server responses. However the motivation in this new version is to reduce overall network congestion and does not particularly address the needs of wireless clients.

5. New Approaches for Radio Webs - The Internet Cellular Access Gateway

As can be seen from the above-mentioned approaches, an additional function is needed to separate the GSM specific radio behavior from the fixed network protocol suite. This can be achieved by establishing a protocol gateway, called Internet Cellular Access Gateway in the following text. While this gateway also undertakes TCP level improvements using a "split connection" as described above, we focus on functions specifically necessary for accessing WWW as a major application in Internet.

5.1 The Gateway Proxy

The Gateway Proxy is located in the Internet Cellular Access Gateway. It has a direct connection to the Internet and it serves all users for a given node of the GSM network. It has three main parts:

The cache. This is a "standard" cache, as provided today by several caching proxies. It need not be limited to HTTP: it can also cache files requested via FTP, gopher or wais.

The distillation engine, which downsizes images or other data transferred via HTTP. This can also include some pre-fetching logic that can improve the quality of the output of the distillation process.

The GSM access module, which connects to the Mobile Proxy over the air link using a stripped-down version of HTTP and performs other operations that are specific to the GSM network.

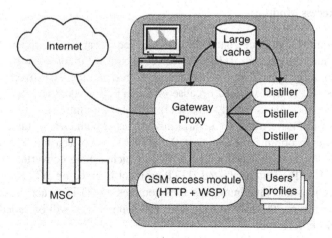

Fig. 4. The Gateway Proxy.

The Cache

The usefulness of a caching proxy grows with the number of users, as it is more likely that some user has already requested the page that a new user wishes to view. A cache in the Gateway Proxy reduces the delays between the Internet Cellular Access Gateway and the servers that can be anywhere on the Internet, because a page that has been requested once does not have to be transferred again. Various proxies exist today and can be used for HTTP, FTP, gopher and more. Almost any of them can be included in the Gateway Proxy with little modification. Among the modifications should be the ability to store both the original and distilled versions of an image.

The Distillation Engine

The distillation engine controls several distillers, which perform the computation-intensive task of reducing the size of specific data types according to the users' and terminals' profiles. A distiller can reduce the resolution or number of colors of images, or convert PostScript to HTML. For example, if the mobile terminal can only display 4 shades of gray and the user wants images to be no larger than 100x100 pixels, these requirements will be taken into consideration in order to transmit images that are as small as possible.

On average, large images can be reduced by a factor of 5 to 10 while still carrying enough information. Small images (2 or 3 KB) are usually transferred without distillation because the expected savings would not be big enough. Text such as HTML or plain text is compressed by a factor of 2 to 3, and decompressed by the Mobile Proxy. The total amount of data saved by the distillation depends on the requested quality and on the contents of the pages (varying amounts of text, small images, large images, sounds, etc.), but the savings are very often above 50%. For example, the homepage of our department is reduced from 157 KB to 21 KB, bringing the transfer time over a GSM link from 3 minutes to less than 30 seconds, including processing time.

The GSM Access Module

HTTP/1.0 requires a separate connection for each request and is inefficient because of TCP connection setup times. HTTP/1.1 solves a part of this problem by allowing a connection to serve more than one request, thus reducing the TCP overhead. However, there is still some overhead because of the HTTP headers and because even a "keep-alive" connection is usually closed after a short idle time (this is configurable, usually a few seconds). Since we control the proxies at both ends of the air link, it is possible to use an optimized version of the HTTP protocol (on top of the Wireless Sockets Protocol, an optimized version of TCP) which will be converted at both ends into standard HTTP. This optimized HTTP protocol is based on HTTP/1.1 with persistent connections and pipelining, and uses compressed HTTP headers, reducing the average request from 200 to 40 bytes. In the following text, it will be called "wireless HTTP" for easier reference.

The GSM access module can also allocate or de-allocate extra time slots on demand if the GSM network and the mobile station support this feature (i.e. HSCSD).

5.2 The Mobile Proxy

The Mobile Proxy is located in the mobile station. It should be as small and simple as possible, so that it does not use too many resources on the terminal, which does not have as much computing power as a non-mobile computer. Simple code will also be more portable to different architectures. Such a proxy provides several features that cannot be provided easily if only one proxy is used in the network:

- Pre-fetching data when the link is idle.
- User interface providing HTML pages for setting the user profile and other parameters related to the GSM link.
- Converting "wireless HTTP" into standard HTTP.

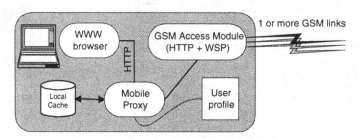

Fig. 5. The Mobile Proxy.

Local cache for pre-fetching data

The Mobile Proxy can pre-fetch data and keep a small cache in which some files can be stored before the WWW browser requests them. Since a normal GSM connection is circuit-switched (until extensions for packet switching such as GPRS are widely available) and the user is paying while the connection is open even if it is not used, the Mobile Proxy can take advantage of the open link to transfer some data that the user is likely to request later. The prediction is done by analyzing the last HTML pages transferred and keeping track of the links that have not been followed yet. The analysis of HTML pages and selection of data to be pre-fetched can be done by either of the two proxies, but the important feature is that the Mobile Proxy can store these files in a local cache once they have been transferred. These speculative transfers will be immediately abandoned in case of a user request. This is facilitated by the use of two proxies because we have full control over the link.

HTML user interface

The Mobile Proxy can also act as a small HTTP server, using HTML pages providing a simple user interface for controlling various parameters of the proxies and the GSM link. Instead of having a separate application for setting these parameters, it is easy to benefit from the user interface offered by the WWW browser. The user's profile, including the capabilities of the terminal and the quality desired by the user for images and other data, is modified using some HTML forms and is transferred to the Gateway Proxy which will give the corresponding instructions to its distillers. Other parameters can also be set, such as the maximum numbers of GSM time slots that the user is willing to allocate (and pay for), if the terminal and the network support this feature.

Wireless HTTP and the GSM access module

When the mobile station establishes a link to the Internet Cellular Access Gateway, its GSM access module will detect if it is talking to a peer which supports the Wireless Sockets Protocol. At a higher protocol level, the Mobile Proxy will check that the Gateway Proxy supports "wireless HTTP" and use it for better performance. If this is not available, the Mobile Proxy will revert to the standard HTTP protocol over TCP.

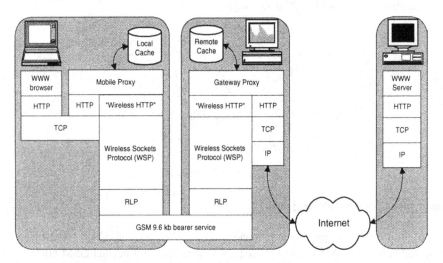

Fig. 6. Using efficient protocols between the client and the gateway.

5.3 Additional features

Concurrently with the pre-fetching between the mobile station and the Gateway Proxy, it could be interesting to do some pre-fetching between the WWW server (on the Internet) and the Gateway Proxy, provided that there is still some spare bandwidth on the Internet link. Data pre-fetched from the WWW server can not only be used to speed up the transmission to the mobile station when the data is requested, but also to give more information to the distillation engine about the images that are included in a page. If the size of the images is already known when a HTML page is passed to the distiller, this information can be included in the HTML code if it was not already provided. The benefit of this is that the WWW browser will automatically scale the (smaller) distilled images to the size of the original and thus preserve the intended layout of the page. If this information is not available before the page is sent to the mobile station and the HTML code does not include size hints for the images, then distilling the images and changing their size will also change the layout of the page

Distilled images have a lower resolution and/or a lower number of colors. Sometimes, the user may want to have all or part of the image at a higher resolution, in order to be able to examine some details that are not visible in the distilled version. This can be achieved in several ways.

The Pythia server in Berkeley inserts a small control button next to each distilled image, which can be used to retrieve more of the image. Although this is simple to implement and use, this can lead to unpleasant results on pages that rely heavily on the exact dimensions of the images and the spacing between them. For example, a graphical button bar in a page can have its layout destroyed if some extra links are inserted between the buttons.

Another solution is to use a (browser-specific) plug-in for loading the images. This plug-in requests the low-resolution image and displays it normally. If the user clicks on the image, the plug-in can request the original image or only a part of it if the user defines a rectangular area in the image. This can be done "in place", without needing to reload the page, since the plug-in controls the display of the image.

6. Glossary

BSC	Base Station Controller
BTS	Base Transceiver Station
DAN	Direct-IP Access Node
GPRS	General Packet Radio Service
HDLC	High-level Data Link Control
HSCSD	High-Speed Circuit-Switched Data service
HTML	Hypertext Markup Language
HTTP	Hypertext Transfer Protocol
IP	Internet Protocol
ISP	Internet Service Provider
IWF	Inter Working Function
LCP	Link Control Protocol
MS	Mobile Station
MSA	Mobile Support Agent
MSC	Mobile Switching Center
PCD	Portable Computing Device
PPP	Point-to-Point Protocol
PSTN	Public Switched Telephone Network
RFC	Request For Comments
RLP	Radio Link Protocol
SLIP	Serial Line Internet Protocol
TAF	Terminal Adaptation Function
TCP	Transmission Control Protocol
WSP	Wireless Sockets Protocol

7. References

[1] V. Jacobson, "Congestion avoidance and control", in SIGCOMM 88, Aug. 1988.

[2] P. Karn and C. Partridge, "Improving Round-Trip Estimates in Reliable Transport Protocols", in SIGCOMM 87, August 1987.

[3] T. Berners-Lee, R. Fielding, H. Frystyk, "Hypertext Transfer Protocol - HTTP/1.0", RFC 1945, May 1996

[4] R. Caceres, L. Iftode, "Improving the Performance of Reliable Transport Protocols in Mobile Computing Environments", IEEE JSAC, 13(5), June 1994.

[5] K. Fall and S. Floyd, "Comparisons of Tahoe, Reno and Sack TCP", Technical Report, Lawrence Berkeley National Laboratory, March 1996.

[6] B. Bakshi, P. Krishna, N. Vaidya, D. Pradhan, "Improving Performance of TCP over Wireless Networks", Technical Report TR-96-014, Dept. of Computer Science, Texas A&M University, May 1996.

[7] H. Balakrishnan, S. Seshnan, E. Amir, R. H. Katz, "Improving TCP/IP Performance over Wireless Networks", Proc. 1st ACM Conf. on Mobile Computing and Networking, November 1995.

[8] A. Bakre, B. R. Badrinath, "I-TCP:Indirect TCP for Mobile Hosts", Technical Report DCS-TR-314, Rutgers University, October 1994

[9] M. Kojo, K. Raatikainen, T. Alanko, "Connecting Mobile Workstations to the Internet over a Digital Cellular Telephone Network", Technical Report C-1994-39, Dept. of Computer Science, University of Helsinki, September 1994.

[10] R. Fielding, J. Gettys, et. al, "Hypertext Transfer Protocol - HTTP/1.1", RFC 2068, January 1997

[11] H. Balakrishnan, V. Padmanabhan, S. Seshan, R. Katz, "A Comparison of Mechanisms for Improving TCP Performance over Wireless Links", to appear, Proc. ACM SIGCOMM '96, Stanford, CA, August 1996.

[12] S. Spero, "Analysis of HTTP Performance Problems", available from http://sunsite.unc.edu/mdma-release/http-prob.html.

[13] D. Tennenhouse, et.al., "A Survey of Active Network Research", IEEE Communications Magazine, January 1997

[14] A. Fox, Eric A. Brewer, "Reducing WWW Latency and Bandwidth Requirements by Real-Time Distillation", Proceedings of the Fifth International World Wide Web Conference, Paris, France, May 6-10, 1996.

[15] V. N. Padmanabhan, J. C. Mogul, "Using Predictive Prefetching to Improve World Wide Web Latency" ACM SIGCOMM Computer Communication Review, July 1996.

[16] M. Mouly, M.-B. Pautet, "The GSM System for Mobile Communications", Mouly/Pautet 1992.

[17] ETSI TC-SMG, "Radio Link Protocol (RLP) for data and telematic services on the Mobile Station", GSM 04.22, May 1994.

An Approach for an Adaptive Visualization in a Mobile Environment

Luc Neumann[1] and Alberto B. Raposo[2]

[1] Computer Graphics Center (ZGDV e.V.)
Rundeturmstraße 6, D-64283 Darmstadt, Germany
neumann@zgdv.de
[2] DCA–FEEC–UNICAMP (State University of Campinas)
Caixa Postal 6101, 13083-970 Campinas, SP, Brazil
alberto@dca.fee.unicamp.br

Abstract. With the evolving availability of wireless communication services and of affordable mobile devices such as notebooks or Personal Digital Assistants, mobile computing is becoming widely accepted and applied. It is one further step towards the vision of information access for anyone, anytime, anywhere. Now, mobile multimedia applications are needed to make the vision more real. However, because of the narrow bandwidth of wireless wide-area networks and the limited resources of mobile devices in comparison to stationary systems, the handling of multimedia data faces severe problems. This leads to a need for effective solutions that enable the interactive handling with multimedia services even over a wireless link.

In this paper we present an approach to optimize the rendering process in terms of response time in a mobile environment that is composed of a mobile client and several stationary servers. We propose an architecture that adapts the rendering tasks to the available resource environment. The main idea is to use the knowledge about the application semantic data, the resource environment, and the user preferences to find a good trade-off between the limitations.

As a first experiment for the adaptation platform, we present a WWW rendering application using VRML 2.0 (Virtual Reality Modeling Language), which filters VRML scenes in order to render only parts selected by the user. It illustrates the handling of application semantic data that can be used to adapt the rendering process.

1 Introduction

The recent development in the number and quality of wireless communication services leads to an increasing interest in mobile computing. Therefore, it is one of the most rapidly growing areas of the computer industry. Mobile users equipped with portable computer such as Personal Digital Assistants (PDAs) or notebooks may access information or services anywhere and at anytime (e.g., tourist information about schedules, hotels or museums). The connection to a stationary server providing remote services and data can be established with a mobile phone or with communication facilities provided by the mobile device itself (like the Nokia 9000 Communicator). This scenario of a mobile application is illustrated in Figure 1.

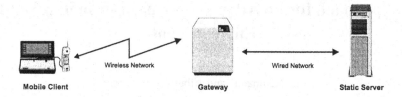

Fig. 1. Mobile Application Scenario.

Mobile users expect to handle all types of multimedia data with mobile devices as with desktop systems. The fulfillment of this requirement will strongly influence the acceptance of mobile computing technology. However, a mobile environment is more restricted than a stationary desktop environment. The restrictions are mainly caused by the limited resources of mobile devices and the narrow bandwidth of wireless Wide Area Networks (WANs). Therefore, it is a challenge to achieve a similar level of interactive multimedia data handling within mobile environments as in desktop environments.

Concepts for an optimized utilization of the available resources within a mobile environment are needed. One approach is to enable mobile applications to adapt to the current resource configuration. For example, to avoid overloaded mobile clients additional operations could be performed on resource-rich stationary servers (e.g. operations to convert the representation of a document). In this paper we propose an architecture that supports adaptation for a rendering application using metainformation such as information about the properties of resources and of data. Furthermore, a first experiment will be presented.

In the next section we will pinpoint the challenging problems associated to mobile computing and present approaches used to overcome them. After that, in Section 3 we will introduce an architecture distributing tasks among the mobile client and stationary servers. This architecture supports application adaptation applying resource dependent identification and delegation of subtasks to fulfill a task. The usage within a mobile rendering scenario and the possible delegation strategies of a rendering task are discussed. Then, the integration of the introduced architecture into the WWW is described. In Section 4 a first experiment for the identified architecture will be presented. It illustrates the usage of application semantic data to reduce the amount of data transferred and to reduce the utilization of the client resources. In Section 5 we come to the conclusions and highlight our future steps.

2 Mobile Computing

A mobile application accessing information in the WWW should be regarded as a distributed application, but this kind of application faces different challenges than an application using a wired stationary environment [2][10]. These new challenges are originated by the heterogeneous, limited and dynamic resources of the mobile environment.

The following properties characterize a mobile environment [7]:

1. Limited resources of the mobile devices in terms of storage, battery power, memory and processing power relative to non-portable devices.

2. Low bandwidth of wireless (WANs). Typical bandwidths of currently available cellular systems are 9.6 Kbps (GSM — Global System for Mobile Communication) or 19.2 Kbps (CDPD — Cellular Digital Packet Data).
3. High costs and low reliability (because of the frequent temporary disconnections) of wireless WANs.
4. Imbalance of resource availability between the mobile device and the stationary servers.
5. Users have no fixed location and can move during a connection.

In order to guarantee the success of highly interactive applications in a mobile environment, it is necessary that the requirements arising from the handling of non-textual bulky data (e.g., graphics, animation) can be fulfilled. This is one of the major challenges of mobile computing.

Comparing available data rates for wireless WANs with the required data throughputs for multimedia applications we can conclude that the available data rates are insufficient to fulfill the requirements of multimedia applications. Therefore, appropriate techniques (e.g., compression, progressive refinement, previewing) are needed for mobile environments. At lower levels of a mobile application Mobile IP [13], wireless TCP [1] and systems like ARTour [6] provide technologies to support mobility of a host[1] and to utilize the bandwidth efficiently (e.g., compressing protocol headers and data). At higher application–oriented layers, the following communication optimization techniques may be applied [15]:

1. *Data compression* increases the effective bandwidth compressing the information to be transmitted. Distillation [3] is an example of data specific compression developed in order to improve response times.
2. *Caching* uses local storage to reduce communication and to cope with voluntary and involuntary disconnections. Coda [11] is an example of a mobile file system that uses caching to reduce bandwidth requirements and to work disconnected. This technique, however, requires powerful storage resources in the mobile device.
3. *Prefetching* tries to anticipate data requests. This technique hides network latency, starting data transmission before the user requires it. However, if the prediction fails, communication and storage resources will be used unnecessarily.
4. *Data reduction* filters the information to be transmitted. This technique is normally used in the context of resource negotiation. Odyssey [11] is a system that uses this technique to provide "application-aware adaptation".

A successful solution should address the following mutually counterproductive aspects:

1. The transferred amount of data has to be as small as possible. This requires that parts of the application data should be processed and stored on the client side.
2. The use of local resources such as processing power and storage space should be the least possible. The client provides a presentation front end and communicates frequently with the server that process computing intensive parts of the application.

---

[1] The "Micro" mobility management such as handoff amongst wireless transceivers is solved by the wireless network system.

An appropriate trade-off between both aspects would overcome the most serious problems of a mobile environment, namely the narrow bandwidth and the client limited resources. In this paper we will introduce an architecture that supports adaptation at application level in regard to the available resource environment. It provides a good balance between the utilization of client resources and the required bandwidth. This allows to achieve a good quality of service within the limits of the available resources.

3 Adaptation Support Platform

In this section we present an architecture that adapts a rendering[2] application to the available resource configuration to optimize a rendering process within a mobile environment. The adaptation is achieved by the distribution of the rendering work (tasks) among the mobile client and the stationary servers using a proxy within the fixed network. The optimization will primarily focus on a fast response to the user even over a narrow bandwidth network. The main approach is to use information about the application semantic data (e.g., structure, table of contents) and the environment resources (e.g., available services) to distribute the tasks. Furthermore, user preferences (e.g., preferences between presentation quality and transfer time) will be used to control the mapping of the tasks.

The usage of a proxy between the client and the server to improve application performance is not new. They are used to reduce the bandwidth requirements [17], to split mobile applications (application partitioning) in a mobile and a static part [16], or to enable clients to negotiate the possible quality of a data representation [4]. The mobile file system Odyssey[11] provides the negotiation for resources without a proxy. However, this requires a specific server. Schill et al. [12] propose an architecture based on DCE with a station manager on each host and a domain manager to manage resource access in dynamic mobile environments. Here, the station manager mediates resource access. A key feature of our approach is a proxy using metainformation about available services within the fixed network and about the application data.

In the following, we introduce first the functional components of the adaptation architecture, discuss possible partitioning strategies for the rendering tasks, and present the establishment of the architecture within the WWW.

3.1 Defining the Scene

It is sure to say that a fixed separation of a mobile application in a client and a server component is feasible, but can overload the client or the server. If we think of the mobile application scenario mentioned above, where the mobile client retrieves some data from the stationary server and visualize it, then such a mobile application can roughly be separated into the following functional components.

1. Display service;
2. Application service (e.g., Rendering service);
3. Data management service.

---

[2] "Rendering" here means the generation of an image from a textual scene description of it.

This means that, for example, a rendering process is separated from the user interface and the data. This type of functional separation is already well known in the area
of business applications.

A straightforward approach to map this application functionality on a multi–server
environment can be to locate the display service on the client, the bulk of the application
service on a server, and the data management service on a third server. This offers different partitioning possibilities, that range from a thin up to a thick client respective server.
Furthermore, the display service can be further separated into two functional components. A presentation service which presents things to the user and an interaction service
which receives user input. Keeping this all together, a mobile application can be defined
as a composition of objects that work together as illustrated in Figure 2.

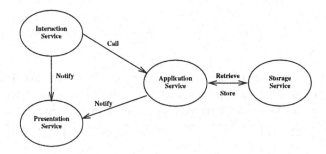

Fig. 2. Functional Decomposition of a Mobile Application.

This functional decomposition of a mobile application represents the basis for the
distribution of the different components and their tasks among the mobile client and the
available servers in the wired network. For the distribution of the tasks we propose an architecture with a so called *Resource and Task Manager (ResTaMan)*. The role of *ResTa-
Man* can be compared with a mediator which is situated between consumer and provider
processes. Within a rendering application *ResTaMan* distributes and controls the rendering process using metainformation. For example, it can use a filter which extracts relevant parts of a scene to distribute the rendering tasks. This is based on an additional
description about the structure of the scene. Figure 3 shows the architecture with the
ResTaMan. In principle, the *ResTaMan* receives a request including user preferences,
client profile and the parameters for the operation from the mobile client. The resource
information base contains the description about the resource configuration of the environment and the metainformation about the requested data. Based on this information
the most suitable resources are identified and the tasks are distributed. The viability of
this architecture is based on the assumption that a mobile application can be viewed as
a collection of objects working together.

3.2 Approaches to Partition Rendering Tasks

In the previous section the functional components of a mobile application and the architecture to distribute tasks have been introduced. However, the decision about the distri-

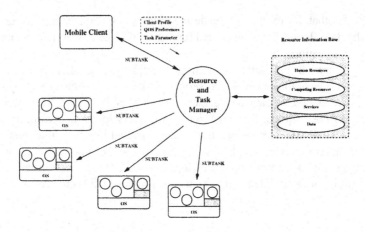

Fig. 3. Functional Components of the Adaptation Architecture.

bution of the tasks based on the application context is still an open issue. For the distribution we are focusing on the application service, that is the rendering service within a mobile rendering application. We are assuming that the presentation service to present images on the screen and the interaction service for the user input are available on the mobile client.

There exist a number of possibilities to partition the rendering work. The resulting task distribution intends to divide the work among several rendering objects to speed up the rendering process. In general, the partitioning of the rendering work can be related to the pixels (rendered image) or the scene content. In the following we introduce two possible partitioning strategies.

Quality Level. One can define different quality of service levels (QoS) that vary in the amount of work items for a rendering task. Then, the entire scene can be rendered in parallel with different qualities from several renderers. The different qualities can refer to:

1. Shading method (wireframe, flat, gouraud, phong, etc.);
2. Texture mapping;
3. Resolution of the image;
4. Level of Detail.

Scene Partitioning. In order to reduce the amount of work items for a subtask, the entire scene will be subdivided in smaller parts. These subparts are rendered in parallel and the results are merged. One obvious approach is to subdivide the pixels among the renderers. Then each of the involved renderer is responsible for a specific subarea of the entire image. If we are using the content of a scene, then we can identify following partitioning strategies.

1. Subparts of a static scene
   The scene is separated into different subscenes. Each subscene contains the objects

of a specific part. For example a scene can be separated into some background sub-scenes describing the surrounding panorama and one foreground scene containing the objects local to the viewpoint. This allows to render realistic panorama images on high end renderer in the fixed network and the foreground objects can locally be rendered on the mobile client. Furthermore, priorities could be assigned to different objects in the world. Then high level objects are essential to the scene and low level objects could be omitted.

2. Subparts of an animation.
The sequence of an animation can be subdivided into N subintervals. They are used by a pool of renderer in the wired network, to generate the subintervals of the animation in parallel. The generated image sequences are combined to a complete video stream and sent to the client. Certainly, one needs a video compression scheme to transmit the sequences. However, the latency time will be reduced even by high quality image sequences.

However, to enable the partitioning of rendering tasks additional knowledge about the structure of the scene is required. Within such a scene description the objects may have names, which depend on their role in the scene, or they are assigned to a specific quality level.

3.3 Establishment of an Adaptive Rendering Facility in the WWW Environment

In order to establish the introduced architecture in the WWW environment, we propose a solution that is based on two main points: the introduction of a semantic content header for the scene description, and the application of Java based Object Request Brokers (ORB)[3]. The ORB allows to integrate the adaptive architecture into the WWW. The Virtual Reality Modeling Language (VRML) [14], which is a file format describing interactive 3D objects and Worlds in the WWW, is used for the scene descriptions. The entire pipeline of the rendering architecture will encompass at least

1. a presentation facility (e.g., located on the mobile client),
2. different renderers on the mobile client and on stationary servers,
3. VRML–analyzer and distribution manager to define and distribute the different tasks, and
4. a composing and synchronization tool for the presentation.

The VRML–analyzer reads the semantic header and the VRML description. The header contains additional information about the content of the scene. It can be regarded as metainformation supporting the definition of tasks to be distributed. Together with an ODP-Trader, which is a yellow page service of the available resources, a good utilization of the rendering resources on the mobile client and in the fixed network can be achieved. Finally, an image composer and synchronizer are necessary to display the result. Figure 4 shows a separation of a rendering process into foreground and background scenes to illustrate the dataflow between the different components. In this case the metainformation describes the foreground and background of the scene.

---

[3] An ORB represents an object bus enabling objects to receive or send requests to local or remote objects.

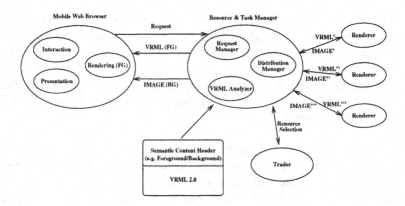

Fig. 4. Distribution of VRML Rendering.

The task for the foreground rendering is done on the mobile client and the background rendering is performed on one ore several renderers on the fixed network. The Request Manager receives the requests from the mobile client, and invokes the VRML analyzer and distribution manager. The distribution manager delegates and schedules the identified subtasks. It can be seen as a master component, that instructs various resources as workers to perform a subtask.

The usage of Java ORBs enables the Web Browser to gain access to arbitrary CORBA–based services. This allows the direct integration of computational services into the WWW. A Java applet can be downloaded to the Web client and serves as CORBA Client accessing services using the Internet Inter–ORB Protocol (IIOP). In our approach it contacts the *ResTaMan* object that represents a meta server from the viewpoint of the client. The entire rendering application consists of a set of interworking objects playing together through the ORB. For the transmission of images or image sequences we will use a stream oriented connection between the Applet and *ResTaMan* that may be established with an IP socket connection. Thus, for the bulk data transfer the stream connection is used and the control operations are transmitted by the IIOP. Figure 5 illustrates the integration of the proposed architecture within the WWW.

4 Object Extraction Method from VRML Scenes

As a first experiment for the adaptation platform presented in the previous section, we developed an application capable of selecting the elements (i.e., geometric objects, light sources, and cameras) of a remote VRML 2.0 world [9]. This application uses the *data reduction* technique to filter the data to be rendered. By presenting only the elements selected, the utilization of resources is reduced and the visualization can be done more efficiently. This filtering technique can be used to extract the relevant parts of a scene for the different renderers, as discussed in Section 3.

Figure 6 shows the strategy used in this application. It is executed in the client as a Java applet [5], downloaded with an HTML page.

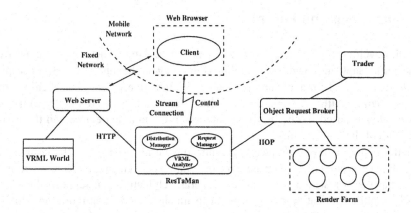

Fig. 5. Integration of the adaptive rendering architecture in the WWW.

The process begins with the establishment of a connection between the client and the application server. Then the URL of a VRML 2.0 world is send to the application server (arrow 1 in Figure 6). The application server will use Java resources to connect the server of the VRML file, read and parse it (arrow 2 in Figure 6). The first output of the application server is a small scene graph document, representing the hierarchy of the elements of the VRML world, which will be sent to the client (arrow 3).

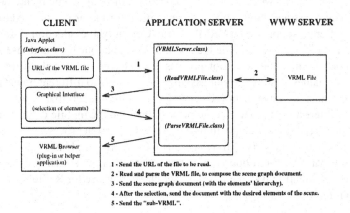

Fig. 6. Strategy to visualize only a subset of a VRML 2.0 world.

Based on the received document, the applet creates an interface adapted to the hierarchical structure of the VRML world. This interface allows the user to select the desired elements of the world. After this selection, a new document is sent to the application server, describing the selected elements (arrow 4 in Figure 6). Using this new document, the application server can create a new VRML 2.0 world from the original one, by extracting only the desired parts of it. This final "sub-VRML" world is then sent to the client (arrow 5), that visualizes the results using any VRML browser.

5   Conclusions and Future

With the development of enabling technologies such as wireless networking and various types of affordable mobile devices, mobile computing has begun to be widely accepted. This leads to a need for applications capable of interactively accessing, manipulating, and visualizing multimedia data in a mobile environment. They are mainly confronted with problems caused by the limited resources on the mobile devices and the narrow bandwidth of the wireless link.

In this paper we have introduced an architecture aiming to improve the visualization in a mobile environment. Furthermore the integration of such an architecture within the WWW has been presented. We defined a general architecture that supports the adaptation of a mobile application in regard to the available resource configuration. This will avoid overloaded mobile clients and allows to achieve the best quality of service within the limits of the system. The main approach is to use application semantic information, information about environment resources and user preferences to divide the tasks to be done.

In order to use this architecture for a mobile rendering architecture we presented different possible partitioning strategies which are related to the quality or the content of a scene. It depends on the goal of the user, which strategy might be used. However, to identify different quality levels or specific subparts of the entire scene additional information about the content of the scene is required. Therefore, as a first step, we have presented an application that is able to select parts of a VRML world and to create sub-VRML worlds. This sub-VRML worlds could be rendered on the mobile client and the stationary server. It demonstrates how application semantic information can be used to identify and handle different subparts of a VRML world. Currently, the selection is done by the user who should have some knowledge about VRML, but our plan is to use semantic headers which automize the extraction process. This header can describe different quality levels, give names to subscenes or assign priority levels to different objects.

At the moment we have achieved a simple adaptation level, that allows to render only parts of interest or up to a specific priority level on the mobile client. Due to the reduced resource requirements this improves the visualization process.

Obviously many issues are still remain open. Currently, implementations of further components to validate the proposed architecture are in progress. This mainly address the Resource Information Base and the task–oriented interface for ResTaMan, allowing to include preferences within a request.

Acknowledgements

This paper is a result of ProBrAl 002/94, a cooperation project between UNICAMP (State University of Campinas) and the Technical University of Darmstadt, sponsored by CAPES (Brazil) and DAAD (Germany). We deeply appreciate the support granted by these institutions. We would like to thank FAPESP and CNPq (Brazil) for the sponsorship of some authors. Special thanks to Dr. Léo P. Magalhães, Dr. Ivan L. M. Ricarte, José M. De Martino, Dr. Rüdiger Strack and Jian Zhang for helpful pointers and clarifying discussions.

References

1. A. Bakre and B.R. Badrinath. I–TCP: Indirect TCP for Mobile Hosts. In *Proc. 15th International Conference on Distributed Computing Systems (ICDCS)*, pages 136–143, Vancouver, British Columbia, May 1995.
2. G. H. Forman and J. Zahorjan. The Challenges of Mobile Computing. *IEEE Computer*, 27(4):38–47, April 1994.
3. A. Fox and E. A. Brewer. Reducing WWW Latency and Bandwidth Requirements by Real-Time Distillation. In *Fifth International World-Wide Web Conference*, Paris, France, May 1996. (http://www5conf.inria.fr/fich_html/papers/P48/Overview.html).
4. A. Fox, S. D. Gribble, E. A. Brewer, and E. Ami. Adapting to Network and Client Variation via On-Demand, Dynamic Distillation. *In Proc. ASPLOS-VII*, October 1996.
5. J. Gosling, B. Joy, and G. Steele. *The Java Language Specification – Version 1.0*. Java Series. Addison–Wesley, 1996. (http://java.sun.com/doc/language_specification/index.html).
6. IBM, WT Mobile Data Communication – ARTe, Heidelberg, Germany. *IBM ARTour – Technical Overview*, 1995.
7. L. Neumann, J. Zhang, and R. Strack. Evaluation of the Existing and Forthcoming Mobile Infrastructure. MOMID Deliverable 2, Computer Graphics Center, March 1996.
8. Object Management Group (OMG). CORBAservices: Common object services specification. Revised Edition 95-3-31, March 1995.
9. A. B. Raposo, L. Neumann, L. P. Magalhães, and I. L. M. Ricarte. Efficient Visualization in a Mobile WWW Environment. Technical Report DCA 001/97, DCA – FECC – UNICAMP, 1997. (ftp://ftp.dca.fee.unicamp.br/pub/docs/techrep/1997/index.html).
10. M. Satyanarayanan. Fundamental Challenges in Mobile Computing. In *15th. ACM Symposium on Principles of Distributed Computing*, Philadelphia, PA, May 1996. (http://www.cs.cmu.edu/afs/cs/project/coda/Web/docs-coda.html).
11. M. Satyanarayanan. Mobile Information Access. *IEEE Personal Communications*, 3(1), February 1996. (http://www.cs.cmu.edu/afs/cs/project/coda/Web/docs-coda.html).
12. A. Schill, B. Bellmann, W. Böhmak, and S Kümmel. System Support for Mobile Distributed Applications . *IEEE Workshop on Services in Distributed and Networked Environments (SDNE)*, pages 124–131, June 1995.
13. J. Solomon. Applicability Statement for IP Mobility Support. RFC 2005, Internet Network Working Group, October 1996.
14. *The Virtual Reality Modeling Language Specification – Version 2.0, ISO/IEC DIS 14772–1*, April 1997. (http://vrml.sgi.com/moving-worlds).
15. T. Watson. Application Design for Wireless Computing. In *Workshop on Mobile Computing Systems and Applications*, Santa Cruz, December 1994. (http://snapple.cs.washington.edu:600/wit/presentations.html).
16. T. Watson, B. Bershad, and H. Levy. Using application data semantics to guide system network policies. In *SOSP'95 WIP Session*, 1995.
17. B. Zenel and D. Duchamp. Intelligent Communication Filtering for Limited Bandwidth Environments. In *Proc. 5th IEEE Workshop on Hot Topics in Operating Systems (HotOS-V)*, Rosario Resort, Orcas Island, Washington, U.S., May 1995. IEE Computer Society Press.

Multimedia Client Implementation on Personal Digital Assistants

Markus Lauff and Hans-Werner Gellersen

TecO - Telecooperation Office - University of Karlsruhe
Vincenz-Prießnitz-Str. 1 - 76131 Karlsruhe - GERMANY
Phone +49 (721) 6902-69, Fax +49 (721) 6902-16
e-mail: {markus, hwg}@teco.uni-karlsruhe.de

Abstract. Their small size and ease of use makes Personal Digital Assistants (PDAs) very attractive for inclusion as clients in mobile multimedia applications. Alas, the cost for integration of PDA in distributed multimedia systems is high, as the resource limitations on PDA require client adaptation, and as software development on PDA is comparatively expensive. We present PocketWeb, a tool facilitating PDA integration in multimedia applications based on the web protocols while freeing application developers from PDA specific software development. We have applied PocketWeb for integration of PDA as multimedia clients in MILLION, a distributed multimedia system developed in an ESPRIT project.

Keywords. Multimedia Client, Personal Digital Assistant (PDA), Mobile Computing, Mobile Multimedia Applications, Multimedia Software Development.

1 Introduction

The new generation of small computing devices such as the Apple Newton MessagePad, the US Robotics Pilot or the Psion Series 3c are currently primarily used in personal stand-alone applications, for example as electronic filofaxes. Now, their small size and ease of use also renders them attractive and potentially useful as clients in networked applications in mobile environments. In fact, a range of applications have already been reported, demonstrating the useful integration of Personal Digital Assistants (PDA) in networked applications, for instance in the health care domain or in field engineering [1,2,3]. The integration of PDA in networked applications, though, remains difficult and expensive for three reasons:

- *Resource Limitations.* PDAs impose limitations on client functionality because of comparatively limited user interface, memory and computing resources. These limitations imply a high adaptation cost for adapting a client to the available resources, for example for development of user interface components that require less screen real estate.

- *Network-level integration.* The facilities for connecting to networks vary with available PDAs, but in general are rather low-level, thus implying additional cost for realizing the connectivity required for a PDA multimedia client.

- *Expensive PDA Software Development.* PDA software development is comparatively expensive for a number of reasons. PDAs in general come with their own development environments [e.g. 4,5,6]. These environments have not yet matured to the efficiencyand reliability available on standard platforms. Further, there is a lack of experience in PDA software development in general, and PDA software developers are hard to find. Last not least, software components developed for standard platforms are difficult to reuse or even to port to PDA, and reusable PDA software components are not yet available to any noticeable degree.

In this paper, we present the tool PocketWeb as a generic solution for integration of PDAs as multimedia clients in mobile multimedia applications. Pocket Web facilitates integration of PDA in applications based on the World-Wide Web protocols. From an end user perspective, PocketWeb is a web browser on Apple Newton MessagePad. From an application developers perspective, though, PocketWeb enables PDA integration in networked applications without requiring software development on PDA, thus reducing the cost of PDA integration effectively. Regarding client adaptation to limited resources on PDA, PocketWeb embodies a number of built-in adaptation strategies based on a thorough analysis of the design space for a multimedia client on PDA.

In the following section we will discuss resource limitations on PDA, and the options for handling them. This discussion leads to consideration of pre-processing of multimedia information for a PDA client. The discussion of the design space for a PDA multimedia client extends to consideration of network-level integration in section 3. In section 4, we will describe the PocketWeb tool and briefly compare it to related work. Finally, in section 5 we will describe the application of PocketWeb to integration of PDA in MILLION, a large distributed multimedia system for the tourism sector currently being developed in an ESPRIT co-funded project [7]. The paper concludes with a brief summary and outlook on further development.

2 Resource Limitations and Multimedia Preprocessing

Today's information system clients such as WWW browsers have lots of built-in modules to display the different types of multimedia information. Apart from the weak computing power, discussed later in this document, PDAs have several other restrictions that makes it difficult to present multimedia information.

These are:
- small display
- limited colour depth
- small memory
- weak computing power
- poor digital audio interface

2.1 Restrictions concerning the display of PDAs

The usual size of displays known from Desktop computers is between 14" to 21" and the size of the display from Laptop computers is about 12". Compared to theses values the size of PDA displays, from about 3" to 6", is very small.

Most of the information available on information systems are thought to be displayed on a display with at most 80 columns. It is no problem to reformat plain text to a smaller display, for example to 50 columns but this is not as easy for tables or multimedia information as pictures or video.

2.1.1 Tables:

Tables need a lot of space because they use additional characters to separate the single columns. This leads in almost every case to a larger width than the available 50 columns on a PDA.

There are several strategies to display a table on the small display:

1. two scrollbars

Usually the user has the possibility to navigate up and down through the document. In addition to this vertical scrolling a horizontal scrolling is used to view the table.

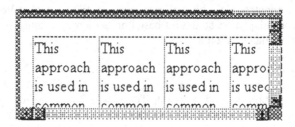

This approach is very inconvenient for the user because it is necessary to scroll for every row from the left to the right margin to view the table. Also it is difficult to overlook the whole table and to remember the position within the table. If the thirst column can be locked, that means the first column is not scrolled and always shown, it is easier to remember the position within the document. The disadvantage is that the space is always used for the first column.

2. reformat the single columns

This approach is used in common HTML viewers. The available space is divided into the number of used columns and the text is reformatted to fit into the column.

This approach is used in common HTML viewers	This approach is used in common HTML viewers	This approach is used in common HTML viewers	This approach is used in common HTML viewers	This approach is used in common HTML viewers

This above table is divided in 5 columns of almost the same width. The single columns are separated with one character as delimiter. So the average size of a single column is about 9 characters. This means that usually only a maximum of two words fit on a single line within the column. A lot of space is lost if only one word fits on the line because the second word is too long. In this case we can force a word break to use the remaining characters. This is shown in the fifth column. Without a forced word break, that makes the document harder to read, the length of the document growth very fast.

3. Don't use a table. Reformat the table to multiple sections.

The idea that leads to this approach is, that a table contains in every row the information that belongs under a given topic to a given point. This is not always applicable but will do for the most cases.

The following example shows the original table

Company	Modell	Price	Size
Apple	MessagePad 130	800 USD	90mm x 140mm x 25mm
Psion	Serie 3c	750 USD	110mm x 70mm x 20mm
US Robotics	Pilot	300 USD	60mm x 80mm x 15mm

and the decomposition to multiple sections

```
Company: Apple
Modell:  MessagePad
Price:   800 USD
Size:    90mm x 140mm x 25mm

Company: Psion
Modell:  Serie 3c
Price:   750 USD
Size:    110mm x 70mm x 20mm

Company: US Robotics
Modell:  Pilot
Price:   300 USD
Size:    60mm x 80mm x 15mm
```

This approach is useful for tables where it is not necessary to compare the different rows.

4. Use floating columns

This approach uses the basic characteristics of a table. A table is a formation with a separation into columns and rows. It is not essential that the columns always start at the same horizontal position. It only must be recognisable for the user to which column the current text belongs.

If the display supports colours a better separation of the columns can be reached by the use of different background colours.

This approach results in a very uncommon structure and is therefore only acceptable on very small displays.

5. Ignore the table

At first sight this approach may look mad or trivial but on a closer look it is at least reasonable to ignore the table construct in some cases. This approach can only be used in addition to another approach to display tables where it is not possible to ignore the table construct.

As We May Think	This article was originally published in the July 1945 issue of "The Atlantic Monthly". As Director of the Office of Scientific Research and Development, Dr. Vannevar Bush has coordinated the activities of some six thousand leading American scientists ...

This leads to the question which table constructs can be ignored? In many cases tables are used to have a nicer presentation of the information. But on a small display this brings the contrary result. Deciding where it is possible to ignore the table or not can only be done by an heuristic algorithm.

As We May Think

This article was originally published in the July 1945 issue of "The Atlantic Monthly".
As Director of the Office of Scientific Research and Development, Dr. Vannevar Bush has coordinated the activities of some six thousand leading American scientists ...

This example shows a typical case where tables are not necessary. The table is only used to separate headers in a pretty way.

In addition to these approaches it is always possible to reduce the size of the font to increase the number of characters per line but this isn't very useful on a PDA because PDAs usually still use small fonts.

2.1.2 Pictures:

The problem with the pictures is, that the differencebetween the PDA and a PC is even more worse than with the before mentioned tables.
The common resolution of a PC display is 800 x 600 pixel or more with at least 256 colours.
For example the Apple Newton MessagePad only has a monochrome display with a resolution of 320 x 240 pixel.

Nevertheless it is necessary to be able to display as many multimedia information as possible in any way because they may contain inalienable information.
On the other hand there are a lot of useless pictures in documents for example pictures displaying bullets or delimiters.

Displaying images can fail because of several limitations (computing power, memory, size of the display) this sections describes how to solve the problems with the small display the other problems are described later in this document.

In every case where the picture has a higher colour depth than the display the image must be dithered before it can displayed.

Similar to the procedure with the tables there are two main strategies to display images.

1. Use two scrollbars
The easiest way to display a large picture is to use a horizontal and a vertical scrollbar to view the picture.
There are two problems occurring with this solution:
• it can be difficult to imagine the complete picture
• the PDA must have enough memory to store the whole picture and enough computing power to scroll through the picture

2. Scaling the image
An obvious solution is to scale the image to fit on the display of the PDA. But depending on the original size and the content of the image the scaled image can be useless and for that reason a fallback to the previous approach should be possible.

3. Ignore the image
This approach is only applicable if the image contains no information, for example bullets or delimiters. A special algorithm has to decide if an image is worth to be displayed or not.

2.2 Restrictions because of the small memory and the low computing power

The common PDA does not have a secondary memory for the storage of programs and data. Therefore all programs and data must be stored in the primary memory. With a usual size of about 2 - 4 MByte the memory of a PDA is a very restricting resource.

2.2.1 Processing and displaying images

For image processing on the PDA there are three things to do:
1. load the image
2. decode the image
3. convert the image

1. load the image

As a matter of course we need the memory to store the image if we want to load it. But this is not always possible. Large images or truecolour images may need to much memory so they can not be stored in the original format. If this happens it is necessary to process the image on-the-fly.
Processing on-the-fly means that the small parts of the image are decoded and converted immediately when they arrive on the PDA and stored in a new format.

2. / 3. decode and convert the image

Decoding a picture from the GIF or the JPG format to a raw pixel-format is a hard job for a PDA. Particularly decoding a JPG image is currently not done in a few seconds.
In addition to the required computing power the image grows in size while it is decoded and more memory is necessary to store the decoded picture.
Therefore the image is dithered (reduce the colourdepth) and scaled before it is stored.

Conclusion loading, decoding and converting images

In many cases it is not possible for the PDA to load, decode or convert a image because of the limited resources. Why should not another machine process the image for the PDA?
The PDA is connected to a network with many powerful machines, why he should not use the computing power and memory from another machine to preprocess the information?
The following section describes several possibilities of preprocessing information for the PDA.

2.3 Preprocess information to eliminate limitations

A PDA is one of the smallest machines in the world of information. It is very comfortable to use a PDA to view as many information as possible. But that is the problem: „as many information as possible„. If the PDA is responsible for decoding, formatting and processing the requested information the limits are reached very soon.
Now it is conceivable that the PDA is able to give some jobs to another machine on the existing information network.
We can now build several models depending on what is done by the PDA and what is preprocessed from another machine. The machine that is responsible for preprocessing is viewed as a proxy.

2.3.1 The PDA only displays prebuild screens

The PDA send the request to a special PDA-Proxy. In addition to the request the PDA sends some information describing his possibilities, for example
- size of the display (320 x 240 pixel)
- resolution (50 dpi)
- number of colours (monochrome)
- digital audio capabilities (no)
- memory available (50 kByte)

The proxy now computes the page, depending on the possibilities of the PDA and sends the result as a bitmap to the PDA.

If the page contains large elements (tables, pictures) the proxy generates a preview and integrates this preview into the page. The user than is able to request the complete objects as an independent page.

Advantages:
- There are no more restrictions concerning the original format of the information. The limits are defined by the memory, the type of the display and the digital audio possibilities of the PDA. The computing power of the PDA is not the restricting factor.

Disadvantages:
- The PDA is not able to request new pages without a proxy
- A page is stored as bitmap and may need a lot of memory depending on the resolution and number of colours from the PDA

Operational area:
- PDA with low computing power
- low resolution and a small number of colours, because the size of the computed bitmap

2.3.2 The PDA can submit supported features with the request

In addition to the request the browser sends his capabilities concerning the different HTML tags to the proxy. The proxy examines the HTML document and converts from the browser unsupported elements to supported elements.

Example 1:
If the browser is not able to format tables the proxy can restructure the table to a preformatted textarea using monospaced fonts.

Example 2:
If the browser can not display inline graphics the proxy can replace pictures that are used as bullets in front of a list with a usual bullet symbol.

Advantages:
- The proxy can replace some elements that are not supported by the browser with an at least readable alternative.

Disadvantages:
- The proxy must parse and modify the HTML document
- If the browser uses heuristical algorithm to remove unsupported elements the information can be corrupted

Operational area:
- The browser on the PDA is able to format standard items and the proxy is especially written to enhance the functionality of the browser

2.3.3 The PDA can submit supported formats with the request

With the request the PDA sends the supported formats or desired properties (size, number of colours) to the proxy. The proxy examines the requested data and tries to convert the information to one of the supported formats.
If the proxy is able to recognise the identification of the browser sent within the header of the request, there is modification of the HTTP necessary. In this case the PDA needs not to sent the desired properties or formats, because the proxy knows these facts from the identification of the browser.

For example:
If the PDA only supports GIF images at a given maximum size and requests a page containing a bigger JPG-coded image, the proxy converts the JPG image to the GIF format and scales the image to the maximum size.

Advantages:
- The PDA is also able to work if the proxy is not available
- The proxy can use standard tools for image conversions the HTML document is not modified
- No modification of HTTP is necessary

Disadvantages:
- The proxy only enhances the functionality of the browser that are originally limited by the number of supported formats
- There is no support for preview items as described above

Operational area:
- The browser on the PDA is able to format standard items and the functionality should be transparent enlarged.

3 Network Connectivity for PDA

To retrieve any information it is necessary that the client is connected to a source where the information is stored, usually this is an information server. In many cases the information is spread over a lot of servers that are connected through a network.

There are several possibilities to connect a PDA to a network. Almost every PDA has other facilities to connect to a network.
For example the Apple Newton MessagePad can connect directly to an AppleTalk network. To use the Newton with an TCP/IP network a serial line to a PPP[1]-server is necessary.
A US Robotics Pilot has no possibility to connect to a network, the Pilot only has a serial connector that can be used to connect to another machine.

---

[1] Point to Point Protocol

A Psion Serie 3c also has only a serial connector with an overlying TCP/IP protocol to connect to a PPP-server or an IrDA infrared channel to connect to another Psion Serie 3c

At the moment there is no PDA that is able to support a direct TCP/IP connection to a network and so in almost every case a serial line is used to connect to an network provider.

With a PDA the usual mobile environment uses a modem to connect through the telephone line to a network provider.

If we use a mobile phone we can speak from a real mobile environment because there are no restrictions we have to make against the environment. Using a mobile phone is surely the most convenient solution but it is also the most expensive one.

A connection is either online or offline.

Online means that we have a physical connection and so we can retrieve new information from our information provider. The opposite to this is offline meaning that there is no possibility to retrieve new information.

A online connection always means maintaining a connection to an service provider and therefore producing costs.

But an access to an information system is only meaningful if the costs are proportionate to the outcome.

Because of this we have developed another interesting connection type besides online and offline that tries to use a few new elements to optimise the ratio between costs and outcome.

We name this approach „semi-online connection,, meaning that we try to present the user the information like having a real online connection.

3.1   „Semi-Online,, Connection

If the standard request is used every information is requested with an own request statement. If the user wants to load the following information he has to open a new connection to do the request.

In many cases it is possible to guess the information the user wants to have and to request the information in advance within the current connection.

Whenever it is possible to modify the original information, it is suitable to add extra information to group the information to blocks which can loaded altogether to reduce the number of necessary connections. For this we have introduced a new HTML tag „REQUIRE". The value of the tag is a page/picture that is required to view/follow the current page. With the help of this tag we can build a tree of information. The spanning tree is called a HTML package and is requested with a single request statement.

4 PocketWeb: A Generic Multimedia Client

As a result of our analysis we have developed „PocketWeb„ as an generic multimedia client on an Apple Newton MessagePad.

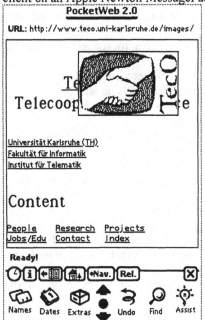

PocketWeb is based on HTML/HTTP and FTP and extensions developed through our before mentioned studies.

The current version 2.1 is available from the TecO.

Features of PocketWeb:

* support of most HTML 2.0 elements
* pictures
* forms
* flexible cache management
* supports download of complete HTML-packages

PocketWeb uses optimized methods for requesting, displaying and storing multimedia objects on the Apple MessagePad.

Some examples on how objects are requested or displayed have been explained in section 2,3. Now we will show how the cache is optimized for performance and an optimum memory usage.

Optimized cache architecture

Instead of storing the information as they arrive the information is preprocessed and than stored in an object oriented cache architecture. This means, that the information within the HTML page is bind to the procedure that is responsible to display it.

This cache architecture allows the information to be displayed very fast without wasting unnecessary memory.

A comparison of the memory usage of cached pages between PocketWeb and the commercial product Nethopper from Allpen has shown, that Allpen uses 2-7 times more memory for caching pages than PocketWeb depending on the number of used HTML elements within the page.

Optimized requests

In the current version PocketWeb works with a special subclient. This subclient is responsible for translating serial requests to TCP/IP requests and for preprocessing information as explained in section 2.

At the moment we are working on a new version of the subclient that acts as an usual proxy on the net and is able to preprocess the information as requested from the client.

In addition to the explained optimizations we are thinking about adding a quality of service module that has been developed at the TecO for mobile clients to access the net.

5   Application of PocketWeb

5.1   Use of PocketWeb Clients in MILLION

We have refined and applied the PocketWeb tool in the ESPRIT project MILLION. The objective of MILLION is to develop a coherent set of tools and services for municipal tourism, including tools for commercial transactions. In the cities of Venice, Bologna and Khania/Crete MILLION system pilots are being installed currently. In these pilots, PDA are integrated for mobile multimedia information access. In the pilot evaluation phase, PDAs will be handed out by tourism agencies to tourists for the duration of their stay. Tourists can download multimedia information according to their personal preferencesand the use the PDA in an offline mode as travel guide. This part of the MILLION project is called VIP and covers all areas of mobile access and transactions within MILLION using PDAs.

With the help of PocketWeb and the ideas we describe in this document we have reached the goal to use a PDA as a multimedia client without adapting and developing new specialized software for every new area.

5.2   Other Parts of MILLION

MILLION BOOK

MILLION BOOK is the fundamental part of MILLION and covers the integration and development of multimedia databases. A main goal of MILLION is the integration of existing databases of any type in a for the user transparent fashion.

MILLION GOLD

The Dutch company Digicash integrate their transaction and electronic cash payment system ecash in the MILLION project to allow secure transactions and payments on the net.

MILLION VOICE

This part of the projects studies the possibilities to add voice services to the MILLION system. For example one area for voice integration in MILLION is the access to the MILLION system through usual telephone using text to speech and voice recognition.

5.3   Further Applications of PocketWeb

Through the universal architecture of PocketWeb it can be reused in many areas to access multimedia information without developing new expensive software for PDAs.

A very early version of PocketWeb [8] has been successful tested within the ViaPerfecta project to access a route optimisation system through a PDA.

6   Conclusion

PDAs are in principal very attractive as clients in mobile multimedia applications with their major advantages being small size, ease of use, and comparatively low cost. Integration of PDAs in distributed multimedia systems, though, implies a range of problems: handling of multimedia in the context of limited resources, connecting

PDA at network-level, and, in order to solve these issues, expensive software development on PDA. We have presented a generic solution to these problems based on the tool PocketWeb. While PocketWeb is primarily perceived as web browser for end users, it actually facilitates integration of PDA in intranet/internet applications based on the web protocols. PocketWeb embodies strategies for handling multimedia on PDA, thus minimising if not eliminating the cost for adaptation of user interfaces to PDA resources. Further PocketWeb uses the http and html protocols, abstracting from low-level communication and granting wide-spread acceptance and applicability. PocketWeb is applied within the ESPRIT project MILLION for integration of PDAs as clients in a multimedia system for the tourism sector. As part of the MILLION system, the PocketWeb-based multimedia clients will be evaluated in pilot installations in Venice, Bologna and Khania/Crete.

Acknowledgements

This work was partially supported by the European Union's ESPRIT programme in Project 20772 MILLION, whose partners are Omega Generation, Gruppo Formula, Digicash, Venis, Intracom, MUSIC and the municipalities of Venice, Bologna and Khania.

References

1. Brian Grimm, Education Research Laboratories Inc., *The Williams Pocket Clinical Consultant, Resident's Medical Reference Library*, Contact: health.wave@quantum.net

2. *PDA for Field Engineers in Railway*, Information Systems Group, Railway Technical Research Institute, 2-8-38, Hikari-cho, Kokubunji-shi, Tokyo 185, Japan.

3. *FieldWorker*, Cindy Park, 551 Millwood, Toronto, Canada, http://www.fieldworker.com/

4. *Newton Developer Kit*, Newton Script Programming Environment, Apple Computer Inc., http://www.newton.apple.com/

5. Psion EPOC/16, *SIBO hardware and software architecture*, Organiser Programming Language, http://www.psion.com/developerreference/develbackindex.html

6. U.S. Robotics Palm OS Software Development Kits, http://www.usr.com/palm/5034.html

7. *Multimedia Interactive Leading Life-giving Initiative On Net*, Esprit Project MILLION, Omega Generation, http://www.omega.it/million/

8. Gessler, S., Kotulla, A., *PDAs as mobile WWW browsers*. Proc. of 2nd International WWW Conference, Chicago, Oct. 1994.

Combining CORBA and ITU-T.120
to an Efficient Conferencing Service[1]

Tobias Helbig, Stefan Tretter

Philips GmbH

Research Laboratories

Weißhausstr. 2

D-52066 Aachen, Germany

{helbig,tretter}@pfa.research.philips.com

Dirk Trossen

Technical University of Aachen

Dept. of Computer Science IV

Ahornstr. 55

D-52056 Aachen, Germany

trossen@informatik.rwth-aachen.de

Abstract: An efficient conference service facilitates the implementation and the run-time control of conference applications. Key features of a conference service are conference management, multicast communication support, application state synchronization, and user data marshalling. The paper compares the features offered by the T.120 standards and the CORBA distribution platform. Since they complement each other, it is discussed how a conference service based on a combination of T.120 standards and CORBA can be realized.

1 Introduction

Conference applications enable the communication and cooperation of widely distributed end-systems. Progress in communication systems and increased knowledge of the functionality of conference applications are the main pushing factors for these applications from a technical point of view. The increasing costs and distribution of institutions and companies are the pulling factors for conference applications from an application's point of view since they promise to be an efficient utility for communication and cooperation.

The key function in conference applications is the multi-point communication among several users. In heterogeneous environments this involves marshalling and demarshalling of user data. These features are crucial for applications such as cooperative working on documents and multi-user games. However, additional functions are needed in most conference applications. Due to concurrent activities, the need arises for coordination and synchronization of distributed program entities and users. Furthermore, functionality to inquire for information about the state and structure of existing conferences may be needed, e.g. to join a conference or to invite other users.

All these features of conference applications are mostly independent of specific scenarios. Therefore it is useful to provide them by a generic system layer. Usage of the generic functionality accelerates and simplifies the implementation of conference applications. The standardization furthermore is the key prerequisite for systems of different vendors to interoperate.

In our paper we aim at evaluating CORBA and the ITU T.120 standards with regard to their applicability for building an efficient conferencing service. The analysis of both platforms shows that their features match very well from an applications point of view. The *International Telecommunication Union's* T.120 standards define protocols for generic conferencing management and multi-point communication among distributed application entities. However, there is neither support for marshalling and demarshalling of user data transferred in a heterogeneous T.120 environment nor are there any

---

1. The work presented in the paper results from a joint project between Philips Research and Technical University of Aachen. The project was funded by Philips.

concepts for typing of user data, object modeling or interface description for application objects.

The latter are the key concepts standardized by the *Object Management Group* (OMG) in the *Common Object Request Broker Architecture* (CORBA). The architecture defines an object model and the means to define interfaces which can be used to automatically derive stubs for marshalling and demarshalling user data. However, the CORBA communication infrastructure does not offer multi-point communication support as needed in conferencing applications nor are there standards for conference management.

Seen from a perspective of features visible to applications, both the T.120 standards and CORBA fit very well together. The challenging question is to find the best way to offer these features to applications by means of an *efficient conferencing service*. To solve the question, we evaluate three different frameworks in terms of implementation overhead and expected run-time performance.

The paper is organized as follows. In section 2, we discuss an application scenario to derive the requirements of conferencing applications. In section 3, an overview of the features of T.120 and CORBA is given. The comparison of frameworks to combine the features to a conferencing service are described in section 4. In section 5 conclusions and related work are presented.

2 Requirements of Conferencing Applications

A number of features are inherent to different conference applications. These features can be offered in a generic way by a conferencing service infrastructure. In this section, we evaluate a simple data conferencing application, a *shared whiteboard application*, with regard to application requirements and the system support functions that can be derived from them. The four essential ones are *conference management, multicast support, state synchronization,* and *user data marshalling.*

2.1 Application Requirements in a Shared Whiteboard Application

In a shared whiteboard application, distributed users work cooperatively on a common document. Additionally, they use an audio conference for speech communication. A typical situation in such a setting is that one of the users wants to write on the whiteboard while he/she is speaking to the other conference members. The interactions need to be *communicated* among the different sites in a *coordinated* and *synchronized* fashion. This requires to enable multicast communication among the users, the whiteboard content has to be refreshed at each site to avoid inconsistencies (*synchronization*). Speech communication and input to the whiteboard can be coordinated by floor control mechanisms such as passing the right for *conductorship*. In heterogeneous settings, the marshalling of user data is mandatory. Otherwise, the data could not be used properly. Another typical situation in conference applications is a user who is establishing a conference or entering the conference while it is already in progress. This requires mechanisms for conference management such as initiating and administrating a conference (inviting or dynamically adding new members to the conference) as well as maintaining some form of database to hold information about the current members of the conference and their capabilities.

2.2 Features of a Conferencing Service

A conferencing service should at least offer features that can be derived from the whiteboard application described in section 2.1. The conferencing service shall enable to efficiently implement and run conferencing applications. In table 1, we give an overview of the essential features which are discussed in more detail afterwards.

Group of Features	Features
Data communication	unicast/multicast, user data marshalling, data priorities, location-transparent addressing, internetworking
Conference management	initiation and administration of conference, database of members and applications
Distributed application support	application state synchronization, conductorship
Standardized applications	whiteboard, audio (video) conferencing

Table 1: Features of a Conferencing Service

Data communication covers all aspects of the exchange of user data among the sites participating in a conference. The communication patterns needed to be covered by a conferencing service range from one-to-one to many-to-many communication. In particular, multicast exchange of messages must be supported by the underlying communication infrastructure. Data priorities are helpful to gain flexibility in prioritizing data transfers. A location-transparent addressing scheme simplifies application development. To increase connectivity in particular in wide area settings, interworking between different types of networks like B-ISDN, ATM or PSTN is required.

Conference management covers all features to set up or terminate a conference, to query for participants, to join or leave a conference as well as to invite others, and to manage security and conference data bases.

Distributed application support covers features to facilitate the operation of distributed applications. This means in particular the provision of mechanisms to synchronize the state of applications to enable a coordinated access to resources, e.g. by applying token mechanisms or a conductorship concept.

Standardized applications: Applications use the features offered by a conference service. Interoperability among applications on different sites requires a standardization of application protocols. Relevant applications in the conference context are, among others, whiteboard, audio and video conferencing and multi-point file transfer.

Since the wide variety of features can not be adequately treated in a single paper, we restrict ourselves in the following to only four essential features, namely *conference management, multicast support, state synchronization,* and *user data marshalling.*

3 Features Offered by T.120 and CORBA

As mentioned before, we aim at evaluating CORBA and the ITU T.120 standards to base a generic conference service on them. This section gives an overview of the features of both approaches.

3.1 Overview of the ITU T.120 Standards

The T.120 standards were defined by the International Telecommunication Union. They provide the means for data communication between two or more multimedia terminals and for managing such communication. The protocols comprise basic functions for managing the multipoint communication infrastructure and application functions. Main parts of the *basic functionality* are provided by the T.122 ([8]) and T.125 ([11])

standards, defining the *Multipoint Communication Service* (MCS) and the T.124 standard, defining *Generic Conference Control* (GCC). The MCS provides full-duplex connections among application entities over a variety of networks as defined in T.123 ([9]). MCS provides multicast connections that are mapped to reliable point-to-point connections via a flexible *channel* concept. It also supports *token resource management* to realize application synchronization. The T.124 standard ([10]) is defined to manage multipoint conferences. It uses the functionality of the MCS. Multipoint connections are mapped into the conference context. The GCC handles the data base of conference and application information.

The *application functionality* of the T.120 standard family is provided by *Still Image Transfer* in T.126 ([12]) and *Multipoint Binary File Transfer* in T.127 ([13]). These standards support the exchange of images and files in the multipoint communication environment of T.120.

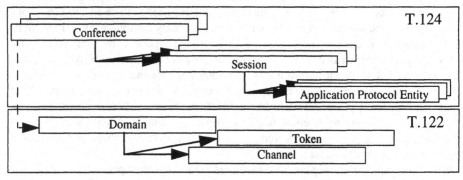

Figure 1: Communication within T.120

In the following, we give some more details about the concepts of the T.120 standards.

3.1.1   Communication Concepts of the T.120

The standard T.124 provides *conferences* subdivided in *sessions*. Within these sessions *application protocol entities* may communicate with each other. Each conference has a one-to-one correspondence with a single MCS *domain*. Within these domains communication takes place via *channels* between applications in a multicast fashion which is realized via reliable unicast connections between the providers. Figure 1 depicts the mappings of the communication context within T.120. Each communication partner within a conference may communicate with each other. The communication within the MCS domains is done in a hierarchical fashion. Connections between well known *providers* (identified by an end system-specific address and a locally unique service access point) are established when rising the domain (e.g. when creating a conference). Thus the topology of a conference is determined at creation of the conference.

Application protocol entities may be attached to providers to use the functionality of the GCC or MCS. If the connection failed because of network errors, the complete subtree below is disconnected. Mechanisms to rebuild the topology are not provided by T.120.

3.1.2   Exchange of User Data

User data within a T.120 context are exchanged as a bitstream via multicast channels. The bitstream is not changed by MCS, i.e. there is no user data marshalling defined. Hence, marshalling of user data has to be done by the applications.

3.2 CORBA

The *Common Object Request Broker Architecture* (CORBA) was specified by the *Object Management Group* (OMG). Its main goal is to provide a standard for the interoperability of distributed objects in heterogeneous environments supporting different computer architectures, operating systems, and programming languages. Therefore, an object model and a language for the implementation independent description of object interfaces, the *Interface Definition Language* (IDL), are introduced. An *object* is an identifiable, encapsulated entity providing services defined as an IDL interface to clients issuing requests. The CORBA standard defines mappings from IDL to different implementation languages that enable the compiler-based creation of specific client and server-side access code to IDL-defined interfaces.

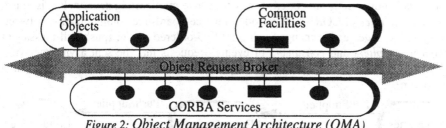

Figure 2: Object Management Architecture (OMA)

While the CORBA standard describes the communication mechanisms needed in a distributed system, it is embedded in the broader scope of the OMG's *Object Management Architecture* (OMA) which is depicted in figure 2. Grouped around the central communication infrastructure are *CORBA services* (like *Naming Service, Event Service, Concurrency Control*) and *Common Facilities* (for user interfaces, information management, system management), that are also specified by the OMG and thought as enhanced support for new application objects.

3.2.1 Communication Concepts in CORBA

This section gives an overview of the general communication concepts of CORBA and shows how the event mechanism can be used for multicast communication.

In CORBA, the communication takes place between a client and an object implementation which may reside at another site using the *Object Request Broker* (ORB). A client sends a (unicast) object request including the server's address and the object name via the ORB to the object implementation which may or may not return results. Figure 3 ([15]) shows the mechanism of requesting a service of an object implementation.

To provide interface specific communication code, the implementation object's IDL interface definition is processed by a compiler. This IDL compiler generates client-side stub code to launch requests and server-side skeleton functions to dispatch incoming requests. The client stubs take and marshal the parameters of object invocations. The marshalled data are sent to the server where the skeleton routines unmarshal the parameters and invoke the server object implementation. Thus marshalling of user data is automatically provided by CORBA. But the connection established by the *ORB core functions* between the client and the object implementation is strictly unicast.

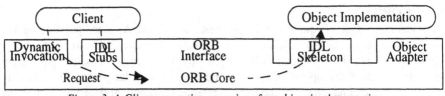

Figure 3: A Client requesting a service of an object implementation

3.2.2 Multicast Communication in CORBA

As mentioned before, communication between objects via the ORB is strictly unicast. Multicast ORBs are neither implemented nor standardized. Thus multicast exchange of user data is not directly supported. A way to realize multicasting nevertheless is to use the *Event Service* of CORBA. It provides a notice-board-like communication between entities. Messages are exchanged using a FIFO-ordered model (push/pull messages) between *suppliers* and *consumers* of events. Requests may be sent asynchronously. Notification is done via blocking (pull-operation) or non-blocking polling (try_pull-operation).

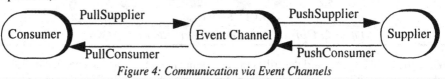

Figure 4: Communication via Event Channels

Multicast communication is realized by creating *Event Channels* as CORBA objects which act as *supplier* and *consumer* of events (Figure 4). Events are exchanged via these channels in a reliable fashion. After creation of an event channel applications may be added as consumers and/or suppliers of events to communicate with other applications via this channel. In the CORBA standard applications are added by the *administrator* of the channel (the creator of the event channel object). This is similar to the private channel model of T.120.

It is not specified how the actual transfer of user data is organized in event channels since CORBA does not standardize the implementation of a service.

3.3 Comparison T.120 and CORBA Features

The section gives a direct comparison of the features of CORBA and T.120 with regard to our evaluation criteria (see table 2). The conclusion is that both systems match very well in their features to combine them for a generic conferencing service.

Conference management: T.124 was defined for conference management. Consequently, the T.120 standards support initiation and management of conferences, e.g. application enrolling to conferences, conference data bases. Conference conductorship is also provided. In contrast, conference management is not in the scope of CORBA. It can, however, be implemented on top of CORBA.

Multicast support: T.122 and T.125 were defined for providing multicast features in conference applications. Consequently, the T.120 standards support flexible multicasting via channels. Multipoint communication between conference participants is directly supported. In CORBA, multicast is not directly supported by the ORB. The *Event Service* of CORBA may be used to realize multicast communication. Implementations like

ORBIX from IONA Inc. use multicast protocols to provide the Event Service but this is a proprietary standard.

Feature	T.120	CORBA
Conference management	Fully supported by GCC (T.124)	Not directly supported
Multicast support	Fully supported by MCS (T.122/T.125)	Restricted support by using the Event Service
State synchronization	Supported by the token concept of MCS (T.122/T.125) and by conductorship of GCC (T.124)	Restricted support (mutual exclusion of method invocations)
User data marshalling	Entirely missing	Fully supported (IDL, stub/skeleton)

Table 2: Comparison of Features

State synchronization: T.122/T.125 offer a very general concept for state synchronization in distributed applications, the token. Furthermore, conferencing-specific synchronization is enabled by the conductorship mechanisms implemented by T.124. In CORBA, state synchronization is restricted to mutual exclusion of the invocation of object methods. Other mechanisms need to be implemented on top of CORBA.

User data marshalling: The T.120 offers no features for user data marshalling. In contrast, it is fully supported by CORBA.

3.4 Conclusion

If we compare the features that are essential for building a generic conference service, CORBA and T.120 match very well. The challenging question is now, how to actually realize the features in a way that

- causes a low implementation overhead
- results in a high run-time efficiency

Consequently, in the second part of our paper we will discuss first studies that show alternatives for implementing a generic conference service that incorporates the superset of the features we discussed above.

4 Frameworks for Realizing a Conference Service

A conference service that combines the above mentioned features of CORBA and T.120 and that is based on CORBA and/or T.120 can be implemented in one of the following three ways:

- Framework 1: Start with an existing CORBA implementation and extend it by the missing features of T.120
- Framework 2: Start with an existing T.120 implementation and extend it by the missing features of CORBA
- Framework 3: Take existing implementations of CORBA and T.120 and combine them

Each of these frameworks will be discussed in the following sections. It will be shown how it can be realized, how much implementation overhead and what run-time behavior can be expected.

4.1   Framework 1: Extend CORBA by T.120-Features

Extending CORBA by T.120 features requires to add conference management functionality, sophisticated multicasting and to extend synchronization mechanisms for application states.

Adding *conference management functionality* can be implemented by one or more *conference servers (CS)* which are used by *conference clients (CC)*. Such a conference server is identified by a name that corresponds to the name of the conference. A trader may be used to decide on which conference server to use. A conference server offers an interface with features alike to GCC (T.124), i.e. it offers functions such as joining a conference, inviting to a conference and conductorship. The conference server is accessed via object invocations. The conference server is responsible to update the conference data base and to allocate *channels* for multicast communication using the *Event Service.*

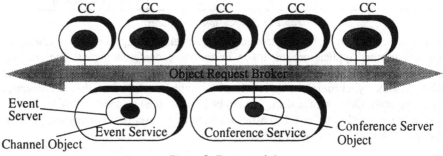

Figure 5: Framework 1

Multicast support is realized by using the Event Service. Event channels are provided by an *event server*. The event server is either accessed directly by conference clients to establish or use channels or indirectly via the conference server. The conference server uses the event server to send control messages and, if requested by conference clients, to send user data via *channel objects. Channel objects* multicast marshalled user data (via unicast connections).

At conference startup an initial multicast channel is established in the event server. The client invokes a conference server method to create a conference. The conference server automatically creates a broadcast event channel, if needed. Client and server enter the channel as consumer and supplier.

Extending *state synchronization* of CORBA may be realized straight-forward by specific synchronization objects that are accessed by the conference clients that need to be synchronized.

4.1.1   Implementation Overhead

The proposed framework 1 extends any existing CORBA implementation (e.g. ORBIX from Iona). The conference server is realized as either a centralized or a distributed database containing conference and application information similar to the conference and application rosters of T.124 ([10]). In case of a distributed data base, the replication protocol can be implemented based on the Event Service. Mechanisms to build multicast groups via the event channels of the Event Service have to be provided by the central conference server similar to the top provider of the T.120 protocols in a star topology.

All these mechanisms need to be implemented. This roughly amounts to a re-implementation of features that already exist in T.120. The implementation overhead is significant.

4.1.2 Run-Time Performance

General performance predictions are difficult since CORBA does not standardize the mechanisms that are to be used for implementations. We hence will concentrate on a CORBA implementation that is available to us, namely ORBIX. In ORBIX the Event Service is realized on top of a multicast protocol (*ORBIX Talk*) which uses IP multicast on network layer. Therefore real multicast is provided. Performance evaluations of the Event Service are currently being investigated but this kind of multicast seems to be insufficient for conferencing.

Some references state bottlenecks in CORBA systems ([17]), such as presentation layer conversions and data copying, server demultiplexing techniques and long chains of intra-ORB virtual function calls. These bottlenecks will also be investigated and evaluated.

4.2 Framework 2: Extend T.120 by CORBA-Features

Extending T.120 by CORBA features requires to add concepts for an appropriate object model to support object orientation, object interface description techniques and stub generators for user data marshalling and demarshalling routines. Based on these extensions, the conference and multicast functions of the T.120 are encapsulated as objects which are used by the applications.

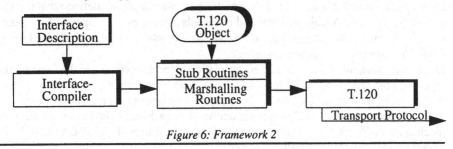

Figure 6: Framework 2

4.2.1 Implementation Overhead

The largest part that needs to be implemented is the interface compiler including the code generator for generating marshalling/demarshalling stubs of user data (Figure 6). ASN.1 marshalling routines are available in the public domain and they are already used to marshal T.120 data internally. However, it is not to be expected that they will be applicable in a CORBA-based environment since an additional format conversion between the T.120 run-time environment and CORBA environment will reduce the performance. Hence, implementing framework 2 leads to re-implementing a CORBA-like interface compiler and stub generator. This is inefficient and expensive with regard to maintenance.

4.2.2 Run-Time Performance

Performance evaluations of the communication infrastructure are described in [3]. Since these measurements were made without user data marshalling, the overhead for marshalling will (slightly) reduce the performance.

4.3 Framework 3: Combine T.120 and CORBA

By adding the features of one system to the other, frameworks 1 and 2 mainly end up re-implementing already existing functions. Consequently, framework 3 aims at combining existing implementations of CORBA and T.120 (see Fig. 7).

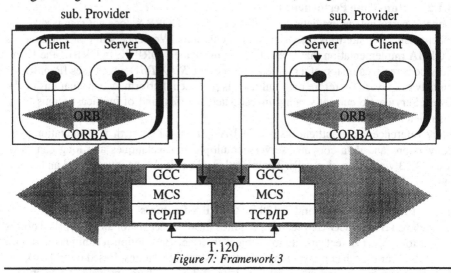

Figure 7: Framework 3

Framework 3 combines the object oriented approach of CORBA with the conference communication architecture of T.120. For this purpose objects are defined that use the standardized CORBA object model and encapsulate services for conference management, multicasting and state synchronization. The objects' interfaces are defined using CORBA's *Interface Definition Language*.

During run-time, instances of T.120 encapsulation objects are created in the same server process as the clients that want to access the services. Access to these objects is a local invocation of their methods. The T.120 encapsulation objects internally implement the standard protocols of T.120. They communicate with each other via the communication mechanisms of the T.120 infrastructure that is separated from the CORBA ORB. With regard to our evaluation criteria this means:

Conference management functionality, multicast support and *state synchronization* are realized by T.120 encapsulation objects that interact via an T.120 MCS and GCC implementation. *User data marshalling* is realized by CORBA mechanisms. This is achieved by passing typed user data via the "ANY" typing concept of CORBA to the T.120 encapsulation objects that transmit the data to other sites.

4.3.1 Implementation Overhead

Framework 3 combines existing implementations of T.120 and CORBA. The implementation overhead is reduced to creating the T.120 encapsulation objects that provide the glue for using T.120 features in a CORBA environment.

4.3.2 Run-Time Performance

The performance of framework 3 for a conference service is comparable to the performance of a pure T.120 implementation. Performance measurements of a T.120 communication infrastructure can be found in ([3]). The overhead caused by user data marshalling via CORBA marshalling/demarshalling mechanisms reduces the available performance. The exact performance will be evaluated as soon as a prototypical implementation is realized.

5 Conclusions and Related Work

Providing an efficient conference service facilitates the implementation and run-time control of conferencing applications. The paper discussed early results of an analysis how to implement a conference service based on a combination of the features offered by the T.120 standards and the CORBA distributed communication platform. A comparison of the features of T.120 and CORBA has shown that both systems match very well since they complement each other. CORBA offers concepts for object modeling and user data marshalling. T.120 is focused on conference management, multicast support and state synchronization.

For an efficient implementation of the features we compared three different frameworks. The first framework adds conferencing functionality to CORBA using the Event Service of CORBA to realize multicast communication. The approach has the disadvantage that this kind of multicast support is insufficient for conferencing. Furthermore, conference management and state synchronization have to be re-implemented.

The second framework adds CORBA functionality to T.120. Components such as an interface compiler and marshalling routines need to be implemented which increases the implementation overhead. Additionally, an object model and service descriptions have to be defined. The expected communication performance is higher than in the CORBA framework because T.120 uses a vertical approach to realize the multipoint conference functionalities. But the additional object oriented functionality would be realized by proprietary products.

The third framework combines the T.120 conferencing and communication architecture and the CORBA object environment with minimum implementation overhead. This is achieved by encapsulating T.120 code into CORBA server objects. The remaining code can be reused from existing implementations: Interface compiler and object model of CORBA are used. The object implementations of the conference service use T.120 implementations. We expect the performance to be comparable to a pure T.120 implementation. We will base our implementation on the framework 3 since we expect the third framework to deliver good performance with low implementation overhead.

Another approach to realize a distributed conference service was developed in the *HORUS* project [16]. There, a description is given how to realize a framework for distributed conference applications with a CORBA compliant interface ([14]). User data marshalling is realized by using CORBA. Multipoint conferences are supported but there is no standardized communication platform similar to T.120.

The IETF (Internet Engineering Task Force) defines conference and multicasting services that can be compared with the features of T.120. They, for example, define a *simple conference control protocol* ([1]), a *simple conference invitation protocol* ([18]) and multicast transport protocols such as RTP (*real-time protocol*) to transport real-time streams or *IP multicast*. However, a detailed discussion of an integration of these approaches with CORBA is beyond the scope of this paper.

A distributed platform comparable to CORBA is the *Distributed Computing Environ-*

ment (DCE) ([2]) of the *Open Software Foundation* (OSF). Services like RPC, Security, Directory, Threads and Distributed File Services are provided. Efficient multicast transfer or conferencing functionality is not supported. Thus there are similar weaknesses as in CORBA.

Our next steps will be to implement a conference service as described in framework 3. We will perform extensive performance evaluations to evaluate performance degradations due to the marshalling/demarshalling routines. Additionally, we will investigate the Event Service as a basis for multicasting in CORBA.

6 References

[1]   Carsten Borman, Joerg Ott, Christoph Reichert (1996): Simple Conference Control Protocol, ftp://ftp.ietf.org/internet-drafts/draft-ietf-mmusic-sccp-00.txt

[2]   DCE Overview, available on http://www.osf.org/comm/lit/TOG-DCE-PD-1296.html

[3]   T. Helbig, D. Trossen (1997): Die ITU Standard-Familie T.120 als Basis für verteilte Mehrbenutzersysteme, Proc. of Kommunikation in Verteilten Systemen, Braunschweig (in german)

[4]   IMTC (1996): IMTC MCS API, ftp://ftp.imtc-files.org/imtc-site/AP-AG/TECH/CD/MCS-API

[5]   IMTC (1996): IMTC GCC API, ftp://ftp.imtc-files.org/imtc-site/AP-AG/TECH/CD/GCC-API

[6]   ITU-T Study Group 8, ITU-T Recommendation T.120 (Draft, March 1995): Data Protocols for Multimedia Conferencing

[7]   ITU-T Study Group 8, ITU-T Recommendation T.121 (Draft, February 1995): Generic Application Template

[8]   ITU-T Study Group 8, ITU-T Recommendation T.122 (1993): Multipoint Communication Service for Audio Graphics and Audiovisual Conferencing

[9]   ITU-T Study Group 8, ITU-T Recommendation T.123 (Draft, June 1994): Protocol Stack for Audiographic and Audiovisual Teleconference Applications

[10]   ITU-T Study Group 8, ITU-T Recommendation T.124 (Draft, March 1995): Generic Conference Control for Audiographic and Audiovisual Teleconference Applications

[11]   ITU-T Study Group 8, ITU-T Recommendation T.125 (Draft, 1993): Multipoint Communication Service Protocol Specification

[12]   ITU-T Study Group 8, ITU-T Recommendation T.126 (Draft, 1995): Still Image Transfer Protocol Specification

[13]   ITU-T Study Group 8, ITU-T Recommendation T.127: Multipoint Binary File Transfer Protocol

[14]   S. Maffeis (1995), Adding Group Communication and Fault Tolerance to CORBA, Proceedings of the 1995 USENIX Conf. of Object Oriented Technologies, Monterey CA, June 1995

[15]   OMG (1993): The Common Object Request Broker: Architecture and Specification

[16]   R. Renesse, K. Birman, S. Maffeis (1996), Horus: A flexible Group Communication System, Communication of the ACM, April 1996

[17]   D.C. Schmidt, A. Gokhale, T. Harrison, G. Parulkar (1997), A High-Performance Endsystem Architecture for Real-time CORBA, IEEE Communications Magazine, Vol.35, No.2, February 1997

[18]   Schulzrinne (1996): Simple Conference Invitation Protocol, ftp://ftp.ietf.org/internet-drafts/draft-ietf-mmusic-scip-00.txt

Supportive Environments for Executing Multimedia Applications

Mario Bochicchio, bomal@ingle01.unile.it, (1)
Paolo Paolini, paolini@elet.polimi.it, paolini@ingle01.unile.it, (1,2)
(1) Telemedia Lab - University of Lecce - Italy
(2) Hypermedia Open Center - Politecnico di Milano - Italy;

Abstract

"Messapia" is a multimedia application conceived to contain digital films, maps, audio and over 1500 high quality (true colors, high resolution) images, requiring large multimedia files to be stored.

Suitable compression techniques where adopted to cope with the problems created by this large multimedia files. The compression's benefits resulted in an global application size fitting the storage capacity of an ordinary CD-ROM. Nevertheless, as a negative side effect, they involved the need of expensive high performance PCs or Workstations, customized for an efficient and rapid decompression and visualization. On the other hand, relatively slow multimedia machines, i.e. '486 at 33 MHz up to Pentium at 133 MHz or equivalent, are an important fraction of the user community, where CD-ROM and Internet applications are mostly "consumed".

In this paper we device a way to build a multimedia application (either on CD-ROM or Internet), using compressed files, so that the slowness of the application, when it is used on an inexpensive machine, is eliminated when the same machine is plugged in a suitable supportive environment. It is very important the fact that the application (or the Web browser) does not need to be modified when plugged in the supportive environment, but its behavior is "adjusted" according to the situation.

Preliminary experiments and examinations of expected performances are presented: the results are very promising and show that supportive environments are feasible, and also they can be very effective in improving the performances of multimedia applications.

1 - Introduction and background

"Messapia" [1] is a multimedia application[2] conceived to contain several high quality multimedia objects, requiring large multimedia files to be stored.

As other modern multimedia applications, Messapia largely rely on compression of files representing images, sound, graphics or alive video.

In this paper we will focus mainly on still images, and their compression techniques, but similar arguments can be applied to other media and other techniques.

1.1 The Application

Many multimedia applications are (relatively) large including thousands of multimedia objects.

---

[1] The Messapi where a pre-Roman population of the south of Apulia, in Italy. Many archeological finds and ruins still testify the history of this people.

[2] Under development at the Telemedia Lab, University of Lecce, Italy, using the Hypermedia Design Model [HDM-1,2,3,4]. The application is being developed in cooperation with the Castromediano Museum of Lecce, and the Archeology Department of the University of Lecce.

Messapia for example, when fully developed[3], will contain more than 1500 high resolution true colors images (6000 images in total) of archeological finds and sites, 30 minutes of interviews in form of digital films, 5 minutes of high quality 3-D virtual reconstruction's of sites and houses (in form of digital films) , maps, texts, audio comments (Italian and English) and music.

Messapia has been designed using the Hypermedia Design Model [HDM- 1,2,3,4], that provides the primitives to model the applications, the directives for its implementation and also an ordered development cycle. According to HDM, an hypermedia applications can be structured in an Hyperbase in-the-large, where the main features of the objects are designed, an Hyperbase in-the-small, where the details of the objects are described, and the Access structures, that allow the user to locate the objects of his/her interest.

The Hyperbase of Messapia consists of three main entity types: Sites (describing relevant places), Objects (describing finding of all kinds) and Topics (Describing subjects as History, Traditions, etc.). A hypertextual Glossary provides additional information, where it is needed. Entity types are structured in components, and a number of Application Link types, provide semantic relationships.

The design in the small takes into account the internal structure of this entities in terms of multimedia slots (images, slide shows, films, audio, ...) and relations between them (synchronization between audio and slide show, text and pictures, ...).

Several access structures (indexes, to random select objects, and guided tours, to browse across objects), support different fruition strategies: from the occasional user to the curious of archeology up to (in some way) the experienced researcher.

Run time tools are also available to support the user in the creation and the customization of indexes and collections, giving to him the possibility to reuse the available multimedia objects for its own purposes (presentations, lessons, ...).

The application, in its final shape will consist of a CD-ROM and a Web-site. The Web site will contain most of the content of the CD-ROM, and additional updates, information about relevant cultural events, references to other relevant sites, etc. The intended audience for Messapia includes schools, students, average well educated museum visitors, well educated general public, etc. Being, in general, non professional, it is likely that inexpensive machine would be mostly used to "consume" the application. It is desirable, however, to maintain an high quality of the pictures and of multimedia in general.

1.2 The technical problems

Referring to the contents of the application, the richness of features of the messapian archeological finds[4] required the storage of about 1500 high quality, big sizes images. This created a serious problem of storage space or, alternatively, of computing power. In fact, assuming an average size of 10^6 pixels per image, at 24 bit per pixel, it is easy

---

[3] In March '97 the design and the software implementation are completed; the content has been developed up to 60%. Completion of the CD-ROM is planned for July '97. Completion of a corresponding Web site (http://labtel.unile.it) is foreseen for December '97.

[4] Most of all earthenware and finely decorated vases.

to calculate that a typical CD-ROM[5] can hold no more than 213 images. On the contrary, adopting a suitable image compression technique, and assuming an average compression ratio[6] of 25:1, about 6000 images can be packed in the same CD-ROM, but the time needed to decompress the images before their actual usage can be not negligible, even on powerful machines. Fig. 1.1 shows, as example, the mean time measured to display a JPEG compressed image ($\approx$20:1 compr. ratio) vs. the image size, when a common 80486 at 33 MHz with a 2x CD drive, or a Pentium at 133 MHz with an 8x CD drive (or a 128 Kbit/s ISDN Internet connection) are used[7].

From fig. 1.1 we can draw that neither the 486 nor the Pentium 133 approach the "10^6 pixel/s region" desirable for a satisfactory interaction with Messapia.

On the other side, in the near future, most of the non-professional uses of multimedia applications, either CD-ROM's or WWW, are likely to be consumed on this kind of relatively slow machines.

Trying to improve the performance of executing high quality multimedia application, using commonly available multimedia machines, is the main topic of this paper.

More in general, compressed multimedia files (and still image files in particular) occupy less space and, also, they are faster to load. Practically speaking, in multimedia applications image files come from two sources: CD-ROM's and the Internet (WWW applications).

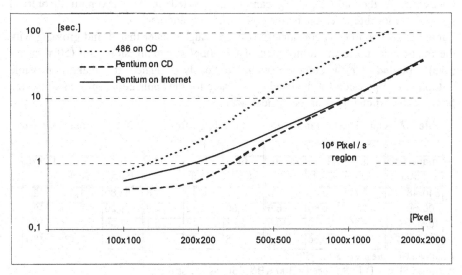

Figure 1.1 Display time vs. image size on common multimedia PCs with JPEG compression

---

[5] With a storage space of 640 Mbytes.

[6] Still images can be compressed with different ratios and different techniques. Popular formats are JPEG [JPEG- 1,2,3], Photo-CD [PHO- 1,2], fractal compression [FRA-1,2] and wavelet compression [WAV-1,2,3].Compression ratio should not exceed 50:1, at least using commonly available software tools, in order to balance performances with acceptable quality.

[7] The typical images used in Messapia where considered to obtain the mean times reported in fig. 1.1. More details about the software and hardware used are reported in the following of the paper.

CD-ROM's have different speed, mostly ranging from 300 KB/s with 350 ms of access time, up to 1200 KB/s with 200 ms of access time. Internet connections, come with all kinds of performances; in this paper we consider medium-end connections, that on practical ground, whatever is the nominal speed, rarely perform better than 128 kbit/s with a latency typically ranging from 20 ms to 1000 ms, depending on the traffic, the protocol etc.

The table 1.2 shows the estimated[8] time to load in memory images of different sizes[9]. It should be noted how the larger pictures can take up to 14 second to be loaded from a CD-ROM (246 sec from the net), if not compressed, and less than one second (less than six seconds from the net) when compressed (50:1). Even half that compression ratio (that is advisable, in order to keep under control the loss of quality) can produce high gains of performances with the reading time. In tab 1.2, T_U can be considered equivalent to the visualization time in the uncompressed case, because after reading only a negligible time is required to send the image from the main memory to the video buffer of the SVGA card. In the compressed case instead, the visualization time is greater than the compressed reading time T_C (see fig. 1.1), because of the time required by the decompression process. In each case the reading time represents a lower limit to the visualization time, in fact T_C can be considered as the visualization time when a very fast CPU (or specialized hardware) is used for decompression. Consequently the ratio T_U / T_C can be interpreted as the maximum theoretical speedup achievable when the image compression is adopted.

Table 1.2 shows, in each considered case, speedups greater than 1 and growing with the image size, up to the limit value of the adopted compression ratio (50:1 in our case). Coupled with the storage savings previously evaluated, this make impossible not to use compression (of the level appropriate for the application), provided that the possible loss of quality is properly dealt with.

Table 1.2 Nominal performance while reading images from CD-ROM's or from the Internet

Image Size	Slow CD			Fast CD			Net		
	T_U	T_C	T_U / T_C	T_U	T_C	T_U / T_C	T_U	T_C	T_U / T_C
[Pixel]	[Sec]	[Sec]		[Sec]	[Sec]		[Sec]	[Sec]	
640x480	3,77	0,76	5,0	1,17	0,42	2,8	57,80	1,35	42,8
800x600	5,50	0,80	6,9	1,60	0,42	3,8	90,20	2,00	45,1
1024x768	8,56	0,86	10,0	2,37	0,44	5,4	147,66	3,15	46,9
1280x1024	13,81	0,96	14,4	3,68	0,47	7,9	245,96	5,12	48,1

Assumed Compression ratio = 50:1;
Assumed Slow CD-ROM speed = 300 KB/s, 350 ms access time;
Assumed Fast CD-ROM speed = 1200 KB/s, 200 ms access time;
Assumed Internet bandwidth = 128 Kbit/s, 100 ms latency;
The "Image Size" column refers to the image dimension as [Hor. Pixel]x[Vert. Pixel];
T_U column refers to the reading time of the uncompressed image;
T_C column refers to the reading time of the compressed image;
T_U/T_C column represents the upper limit to the speedup when using a compression technique.

[8] We found that, in the case here considered, an acceptable approximation for the loading time is given by the linear model: $T_{load}=2(access\_time)+speed * (image size in bytes)$.
[9] The typical screen dimension available on most SVGA adapters are considered.

Unfortunately, the normal multimedia PCs are very slow, when adopting standard techniques, in decompressing the large images here considered. Consequently, speedups smaller than 1 (i.e. slowdown) are normally achieved on these PCs.

When a standalone PC is considered , the only solution (waiting for the next generation of super-fast algorithms) is the hardware upgrade. When instead a networked PC is considered and a suitable computing resource is available on the network, a distributed decompression technique can be usefully exploited. This is the case of most universities, schools, museums and companies.

In the following of this paper a such distributed technique is proposed to speedup the decompression and visualization process, constituting so a supportive environment for the execution of high quality multimedia objects on the current generation of multimedia PCs.

Our idea of supportive environment is the following: the user employing his personal machine, (even a laptop or transportable) may get slow execution of high quality multimedia applications (CD-ROM's or WWW applications). Plugging the same machine in a supportive environment, without any change to the application itself, should provide much better performances.

A supportive environment, therefore, allows the upgrading of the performances of multimedia application in an not-intrusive manner.

In section 2 we propose a simple architectural solution for a supportive environment, and we try to reason about the expected performances; in section 3 we show the results of some measurements, providing very promising results; in section 4 we draw the conclusions and sketch the future directions for work; 5 is for bibliography and references.

2 - The proposed solution

Our architectural solution for a supportive environment (SE in the following) is quite simple: a network environment where the user machine, executing a multimedia application, is plugged in.

When the machine is not connected to the environment, it reads a compressed file (from CD-ROM or Internet), decompress it and uses it. When the machine is connected to the environment, after a compressed image is fetched, instead of being decompressed locally, it is sent to a "decompression engine" for decompression and possible reformatting, and then received back "ready to use". The "engine" could be a PC equipped with a suitable hardware accelerator, an high performance UNIX workstation or a cluster of powerful machines running in parallel [HPCN- 1,2], for more demanding jobs. A simple example of SE is depicted in figure 2.1.

The key point is that the machine executing the application, and the application software, should be unaware of the complexity of the supportive environment. This machine, in fact, simply requires a decompression service to the available networking resources and takes back the decompressed image when available. A suitable decompression software driver must be implemented for the purpose, capable to detect the presence of a fast decompression resource available on the LAN. This software driver, when requested from a multimedia application or from a Web

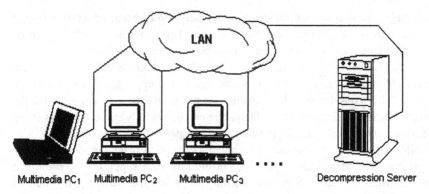

Figure 2.1: A distributed environment for multimedia applications

browser, should tests the presence of the decompression server in order to determine if the decompression can be done remotely or locally. The test process should be not performed at each decompression request, but only when the requesting application starts or in case of network disconnection, reconnection or time-out.

The network traffic generated by this kind of software driver on a dedicated LAN could be represented as a short burst (the transmission of the compressed file) followed by a pause (needed for the decompression) and by a successive long burst (the transmission of the uncompressed image). The global duration of this process should not exceed a few seconds.

Because of the performance degradation due to the "collision" between two or more decompression requests, the number of clients per each available server, and the network segmentation, should be carefully considered when a such supportive environment is designed.

Let us call "PC" the multimedia PC executing the application, and "W" the workstation (or cluster of computing resources) implementing the decompression service.

The total time to visualize an image when the PC is not connected to the SE (operating as a stand-alone machine) is given by the formula

$$t_{visualize} \cong t_{read} + t_{PC\text{-}dec} \qquad \textbf{(formula-1)}$$

where t_{read} is the time required to transfer the compressed file from the CD-ROM (the net) to the memory of the PC, while $t_{PC\text{-}dec}$ is the time needed to the PC in order to decompress the image. The actual time needed to copy the decompressed image from the main memory to the video buffer is considered to be negligible, being of an order of magnitude smaller than the other times.

When the PC is connected to the SE, instead, the total time to visualize an image is given by the formula:

$$t_{SE\text{-}visualize} \cong t_{read} + t_{send} + t_{receive} + t_{W\text{-}dec} \qquad \textbf{(formula-2)}$$

where t_{read} is the time required to transfer the compressed file from the CD-ROM (the net) to the memory of the PC, t_{send} is the time to send the compressed file from the PC to the decompression server, $t_{receive}$ is the time needed to receive the decompressed file back to the PC, and $t_{W\text{-}dec}$ is the time needed to the server to decompress the file.

The possible gain of performance is given by the formula

$$t_{gain} = t_{visualize} - t_{SE\text{-}visualize} \cong (t_{PC\text{-}dec} - t_{W\text{-}dec}) - (t_{send} + t_{receive}) \quad \textbf{(formula-3)}$$

while the speedup is

$$S = t_{visualize} / t_{SE\text{-}visualize} \quad \textbf{(formula-4)}$$

The positive factor of the gain is given by the faster decompression of the server; the negative factor is given by the need of sending the compressed file to the server, and receive the decompressed file back to the PC.

The visualization time could be reduced, considering the involved hardware, by concurrently performing the operations of reading, sending, decompressing and receiving back. This is possible because of the nature of many modern compression algorithms, that make possible to start the decompression and visualization process before the complete acquisition in memory of the compressed data. In this hypothesis, and considering an half duplex communication channel (e.g. ethernet), the following formula define the minimum theoretical time to visualize an image:

$$t_{SE\text{-}visualize\text{-}MIN} \cong \max(t_{read}, t_{send} + t_{receive}, t_{W\text{-}dec}) \quad \textbf{(formula-5)}$$

In a similar manner we can consider the possibility of executing in parallel the operations of reading and decompressing on the standing alone machine (a suitable use of the DMA or the Bus Mastering capabilities of many modern disk controllers make possible this):

$$t_{visualize\text{-}MIN} \cong \max(t_{read}, t_{PC\text{-}dec}) \quad \textbf{(formula-6)}$$

The two ratios: $\varepsilon_{SE} = t_{SE\text{-}visualize\text{-}MIN} / t_{SE\text{-}visualize}$ and $\varepsilon_{PC} = t_{visualize\text{-}MIN} / t_{visualize}$ represents the degree of parallelism (efficiency) achieved by the implementation, with the value being smaller than 1, or equal to 1 when the theoretical perfect parallelism is actually achieved.

Since the formula-5 is based upon the maximum of a function, a balanced implementation should keep all the three factors of the same order of magnitude. In fact there is no point in trying to stress one of the factors, having an unbalanced implementation, with no actual gain in performances. This can be expressed by the relationship:

$$t_{SE\text{-}visualize\text{-}MIN} \cong t_{read} \cong (t_{send} + t_{receive}) \cong t_{W\text{-}dec} \quad \textbf{(formula-7)}.$$

useful for system dimensioning purposes, when the cheapest possible SE is required for a given visualization time.

In the next section we will use the above formulas in order to test the feasibility of an effective supportive environment.

3 - Testing the solution

We tested in our laboratory the actual feasibility and effectiveness of a supportive environment, as the one shown in figure 2.1, and we tried also to estimate the performances of a multimedia application, before actually building it.

We considered two multimedia machines, indicated as PC/33[10] and PC/133[11], respectively. We used as a decompression server a RISC workstation[12]. The connection was a 10 Mbit/s thin ethernet completely available to the experiment. We purposely did not use a Fast Ethernet (100 Mbit/s) or a switched network, in order to test our hypothesis on a very common and inexpensive environment.

We considered four sizes of true color images, i.e. 640x480, 800x600 1024x768 and 1280x1024 pixels; obviously the images was stored at 24 bit/pixel, but they was reproduced at 8, 16 or 24 bit/pixel, depending on the hardware available on the visualization machine. We also considered two different compression formats with an equivalent visual quality for the images used for the experiment: JPEG (at the compression ratio 20:1), and CREW [WAW-3] (at the compression ratio 50:1). The first formats is well known and commercially available. CREW is based on wavelet compression, that seem to be very promising in the near future; it has been selected among the possible candidates for ISO/IEC standardization. We did not use Fractal or Photo-CD compression because of its proprietary nature.

As far as JPEG is concerned, the time for decompression has actually been measured for both PCs and the decompression server. The performances of CREW have been estimated in accordance with the declarations of the authors.

The performances of decompression for JPEG have been measured using the version 6.0 of the source code made available by the Independent JPEG group [JPEG-3].

Under this hypotheses the measured performances of the JPEG where more or less linear with respect to the number of pixels. In particular we measured a rough average of 50 µs/pixel, as the JPEG decompression rate on the PC/33, and 9.5 µs/pixel on the PC/133. The performance ratio when decompressing, between the RISC and the PC/33, was about 25:1, and this is a key observation of our experiment. Referring to formula-3 this tremendous gain for decompression overbalances the time for communication (send and receive), and makes our overall architecture worthwhile.

The estimated decompression time for CREW is about 20% higher than the one for JPEG.

As far as communication is concerned, we need to efficiently connect Windows based PCs to a UNIX workstation. The obvious choice for the protocol has been TCP/IP and the multiplatform "Berkeley Sockets" support library[13] has been used for the actual implementation.

---

[10] The CPU is a 80486DX33 with a 16 bit ISA bus, 8 MB of RAM, ethernet NE2000 compatible, double speed SCSI CD-ROM, a 16 bits ISA graphic card based on the TsengLab ET4000 used at 1024x768 with 256 colors, audio card, running Windows 3.11 for workgroups.
[11] The CPU is a Pentium 133 with a 32 bit PCI bus, 16 MB of RAM, 3Com 3C590 PCI ethernet, eight speed EIDE CD-ROM, a PCI graphic card based on the S3 Trio 64V+ used at 1280*1024 with 64K colors, audio card, running Windows 95.
[12] An IBM RISC 3000/360 with 64MB of RAM and a theoretical peak power of 100 MFlops.
[13] The TCP/IP suite "LAN WorkPlace" from Novell was used for the PC/33 under Win 3.11 for Workgroups; the 32 bit TCP/IP from Microsoft, native for Win95, was used for the PC/133. Visual C++ was used as development environment. Various deviations from the

The performances of communication have been measured and interpreted using the model [HPCN- 1,2] given by the formula:

$$T_{comm} = \alpha + \beta * L \qquad \text{(formula-8)}$$

where L is the length of the message, α is the latency time and β express the linear dependency of the communication time upon the message length.

Measuring then the time needed to transmit packets of various dimension we obtained for α the average values of 57 milliseconds and 31 milliseconds, for the PC/33 and the PC/133 respectively. We also obtained for $1/\beta$ the average values 220 KB/s[14] and 650 KB/s[15] respectively.

In order to compute the overall performances we need to apply the formulas introduced in section 2. We have to consider two different situations:

1- Compressed stand-alone: the images are compressed, and decompression is performed on the machine executing the application.

2- Compressed, using the network as a supportive environment: the images are compressed, and the supportive environment provided by the networked server is used to decompress the images.

For both situations, according to the formula of section 2, we have to consider the worst case, where all the operations are performed in sequence, and the best case, where a maximum parallelism is achieved. The best performances are not attainable, in practice, but they still represent a good point of reference.

Table 3.1 Maximum theoretical speedup on the proposed supportive environment

	Img. Size	Tread	Tpc-dec	Tw-dec	Tsend	Treceive	Tse-v	Tse-v-min	Tsa-v	Tsa-v-min	S
	[pixel]	[s]	[s]	[s]	[s]	[s]	[s]	[s]	[s]	[s]	
486/33	640x480	0,88	15,36	0,61	0,27	1,45	3,22	1,72	16,24	15,36	8,93
JPEG 20:1	800x600	0,99	24,00	0,96	0,38	2,24	4,57	2,62	24,99	24,00	9,15
CD 2x	1024x768	1,17	39,32	1,57	0,59	3,63	6,97	4,22	40,49	39,32	9,31
256 colors	1280x1024	-	-	-	-	-	-	-	-	-	-
Pentium/133	640x480	0,44	2,92	0,61	0,10	0,98	2,13	1,1	3,36	2,92	2,71
JPEG 20:1	800x600	0,47	4,56	0,96	0,14	1,51	3,08	1,6	5,03	4,56	2,76
CD 8x	1024x768	0,51	7,47	1,57	0,21	2,45	4,74	2,7	7,98	7,47	2,81
64 K colors	1280x1024	0,58	12,45	2,62	0,33	4,06	7,60	4,4	13,03	12,45	2,83

Tread = reading time for the compressed file;

Tpc-dec = CPU time needed to decompress the image locally (on the PC CPU);

Tw-dec = CPU time needed to decompress the image remotely (on the RISC CPU);

Tsend = time needed to send the compressed file from the PC to the RISC;

Treceive = time needed to receive back the compressed file to the PC;

Tse-v = global visualization time on the supportive environment, sequential (worst) case;

Tse-v-min = global visualization time on the supportive environment, parallel (best) case;

Tsa-v = global visualization time on the stand alone PC, sequential (worst) case;

Tsa-v-min = global visualization time on the stand alone PC, parallel (best) case;

S = maximum theoretical speedup = Tsa-v-min / Tse-v-min.

---

standard Berkeley Sockets was observed for both the packages, making the performance tuning not straightforward.

[14] The networking hardware on the PC/33 was a 16 bit NE2000 compatible card.

[15] The networking hardware on the PC/133 was a 3C590 combo card, with a 32 bit PCI Bus.

In table 3.1 are shown the mean values of the terms[16] needed to compute the overall performances for the different picture sizes and the different PC's.

From the data of table 3.1 we can draw that, when adopting the JPEG, the supportive environment can theoretically increases the performances of the application, by more than 9:1 for the 486 and nearly 2,8:1 for the Pentium. No relevant difference on the visualization times are achieved when adopting CREW instead of JPEG in CD-ROM based applications.

A preliminary implementation of the proposed technique gives for S an interesting 7:1 for the PC/33 and a 1.9:1 for the PC/133. Overall the supportive environment seems to provide an effective boost to performances.

Figure 3.2 depicts in a graphical manner the estimated performances, plotting the worst case and the best case times, as a functions of the image sizes, for the PC/33 and for the PC/133.

It should be noted how the supportive environment improves its gain, with respect to the stand-alone solution, as the size of the picture increases,.

The difference between worst and best case is relevant, for the supportive environment, only for very large pictures, and is irrelevant in each sequential case.

Table 3.3 refers to the comparison[17] between JPEG and CREW on Internet. We can see as the Pentium is fast enough to decompress the data while they arrives, so no speedup is possible using the supportive environment.

On the contrary, when the CREW is adopted, thanks to its superior compression ratio (at an equivalent visual quality) the time needed to read the compressed file from the Internet is smaller than the stand-alone decompression time Tpc-dec, so that the supportive environment can be usefully exploited.

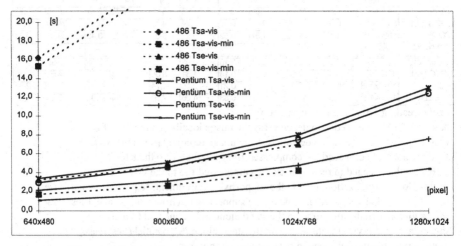

Figure 3.2 Overall performances: worst case and the best case times

---

16 The terms Tread, Tpc-dec, Tw-dec, Tsend and Treceive where measured on a subset of 20 images from the "Messapia" collection. Formulas 2, 5, 1 and 6 where respectively applied for evaluating Tse-v, Tse-v-min, Tsa-v and Tsa-v-min.

17 The terms Tread, Tpc-dec, Tw-dec, Tsend and Treceive where measured for the JPEG case, and estimated for the CREW case in accordance with the declarations of the authors.

Table 3.3 Comparison between JPEG and CREW on Internet

	Img. Size	Tread	Tpc-dec	Tw-dec	Tsend	Treceive	Tse-v-min	S
	[pixel]	[s]	[s]	[s]	[s]	[s]	[s]	
Pentium/133	640x480	3,08	2,92	0,61	0,10	0,98	3,08	1,0
JPEG 20:1	800x600	4,70	4,56	0,96	0,14	1,51	4,70	1,0
Internet	1024x768	7,57	7,47	1,57	0,21	2,45	7,57	1,0
64 K colors	1280x102	12,49	12,45	2,62	0,33	4,06	12,49	1,0
Pentium/133	640x480	1,35	3,50	0,74	0,06	0,98	1,35	2,6
CREW 50:1	800x600	2,00	5,47	1,15	0,08	1,51	2,00	2,7
Internet	1024x768	3,15	8,97	1,89	0,10	2,45	3,15	2,8
64 K colors	1280x102	5,12	14,94	3,15	0,15	4,06	5,12	2,9

Tread = reading time for the compressed file;
Tpc-dec = CPU time needed to decompress the image locally (on the PC CPU);
Tw-dec = CPU time needed to decompress the image remotely (on the RISC CPU);
Tsend = time needed to send the compressed file from the PC to the RISC;
Treceive = time needed to receive back the compressed file to the PC;
Tse-v-min = global visualization time on the supportive environment, parallel (best) case;
S = maximum theoretical speedup = Tsa-v-min / Tse-v-min.

4 - Conclusions and future work

The overall results of our experimentation's induce the following observations:

- Compressed images, thanks to the supportive environment, can always perform better than uncompressed images; coupled with the storage saving, it is impossible not to use compression (of the level appropriate for the application), provided that the possible loss of quality is properly dealt with.
- The gain in performance offered by the supportive environment is significant, in general.

In practice, for Messapia, the supportive environment will work like this:

- the user, utilizing CD_ROM or accessing the Web site with its own inexpensive slow machine will get (relatively) slow performances
- a supportive environment will be set up in the Castromediano museum, so that the same CD-ROM, or the Web site will be accessed more efficiently;
- in any environment where a LAN and a performing machine is available the user may (easily) setup a supportive environment, to speed up access.

From the application developer point of view the advantage consists in having the same multimedia application (of limited size, since compression is used), performing differently, according to the environment, instead of having to develop two different applications.

In the near future we foresee this developments for our work:

1. while completing Messapia, we will compress most of its pictures (JPEG and wavelet), in order to squeeze it in a CD-ROM and in order to build a small-size, fast Web site.
2. A straightforward supportive environment will be available both for a CD-ROM and for the Web Site.
3. A more sophisticated supportive environment, capable of exploiting the parallelism of clusters of workstations, will be experimented for very large pictures, for compressed video and for virtual reality.

4. We will experiment advanced compression schemes (based on wavelet), in order to verify the possibility of using parallelism "within" the decompression algorithms, and not just around them.

5. A generalization of the proposed approach to other type of multimedia clients, servers, LANs and services will be developed. Particular attention will be addressed to the emerging "fast ethernet" standard (LANs), and to the "simulation on demand" and "virtual reality" fields (services).

5 - Bibliography

[HDM-1] F.Garzotto, P.Paolini and D.Schwabe, "HDM - A Model Based Approach to Hypermedia Application Design", ACM Transactions on Information Systems, 11, 1 (Jan. 1993), 1-26

[HDM-2] U.Cavallaro, F.Garzotto, P.Paolini and D.Totaro, "HIFI: Hypertext Interface for Information Systems", IEEE Software 10, 6 (Nov. 1993), 48-51.

[HDM-3] F.Garzotto, L.Mainetti and P.Paolini, "Hypermedia Application Design: A Structured Approach", In J.W.Schuler, N.Hannemann and N.Streitz Eds., "Designing User Interfaces for Hypermedia", Springer Verlag, 1995.

[HDM-4] F.Garzotto, L.Mainetti and P.Paolini, "Navigation in Hypermedia Applications: Modeling and Semantics", Journal of Organizational Computing (in press)

[HPCN-1] V.S.Sunderam, "PVM: a framework for parallel distributed computing", Concurrency: Practice and Experience, Vol. 2, Dec. 1990, pp.315-339.

[HPCN-2] G.A.Geist and V.S.Sunderam, "Network based concurrent computing on the PVM system", Concurrency: Practice and Experience, Vol. 4, June 1992, pp.293-311.

[FRA-1] A.Jacquin, "A Fractal Theory of Iterated Markov Operators with Applications to Digital Image Coding", Ph.D. thesis, Georgia Institute of Technology, 1989.

[FRA-2] Y.Fischer, "Fractal Compression Theory and Application to Digital Images", Springer Verlag, New York, 1994

[JPEG-1] G.Wallace, "Overview of the JPEG (ISO/CCITT) still image compression standard, Image Processing Algorithms and Techniques", proceedings of the SPIE, vol. 1244, Feb. 1990, pp. 220-233.

[JPEG-2] A.Léger, T.Omachi, G.Wallace, "The JPEG Still Picture Compression Algorithm", Optical Engineering, vol. 30, n. 7, July 1991, pp. 947-954.

[JPEG-3] T.Lane et al., "Independent JPEG Group's free JPEG software: Using the IJG Library", on Internet {ftp://ftp.uu.net/graphics/jpeg/jpegsrc.v6.tar.gz}, December 1995.

[PHO-1] "The official Photo-CD Handbook", Verbum Books, San Francisco, California.

[PHO-2] J.Larish, "Photo CD: Quality Photos at Yours Fingertips", Micro Publishing News, Torrance, California.

[WAV-1]E.Simoncelli, E.Adelson, "Subband Transforms" in Subband Coding (J.Woods, ed.), ch. 4, Norwell, Massachusetts, Kluwer Academic Publishers, 1990.

[WAV-2] A.Lewis and G.Knowles, "Image Compression Using the 2D Wavelet Transform", IEEE Trans. Image Processing, vol. 1, pp. 244-250, April 1992.

[WAV-3] A.Zandi, J.Allen, E.Schwartz and M.Boliek, "CREW: Compression with Reversible Embedded Wavelets", in IEEE Data Compression Conference, Snowbird, Utah, pp. 212-221, March 1995.

Patterns for Constructing CSCW Applications in TINA

Eckhart Koerner
Université de Liège
Institut d'Electricité Montefiore B28, B- 4000 Liège, Belgium
Email: koerner@montefiore.ulg.ac.be

Abstract

There is a convergence between the idea of design patterns and the goal to achieve reuse of component design in the Telecommunications Information Networking Architecture (TINA). Patterns can be specified using graphical and syntactical notations proposed by the TINA Consortium (TINA-C) for the information and computational viewpoint languages of the Open Distributed Processing Reference Model (ODP-RM). In this article four design patterns are proposed that efficiently support Computer Supported Collaborative Work (CSCW) applications in TINA: voting, floor control, role management and group integrity criteria. The design is done with the goal of high configurability in mind. It is planned to implement the patterns to achieve an extension of a prototype of a TINA multimedia communication service. This service is thus to evolve to a collaboration service. Eventually, the implementation of the patterns should become part of an object-oriented framework for TINA applications.

Keywords

TINA, CSCW, design patterns, multimedia collaboration service

1. Introduction

The software architecture proposed by TINA-C provides substantial assistance to designers of telecommunications services including multimedia, multi-party and mobility features. In TINA operating system and network technology aspects are hidden to the designers by building a technology-independent Distributed Processing Environment (DPE) on top, in which the telecommunications applications can run [TN-ARC].

The TINA specifications have already been successfully implemented in the TANGRAM project at GMD Fokus in Berlin (see figure 1) [DEE+96]. The application layer is partitioned into service, resource and element layers. The two lower layers provide an abstraction of network connections to the service layer in the form of stream bindings. In TANGRAM, this segment has been implemented by writing object wrappers for components from a former prototype of a multimedia teleservice [BK-MMC]. The service layer has been segmented and designed according to the TINA Service Architecture [TN-SA]. The Service Session Segment provides general procedures for service session control based on the concept of a Session ("State") Graph. The following functional aspects are treated in this segment:
• party attachment to a service,
• stream binding control,
• explicit control of the use of resources (access and ownership control).

The specific logic for the actual application is encapsulated in the Service Segment. In TANGRAM, a Multimedia Communication Service (T-MMCS) has been realised as a sample application. It allows multiple parties to exchange audio-visual information in a service session. The T-MMCS runs on the TANGRAM-DPE which is built out of interconnected CORBA 2 [OMG-C2] processing platforms. Sun Solaris is used as computing support. The Internet Inter-ORB protocol stack (IIOP) over ATM provides the communications environment.

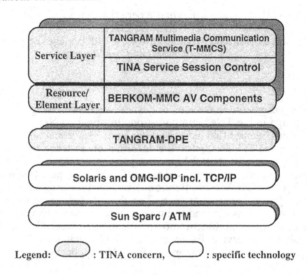

Legend: ⬭ : TINA concern, ⬭ : specific technology

Figure 1: TANGRAM and the TINA environment

In this article an evolution path for the T-MMCS towards a sophisticated multimedia collaboration service is shown. A requirements analysis for this evolution has already been presented in [Koe96]. As suggested by TINA-C we use the viewpoint separation of the ODP-RM [ITU-X901] as the main structuring means for the service specifications. TINA has concretised the abstract viewpoint languages for the information and computational viewpoint by graphical and syntactical notations:

- OMT [Rum91] as a graphical and qGDMO/qGRM as a syntactical notation for the information model [TN-IMC],
- a non-standard extension of OMT as graphical and the Object Definition Language TINA-ODL [TN-ODL] [TN-CMC] as a syntactical notation for the computational viewpoint. TINA-ODL has been defined as a strict superset of CORBA IDL. The major added value of ODL consists in the definition of syntax for stream interfaces with QoS. Besides, computational objects with multiple interfaces can be defined. The computational viewpoint is further supported by the documentation of application scenarios in object interaction diagrams.

While the TINA approach focuses on design reuse, the TANGRAM project also wants to achieve implementation reuse. The idea of object-oriented frameworks is applied for that purpose [EFS+97]. An object-oriented framework consists of a set of prefabricated object services with semi-complete implementations in a chosen programming language. The services are glued together according to architectural concepts and rules. Subsequently, the implementations can be completed by inheritance and subtyping to achieve a solution to a given problem. In the case of TINA the goal is to construct a framework for telecommunications specific applications with the Service Architecture as the glue for the different components. Patterns aid the development of frameworks. A set of patterns can be viewed as an abstract description of a framework that documents its form and contents [Sch95]. The notion of design pattern can be matched in TINA with the documentation of the information and computational model of successful solutions to problems that arise in telecommunications. For instance, the following design patterns are units of the TANGRAM framework:

- subscription to a telecommunication service,
- creation and instantiation of a service session,
- establishment and usage of a stream connection (see AV components in figure 1).

In the next four sections additional design patterns are proposed that specifically support CSCW applications: voting, floor control, role management and group integrity criteria. In the final section some concluding remarks are made.

2. The Voting Pattern

Voting is an essential feature in CSCW applications to support group decision making. The use of voting in TINA applications may be twofold. First, it is an essential ingredient in the flexible negotiation of session configurations and operations. For example, parties may want to vote about the joining of another party to the session. Secondly, it may be made available to the parties whenever they want to use it as a tool to reach a decision in their collaboration.

Figure 2 depicts an abstract information model for voting. A *Voting_Group* is associated to a *Voting*. The interaction between instances of the two classes is governed by the *Voting_Control_Rules*. For instance, the *decision_rules* attribute defines the counting method to be applied, such as minimum, simple majority, absolute majority, consensus and unanimity. A *Voting* is composed of *Vote* objects, as well as a *Voting_Group* is an aggregation of *Voter* objects. *Voter* and *Vote* are related with optional-to-optional cardinality. This relation first of all expresses that each voter has one vote. Further, in a secret *Voting* the association between a *Voter* and a *Vote* may be removed as soon as the *Vote* has been given.

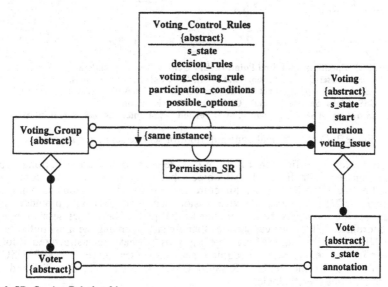

Legend: SR: Session Relationship

Figure 2: OMT static information model for voting

The *Permission_SR* class is needed to configure the monitoring mode of the *Voting*. In an open *Voting* read access to a *Vote* object including the link to the *Voter* object is granted and announcements about individual votes may be sent out to the *Voting_Group* members. In a secret *Voting* the access to the links between *Voter* and *Vote* objects has to be denied. Read access to the *Vote* objects may still be granted, in order to allow the inspection of annotations given as part of a *Vote*.

Two kinds of voting can be distinguished:
- Yes-No Voting: voters decide about the acceptance or rejection of a single proposal,
- Alternative Voting: voters decide about the adoption of one among several proposed alternatives.

The abstract *Voting*, *Vote* and *Voting_Control_Rules* classes are specialised into concrete classes for these types of voting. In particular, the attribute characterising the result of a voting is defined as part of the specialisation of the *Voting* class. The actual choice taken by a voter is defined as an attribute in the class derived from *Vote*. The *Voting_Group* and *Voter* classes can be concretised by mapping them onto the *Party* and *Party_Session_Member_Group* classes in the TINA Session Graph.

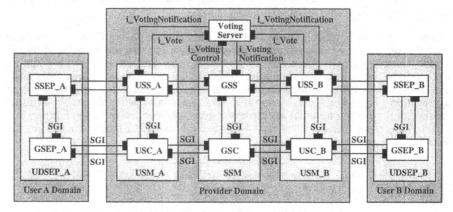

Legend: SSEP: Specific Session End Point, GSEP: Generic Session End Point,
          UDSEP: User Domain Session End Point, USS: User Service Segment,
          USC: User Session Control, USM: User Service Session Manager,
          GSS: Global Service Segment, GSC: Global Session Control,
          SSM: Service Session Manager, SGI: Session Graph Interface, ■ : server interface

Figure 3: Computational configuration for service session management

The information specification of voting is related to the computational object configuration shown in figure 3. All objects but the voting server are components suggested by the TINA Service Architecture. Each party has a session end point in the user domain (UDSEP) and a user service session manager (USM) in the provider domain that manages the service aspects particular to that party. The service session manager (SSM) coordinates all service capabilities that are shared among the users in the service session. UDSEP, USM and SSM are object groups that have been structured according to the service and service session segment separation (see figure 1). GSEP, USC and GSC hence provide the generic Session Graph Interface. SSEP, USS and GSS encapsulate service specific logic.

A voting server makes sense as a decentralised Session Support Object (SSO) in this configuration. Firstly, it can be used as a group decision support tool independently of session management. Secondly, when used to reach an agreement about session management operations the interaction and coordination with the SSM objects only takes place at the start and end of a voting.

The computational interfaces of voting are:

- *i_votingControl*: This interface allows a client to manage votings. Basic operations are supplied to define, modify and remove a voting and associated voting control rules. Query operations are provided to get the configuration of a voting. References to voting groups can be exchanged. Subsequently, they can be associated to a voting. State management operations are provided to drive the voting from the defining to the open state and from the open to the closed state. The latter operation can be used to cancel an ongoing voting,
- *i_vote*: This interface is used to submit a vote. Further, this interface can be used to query the decision of a voter. For secret voting a query is defined that allows to browse the annotations of individual votes without seeing the association to the corresponding voter,
- *i_votingNotification*: This interface serves for receiving reports of ongoing voting activities. This includes the report about an imminent voting with associated voting control rules through which all voters receive the necessary information to take part in the voting. Other notifications are defined to inform about an individual vote, to indicate the result of a voting and to notify about the cancellation of a voting.

A voting server object supports the *i_votingControl* and the *i_vote* interfaces as a server. It is a client of the *i_votingNotification* interface.

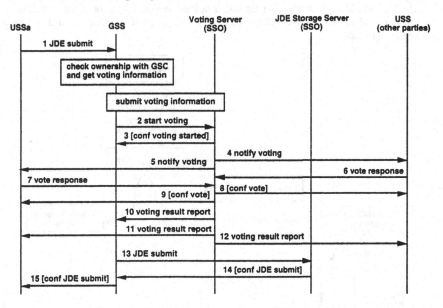

Figure 4: Voting about a service specific operation

Figure 4 exemplifies the usage of voting in the negotiation of a session management operation. A party submits a document to a Joint Document Editing (JDE) storage server as an outcome of a collaborative session. After invocation of the operation the GSS checks with the GSC whether a voting for this operation is required. The GSC has the knowledge about the owners of the document which need to agree and thus constitute the voting group. The case of a voting is assumed. The GSS then submits the voting group, voting control rules and permission session relationship information to the voting server and requests the start of the voting. A secret voting is assumed so that no reports about the individual decisions are sent. The voting server informs the parties of the imminent vote and collects the decisions. The result is sent to the GSS and to the parties via the USS objects. An acceptance of the operation is assumed. The GSS finally invokes the operation at the JDE storage server and confirms the operation to the invoker.

3. The Floor Control Pattern

Floor control is used to attribute the right to perform a particular activity in a collaborative service session. For an audio-visual conference a token holder is attributed the right to provide audio-visual input. In a shared application token passing is used for maintaining the consistency of multimedia documents shown among parties. A token holder has the right to manipulate shared windows on which the documents are shared with all the parties. He can accomplish such operations as opening and closing shared windows, loading documents to a shared window, scrolling documents, and inputting multimedia information to documents. Several tokens may be linked together such that they are always passed in combination.

Floor control is governed by token control rules. Three basic floor passing modes are defined [BK-MMC]:
- in the Automatic mode the floor is passed automatically among parties according to one of several possible automatic assignment rules,
- in the Master mode the decision about the next token holder is taken by a master. If the master releases the token, automatic assignment is enabled until the next token assignment,
- in the Tokenholder mode the current token holder designates the next token holder. If the tokenholder releases the token, automatic assignment is enabled until the next token assignment.

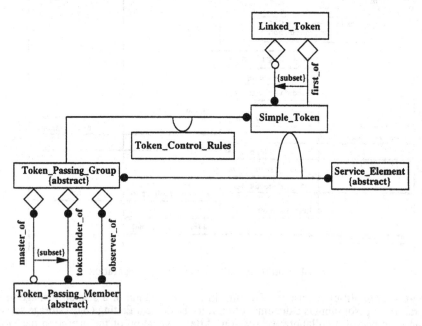

Figure 5: Floor control information model

The information modelling of floor control is shown in figure 5. A token is modelled as a relationship class that controls the association between a *Token_Passing_Group* and a *Service_Element* such as a video conference or a shared application. Three roles are distinguished for floor control. The master role is optional as it may be unassigned in the automatic and tokenholder mode. A tokenholder is entitled to become holder of the token. Observers are merely monitoring the token passing. It should be noted that the three roles in token passing are particular to that component and have to be associated to roles in an actual TINA application. For instance, in a conference application the master role may be related to a chairman, the tokenholder role to the speakers and the observer role to the attendees. The interaction between the *Token_Passing_Group* and a simple or linked token is governed by an instance of the *Token_Control_Rules*. The *Token_Control_Rules* object notably establishes the automatic, master or tokenholder mode. The aggregation of *Simple_Token* instances into a *Linked_Token* object expresses the propagation of operations from the *Linked_Token* to the contained *Simple_Token* objects. The first *Simple_Token* that is linked determines the *Token_Control_Rules* adopted for the *Linked_Token*.

In the computational viewpoint it could be envisaged to coordinate floor control by a Concurrency Access Control Server (CACS) object as a decentralised TINA SSO. However, a CACS server would necessitate much interaction and coordination with the GSS to keep consistency. The GSS would have to communicate the knowledge about the token passing group to the CACS, based on the current session membership. Besides, token passing typically maps onto the activation or establishment of underlying connectivity resources via the Session Graph Interface at the GSC. Consequently, the token passing interactions will preferably be executed by instantiating adequate operational interfaces for floor control at the USS and the GSS.

4. The Role Management Component

Roles have been widely used for modelling the authority and responsibility within organisations. Such organisational structures may also be reflected in collaboration sessions. Role management has been studied in computer systems mainly for role-based access control [SaC96]. Such role-based systems benefit from a logical independence of user-to-role assignment and the specification of authorisations associated to that role. Figure 6 shows this separation in the information model. A *Role* instance is characterised by a name and several role constraints. Roles can be assigned to individual parties or a group of parties which are abstracted into a *Controller* class. In the general case, a *Controller* may be assigned several roles concurrently. This association is constrained by mutually exclusive roles and prerequisite roles. The maximum number of parties in a role is fixed in the *cardinality_constraints* attribute. A *Controller* has control capabilities towards resources, abstractly represented by the *Controlled* class. The set of control capabilities then characterises the role of the *Controller*. In the case of role-based access control the *Control_SR* class is concretised by permission session relationships.

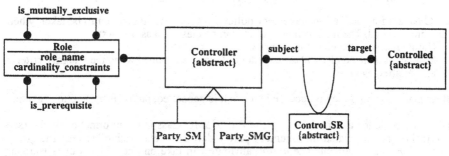

Legend: SM: Session Member, SMG: Session Member Group

Figure 6: Information model for role management

In the computational model interfaces are defined that provide operations to associate a *Controller* to a *Role* and to establish control session relationships between a *Controller* and a *Controlled* object. With that model roles can thus be configured very flexibly.

5. Group Integrity Criteria

Group Integrity Criteria (G-IC) specify conditions on the membership in a service session. The G-IC concept has first been introduced in the context of the OSI architecture [MLB+94], but can be adapted to TINA. The requirement for G-IC can be illustrated by the following simple scenario: an end-user initiates a multi-party voice call and invites five people to the service session. Only one of the called end-users is willing to accept the invitation. It is most likely that the caller will prefer not to establish the audio binding, because he is motivated to save money. He will rather shutdown the session. While this example only illustrates the use of G-IC at session establishment, they may also need to be monitored during session execution. For example, a business meeting may require the permanent telepresence of "at least four managers".

G-IC can be expressed as a combination of the following conditions:
• Minimum: A condition that specifies the minimum number of parties such that the G-IC are satisfied,
• Maximum: A condition that specifies the maximum number of parties required to consider that the G-IC are satisfied,
• Key member: A party necessary to consider that the G-IC are satisfied.

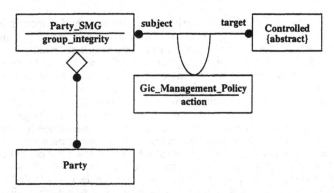

Figure 7: Session Graph object types related to group integrity criteria

G-IC have an associated management policy that specifies the action to be taken when they are violated. The management policy can be classified as follows:
• Hard: an associated element is removed,
• Soft: an associated element is suspended. It can be resumed when the integrity criteria are satisfied again.

If no management policy is specified an event is generated, but no further action is taken.

G-IC are modelled as an attribute of the *Party_SMG* class in the information model (see figure 7). The associated management policy defines a relationship between the group and a *Controlled* element. The *action* attribute can take on one of the values Hard or Soft. The *Controlled* class can be concretised by, for example, a class representing the service session or representing a service element.

It should be noted that G-IC cannot only be specified for a party group that represents the session membership as a whole. For example, in a collaboration session party groups involved in several shared applications could have disjoint membership. Then G-IC may be specified for each of them separately. G-IC may conflict with cardinality constraints expressed for roles (see figure 6). This should be considered by the service designer.

In the computational viewpoint specific operations to manipulate G-IC can be defined. G-IC may also be passed by extension of existing operations. Figure 8 shows the use of G-IC in relation to a service specific invitation procedure. It is assumed that user A has created the service session and initiates an invitation at the SSEPa with users B and C as key members. Further, it is assumed that a G-IC management policy is established that enforces the shutdown of the service session if the G-IC are not satisfied. SSEPa forwards the invitation to USSa which will check the access rights of user A to invoke this operation. As soon as the check succeeds, the request is forwarded to the GSS. The GSS sends a separate invitation to the concerned user agents (UAb and UAc). When UAb rejects the GSS recognises that the G-IC cannot be satisfied. Consequently, the GSS cancels the invitation to UAc which has not responded yet. Then the GSS coordinates the orderly session shutdown.

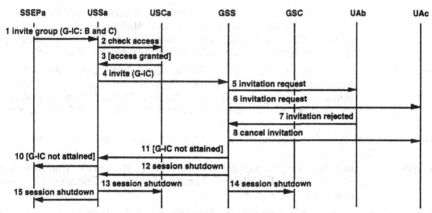

Figure 8: Scenario of failing session invitation

6. Conclusion

The patterns have been designed to achieve a high degree of configurability. In voting and floor control the coordination among parties is governed by control rules. These have been explicitly modelled and can be chosen among a large number of possible configurations. Similarly, in role management and management of group integrity criteria policies have been identified that allow to flexibly influence and modify the behaviour of the components.

This property of high configurability also distinguishes this design from former multimedia collaboration teleservices like the BERKOM MMC [BK-MMC]. Besides, voting and group integrity criteria are features that have not been distinctively offered by comparable systems. Roles have typically been defined statically with fixed authorisations.

All presented features are complementary to the TINA Session Graph. The information models are basically independent of TINA and may hence also be interesting in another architecture or market segment. The computational model necessarily takes the offered TINA components into account. The patterns should be useful to build a general-purpose TINA collaboration service. For such an application, a high number of configuration options and a high degree of dynamic configuration changes during a service session may be preserved. Such an application can then be further specialised for specific segments like business meetings, tele-education and medical diagnostics using a subset of the configuration options. The frameworks approach should be very advantageous to drive such an approach.

Acknowledgements

The presented work has been carried out in the framework of a contract between the University of Liege, Belgium and GMD Fokus, Germany. The author would like to express his gratitude to the members of the DEIMOS and TANGRAM projects at GMD Fokus for the excellent cooperation.

References

[BK-MMC] BERKOM Project, *The BERKOM Multimedia Collaboration Teleservice, Release 3.1*, Technical Report, December 1993

[DEE+96] M. Khayrat Durmosch, Christian Egelhaaf, Klaus-Dietrich Engel, Peter Schoo, *Design and Implementation of a Multimedia Communication Service in a Distributed Environment based on the TINA-C Architecture*, Aachen Workshop on Trends in Distributed Systems, October 1996

[EFS+97] K.-P. Eckert, M. Festini, P. Schoo, G. Schuermann, *TANGRAM: Development of Object-oriented Frameworks for TINA-C Based Multimedia Telecommunication Applications*, Proc. of Third International Symposium on Autonomous Decentralised Systems (ISADS97), April 1997

[ITU-X901] ISO IS 10746-1 I ITU-T Rec. X.901, *ODP Reference Model: Overview*, June 1995

[Koe96] Eckhart Koerner, *Group Management for a Multimedia Collaboration Service*, Proc. of EUNICE96 Summer School on Telecommunication Services, September 1996

[MLB+94] L. Mathy, G. Leduc, O. Bonaventure, A. Danthine, *A Group Communication Framework*, Proc. of Broadband Islands Conference (BRIS94), 1994

[OMG-C2] OMG, *The Common Object Request Broker: Architecture and Specification*, Revision 2.0, July 1995

[Rum91] J. Rumbaugh et al., *Object-Oriented Modelling and Design*, Prentice Hall, 1991

[SaC96] R. Sandhu, E. Coyne, *Role-Based Access Control Models*, IEEE Computer, February 1996

[Sch95] Douglas C. Schmidt, *Using Design Patterns to Develop Reusable Object-Oriented Communication Software*, Communications of the ACM, Vol. 38, No. 10, October 1995

[TN-ARC] TINA-C, *Overall Concepts and Principles of TINA*, Archiving Label: TB_MDC.018_1.0_94, February 1995

[TN-CMC] TINA-C, *Computational Modelling Concepts*, Archiving Label: TB_A2.HC.012_1.2_94, February 1995

[TN-IMC] TINA-C, *Information Modelling Concepts*, Archiving Label: TB_EAC.001_1.2_94, April 1995

[TN-ODL] TINA-C, *TINA Object Definition Language (TINA-ODL) Manual Version 2.3*, Archiving Label: TR_NM.002_2.2_96, July 1996

[TN-SA] TINA-C, *Definition of Service Architecture Version 4.0*, Archiving Label: TB_RM.001_4.0_96, December 1996

Performance Evaluation of the Fuzzy Policing Mechanism for Still Picture in ATM Networks

Leonard Barolli, Kuninobu Tanno and Ardian Greca

Yamagata University, 4-3-16 Jounan, Yonezawa, Yamagata 992, Japan

Abstract. The Asynchronous Transfer Mode (ATM) technique has been accepted as a basis for future B-ISDN networks. In ATM networks, all information is packetized and transfered in small packets of fixed length, called cells. The packetized information transfer, without flow control between the user and the network and the use of statistical multiplexing, results in a need of a Policing Mechanism (PM) to control the traffic parameters of each virtual connection in order to guarantee the required Quality of Service (QoS). Policing of the peak cell rate is generally not complex. The control of the mean cell rate is more difficult, but is intended to improve the link utilization when it has to handle bursty traffic sources. Conventional PMs, such as the Leaky Bucket Mechanism (LBM) and the Window Mechanisms (WMs), are not well suited to the bursty nature of sources supported by ATM networks, therefore intelligent PMs are needed. In this paper, we propose a Fuzzy Policing Mechanism (FPM) to police the mean cell rate of the still picture source. We consider the case when the peak cell rate of the still picture source is not controlled separately by other PMs. The performance evaluation via simulations shows that the FPM efficiently controls the mean cell rate of the still picture source. The FPM has better responsiveness and selectivity characterisics than the LBM.

1 Introduction

ATM networks are proposed as a transfer mode for future B-ISDN. ATM networks will provide the flexibility to cope with a wide range of services, characterized by different traffic behaviors. The services will share the same ATM transport network, where packets of fixed length belonging to different calls are multiplexed together, and transmitted between user and network without flow control. Due to the characteristics of ATM networks, a source (user) once accepted by Connection Admission Control (CAC) may exceed the negotiated parameters, therefore PMs or Usage Parameter Control (UPC) are needed.

The proposed parameters for monitoring source traffic characteristics are the mean cell rate, the peak cell rate or the peak burst duration [9]. Policing of the peak cell rate is generally not complex and can be achieved by using a cell spacer or other PMs [10]. Monitoring the mean cell rate is more difficult, but is intended to improve the link utilization when it has to handle bursty traffic sources.

So far, some performance evaluations of the conventional PMs for ATM networks have been carried out [1], [2], [3]. The conventional PMs such as the LBM and the WMs can't efficiently monitor the mean cell rate of bursty sources. Most of these PMs suffer from serious shortcomings. Some are simple but include many approximations and assumptions that are hard to justify. Others include complicated mathematical solutions that may not be feasible for real time implementation. The WMs of traditional packet switched networks are not well suited to the bursty nature of sources supported by B-ISDN, and the LBM in the case of mean cell rate control requires a very high counter threshold to obtain an acceptable cell loss probability. This means that very long times are necessary to detect a violation of the mean cell rate. Therefore, new adaptive PMs are needed to control efficiently the mean cell rate of bursty sources in ATM networks.

The difficulty of characterizing a policer accurately, if conventional methods and models are used, led us to explore alternative solutions based on soft computing techniques, especially in the field of fuzzy systems. Fuzzy set theory has been accepted in literature as a robust mathematical framework for dealing with certain forms of imprecision that frequently occur in decision making environments, but for which the probability calculus is inadequate. Such imprecision is inherent in diverse ATM environments with bursty nature of sources. In practical situations the mean arrival rate and the mean service rate are frequently fuzzy, i.e., they can't be expressed in exact terms.

Fuzzy systems promise to offer a rich language for traffic control by providing soft and flexible policing action, characterizing imprecise quantities (e.g., mean arrival cell rate, mean silence duration), and capturing linguistic, rule based control strategies. The philosophy on which the FPM is based exploits the fuzzy logic capability to deduce a system model on the basis of linguistic variables, fuzzy sets and fuzzy inferences. The rules are expressed in approximate terms, but at the same time corresponding to an expert description. This allows the rules to be translated into a rigorous fuzzy inferential system, the performance of which is very close to the ideal characteristics required for a PM.

In this paper, we propose a FPM to police the mean cell rate of the still picture source in ATM networks. The organization of this paper is as follows. The source and system models will be given in Section 2. In Section 3, the FPM design will be presented. Simulation results are discussed in Section 4. The implementation issues of the FPM will be treated in Section 5. The conclusions are given in Section 6.

2 Source and System Models

We assume for the cell arrival process pattern a bursty source as shown in Fig.1(a). This on-off source is considered as the worst case traffic pattern [1], [2]. Each burst has a duration *mbd* (mean burst duration) random variable and a cell rate of *pcr* cell/s (peak cell rate). The duration of inactive (silence) period is the random variable *msd* (mean silence duration). The source is characterized by the following set of parameters: the peak (burst) cell rate [*pcr*]; the mean burst duration [*mbd*]; the mean silence duration [*msd*]; the source burstiness [*sb = (mbd+msd)/mbd*]; the mean burst length in cells (or burst cell number) [*bcn = pcr · mbd*]; the mean source cell rate [*mcr = bcn/(mbd+msd)*]; and the mean cycle duration [*mcd = mbd+msd*].

The system model has three parts (see Fig.1(b)): the Fuzzy Logic Controller (FLC), the subtractor and the counter.

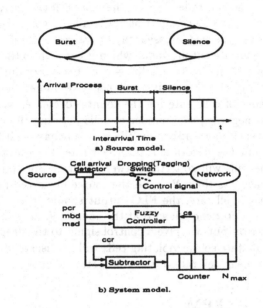

Fig. 1. Source and system models.

The FLC is the major component in the proposed FPM. The main function of the FLC is to control the source short-term behavior. The components of the FLC are the fuzzifier, fuzzy rule base, inference engine and defuzzifier. The structure of the FLC is shown in Fig.2.

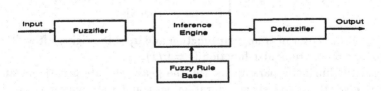

Fig. 2. FLC structure.

The counter's main function is to control the long-term behavior of the source. If the cell arrival number exceeds a predefined number (the maximum value of the counter) the FPM will act and will discard or tag the exceeded cells. The counter state is expressed as: $cs = cs_0 + bcn - ccr$, where cs_0 is the counter state (in cells) at the starting moment, bcn is the number of cells in a burst, and ccr is the output of the FLC which indicates the number of cells that counter state should be changed.

The FPM works in this way: the detector counts the number of cells going to the network and at the same time going to the FLC and the subtractor. The parameters of the controlled source: the peak cell rate pcr, the mean burst duration mbd, the mean silence duration msd and the counter state cs, are the input parameters for the FLC. Based on the values of input parameters, the FLC gives an appropriate output value, which it enters into the subtractor. The subtractor carries out the operation $bcn - ccr$. The fuzzy output membership function ccr (see Fig.4) has positive and negative values. If the ccr value is positive, the number of cells entering the counter decrease, on the other hand, if the ccr value is negative, the number of cells entering the counter increase. FPM monitors at each moment of time the source behavior. If the source mean cell rate is less than the negotiated mean cell rate, the dropping switch is in on-state, so all the cells will go to the network. During this time FLC keeps the counter state always less than N_{max}. If the source mean cell rate is more than the negotiated mean cell rate, the FLC output should increase the number of cells entering the counter enough so that the counter state would be more than N_{max}. The counter output will give a control signal to the dropping switch and change it to the off-state so the violating cells will be discarded. In this way, the FPM controls the mean cell rate of the still picture source.

3 Design of the FPM

The selection of shape and position of the membership function is a very important issue of the FPM design. During many experiments we found that the triangular membership function is more appropriate for our system, because it is easy to tune the membership functions and the error is smaller compared with other membership function shapes. The function $f(x, x_0, a_0, a_1)$ for triangular shape is defined (see Fig.3) below:

$$f(x, x_0, a_0, a_1) = \begin{cases} \frac{x - x_0}{a_0} + 1 & \text{for } x_0 - a_0 < x \leq x_0 \\ \frac{x_0 - x}{a_1} + 1 & \text{for } x_0 < x \leq x_0 + a_1 \\ 0 & \text{otherwise} \end{cases}$$

where x_0 is the center of triangular function and a_j is the right/left width of the monotonic part of triangular function ($j = 0/1$).

The input linguistic parameters are: the peak cell rate pcr, the mean burst duration mbd, the mean silence duration msd and the counter state cs. The output linguistic parameter is the controlled cell rate ccr that enters into the counter. The term sets of pcr, mbd, msd and cs are defined respectively as:

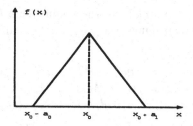

Fig. 3. Triangular membership function.

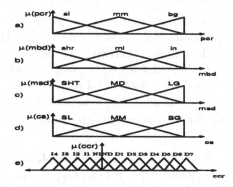

Fig. 4. Membership functions for a) *pcr,* **b)** *mbd,* **c)** *msd,* **d)** *cs* **and e)** *ccr.*

$$T(pcr) = \{small, medium, big\} = \{sl, mm, bg\};$$
$$T(mbd) = \{short, middle, long\} = \{shr, mi, ln\};$$
$$T(msd) = \{SHORT, MIDDLE, LONG\} = \{SHT, MD, LG\};$$
$$T(cs) = \{SMALL, MEDIUM, BIG\} = \{SL, MM, BG\}.$$

The set of the membership functions associated with *pcr*, $T(pcr)$, are denoted by $M(pcr) = \{\mu_{sl}, \mu_{mm}, \mu_{bg}\}$, where $\mu_{sl}, \mu_{mm}, \mu_{bg}$ are the membership functions for *sl, mm, bg,* respectively. They are given by (see Fig. 4 (a)):

$$\mu_{sl}(pcr) = f(pcr, sl_c, sl_{w0}, sl_{w1});$$
$$\mu_{mm}(pcr) = f(pcr, mm_c, mm_{w0}, mm_{w1});$$
$$\mu_{bg}(pcr) = f(pcr, bg_c, bg_{w0}, bg_{w1}).$$

$M(mbd) = \{\mu_{shr}, \mu_{mi}, \mu_{ln}\}$ are the membership functions for the term set of *mbd*. The membership functions μ_{shr}, μ_{mi} and μ_{ln} are given by (see Fig. 4 (b)):

$$\mu_{shr}(mbd) = f(mbd, shr_c, shr_{w0}, shr_{w1});$$
$$\mu_{mi}(mbd) = f(mbd, mi_c, mi_{w0}, mi_{w1});$$
$$\mu_{ln}(mbd) = f(mbd, ln_c, ln_{w0}, ln_{w1}).$$

$M(msd) = \{\mu_{SHT}, \mu_{MD}, \mu_{LG}\}$ are the membership functions for the term set of msd. The membership functions μ_{SHT}, μ_{MD} and μ_{LG} are given by (see Fig. 4 (c)):

$$\mu_{SHT}(msd) = f(msd, SHT_c, SHT_{w0}, SHT_{w1});$$
$$\mu_{MD}(msd) = f(msd, MD_c, MD_{w0}, MD_{w1});$$
$$\mu_{LG}(msd) = f(msd, LG_c, LG_{w0}, LG_{w1}).$$

The membership functions for the term set cs are $M(cs) = \{\mu_{SL}, \mu_{MM}, \mu_{BG}\}$, and μ_{SL}, μ_{MM} and μ_{BG} are given by (see Fig. 4 (d)):

$$\mu_{SL}(cs) = f(cs, SL_c, SL_{w0}, SL_{w1});$$
$$\mu_{MM}(cs) = f(cs, MM_c, MM_{w0}, MM_{w1});$$
$$\mu_{BG}(cs) = f(cs, BG_c, BG_{w0}, BG_{w1}).$$

We define the term set of the output linguistic parameter $T(ccr)$ as {Increase4, Increase3, Increase2, Increase1, Not Increase Not Decrease, Decrease1, Decrease2, Decrease3, Decrease4, Decrease5, Decrease6, Decrease7}. We write for short as $\{I4, I3, I2, I1, NIND, D1, D2, D3, D4, D5, D6, D7\}$, where I2 increases more than I1 and D2 decreases more than D1 and so on.

The term set of output membership functions, are denoted by $M(ccr) = \{\mu_{I4}, \mu_{I3}, \mu_{I2}, \mu_{I1}, \mu_{NIND}, \mu_{D1}, \mu_{D2}, \mu_{D3}, \mu_{D4}, \mu_{D5}, \mu_{D6}, \mu_{D7}\}$, which are given by (see Fig.4 (e)):

$$\mu_{I4}(ccr) = f(ccr, I4_c, I4_{w0}, I4_{w1});$$
$$\mu_{I3}(ccr) = f(ccr, I3_c, I3_{w0}, I3_{w1});$$
$$\mu_{I2}(ccr) = f(ccr, I2_c, I2_{w0}, I2_{w1});$$
$$\mu_{I1}(ccr) = f(ccr, I1_c, I1_{w0}, I1_{w1});$$
$$\mu_{NIND}(ccr) = f(ccr, NIND_c, NIND_{w0}, NIND_{w1});$$
$$\mu_{D1}(ccr) = f(ccr, D1_c, D1_{w0}, D1_{w1});$$
$$\mu_{D2}(ccr) = f(ccr, D2_c, D2_{w0}, D2_{w1});$$
$$\mu_{D3}(ccr) = f(ccr, D3_c, D3_{w0}, D3_{w1});$$
$$\mu_{D4}(ccr) = f(ccr, D4_c, D4_{w0}, D4_{w1});$$
$$\mu_{D5}(ccr) = f(ccr, D5_c, D5_{w0}, D5_{w1});$$
$$\mu_{D6}(ccr) = f(ccr, D6_c, D6_{w0}, D6_{w1});$$
$$\mu_{D7}(ccr) = f(ccr, D7_c, D7_{w0}, D7_{w1}).$$

Based on the above linguistic description of input and output parameters we make a Fuzzy Rule Base (FRB). The FRB forms a fuzzy set of dimensions $|T(pcr)| \times |T(mbd)| \times |T(msd)| \times |T(cs)|$, where $|T(x)|$ is the number of terms on $T(x)$. So, there are 81 rules in the FRB. The FRB of the membership functions is shown in Table 1. The control rules have the following form: IF "conditions" THEN "control action". The condition part has four input linguistic parameters and the control action is the output linguistic parameter. The input and output linguistic parameters are expressed by membership functions shown in Fig. 4. The fuzzy rule base is made based on the range of input and output parameters. Statements on conditions take forms such as "the pcr is small", "the mbd is middle", "the msd is long", or "the cs is big". Likewise, statements on control action might be "increase the ccr" or "decrease the ccr".

Table 1. FRB.

Rules	pcr	mbd	msd	cs	ccr	Rules	pcr	mbd	msd	cs	ccr
0	sl	shr	SHT	SL	I3	41	mm	mi	MD	BG	D3
1	sl	shr	SHT	MM	I2	42	mm	mi	LG	SL	D1
2	sl	shr	SHT	BG	NIND	43	mm	mi	LG	MM	D4
3	sl	shr	MD	SL	I2	44	mm	mi	LG	BG	D5
4	sl	shr	MD	MM	NIND	45	mm	ln	SHT	SL	I2
5	sl	shr	MD	BG	D1	46	mm	ln	SHT	MM	I1
6	sl	shr	LG	SL	NIND	47	mm	ln	SHT	BG	NIND
7	sl	shr	LG	MM	D1	48	mm	ln	MD	SL	D1
8	sl	shr	LG	BG	D2	49	mm	ln	MD	MM	D2
9	sl	mi	SHT	SL	I3	50	mm	ln	MD	BG	D3
10	sl	mi	SHT	MM	I1	51	mm	ln	LG	SL	D2
11	sl	mi	SHT	BG	D1	52	mm	ln	LG	MM	D4
12	sl	mi	MD	SL	NIND	53	mm	ln	LG	BG	D5
13	sl	mi	MD	MM	D2	54	bg	shr	SHT	SL	I4
14	sl	mi	MD	BG	D3	55	bg	shr	SHT	MM	I2
15	sl	mi	LG	SL	NIND	56	bg	shr	SHT	BG	NIND
16	sl	mi	LG	MM	D2	57	bg	shr	MD	SL	NIND
17	sl	mi	LG	BG	D3	58	bg	shr	MD	MM	D2
18	sl	ln	SHT	SL	I4	59	bg	shr	MD	BG	D3
19	sl	ln	SHT	MM	I2	60	bg	shr	LG	SL	D4
20	sl	ln	SHT	BG	NIND	61	bg	shr	LG	MM	D6
21	sl	ln	MD	SL	I1	62	bg	shr	LG	BG	D7
22	sl	ln	MD	MM	D2	63	bg	mi	SHT	SL	I3
23	sl	ln	MD	BG	D3	64	bg	mi	SHT	MM	I1
24	sl	ln	LG	SL	D4	65	bg	mi	SHT	BG	D1
25	sl	ln	LG	MM	D5	66	bg	mi	MD	SL	NIND
26	sl	ln	LG	BG	D6	67	bg	mi	MD	MM	D1
27	mm	shr	SHT	SL	I3	68	bg	mi	MD	BG	D4
28	mm	shr	SHT	MM	I2	69	bg	mi	LG	SL	D2
29	mm	shr	SHT	BG	NIND	70	bg	mi	LG	MM	D3
30	mm	shr	MD	SL	I1	71	bg	mi	LG	BG	D5
31	mm	shr	MD	MM	D2	72	bg	ln	SHT	SL	I3
32	mm	shr	MD	BG	D3	73	bg	ln	SHT	MM	I2
33	mm	shr	LG	SL	D1	74	bg	ln	SHT	BG	D1
34	mm	shr	LG	MM	D2	75	bg	ln	MD	SL	NIND
35	mm	shr	LG	BG	D3	76	bg	ln	MD	MM	D3
36	mm	mi	SHT	SL	I3	77	bg	ln	MD	BG	D4
37	mm	mi	SHT	MM	I2	78	bg	ln	LG	SL	D2
38	mm	mi	SHT	BG	NIND	79	bg	ln	LG	MM	D5
39	mm	mi	MD	SL	NIND	80	bg	ln	LG	BG	D7
40	mm	mi	MD	MM	D2						

4 Simulation Results

The evaluation of the FPM is based on the ATM Forum definition of the UPC mechanism transparency, which can be defined by the accuracy with which the UPC mechanism approaches the ideal mechanism [11].

We make the following definitions for the simulations.

- **Violation probability.**
  This is the ratio of the number of discarded cells to the number of cells sending from the source to network.
- **Number of emitted cells.**
  This is the number of cells that the source sends to the network.
- **Normalized parameters.**
  These are the ratios $pcr/pcr0$, $mbd/mbd0$, and $msd0/msd$.

We generate the burst and silence periods in an independent way. The distribution functions are exponential and density functions f(.) are expressed as: $f_B(b) = 1/mbd * \exp^{-mbd*b}$ and $f_S(s) = 1/msd * \exp^{-msd*s}$ for burst and silence, respectively. The peak cell rate is generated by the uniform distribution function.

For simulation purposes, we choose the still picture source which is known as a prototype of an on-off source. The parameters of still picture source are: $pcr0$ = 2 Mb/s = 4716 cell/s; $mbd0$ = 500 milliseconds (ms); $msd0$ = 11000 ms; $bcn0$ = 2358 cells; and $mcr0$ = 87 kb/s = 205 cell/s. The values for input and output linguistic parameters are assigned as shown in Table 2.

The characteristic of violation probability versus the number of emitted cells for different mbd and msd is shown in Fig.5. The mbd and msd are kept constant. The increase in the mean cell rate has been achieved by increasing the pcr. For mbd 500 ms and msd 11 seconds (sec), mbd 700 ms and msd 14 sec, mbd 650 ms and msd 7 sec, the violation probability is zero until the mean cell rate is 2534 cells, 9555 cells, and 12965 cells, respectively. At these points the mean cell rate value is 205 cell/s. After these points, the FPM starts to discard the violating cells. The violation probability increases gradually and afterwards increases slowly. The characteristic for mbd 500 ms and msd 11 sec increases rapidly at beginning. This happens because the number of emitted cells is small. The violation probability is lower for mbd 500 ms and msd 11 sec.

Fig.6 shows the characteristic of violation probability as a function of number of emitted cells for different pcr and msd. At the beginning, we keep the pcr 4000 cell/s and msd 13 sec constant. The increase in the mean cell rate has been achieved by increasing the mbd. The violation probability is zero until the number of emitted cells becomes 1650 cells. At this point the mean cell rate becomes 205 cell/s. The violation probability increases gradually and afterwards increases very slowly. Next, we change the pcr to 5500 cell/s and the msd to 7 sec. The violation probability is higher than the first pattern. Finally, the pcr and the msd are selected 4716 cell/s and 11 sec, respectively. From the three patterns the violation probability is higher for the second pattern. This is because by increasing the pcr and decreasing msd, the violation probability increases.

Table 2. Assignment of values for input and output linguistic parameters.

pcr		
$sl_c = 0$	$sl_{w0} = 0$	$sl_{w1} = 4716$
$mm_c = 4716$	$mm_{w0} = 4716$	$mm_{w1} = 4716$
$bg_c = 9432$	$bg_{w0} = 4716$	$bg_{w1} = 0$
mbd		
$shr_c = 0$	$shr_{w0} = 0$	$shr_{w1} = 500$
$mi_c = 500$	$mi_{w0} = 500$	$mi_{w1} = 500$
$ln_c = 1000$	$ln_{w0} = 500$	$ln_{w1} = 0$
msd		
$SHT_c = 0$	$SHT_{w0} = 0$	$SHT_{w1} = 11000$
$MD_c = 11000$	$MD_{w0} = 11000$	$MD_{w1} = 11000$
$LG_c = 22000$	$LG_{w0} = 11000$	$LG_{w1} = 0$
cs		
$SL_c = 0$	$SL_{w0} = 0$	$SL_{w1} = 1175$
$MM_c = 1175$	$MM_{w0} = 1175$	$MM_{w1} = 1175$
$BG_c = 2350$	$BG_{w0} = 1175$	$BG_{w1} = 0$
ccr		
$I4_c = -3000$	$I4_{w0} = 750$	$I4_{w1} = 750$
$I3_c = -2250$	$I3_{w0} = 750$	$I3_{w1} = 750$
$I2_c = -1500$	$I2_{w0} = 750$	$I2_{w1} = 750$
$I1_c = -750$	$I1_{w0} = 750$	$I1_{w1} = 750$
$NIND_c = 0$	$NIND_{w0} = 750$	$NIND_{w1} = 750$
$D1_c = 750$	$D1_{w0} = 750$	$D1_{w1} = 750$
$D2_c = 1500$	$D2_{w0} = 750$	$D2_{w1} = 750$
$D3_c = 2250$	$D3_{w0} = 750$	$D3_{w1} = 750$
$D4_c = 3000$	$D4_{w0} = 750$	$D4_{w1} = 750$
$D5_c = 3750$	$D5_{w0} = 750$	$D5_{w1} = 750$
$D6_c = 4500$	$D6_{w0} = 750$	$D6_{w1} = 750$
$D7_c = 5250$	$D7_{w0} = 750$	$D7_{w1} = 750$

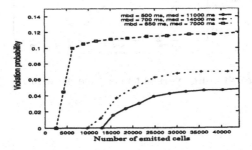

Fig. 5. Violation probability versus the number of emitted cells for different *mbd* and *msd* values.

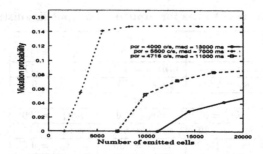

Fig. 6. Violation probability versus the number of emitted cells for different *pcr* and *msd* values.

Fig.7 shows the characteristic of violation probability versus the silence duration for different *pcr* and *mbd*. At the beginning, we keep the *pcr* 3500 cell/s and *mbd* 300 ms constant. The increase in the mean cell rate has been achieved by decreasing the *msd*. The violation probability is zero from 22 sec until the *msd* value becomes 16 sec. At this point the mean cell rate becomes 205 cell/s. With decreasing the value of *msd* the violation probability increases. For *pcr* 8500 cell/s, *mbd* 250 ms and *pcr* 5000 cell/s, *mbd* 650 ms the violation probability value is zero until the *msd* becomes 10 sec and 6 sec, respectively. The violation probability is lower for *pcr* 3500 cell/s and *mbd* 300 ms.

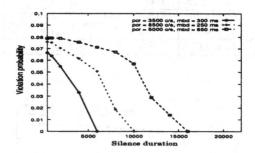

Fig. 7. Violation probability versus the silence duration for different *pcr* and *mbd* values.

The performance characteristics of the FPM and the LBM are shown in Fig.8. For the FPM, the violation probability is zero until the normalized value of *pcr*, *mbd* and *msd* becomes 1. At this point, the mean cell rate becomes 205 cell/s. After this point, the FPM starts to discard the cells which violate the negotiated parameters. The violation probability increases steeply and afterwards increases very slowly. The LBM starts to discard the cells before the normalized value of *mbd* becomes 1. Whereas, the FPM starts to discard the cells after the normalized parameters value becomes 1. This means the FPM has a better responsiveness characteristic than the LBM. The performance characteristics of the FPM are closer to the ideal characteristic than the LBM. This means the FPM has a better selectivity characteristic than the LBM.

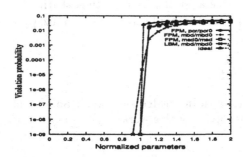

Fig. 8. Performance characteristics.

5 Implementation Issues

We consider as a fuzzy chip for implementation of the FPM the parallel architecture proposed in [14]. The speed of this fuzzy chip is about 77000 Fuzzy Logic Inference per Second (FLIPS), if one FPU is operating. The processing speed depends on the statistical characteristics of the source to be controlled. We consider the inter arrival cell time t_c 12 ms [9]. This is the maximum time limit by which the FPM has to infer the output. The FPM can begin processing rules when the input variables are all available, which does not happen until the end of the cycle. If we denote the processing time of the FPM with t_f, the value of t_f should be smaller than t_c in order to prevent a cell arriving at the beginning of the new cycle to escape control action. Based on this fuzzy chip, the FPM is capable of policing almost a thousand sources. This results in an improvement of the FPM hardware exploitation when the FPM is used to police a set of sources in time sharing. It should be noted that, the FRB of the FLC is more complex than in the case when the peak cell rate is considered constant, but we need only a fuzzy chip, a counter, and a subtractor to implement the FPM. Otherwise, the conventional PMs need another PM to control the peak cell rate.

6 Conclusions

In this paper, we proposed a FPM to police the mean cell rate of the still picture source in ATM networks. The FPM behavior was investigated by simulations.

Performance evaluation via simulations shows:

- The FPM efficiently monitors the mean cell rate of the still picture source.
- The FPM has a good dynamic responsiveness to parameter variations.
- The FPM has a better selectivity and responsiveness characteristics than the LBM. The FPM selectivity characteristics are very close to the ideal characteristic required for a PM.
- The violation probability depends on the source pattern.
- The FPM can police a set of sources in time sharing.
- It is easy to implement in hardware by using a fuzzy chip control processor, a subtractor, and a counter.

Acknowledgment

The authors gratefully acknowledge the Japan Society for the Promotion of Science (JSPS) for supporting this work.

References

1. Rathgeb, E.: Modeling and Performance Comparison of Policing Mechanisms for ATM Networks. IEEE J. Selec. Areas Comm. 9 No. 3 (1991) 325–334
2. Butto', M., Cavallero, E. and Tonietti, A.: Effectiveness of the "Leaky Bucket" Policing Mechanism in ATM Networks. IEEE J. Selec. Areas Comm. 9 No. 3 (1991) 335–342
3. Yamanaka, N., Sato, Y. and Sato, K.: Performance Limitation of the Leaky Bucket Algorithm for ATM Networks. IEEE Trans. on Comm. 43 No. 8 (1995) 2298–2300
4. Barolli, L. and Tanno, K.: A Fuzzy Approach for Source Policing in ATM Networks. Proc. ICOIN'96 Korea (1996) 482–490
5. Barolli, L. and Tanno, K.: A Fuzzy Policing Mechanism for Still Picture in ATM Networks. Proc. LCN'96 USA (1996) 40–47
6. Barolli, L. and Tanno, K.: A Fuzzy Policing Mechanism for Packet Voice in ATM Networks. Proc. IEEE ICCS'96/ISPACS'96 Singapore (1996) 124–128
7. Barolli, L. and Tanno, K.: Effectiveness of the Fuzzy Policing Mechanism for ATM Networks. IEICE Trans. on Inf. and Syst E80-D No.5 (1997) 608–613
8. Barolli, L. and Tanno, K.: Policing Mechanism in ATM Networks Using Fuzzy Set Theory. Trans. on Inf. Process. Soc. of Japan (IPSJ) 38 No.6 (1997) 1103–1115
9. De Prycer, M.: Asynchronous Transfer Mode. Solution for Broadband ISDN. Ellis Horwood Second Edition (1993)
10. Guillemin, F., Boyer, P., Dupis, A. and Romoeuf, L.: Peak Rate Enforcement in ATM Networks. Proc. IEEE Infocom'92 (1992) 6A.1.1–6A.1.6
11. The ATM Forum : ATM User-Network Interface Specification. Version 3.0 Englewood Cliffs New Jersey (1993)
12. Chang, C., and Cheng, R.: Traffic Control in a ATM Network Using Fuzzy Set Theory. Proc. Infocom'94 (1994) 9c.1.1–9c.1.8
13. Hirota, K.: Industrial Applications of Fuzzy Technology. Springer-Verlag (1993)
14. Catania, V. and Ascia, G.: A VLSI Parallel Architecture For Fuzzy Expert Systems. International J. of Pat. Recogn. and AI 9 No. 2 (1995) 421–447

Implementation of an Audio/Video Conferencing Application over Native ATM

Torsten Braun
IBM European Networking Center
Vangerowstr. 18, D-69115 Heidelberg
Phone: +49 6221 59-4352, Fax: +49 6221 59-3300
E-mail: braun@heidelbg.ibm.com

Andreas Reisenauer
IBM PS Telecommunication & Media
Lyoner Str. 13, D-60528 Frankfurt
Phone: +49 69 6645-2662, Fax: +49 69 6645-3848
E-mail: andreas_reisenauer@de.ibm.com

Abstract. Distributed Multimedia Applications have high Quality-of-Service (QoS) requirements regarding throughput, delay, and delay jitter. Current communication systems do not provide QoS to the applications even in the cases where the network technology may provide QoS guarantees. This paper describes a Internet audio/video conferencing application on top of a native ATM communication system. The UDP/IP communication system has been substituted by AAL5/ATM. The paper describes how the applications can now benefit from the QoS capabilities of ATM and the required modifications of the audio/video conferencing applications in order to use the native ATM service.

1 Introduction

Distributed Multimedia Applications have high quality-of-service requirements concerning bandwidth, delay, and delay jitter. Legacy networks such as LAN technologies, e.g. Ethernet, Token Ring, and FDDI, or wide area networks such as X.25 do not provide quality-of-service to the application or to the communication system on top of the network. The ATM network technology overcomes this drawback and is able to offer QoS guarantees to the communication system. However, today's communication systems running on top of ATM such as LAN emulation [2] or classical IP over ATM [10] do not provide the full QoS capabilities of ATM to the application. Therefore, Internet applications based on UDP/IP or TCP/IP do often not benefit from ATM.

One approach to provide QoS to Internet applications is to extend the Internet protocol stack by the Resource Reservation Protocol (RSVP) [5]. Applications can then specify their desired QoS and deliver it to the RSVP entity, which in turn has to map the application QoS parameters to ATM parameters. In that case, the RSVP implementation has to interact with the underlying classical IP over ATM [6][7] or LAN emulation implementation. The RSVP implementation has to deliver the ATM QoS parameters to be used for ATM connection establishment to the classical IP over ATM or to the LAN emulation implementations respectively. These have then to establish ATM connections with the specified QoS and have to map IP packets to the established ATM connections.

The advantage of the RSVP approach is that the application must only slightly be changed. Only function calls must be inserted that allow to deliver the application QoS parameters via the RSVP API to the RSVP implementation. The socket calls to establish TCP connections and to send or receive application data via TCP or UDP remain unchanged.

The drawbacks of the RSVP approach depend on the limitations of the classical IP subnet model. No direct ATM connections are possible between ATM systems belonging to different logical IP subnets. Data streams on application level are mapped to the datagram-oriented IP service and the IP datagrams have to be mapped to ATM connections. Moreover, the application QoS parameters must be expressed using the RSVP terminology. The RSVP QoS parameters are then mapped to ATM parameters. Most of the Internet real-time applications are based on UDP. However, UDP and IP do not provide any functions which are necessary for real-time applications but they introduce additional protocol processing.

Because real-time applications do not require the functionality provided by UDP/IP or TCP/IP they can also run directly on top of an ATM service. The ATM API allows the application to specify the ATM QoS parameters. This allows a direct mapping of application QoS parameters to ATM QoS parameters without intermediate parameters such as RSVP parameters as mentioned above. Furthermore, the application data streams can be directly mapped to ATM connections while in IP over ATM communication systems, application data streams are broken into IP packets, which must be filtered and mapped to ATM connections on lower levels of the communication system. Another advantage of native ATM applications is that direct ATM connections can be set up between two ATM end systems even in the case they belong to different logical IP subnets.

The paper describes the implementation of vic and vat over native ATM. Vic (video conferencing tool) and vat (visual audio tool) are the most popular audio/video conferencing tools being used in the Internet. They usually run on top of UDP/IP, but they do not provide any interface for QoS selection. The paper describes the required modifications to implement vic and vat on top of an native ATM socket interface instead of an UDP/IP based datagram socket interface. Further modifications have been required to provide ATM QoS selection to the user.

2 ATM Sockets

AIX version 4.2 provides extensions to the socket programming interface to enable applications to access ATM services directly (cf. Fig. 1.). The existing socket calls were slightly modified to provide the necessary functionality, leaving the basic socket model unchanged [9]. A new address family (AF_NDD) was added to AIX 4.2 to enable applications to run directly on top of physical networks without any protocols in between. Fig. 2. shows the sequence of socket calls required to establish an ATM connection from a client to a server. The *bind* call is used to associate an ATM socket with a specific network adapter after the socket has been created using the *socket* call. ATM sockets that wish to establish Switched Virtual Circuits (SVCs), are required to set parameters that will be used by the signalling software and passed on through the ATM network. These parameters define the characteristics of the traffic that the application intends to send or receive. Additional parameters carry information intended for higher layers, i.e., a means for addressing different entities located on a signal ATM end point. These parameters can be written or read with the *setsockopt* and *getsockopt* calls respectively.

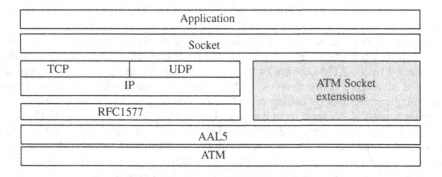

Fig. 1. : ATM Socket Extensions

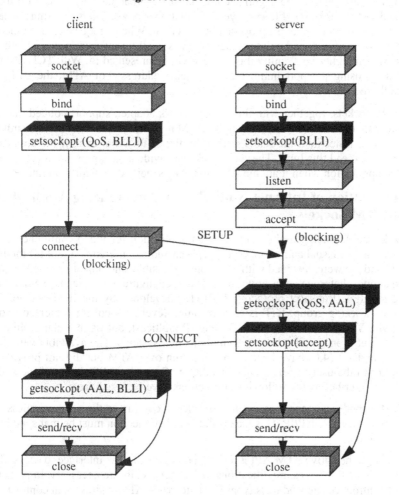

Fig. 2. : ATM connection establishment with native ATM sockets

The *connect* call is used to initiate a SVC connection. The application specifies the ATM address of the remote end point. The signalling parameters that were set with the *setsockopt* calls are used to construct the signalling message elements and attempt to establish the connection. The *connect* call will block until the connection is established or rejected by the ATM network or the remote station. On the server side, sockets are enabled to receive connections with the *listen* call. New sockets are created from incoming connections by calling *accept* on the listening socket. If an incoming connection is present, *accept* returns a new socket. The application may wish to check the connection parameters that were specified by the initiator with *getsockopt* and then either fully establish the connection, attempt to establish with negotiated parameters, or fully reject the connection. The application calls *setsockopt(accept)* to indicate to the signalling software that this connection is to be fully established.

Among the parameters to be negotiated are the BLLI (broadband higher layer information) and BHLI (broadband lower layer information) values. The server must select a BLLI out of up to three values proposed by the client. When a request for a connection initiated from a remote end point is received, the signalling software needs a method to choose which higher level entity this request will be presented to. With TCP/IP, this is handled by using the port number of the socket address. For ATM, the BLLI and optionally the BHLI are used.

The ATM sockets, together with the ATM network adapter, support policed rate connections. On a policed rate connection, the ATM network adapter will not transmit data to the switch faster than the rate specified by the application, even if the application attempts to exceed this limit. The ATM sockets provide a back pressure mechanism to alert the application when it attempts to exceed its specified transmission rate.

3 Adaptation of Internet Audio/Video Conferencing Applications to ATM Sockets

The video conferencing tool (vic) [12] is an application for video conferencing over the Internet. The visual audio tool (vat) [14] is an audio conferencing tool for the Internet. Vic and vat were designed with a flexible, extensible, and object-oriented architecture to support heterogeneous environments and configurations. Vic and vat are based on the real-time transport protocol (RTP) [13] developed by the IETF Audio/Video Transport working group. RTP is an application-level protocol implemented entirely within vic. Vic usually runs over UDP and IP multicast, but its modular architecture allows to put it on top of arbitrary communication systems. The available versions of vic (2.7) and vat (4.0) have been prepared to run over ATM but did not provide any possibility to the user to select guaranteed QoS. Several modifications of vic and vat which are described in the following subsections have been required:

- ATM provides a connection-oriented service in contrast to the connectionless datagram service of UDP/IP. This means that ATM connection must be managed by the application.

- Although the native ATM API is a socket interface like the interface to access TCP/IP or UDP/IP services, modifications to the socket calls are necessary. In particular, ATM introduces new address types (20 bytes wide ATM addresses in contrast to 32 bit wide IP addresses). Also the end point addresses are different. UDP and TCP use so-called ports identified by a 16 bit wide port number. The port equivalent of ATM addresses can be seen in the BLLI and BHLI values.

- A significant improvement of the implementation described in this paper is the QoS support of the communication system. In order to make use of this service, the application should be able to select the desired QoS parameters. Because vic and vat did not provide an user interface to specify QoS parameters to be supported, an appropriate user interface had to be designed and integrated into the vic and vat applications.

3.1 Protocol Stack and Address Selection

The introduction of native ATM support has been simplified by the modular implementation structure of vic and vat. Basically, a new network class for the native ATM sockets had to be implemented. The implementation allows to select whether the UDP/IP or the AAL5/ATM communication system shall be used. At the beginning of a communication a menu appears asking the user which communication system to be used (Fig. 3.). The user has the choice between the UDP/IP or the AAL5/ATM communication system. Dependent on the user's choice, the objects of the IP or the ATM network class are invoked from the application.

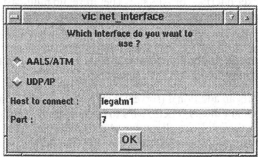

Fig. 3. : vic network initialization menu

The combination of both, the UDP/IP and the AAL5/ATM communication system allows the user to use the AAL5/ATM communication system to connect to ATM attached hosts, while the UDP/IP communication system can be used for legacy LAN attached end systems or if the two ATM end systems are interconnected by a non-ATM network. In addition, the menu asks the user which host the application should connect to. If UDP/IP is used, the host address must be either an IP address or a IP host name. The port specifies the destination port.

If AAL5/ATM is selected the host name can be either an ATM address or an IP host name if the IP host name is an ATM attached host belonging to the same logical IP subnet. In the latter case, the ATM address is resolved using the ATM address resolution protocol (ATM ARP). An ICMP packet is sent to the IP/ATM host which triggers an ATM connection setup by the classical IP over ATM implementation. After transmitting the ICMP packet, the local ATM ARP table contains an address entry mapping the IP address of the destination to an ATM address. This ATM address is then used for the native ATM connection establishment. The use of NHRP (NBMA Next Hop Resolution Protocol) [11] would even allow to resolve ATM addresses of hosts belonging to different logical IP subnets. Another possibility to map host names to ATM addresses is the ATM name service (ANS) mapping host names to ATM addresses. In the ATM case, the end point identifier consists of the BLLI value to be used for the ATM connection setup.

3.2 ATM Connection Management

In the UDP/IP communication stack, no communication is required between the communicating hosts when the applications are starting. Only the sockets to be used must be created and eventually bound to an address. The communication system is then ready and able to send and receive application data.

In the ATM case, however, an ATM connection must be established, before data can be exchanged between applications. The connection setup should be done in advance. Otherwise the connection must be established for the first data packet to be exchanged. In that case, the first data packet can only be sent after a successful ATM connection establishment.

In order to avoid delays we implemented the connection establishment procedure in advance, i.e. the application tries to connect to the destination host after it is started. As soon as the peer application has also started, the ATM connection between the two end systems is established. Point-to-multipoint ATM connections are not yet supported by the native ATM socket interface of AIX 4.2. Therefore, no multicast communication is supported. Point-to-multipoint connections are uni-directional. Similar to that, the established point-to-point connections are only used for uni-directional data transfer. This simplifies migration to point-to-multipoint connections as soon as the ATM socket interface supports it. Point-to-point communication between two hosts requires two uni-directional point-to-point ATM connections.

Each application starts an additional (server) thread waiting for incoming SETUP messages (Fig. 4.). The main thread (client) tries to connect to the peer server. Both applications start a client thread. Each client thread establishes an ATM connection to the corresponding peer server thread. So, two point-to-point ATM connections are established for data transfer.

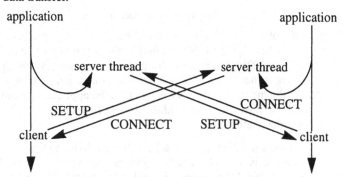

Fig. 4. : Connection establishment handling

Vic and vat are using the real-time transport protocol (RTP) to exchange real-time data (Fig. 5.). RTP adds time-stamps, A/V coding information etc. to the audio or video data packets. RTP has an integrated control protocol called RTCP (real-time control protocol), which is used to exchange control information such as status or QoS status reports. An additional ATM connection pair is required to exchange RTCP packets. In an UDP/IP environment, RTCP uses different port numbers than RTP, while different BLLI values are assigned to the RTCP connection in the ATM socket case. In total, four unidirectional ATM connections are established.

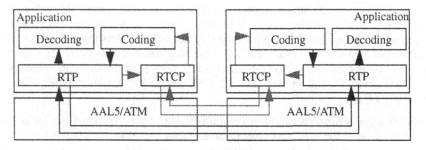

Fig. 5. : RTP and RTCP integrated into A/V applications over AAL5/ATM

Transmission

| Rate Control | | 23 f/s | 2.4 Mb/s |

Transmit
Release

17030 kbp.

26 fps

Encoder

Device...
Port...
Options...

nv · jpeg · raw
nvdct · h261 · raw1
cellb · bvc · raw2

small
· normal
large

Quality
77

Display

Options...
Tile...
Extend

· Ordered · Quantize
· Error Diff · Gray

Gamma: 0.7 [8-bit]

Session

Dest: 39.99.99.99.99.99.0.0.99.99.1.1.4D.20.45.43.20.20.0. BLLI/BHLI: 7/91 ID: 843256221

Name: Andreas Reisenauer

Note:

Key:

Global Stats Members Redial

Dismiss

Fig. 6. : vic transmission control menu

3.3 Quality-of-Service Selection

After starting the vic or vat application two ATM connection pairs using UBR service
are established among the two communicating ATM end systems. One connection pair
is for RTP data exchange, the other ATM connection pair is used for RTCP message
transfer. The user has the possibility to change the RTP connection parameters if the
UBR service is not sufficient for the application requirements. The user invokes the
transmission control window by pushing the "Menu" button of the vic and vat main
window. The user can initiate to change the QoS by pushing the "Redial" Button in the
transmission control window (Fig. 6.).

A new QoS parameter window appears and allows the user to specify the desired ATM QoS (Fig. 7.). The user has the choice among several service types and parameters which are shown in Fig. 7. To specify the bit rate values, sliders are used allowing the user to specify up to the maximum value the ATM interface supports. The specified QoS parameters are applied to both ATM connections used for application data transfer. The current implementation assumes UNI 3.1/3.0 [3]. ABR should be supported with UNI 4.0 [4].

After all settings are done the user can push the "really redial now" button and new connections with the modified QoS parameters are established. After the new ATM connections are established, the old ones are torn down and data are sent over the new connection. This allows the on-the-fly change from the old to the new ATM connections. If the old connection would be torn down before the new one has been established, this would lead to an interrupt of the audio/video communication. However, with the selected approach, for a short time period two data connections are established for each direction. This requires that a new connection uses new BHLI values to specify the communication end points. The BHLI values for a certain session are taken from a predefined set of consecutive BHLI values.

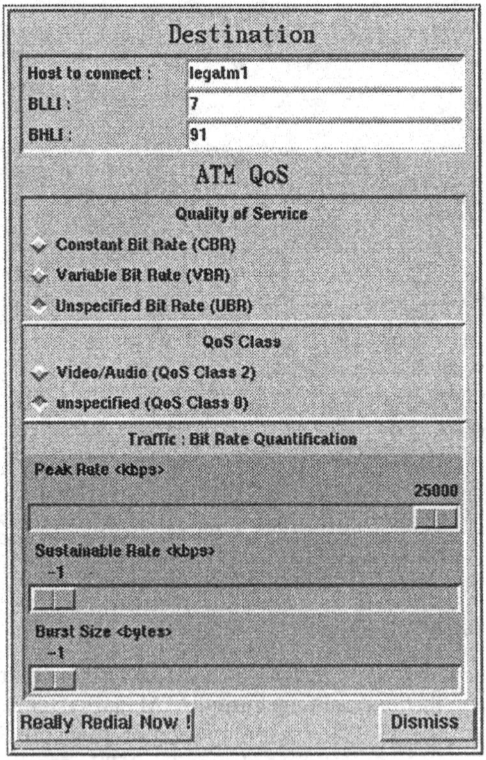

Fig. 7. : Native ATM Socket QoS Parameter Menu

4 Evaluation

Compared to a scenario where vic and vat are running over an IP router network without quality-of-service reservation, the achieved video and audio quality has been

improved significantly. Running vic and vat over the native ATM sockets with sufficient resource reservation results in loss-free transmissions. The delay mainly depends
on the delay due to packetization of the data, in particular if audio samples have to be
packetized into AAL5 frames. Also the delay jitter was very low such that users did
not notice any quality degradation.

Although we are running vic and vat only in a local ATM environment, we expect that
the approach also scales for wide area ATM networks. This is in contrast to legacy
LAN environments. While real-time audio/video conferences are running well over IP
router networks, significant quality degradation can usually be noticed when running
these tools over a heavily loaded IP router network without resource reservation capabilities. Although vic and vat are designed to be adaptive, their adaptivity is obviously
limited when the available bandwidth is below a certain threshold value. In our opinion resource reservation is necessary to run real-time applications with a reasonable
quality.

5 Conclusions

The native ATM socket API of AIX version 4.2 allows to offer the QoS capabilities
provided by ATM to applications. In particular, audio/video server and conferencing
products already use the native ATM interface. This paper presented an example for
Internet real-time applications which have been modified and enhanced in order to use
the ATM's QoS support.

The discussed implementation does not yet support multicast. Multicast functionality
requires the ability of the native ATM socket implementation to support point-to-
multipoint ATM connections. A receiver-oriented approach to join point-to-multipoint
VCs as defined by ATM UNI 4.0 will simplify this task, because the IP multicast service on which vic and vat are running provides also a receiver-oriented approach to join/
leave multicast groups.

Support of new service types such as ABR (available bit rate) is another issue for
future work. Especially the ABR service could be very interesting in the proposed scenario since ABR are expected to be cheaper for the user than CBR or VBR. In the
ABR case, there would be two adaptation mechanisms implemented within the communication system. The first mechanism is implemented on ATM level, while the second mechanism is implemented by RTP on top of the socket interface. Both RTP and
ABR try to adapt to dynamic network overload situations. It is still not clear whether
and how the two adaptation mechanisms have impact on each other.

A drawback of the native ATM sockets compared to the RSVP over classical IP over
ATM approach, is that all the ATM connection handling must be performed within the
application running in user space. In the RSVP over classical IP over ATM approach,
ATM connection handling is done in the AIX kernel. If the user desires a new QoS for
the communication, a new ATM connection must be established and the old one has to
be released. A renegotiation mechanism as discussed within the ATM Forum could
significantly reduce the connection handling overhead within the application.

A major problem of the native ATM sockets is that it is currently only applicable in
homogeneous ATM environments. One approach to allow the communication among
an Ethernet attached and an ATM attached end system is the cells in frames (CIF) concept [8]. CIF emulates ATM services on top of the LAN MAC layer. Several ATM
cells of a VC are packed into a LAN MAC frame and sent from an Ethernet attached

host to a kind of LAN/ATM switch, which restores the ATM cells from the Ethernet packets and transmits them over an ATM VC to an ATM end system. This concept only allows interoperability of Ethernet and ATM end systems if no router is between the two systems. Therefore, we think that native ATM sockets will not substitute UDP/ IP stacks, but may enhance Internet protocol stacks.

Another possibility to provide interoperability between ATM and IP systems is the implementation of a RTP gateway such as vgw [1] which already connects non-multicast capable end systems to a multi-party conference based on IP multicast. In a similar way ATM end systems could be interconnected with IP multicast systems.

Another open issue is how to announce MBone sessions to ATM systems. A first approach is to use IP to transfer the session parameters, e.g. the ATM parameters, and ATM/AAL5 to exchange the audio/video data streams.

6 References

[1] E. Amir, S. McCanne, H. Zhang: An Application Level Video Gateway, ACM Multimedia, November 1995, San Francisco, pp. 511-522

[2] ATM Forum Technical Committee: LAN Emulation over ATM, Version 1.0, January 1995

[3] ATM Forum Technical Committee: User-Network Interface Specification Version 3.1, September 1994

[4] ATM Forum Technical Committee: User-Network Interface (UNI) Signalling Specification, Version 4.0, April 1996

[5] R. Braden, L. Zhang, S. Berson, S. Herzog, S. Jamin: Resource ReSerVation Protocol (RSVP) - Version 1 Functional Specification, Internet Draft, May 1996

[6] T. Braun, S. Giorcelli: Quality-of-Service Support for IP flows over ATM , in: M. Zitterbart (ed.): "Kommunikation in Verteilten Systemen", Informatik aktuell, Springer-Verlag, 1997

[7] T. Braun, H. Stüttgen: Implementation of an Internet Video Conferencing Application over ATM, IEEE ATM '97 Workshop, May 26-28, 1997, Lisboa, Portugal

[8] Scott W. Brim: Cells In Frames Version 1.0: Specification, Analysis, and Discussion, http://cif.cornell.edu/specs/v1.0/CIF-baseline.html, 21 October 1996

[9] W. J. Hymas, H. Stüttgen, S. Sharma, S. Wise: Socket Extensions for Native ATM, IEEE ATM '96 Workshop, San Francisco, August 26-28, 1996

[10] M. Laubach: Classical IP and ARP over ATM, RFC 1577, January 1994

[11] J. Luciani, D. Katz, D. Piscitello, B. Cole: NBMA Next Hop Resolution Protocol (NHRP), Internet Draft, March 1997

[12] S. McCanne, V. Jacobson: vic: A Flexible Framework Framework for Packet Video, ACM Multimedia '95

[13] H. Schulzrinne, S. Casner, R. Frederick, V. Jacobson: RTP: A Transport Protocol for Real-Time Applications, RFC 1889, January 1996..

[14] vat - LBNL Audio Conferencing Tool, http://www-nrg.ee.lbl.gov/vat/

A Native ATM API Suited
for Multimedia Communication

Stefan Dresler, Markus Hofmann, Claudia Schmidt, Hajo R. Wiltfang
University of Karlsruhe, Institute of Telematics
Zirkel 2, 76128 Karlsruhe, Germany
Phone +49 721 608 [6397, 6413, 6408, 6406], Fax +49 721 388097
E-mail: [dresler,hofmann,schmidt,wiltfang]@telematik.informatik.uni-karlsruhe.de

Abstract

ATM, the asynchronous transfer mode, was developed to meet two main goals. Firstly, it provides a flexible transmission technology. Secondly, it offers a set of new services to the user [Part94]. In order to take advantage of ATM's strengths, one might want to use its services directly. To be able to do so, in particular to be able to integrate access to ATM services into own applications, one is in need of a powerful ATM application programming interface (API). In this paper, we first present the design and implementation of such an ATM API for Digital Alpha workstations running Digital UNIX. Moreover, we describe how the basic ATM service is enhanced by setting an implementation of SandiaXTP (Xpress Transport Protocol) on top of it. Finally, we give an overview over a monitor for this ATM environment designed to compute quality of service (QoS) parameters from protocol information, detect QoS violations by comparing them to desired values, and display the results graphically.

1 Introduction

During the last years, computer communication has been experiencing drastic changes from simple text-based exchange to advanced communication facilities including audio and video streams. Communication technology has drastically evolved towards fiber optic networks and enhanced transmission technologies, such as SONET/SDH. Based on these developments, very high speed networks are becoming reality. Moreover, emerging networks typically based on the ATM technology (e.g., B-ISDN) are characterized by their service-integrated nature. This makes them suitable to support forthcoming applications. Additionally, group communication is considered to be a basic requirement for upcoming application scenarios.

However, typically the new services offered by these networks are hidden by traditional protocols residing between applications and network services. For example, using IP-over-ATM [Lau94] or LAN emulation [ATMF95a] reduces ATM to a traditional link layer protocol and hides the new functionality. There exist two possibilities for an application to take advantage of the services offered by ATM and its adaptation layers. Firstly, a native ATM service interface provides applications direct access to the functionality of the ATM Adaptation Layers. The ATM Forum has defined a semantic description of such an ATM application programming interface (ATM API) [ATMF95b]. Secondly, using forthcoming communication protocols on top of ATM provides applications with an enhanced service, such as group management, end-to-end quality of service support, or several reliability levels and simultaneously hides network-related problems from them. In order to ensure the achievement of the requested QoS, QoS maintenance functions are needed inside the communication system. Consequently, the achieved QoS values have to be monitored during operation. A QoS monitor [ScBl96] is able to compute QoS parameters from protocol information and detects QoS violations by comparing monitored parameters to the parameters initially agreed upon in the service contract. Especially in high performance scenarios (e.g., ATM networks), QoS monitoring plays an important role to measure the performance actually achieved [WiSc97].

In this paper, the design and implementation of a native ATM API for Digital Alpha workstations running Digital UNIX is described [Kle96]. In section 2 and 3, the design and implementation of the API is explained. Section 4 gives an overview on applications deploying the functionality of the implemented API. Firstly, a protocol stack offering services for forthcoming applications is presented. It comprises XTP over AAL5/ATM and thus offers an enhanced communication service on top of ATM. We describe the adaptation of SandiaXTP [SGC94] to the developed ATM API. The second part introduces a QoS monitor that is able to collect QoS parameters at different service interfaces. The paper concludes with a summary of the achievements.

2 System Environment

Basis for the developments described in this paper was Digital UNIX (formerly OSF/1) on Digital Alpha workstations. To make the explanations of following chapters easier, two concepts should be explained in this chapter that are not available on all UNIX platforms.

The first concept is the notion of *dynamically loadable modules*. This is a feature of Digital UNIX in its version 3.2D which supports modules to be loaded to the kernel, configured, and unloaded at runtime.

The second mechanism that heavily influenced the design of the API is the so-called *Connection Management Module (CMM)*. In Digital UNIX 3.2, the ATM subsystem is grouped around the CMM as a core module (see Figure 2). This architecture introduces flexibility in that it sets in relation three different module classes which are regarded as orthogonal: *device drivers*, *signaling modules*, and *protocol convergence modules*. Depending on the needs of applications, modules of each class can be added to the CMM, providing additional functionality.

As usual in operating systems, the *device drivers* abstract from the methods of how the data to be transmitted is put onto and picked from the medium. Protocol convergence modules and signaling modules connected to the CMM can request the specific abilities of device drivers. In order to support different signaling protocols, it is possible to register several *signaling modules* with the CMM. Signaling channels are treated by the CMM like regular (switched or permanent) virtual channels. Switches Virtual Channels (SVCs) are established and released by the signaling modules. Without such a module, only Permanent Virtual Channels (PVCs) can be used.

354

Protocol convergence modules, finally, serve to bridge the gap between the CMM and other modules, possibly applications in user space. The implementation of IP-over-ATM that is shipped with the CMM may serve as an example for a convergence module. Convergence modules can also initiate connection establishments and releases. The kernel part of the API is realized as such a module.

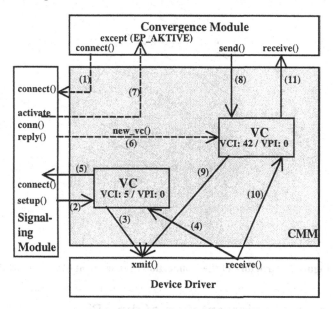

Figure 1: Communication paths in the CMM

The establishment of an active (i.e., outgoing) connection may serve as an example (see Figure 1) for the coordination work of the CMM. It is assumed that all modules have already registered with the CMM.

In a first step, the convergence module chooses an appropriate so-called Physical Point of Attachment (PPA), requests from the CMM memory for the address of the peer entity, for the service structure, and for information elements. The convergence module then calls the CMM function atm_cmm_connect() (cf. (1) in Figure 1). The signaling module uses the signaling channel (VPI 0, VCI 5) to ask the switch for the establishment of a connection (2); the request is forwarded by the CMM (3).

Successful completion of the establishment of the connection is signaled by the switch (4). The VPI/VCI values in the header of the (signaling) message indicate that the CMM has to forward the message to the signaling module (5). Up to now, the CMM has no knowledge of the newly established connection. To make the CMM aware of it, the signaling module calls the function atm_cmm_reply() with the appropriate path and channel identifiers (6). The connection can be used after a final call of atm_cmm_activate_conn() by the signaling module (7). This call is forwarded to the convergence module, causing an exception in the convergence module.

For regular data transmissions, the signaling module is not needed. Data passed from the convergence module to the CMM (8) is forwarded to the appropriate device driver (9). Similarly, incoming data from a device driver (10) is passed to the registered convergence module (11) by the CMM, based on the VPI/VCI information.

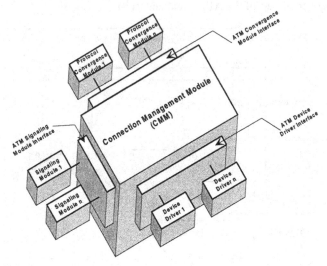

Figure 2: Structure of the Connection Management Module (CMM)

3 Design and Implementation of the API

The semantic description of an API for access to ATM services given by the ATM Forum [ATMF95b] has been implemented on several architectures like those for the UNIX derivate Linux [Alm95] and for PCs running MS-DOS or the Plan 9 operating system and Fore adapters [KeSa94]. The latter has been ported to FreeBSD [JaKe96]. [ShKe94] also deals with the design and implementation of operating system and signaling support for native-mode applications. [AKS96] gives an example of a transport service offered on top of ATM AAL 5. The Winsock 2 Application Programming Interface [MIJ95], which offers a socket interface to applications running under Microsoft Windows, was extended to support native ATM services [WIJ96]. Fore Systems, Inc., offers an API which supports the proprietary SPANS signalling protocol, which differs from the UNI specification of the ATM Forum. A comparison of the communication performance of four APIs (Fore Systems' ATM API, BSD socket programming interface, Sun's Remote Procedure Call (RPC), and the Parallel Virtual Machine (PRM) message passing library) can be found in [LHD+95]. However, up to now there was no implementation of a native ATM API available for Digital UNIX.

The ATM API described in this paper supplements the environment of Digital UNIX. The adapters used are Digital's ATMworks 750. In particular, the API interacts with the Connection Management Module (CMM) described above. The implementation of the ATM API was realized using the programming language C.

3.1 Interface Functions

An application accesses the functionality provided by the API through function calls ("downcalls", for commands and requests to the system) and callback functions ("upcalls", for indications to the application, initiated by the system). Most of the interface functions are directly derived from corresponding primitives of the semantics description of the ATM Forum [ATMF95b]. The communication between kernel and user space parts of the API uses the socket interface, however. This is similar to the Fore API [Fore93], which either uses the BSD socket interface or the System V streams interface as communication channel between API and device driver. Our approach does not emulate a regular socket interface.

Downcall functions provided by the API can be grouped into certain categories. One category contains functions that are concerned with local tasks like the creation or deletion of a local connection end point (CEP) identifier (e.g., ATM_associate_end-point()). Other categories are establishment, release and abortion of outgoing or incoming connections, resp. (e.g., ATM_connect_outgoing_call(), ATM_wait_on_incoming_call(), and ATM_call_release()). Data is send by ATM_send_data(). Furthermore, there are functions that support multicast communication, e.g. to add or drop leaves of the multicast tree. In addition to that, there are support functions that handle addresses, set or request connection attributes, or display error messages. Finally, there are functions that deal with the way indications and confirmations are handled at the interface (e.g., ATM_process(); described in the following section).

The application may register with the API callback functions for upcalls, which may also be grouped into classes. There are functions to notify the application of an incoming call or of data received. Furthermore, there are callbacks that confirm the completion or report the rejection of operations like adding a party to a multicast connection. It is also reported when a connection is ready to handle data to be sent. Finally, there is a function that indicates the release of a call.

The API provides functions for multicast connections. In Digital UNIX 3.2 the CMM does not support this functionality, however, so that the functions cannot be called by an application yet.

3.2 Implementation Overview

As mentioned, the kernel part of the API was implemented as a convergence module that interacts with the CMM. Since the CMM is located in the kernel, the API has to cross the borders between kernel space and user space. One option would have been the introduction of new system calls. This might have resulted in a more efficient implementation, and an application would not have to link a dedicated library. It would also have meant a lot of implementation work, though. Furthermore, problems could arise with new operating system releases, and realization as a loadable module would have been impossible, thus slowing down the development. For these reasons, it was decided to take a different approach.

Service primitives as described above are not mapped to system calls, but to messages that are passed through a socket interface. This allows for easy extensibility of the functionality of the API. To ease communication between kernel and user space, a new protocol family (PF_ATM) of sockets was defined with a single protocol (SOCK_RAW).

The implementation of the API is based on a single execution thread. To avoid having to block an application whenever e.g. it is waiting for an incoming call, all downcalls of the API were realized as non-blocking functions. In order to enable the API to perform upcalls to the application, it has to be given the execution thread. This is done by calling the downcall function ATM_process() with the corresponding CEP value. Depending on the indication waiting, the API then calls the respective upcall function, thus acting as a dispatcher. An application that favors asynchronous calling of upcall functions needs to fork another thread that blocks until the socket associated with the CEP indicates an event.

The API is mainly contained in the six modules indicated in Figure 3. The module atmifc_lib.c contains the library functions a user has to link to an application. These functions compile to libraries that can be linked statically or dynamically, resp. All other modules are located in kernel space. They serve different purposes, from handling the communication with the application and the CMM, resp., management of data structures and connection attributes to auxiliary functionality.

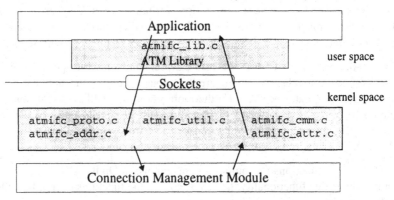

Figure 3: Module structure of the implementation

3.3 Performance Evaluation

In order to assess the performance of the implementation, a comparison against the IP-over-ATM implementation shipped with the operating system as well as against the theoretical limit is given. The tests both with the native interface and over TCP were controlled by a graphical test program which also served as an example application.

Digital's ATMworks 750 adapters offer a line speed of 155.5 Mbit/s. Subtracting protocol and management information that is transmitted, the ATM layer offers a service of 135.63 Mbit/s. With a PDU size of 9kByte, this value is not significantly reduced, because the overhead for packet headers is negligible.

Using this PDU size of 9kByte, the IP-over-ATM implementation yields a performance of around 60 Mbit/s. The native ATM API, which is in a prototypical state and has not yet been optimized for speed, achieved a throughput of 85 to more than 124 Mbit/s if used for a connection between two different machines. The API thus meets our expectations of a high performance implementation. The lower performance of IP-over-ATM is due to its higher protocol processing costs.

If used for a connection to an ATM switch and back to the same machine, the throughput reaches about 40 Mbit/s. In this case, the kernel part of the API consumes about 90% of the processor cycles.

4 Practical Experience with the API

The developed native ATM API has been used to implement the Xpress Transport Protocol on top of ATM. The bearer service provided by ATM is improved by a powerful transport level protocol including QoS support for multimedia applications. Furthermore, QoS monitoring is applied at the ATM API in order to measure the achieved quality of service at AAL level.

4.1 XTP on Top of ATM

Current networks mostly use UDP or TCP as the transport layer in combination with IP-over-ATM to transfer data across ATM-based networks. While this approach permits fast integration of ATM technology in existing networks, it is neither possible to provide any quality of service (QoS) support nor efficient multicast communication. The QoS guarantees supported by ATM are lost by IP, since IP-over-ATM transmits data of multiple transport connections between two peer entities across a single ATM channel. Moreover, the service user cannot directly utilize any QoS support since the interfaces of TCP and UDP do not provide any service parameters. Besides the lack of QoS support, TCP can perform poorly over plain ATM [RF95]. The loss of any one cell, for example, results in a loss of the complete TCP packet and, therefore, in a reset of the congestion window. This is why TCP significantly decreases its average transfer rate due to the loss of a single ATM cell.

A variety of new transport level protocols have been designed to overcome the primary drawbacks of TCP and UDP. The Xpress Transport Protocol (XTP), for example, supports a variety of applications ranging from simple data transfer to multimedia distribution with real-time requirements. Additionally, it supports guaranteed quality of service and multicast communication. Novel protocol mechanisms have been integrated to support efficient data exchange in large-scale, heterogeneous communication groups [Hof96]. XTP protocol functions, such as error and traffic control, can be tailored to application specific requirements. XTP also defines a traffic descriptor on order to negotiate QoS parameters among communication participants.

SandiaXTP is an object-oriented user-space implementation of the Xpress Transport Protocol. It was developed at the Sandia National Laboratories and is realized as a user-space daemon written in C++. In principle, SandiaXTP operates on top of any data delivery protocol. Possible protocols include IP, UDP, FDDI, Ethernet, and

XUNET. However, the current version of SandiaXTP only supports IP and UDP. To take full advantage of the ATM technology, the implementation has been adapted to operate directly on top of our native ATM API.

The design of SandiaXTP has been guided by the object-oriented programming paradigm. It is based on several abstract, hierarchical classes that are applicable to most transport protocols. This collection of generic base classes is named *Meta-Transport Library (MTL)*. It defines functions and data structures common to all transport protocols. The sending of data units, for example, is not a protocol specific function since the data delivery service does not care about the format of data units to be sent. Therefore, the *send(...)* method is defined in the delivery service of MTL and can be used by every derived protocol implementation without any modifications. Protocol dependent issues, such as evaluation of specific protocol parameters or the concrete data unit format, are defined in derived classes. The XTP protocol itself, for example, is realized by the *SandiaXTP* class collection.

The integration of SandiaXTP into a native ATM protocol suite consists of two major tasks:

- Implementation of a native ATM delivery service
- Mapping of the XTP traffic specifier to ATM QoS parameters

The generic delivery service defined in the MTL library currently supports connectionless IP and UDP. Its functionality is not suitable to be used in conjunction with connection-oriented networks such as ATM. Therefore, it was necessary to design another delivery service providing support for the management of underlying connections. It has been named *NAS_delivery_service*. While sending data units is a task within MTL, the mapping of QoS parameters is protocol dependent. Therefore, an XTP specific delivery service, named *XTP_delivery_service*, has been derived from a newly designed NAS_delivery_service. This delivery service has knowledge about protocol specific QoS parameters and, therefore, is able to map them on corresponding ATM signaling parameters.

XTP supports several sets of QoS parameters. The traffic format [XTP95], for example, comprises throughput (number of TSDUs per second), delay (end-to-end delay of one TSDU in milliseconds), maximum TSDU size in bytes, and reliability class. XTP guarantees the negotiated quality of service for the lifetime of a connection according to the guarantee class selected by the user at connection establishment time. XTP differentiates between four guarantee classes: Class 0 (best effort, no indication of QoS violation), Class 1 (best effort, indication of QoS violation), Class 2 (guaranteed, no indication of QoS violation), and Class 3 (guaranteed, indication of QoS violation). The guarantee classes are derived from two distinct concepts: the *probability* that the negotiated QoS is actually provided by the transport service, and the *indication* to the transport user that the QoS has been violated.

On the other hand, ATM provides parameters such as peak cell rate (PCR), sustainable cell rate (SCR), maximum burst size, best effort indicator, and maximum CPCS-SDU size.

To illustrate the mapping of QoS parameters, the translation of XTP throughput to ATM cell rate will be explained. The following variables are used:

- CSDUsize - *maximum size of a CPCS-SDU* in bytes
- ASDUsize - *size of an ATM-SDU* in bytes (using AAL5, this will be 48 bytes)
- CPCIsize - *size of the CPCS Protocol Control Information* in bytes (8 bytes)
- TPCIsize - *size of XTP header* in bytes
- TTPT - *XTP throughput in TSDUs per second*

A complete CPCS-SDU is encapsulated in a single CPCS-PDU, which is segmented into N ATM cells by the Segmentation And Reassembly Layer (SAR). The value of N is given by

$$N = \left\lceil \frac{CSDUsize + CPCIsize}{ASDUsize} \right\rceil = \left\lceil \frac{CSDUsize + 8\ bytes}{48\ bytes} \right\rceil \tag{1}$$

Therefore, sending a single TSDU results in the transfer of N ATM cells. The peak cell rate corresponding to a given XTP throughput thus is calculated to:

$$PCR = TTPT \cdot N = TTPT \cdot \left\lceil \frac{TSDUsize + TPCIsize + CPCIsize}{ASDUsize} \right\rceil \tag{2}$$

Vice versa, XTP throughput is determined by a given PCR:

$$TTPT = \frac{PCR}{N} = \frac{PCR}{\left\lceil \frac{TSDUsize + TPCIsize + CPCIsize}{ASDUsize} \right\rceil} \tag{3}$$

According to formula (2), the initiator of a connection calculates the peak cell rate based on the user given maximum TSDU size and the XTP throughput. The result is used to signal an appropriate ATM peak cell rate. If the desired value is not supported by the network, the initiator calculates the supported XTP throughput based on the return value of ATM according to formula (3) and informs the service user about it.

4.2 Monitoring QoS Parameters of Communication Protocols

The presented ATM API allows to monitor ATM connections at the AAL level. For QoS monitoring of selected connections, a flexible QoS monitor for high performance protocols has been developed and implemented [ScBl96]. The QoS monitor can be applied to different levels of a communication system where it is able to measure QoS parameters of the underlying level. In the past, we tested our QoS monitor with the XTP protocol over ATM where it monitored XTP parameters. For the work presented in this paper, the QoS monitor has been placed beside the described ATM API in order to monitor AAL5-related QoS parameters.

4.2.1 Architecture and Functionality of the QoS Monitor

In order to keep up with the high data rates of high performance protocols, the monitor is placed in an autonomous entity which communicates asynchronously with the monitored protocol. Currently, we are able to monitor the QoS parameters throughput, delay, jitter, and several error-related parameters. Subsequently, the QoS monitor computes recent QoS parameters from information provided by the monitored protocol entity and checks for QoS violations. Possible reactions are a periodical report, a report on demand, an immediate report of QoS violations or of warnings.

The QoS monitor comprises three entities that are able to work independently. They are grouped around a *QoS-MIB* (Management Information Base) where all QoS-related information is stored [ScB196]. For each monitored data stream, the QoS-MIB includes the QoS limits of the service contract as well as measured values (e.g., the current, minimum, and maximum value). The core of the QoS monitor is a *monitor entity* which computes QoS parameters from protocol information, updates QoS parameters in the QoS-MIB, and detects QoS violations. Furthermore, a powerful *presentation entity* based on X/Motif is able to provide a graphical presentation of the monitored parameters, and an *SNMP agent* integrates the monitoring results into classical network management. The described QoS monitor has been implemented under the operating system Digital UNIX V3.2d on Alpha workstations using so-called threads (light weight processes) for each entity.

4.2.2 Monitoring the ATM API

In the special case of ATM, the QoS monitor was configured to operate on top of the designed ATM API. Here, QoS parameters of selected ATM connections can be monitored at AAL level. In order to integrate all required control information into the regular data flow, a new SSCS (Service Specific Convergence Sublayer) for AAL5 was defined [WiSc97]. This *QM-SSCS (QoS Monitoring-SSCS)* operates on top of the AAL5 CPCS sublayer and adds a four byte sequence number as well as an eight byte time stamp to each user data unit (Figure 4). With the help of this additional information, all useful QoS parameters at AAL level can be computed. The sequence number is used at the receiver side to detect the loss of AAL5 CPCS data units in order to calculate the error-related parameters. For computation of time-related parameters such as delay and jitter, the time stamp describes exactly the time when the data unit was sent. The required synchronization of the local times at the sender and receiver can be realized by using synchronization protocols like NTP (Network Time [Mill92]). Finally, throughput parameters can be calculated by using the local time at the receiver and the length field in the AAL5 CPCS trailer.

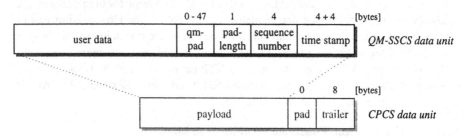

Figure 4: QM-SSCS for AAL5 QoS Monitoring

Additionally, padding is realized at the SSCS sublayer with two fields in the QM-SSCS data unit. The field qm-pad is used instead of the CPCS pad in order to align the length of the resulting CPCS data unit to a 48 byte boundary. Thus, using the QM-SSCS, the length of the CPCS pad field is always zero and monitoring information is located in fixed positions of the last ATM cell. SSCS sublayer padding is not required for AAL layer monitoring, but, for monitoring ATM connections on switches, it will be very useful having all monitoring information in one cell that can be easily identified (ATM header payload type = '1') [Wilt97].

5 Conclusion

The implementation of the ATM API described is fully integrated into the architecture of Digital's Connection Management Module (CMM). This implementation provides an overlying application with the full throughput offered by the ATM layer. Using the API, an implementation of XTP was set on top of ATM. This transport layer offers enhanced services with corresponding service guarantees to applications. Violations of the service contract are detected by a monitor that displays its data using a graphical user interface. The monitor is able to operate at the transport layer as well as at the ATM API presented.

6 Acknowledgements

The authors thank Jochen Klein for his excellent work of implementing the API [Kle96] and drawing some of the figures.

References

[AKS96]   R. Ahuja, S. Keshav, H. Saran: *Design, Implementation and Performance of a Native Mode ATM Transport Layer*; Proc. IEEE INFOCOM, March 1996, or via http://www.cs.att.com/csrc/keshav/papers.html

[Alm95]   W. Almesberger: *Linux ATM API*, Draft, version 0.2, EPFL, LRC, Switzerland, http://lrcwww.epfl.ch/linux-atm/, June 20, 1995

[ATMF94]   The ATM Forum: *ATM User-Network Interface Specification 3.1*; ATM Forum af-uni-0010.002, September 1994

[ATMF95a] The ATM Forum: *LAN Emulation over ATM, Version 1.0;* ATM Forum af-lane-0021.000, January 1995

[ATMF95b] The ATM Forum: *Native ATM Services: Semantic Description*; ATM Forum af-saa-0048.000, February 1996

[DEC95]   Digital Equipment Corporation: *DEC OSF/1: Asynchronous Transfer Mode*; Digital Equipment Corporation (AA-QDP5B-TE), 1995

[Fore93]   Edoardo Biagioni, Eric Cooper, and Robert Sansom (Fore Systems Inc.): *Designing a Practical ATM LAN;* IEEE Network, March 1993

[Hof96]   M. Hofmann: *Adding Scalability to Transport Level Multicast*; Proceedings of Third COST237 Workshop, Barcelona, Spain, November 1996

[JaKe96]   A. Jain, S. Keshav: *Native-mode ATM in FreeBSD: Experience and Performance;* Proc. NOSSDAV'96, April 1996, or via http://www.cs.att.com/ csrc/keshav/papers.html

[KeSa94]   S. Keshav, H. Saran: *Semantics and Implementation of a Native-Mode ATM Protocol Stack;* AT&T Bell Laboratories Technical Memorandum, 1994, or via http:// www.cs.att.com/csrc/keshav/papers.html

[Kle96]   J. Klein: *Design and Implementation of an ATM Application Programming Interface,* Diploma Thesis (in German), Institute of Telematics, University of Karlsruhe, Germany, February 1996

[Lau94]   M. Laubach: *Classical IP and ARP over ATM,* Request for Comments 1577, Internet Engineering Task Force (IETF), January 1994

[LHD+95]   M. Lin, J. Hsieh, D.H.C. Du, J.P. Thomas, J.A. MacDonald: *Distributed Network Computing over Local ATM Networks,* IEEE Journal on Selected Areas in Communications Special Issue of ATM LANs: Implementations and Experiences with an Emerging Technology, 1995

[MIJ95]   Microsoft, Intel, and JSB Corporations: *Windows Sockets 2 Application Programming Interface,* Revision 2.0.6, February 1, 1995 (Draft)

[MIJ96]   Microsoft, Intel, and JSB Corporations: *ATM-Specific Extensions,* Annex to the Winsock 2 Manual

[Mill92]   D. L. Mills: *Network Time Protocol (Version3) - Specification, Implementation and Analysis;* Request for Comments 1305; March 1992

[RF95]   A. Romanow, S. Floyd: *Dynamics of TCP Traffic over ATM Networks;* IEEE Journal on Selected Areas in Communications, May 1995.

[ScBl96]   C. Schmidt, R. Bless: *QoS Monitoring in High Performance Environments;* Proceedings of the Fourth International IFIP Workshop on Quality of Service - IWQoS'96; Paris, France; March 6-8, 1996

[SGC94]   W.T. Strayer, S. Gray, R.E. Cline: *An Object-Oriented Implementation of the Xpress Transfer Protocol;* Proceedings of the second International Workshop on Advanced Communications and Applications for High Speed Networks (IWACA), Heidelberg, Germany, September 1994.

[ShKe94]   R. Sharma, S. Keshav: *Signaling and Operating System Support for Native-Mode ATM Applications;* ACM SigComm, August 1994, via http://www.cs.att.com/ csrc/keshav/papers.html

[XTP95]   XTP Forum: *Xpress Transport Protocol Specification, Version 4.0;* XTP Forum, Santa Barbara, CA, March 1995

[Wilt97]   H. R. Wiltfang: *An Approach to QoS Monitoring in ATM Networks;* in Proceedings of the European Conference on Networks & Optical Communications; Antwerp, Belgium; June 17-20, 1997

[WiSc97]   H. R. Wiltfang, C. Schmidt: *QoS Monitoring for ATM-based Networks;* in Proceedings of the International Conference on Management of Multimedia Networks and Services; Montreal, Canada; July 8-10, 1997

A Model for Collaborative Services in Distributed Learning Environments

Volker Hilt, Werner Geyer
University of Mannheim
{hilt, geyer}@pi4.informatik.uni-mannheim.de

Abstract: Synchronous collaborative work environments, which are mainly based on video-conferencing systems, suffer a lack of human *communication channels* and *social awareness* because mostly only audio, video, and joint editing of documents are supported. *Collaborative services* such as *floor control*, *session control*, and *telepointers* provide additional communication mechanisms to support persons co-working through computers. We propose a *collaborative services model* (CSM) for distributed learning environments. Our object-oriented approach includes floor and session control mechanisms and policies. Due to the realization of collaborative services as operations on the model, no specific network collaboration protocol is required. We present an optimistic synchronization scheme which provides consistency for the distributed model.

1   Introduction

1.1   Motivation

Advances in multimedia technologies and high-speed networks lead to new types of teaching and learning. Various digital media may be integrated and distributed via networks, such that they are available in arbitrary places and at arbitrary times (*independency of space and time*). *Teleteaching* denotes the geographical distribution of teachers and students who are connected via fast networks and who communicate synchronously. Compared to traditional distance learning this allows for a higher degree of interactivity and usage of new instructional media such as digital animations or simulations. In spite of this improvement, teleteaching suffers a lack of communication channels in comparison to traditional classroom instruction. *Social protocols* or *rules* control the human interaction and the course of instruction within a classroom. These communication mechanisms include for instance putting-up hands, giving rights to talk or to write on a blackboard, setting-up work groups and reference pointing. They are difficult to reproduce in remote situations because today's synchronous learning environments, which are mainly based on video-conferencing systems, mostly provide only audio, video, and joint editing of documents. A major goal in developing learning environments is to reproduce, as far as possible, the traditional classroom situation by supplying computer support for controlling remote instruction and human interaction.

1.2   Collaborative Services

The lack of communication channels and *social awareness* cannot just be found in teleteaching but generally in all *computer supported collaborative work* (CSCW) systems. *Collaborative services* (CS) provide mechanisms to support the communication of persons through computers. Basic services are *floor control*, *session control*, and *telepointers*.

Floor Control

Floor control realizes concurrency control for interactive, synchronous cooperation between people by using the metaphor of a floor. A floor denotes the temporary permission to access and manipulate resources (e.g. a shared drawing area or a video channel). The owner of a floor at a certain point of time - called the *floor holder* - is equipped with well defined rights on a certain floor [Dom97]. In a session, floor control mainly has the coordinating tasks of reducing non-determinism, setbacks, redundancy, and inconsistencies, balancing contributions among session participants in a fair manner, regulating the collaboration by a predictive and binding protocol and promoting inter-group awareness, cohesiveness, and integrity [Dom97].

Crowley et al. [Cro90] distinguish between floor control *policy* and *mechanism*. The policy determines the strategy of achieving, holding, and releasing floors while the mechanism provides basic low-level functionality for the implementation of floor control. The best-suited strategy for a certain (instructional) situation depends on the task and size of the group and on the type of interaction in the group; consequently, floor control has to be adaptable [Gut95]. Fluckiger et al. [Flu95] identify four basic floor control policies:

1. *No control*: Each group member can access common resources without control.
2. *Implicit floor control*: This strategy implicitly grants a floor to a group member as soon as he or she starts to use a certain resource. The resource is then locked for other members. A certain time after the floor holder has stopped using the resource the floor is automatically released.
3. *Explicit floor control*: The floor is granted to a participant or is released from a participant only on his or her specific request, for instance, by pressing a dedicated button. The requests of other participants while the floor is locked have to be queued and displayed to the group. Generally, these requests are then served on a first-in-first-out basis.
4. *Chair control*: One person becomes chairperson or moderator (e.g. the teacher in a teleteaching session) of the collaborative group. He or she can grant or withdraw the floor for a certain resource at any time. Pending floor requests of participants have to be queued and monitored to the chairperson.

Furthermore, Dommel et al. [Dom97] classify policies in *queuing* and *non-queuing* depending on whether requests are queued or served immediately. A single floor may be granted to one participant only (*mutually-exclusive*) or to several participants simultaneously (*selective*)[1]. Further classifications can be found in [Dom97].

Session Control

Session control denotes the administration and coordination of multiple sessions with its participants and media [MMU96]. It comprises initiation, pause, resume, and stop of sessions. Membership support includes creation, joining, withdrawing, inviting, excluding etc. of participants [Dom97]. Session control provides social awareness in

---

[1] Dommel et al. [Dom97] denote the policy of assigning k floors to n participants as *mutually-selective*. In contrast to their terminology, we use the term *selective* to express that k participants, $k \leq n$, may hold the same floor.

workgroups because members gain knowledge about other members and their status in the session. Floor control relies on session control because a floor grants resource access to a certain participant who is a member of a session. *Hierarchical* session management allows for the creation of subsessions within a workgroup.

Telepointers

To ease tracking the direction of the lecture, telepointers realize a common point of reference by providing shared pointers. For a detailed insight of telepointers refer to [Nak93]. In our collaborative services model, telepointers are managed as resources.

1.3 Context of the Model

The collaborative services model we propose in this paper has been motivated by our work in the *TeleTeaching* project Mannheim-Heidelberg. The project realizes three different modes of teaching and learning. The three modes are characterized by their scope of distribution, interactivity, and individualization of the learning process. In the *Remote Lecture Room* (RLR) scenario lecture rooms are connected via a high speed network and courses are exchanged synchronously and interactively between the universities of Heidelberg and Mannheim. *Remote Interactive Seminars* (RIS) describes a more interactive type of instruction. Small groups of participants are distributed across a few seminar rooms which are also connected by a network. The *Interactive Home Learning* (IHL) scenario is aimed at the maximization of the geographical distribution degree of all class participants. Each student learns asynchronously as well as synchronously at home in front of his PC. For a thorough description of the project see [Gey97].

Currently, we are using the MBone video-conferencing tools and the Internet for remote lecturing. The tools prove to be not sufficient for the purpose of teleteaching because they are not powerful enough to support team work. They are also not flexible enough for the use of media and they do not provide an integrated user interface. Therefore, we are currently developing a novel, integrated teaching software, called *digital lecture board* (dlb). In the context of the dlb, we designed and implemented the collaborative services model presented in this paper.

1.4 Terminology

There is no commonly accepted framework for collaborative services, although floor control and session control have been well-known concepts for a long time. Based on [Dom97] we are using the following terminology: The infrastructure for groupwork (e.g. teleteaching) is provided by *sessions*, which can be divided into *subsessions* or combined in *supersessions*. A session denotes an on-line aggregation of *participants* co-working synchronously on *shared resources*. Resources in multimedia systems could be, for instance, files, devices, user interface widgets, graphical objects, video and audio streams etc.[2] Participants are associated with certain *roles*. A role is a set of *privileges* [Ell91]. A multimedia environment allowing remote collaboration with multiple resource types is called a *collaborative environment*.

---

[2] Note that we use the term resource in a broader sense for all required collaboration objects.

2   Related Work

The IETF working group MMUSIC develops mechanisms mainly for session control. Session control for *light-weight* sessions simply provides membership information whereas *tightly-coupled* sessions allow for the realization of collaborative services. So far, a standard session control mechanism for tightly-coupled sessions does not exist in the Internet [Han96]. Bormann et al. [Bor96] propose the realization of collaborative services by using a specific protocol (Simple Conference Control Protocol, SCCP). The protocol assumes the existence of a session state representation at each participants site. SCCP uses centralized components and relies on transport services which provide global ordering of messages.

Dommel et al. [Dom97] introduce a notation for the representation of groupwork environments. Sessions may be structured hierarchically, side talks are possible, participants may take different roles in a session, and resources are assigned to floors. The work mainly provides a basic survey of floor control but no specific realization.

Shenker et al. [She95] present a more abstract framework for the management of the session state. For the representation of sessions, they use state variables and define abstract operations. Moreover, protocols, mainly for assuring consistent state transitions in the distributed model, are discussed.

The BERKOM Multimedia Collaboration Service (MMC) [MMC96] provides a complete conferencing architecture including session control and simple floor control for a shared application tool. The conference control concept is based on centralized components. JVTOS [Der93] provides a system level platform for multimedia based telecooperation in heterogeneous environments. Course-grained floor control is realized for an application sharing component (i.e. the complete application is regarded as a resource). The T.120 standard [ITU96] describes a protocol hierarchy to support teleconferencing in public networks such as ISDN. Session control and a token mechanism are based on a centralized concept. The standard does not consider floor control.

Existing work in this area mainly focused on session control. There is little about floor control and its realization in combination with session control. Most of the approaches propose heavy-weighted architectures where specific protocol layers provide session control, ordering, token mechanisms, and other services on which e.g. floor control can be realized. Other approaches simply design collaborative services protocols while not being easily capable of handling the collaboration state and consistency. Consequently, the application has to manage the state, for instance, in variables or tables, which introduces a high complexity.

Our work mainly inspired by [Dom97] and [She95] differs in that we combine fine-grained floor control and session control for tightly-coupled sessions in a single collaborative services model. Additionally, the object-oriented model performs consistency checks based on specific consistency rules, and includes different collaborative services policies, while the separation of mechanism and policy is still provided. We do not need specific protocols for realizing floor control and session control because operations are executed by the model and not by a protocol. This light-weighted approach provides high scalability, does not rely on complex conferencing architectures, and can be easily incorporated into an application. The model has been designed for teleteaching applications but is not limited to them.

3   Collaborative Requirements

The three teleteaching modes (RLR, RIS, IHL) - in regard to common forms of university learning - should comprise of both traditional types (e.g. didactic teaching) and cooperative types of instruction (e.g. jigsaw). Modern types of face-to-face instruction should basically serve as an example for remote instruction. In our analysis [Hil96] we derived the following standard situations that should be supported by all learning environments:

- Joining a session at the beginning
- Joining a running session
- Leaving a session earlier
- Leaving a session at the end
- Removing a participant from the session (e.g. by the moderator)
- Putting-up hands (signalling)
- Putting-up hands recursively (for replies)
- Drawing back signals
- Removing signals by the moderator or teacher
- Selecting participants (with and without signal)
- Selective granting of rights (floors)
- Selective removing of rights (floors)
- Returning own rights
- Pointing to shared instructional materials

Cooperative learning environments should additionally be able to cope with the following situations:

- Creating sub-groups
- Joining running sessions (sub-groups)
- Closing sub-groups
- Private discussions and co-operation
- Creating moderators for sub-groups
- Surveying sub-groups and participants (super user role of the moderator)
- Granting of rights (floors) for working areas and materials to sub-groups
- Accessing working areas and materials without a moderator in order to enable internal group discussions

4   Collaborative Services Model

4.1   Overview

We propose a collaborative services model which implements enhanced floor control and session control in order to realize the described service requirements. The object-oriented model not only holds the required state information but also carries out the collaborative services. While executing services, consistency rules are checked. Our model provides both mechanisms and policies so as to realize its collaborative services. Each participant of a session holds a local copy of the model. The distributed copies are held consistent by applying an optimistic synchronization scheme (see 5.2). This guarantees a good responsiveness, typically required by CSCW applications.

The identified services (see 3) are similar in that they require extensive state information about the session (*collaboration state*). Floor control and session control use state information which overlaps considerably. In order to avoid redundancy and to ease consistency, we manage the collaboration state in a single model. The CSM describes the collaboration state by using *state elements* (*objects*) between which *relationships* are built during a session. Objects are either a mapping of real world objects, like *participants* and *resources*, or abstractions such as *floors* and *sessions*. A participant may be involved into a certain session and may have the floor for a certain resource. We are using specific floor objects to store the status of a floor, its access permissions to resources, and the applied policy. A floor establishes a relationship between a resource, a participant, and a session, and regulates the repeated, concurrent access to shared resources[3] during synchronous groupwork.

Collaborative services are realized by executing *operations* on objects. Thus, using CS always changes the collaboration state, except for simple state queries. The *semantics* of a service is defined by its changes to a given collaboration state (i.e. the required operations needed in order to provide the service). Granting a floor, for instance, always includes the construction of a relationship between the floor and a participant. The operation must check whether or not a relationship between the floor and the other participants already exists.

Implementing CS as operations on the collaboration state allows for performing different service provision strategies. The strategies determine the rules for the execution of state changing operations (*collaborative services policies*). Being independent of the strategies, our model provides the consistent management of objects and relationships, and furthermore, the consistent distribution of the collaboration state in a network (*collaborative services mechanisms*).

4.2   Objects of the Model

The objects of the CSM contain both information for the representation of the collaboration state and data with informative character only. The stored information is universal and independent of a specific application. Figure 1 depicts the CSM in Coad/Yourdan notation.[4]

The container class *CollaborativeLearning* represents the totality of all involved objects. A *Session* object describes the composition of a workgroup. The object may recursively contain several sub-sessions required for cooperative learning. The following relationships may be established with a *Session*:
- A *Session* may contain arbitrary participants by forming relationships with *Participant* objects.
- *SharedResource* objects can be assigned to a *Session* using a *Floor* object.

The following attributes characterize a *Session* [Dom97], [Han95]: *Name, URI (Uni-*

---

[3] All resources used for collaboration and included in the model are considered as *shared*.

[4] Internal object methods are omitted in Figure 1 in order to provide a better readability.

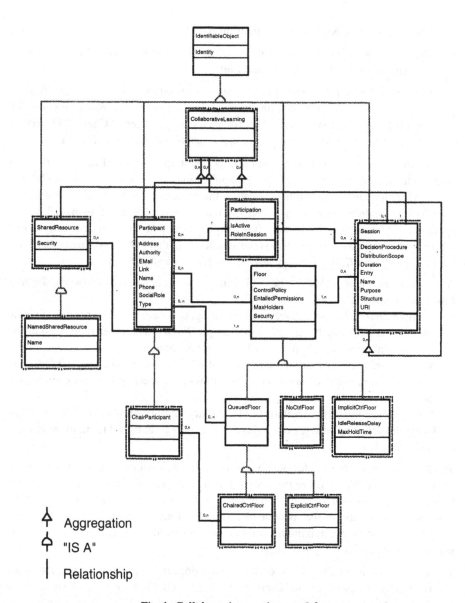

Fig. 1: Collaborative services model.

form Resource Identifier), *Structure, Purpose, Duration, Entry, Decision Procedure,* and *Distribution Scope. Purpose*, for instance, could carry values such as, "lecture", "examination" etc. For a thorough description of these attributes refer to [Hil96].

The *Participant* object represents a potential participant. Since a moderator or teacher has additional rights, we derive a *ChairParticipant* object. The following relationships are possible:

• A *Participant* may participate in arbitrary *Sessions,* using a *Participation* object.

- A *Participant* may be floor holder of arbitrary *Floors*.

The object contains the following attributes [Dom97], [Bor96]: *Name, Address, Phone, E-Mail, Type, Social Role, Authority,* and *Link*.

The participation of a *Participant* in a *Session* is realized by a (1:1) relationship using the *Participation* class. This relationship is characterized by the *RoleInSession* and *IsActive* attributes. *RoleInSession* can be, for instance, "Lecturer", "Chair", "Observer" etc. *IsActive* determines whether a participant is actively or passively involved.

A *SharedResource* object denotes the abstract counterpart to a real resource that has been declared as a shared resource by the application during a session. All collaborative services are carried out in the model on the place-marker *SharedResource*, i.e. it reflects the resource's state only, while the real resource has to be managed by the application. Clearly, the application has to query the state prior to accessing a real resource. A *Shared Resource* object is the smallest unit that can be assigned to a *Floor*, but multiple *SharedResources* may also be assigned to a single *Floor*.

A *Floor* object can establish the following (*n:m*) relationships between *SharedResource, Session,* and *Participant*:
- Multiple *SharedResource* objects may be assigned to a single *Floor* just as a single *SharedResource* may be assigned to multiple *Floors*. The latter requires that the access permissions amongst these *Floors* are not conflicting.
- A *Session* may hold an arbitrary number of *Floor* objects and a single *Floor* may be used in multiple *Session* objects. This allows for accessing resources across sessions.
- *Participant* objects may be floor holders of an arbitrary number of *Floor* objects. Using the *selective* policy, even a single *Floor* may be assigned to multiple *Participant* objects.

The *Floor* object carries the following attributes: *Control Policy, Entailed Permissions, MaxHolders,* and *Security. Control Policy* may take the values "NoControl", "ImplicitControl", ExplicitControl", and "ChairedControl". The *Entailed Permissions* object defines access rights on the associated resource ("Read", "Write", "Execute" etc.). *MaxHolders* limits the maximum number of floor holders. *Security* may limit the usage of the *Floor* to a single *Session*.

The class *Floor* is an abstract base class which does not implement a concrete floor control policy but provides mechanisms needed by all *Floor* objects. Also, *Queued Floor* is abstract however it enhances *Floor* with queuing mechanisms in order to realize queuing policies. In the model, policies are implemented by deriving specialized *Floor* classes enriched with algorithms which set up relationships to *Participant* objects considering given criteria. The classes *NoCtrlFloor, ImplicitCtrlFloor, ChairedCtrlFloor,* and *ExplicitCtrlFloor* realize policies described in 4.4.

4.3 Consistency Rules

Consistency rules subject possible relationships between objects to conditions so as to avoid conflicting situations and to provide a well defined state of the model at any time. Our CSM defines four basic rules:

C1: *SharedResource* may only have multiple *Floor* objects if the floors are compatible[5] with each other.

C2: A *Floor* can only exist if it is at least assigned to one *SharedResource* and one *Session*.

C3: *Participant* objects may only be floor holders if they are members in at least one *Session* the related *Floor* is assigned to.

C4: *Participant* objects may only become floor holders if the maximum number of floor holders associated with a *Floor* is not exceeded.

C1 and C2 concern the relationship *Shared Resource-Floor-Session*. C1 constrains the number of floors for one *SharedResource* so that floor control is not by-passed by simply creating a new *Floor* object.

C2 manifests the purpose of a *Floor* because it allows accessing shared resources in a *Session*. A *Floor* object without a *SharedResource* object and a *Session* object is useless.

C3 provides a consistent relationship between *Participant-Floor-Session*. If *Participant* objects could access shared resources, which are independent of their *Session*, the assignment of *SharedResource* objects to *Session* objects would have no effect. With C3, *Participants* may only use *SharedResources* assigned to their own *Session* by a *Floor*. This allows, for instance, that different *Sessions* can hold different audio resources in order to permit separated discussions. Moreover, when leaving a session, participants are forced to drop their floor holder role so as to avoid locking resources. If a *SharedResource* is separated from a *Session*, all *Participants* of this *Session* (not being in another *Session* including this resource) have to drop the floor holder role for this *SharedResource*.

C4 regulates concurrent access to *SharedResource* objects.

4.4 Collaborative Services Policies

Taking into account the above consistency rules, different floor control policies can be realized. C4 defines a maximum number of floor holders. Setting it to one provides a *mutually-exclusive*, setting it to n, $n>1$, a *selective* access. The *"no control"* policy can be, for instance, realized by setting an adequately large limit (e.g. the number of participants in the session).

The policies *"no control"*, *"implicit control"*, *"explicit control"*, and *"chair control"* can be distinguished by different state automata (see [Hil96] for a detailed description of the state automata).

The classes *NoCtrlFloor* and *ImplicitCtrlFloor* are *non-queuing* while *ExplicitCtrlFloor* and *ChairedCtrlFloor* provide *queuing*. All policies (except of *"no control"*) may either be *mutually-exclusive* or *selective*.

Also for sessions, different policies can be selected. The *Structure* attribute, for instance, allows hierarchical sessions while the *Security* attribute limits floor usage to a single Session. Further application specific strategies can be defined.

---

[5] Floors are considered compatible if the attached access permissions are not conflicting.

4.5 Accessing Services

Collaborative services are provided to the user by an application programming interface (API). All API calls are listed in Table 1.

CreateParticipant	*RemoveFloorFromResource*
DeleteParticipant	*RequestFloor*
CreateSession	*ReleaseFloor*
DeleteSession	*RequestChairRole*
CreateSharedResource	*ReleaseChairRole*
DeleteSharedResource	*ChairedFloorGive*
JoinSession	*ChairedFloorGet*
InviteParticipant	*AttachSubsession*
LeaveSession	*DetachSubsession*
CreateFloor	*GetObjectState*
RemoveFloorFromSession	*SetObjectState*

Table 1: CSM API calls.

By querying the states of objects through the API, the application can read the result of collaborative services. In order to allow for a clear mapping between application objects and CSM objects (which are place markers for real world or user interface objects), the class *IdentifiableObject* provides a unique ID for each CSM object. Whenever a new object is created in the CSM by using the API, the ID is returned to the application. The application itself is then responsible for distributing this ID to all co-operating applications in the network. Each application instance is now able to access its own CSM by using this ID.

5 Implementation Issues

5.1 Distributed Architecture

We have chosen a distributed, replicated architecture for our implementation of the CSM (see Fig. 4). This approach meets the following requirements for synchronous groupware systems [Ell91], [Borg95]. *Short response times* for read- and write-operations are achieved because each participant holds a local copy of the CSM. Without a centralized state-server, *minimal notification times* are attained because operations are transmitted directly to participants. Since we solely send change operations instead of the complete collaboration state, only *low bandwidth* is required. High *robustness* is achieved, due to the replication of the collaboration state and the absence of a "single point of failure". Moreover, a distributed, replicated approach allows for better *spacial scaling* of user groups [Dom97].

The realization of the CSM in a CS *agent*, separated from an application, avoids redundancy when multiple local applications are using collaborative services. An application requests a specific collaborative service by issuing the appropriate operation to its local agent using the *State Control Protocol (SCP)*. The agent carries out the operation on its CSM copy. After successful execution, the operation is distributed to all other agents in the network by the *State Interchange Protocol (SIP)*. The *synchronization*

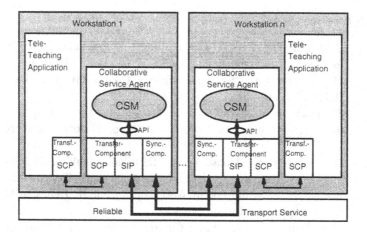

Fig. 2: Communication architecture.

component coordinates concurrent operations in order to ensure consistency amongst the distributed CSM copies. This modular architecture allows for using arbitrary synchronization protocols.

5.2 Optimistic Synchronization

An *optimistic* synchronization scheme detects and resolves conflicts instead of avoiding them. Applying this approach to the CSM means that operations on the model are executed optimistically (without considering possible conflicts), and are reversed in the case of conflicts (*reversible execution* [Ell91]). This approach allows for a maximum concurrency because actually, only contenting operations are reversed. The distributed data of the model has an ephemeral character, i.e. it is only valid during the period of collaboration. Thus, the requirement of 100% fault tolerance, needed, for instance, for databases, can be relaxed in order to achieve a higher responsiveness, which is more important for distributed learning environments than absolute fault tolerance. Analyses of the three different learning modes (see 1.3) indicate that irreversible conflicts will rarely arise and are tolerable for distributed learning. For a detailed description of our implemented synchronization scheme see [Hil96].

6 Conclusion

We have presented an approach for realizing collaborative services in the context of distributed learning environments. Based on an analysis of collaborative service requirements, we developed an object oriented collaborative services model. The lightweighted approach, which follows the paradigm of mechanism and policy, includes fine-grained floor control and session control for tightly-coupled sessions. We do not depend on specific protocols for realizing collaborative services in distributed workgroups because operations are executed by the model and not by a protocol. We implemented the CSM with an optimistic synchronization scheme which offers a high responsiveness in a distributed architecture. Further research will focus on the development of a suitable user interface representation for the CS provided by our model and the integration into the digital lecture board.

7 References

[Borg95]   U.M. Borghoff et al.: "Rechnergestützte Gruppenarbeit, Eine Einführung in verteilte Anwendungen". Springer Verlag, Berlin, Heidelberg, 1995.

[Bor96]   C. Bormann et al.: "Simple Conference Control Protocol". Internet draft, IETF, draft-ietf-mmusic-sccp-00.txt, June 1996.

[Cro90]   T. Crowley et al.: "MMConf: An infrastructure for building shared multimedia applications". In *Proc. CSCW'90*, p. 637-650, ACM, New York, 1990.

[Der93]   G. Dermler et al.: "Constructing a Distributed Multimedia Joint Viewing and Tele-Operation Service for Heterogeneous Workstation Environments". Proc. of the Fourth Workshop on Future Trends in Distributed Systems, pp. 8-15, Lisbon, 1993.

[Dom97]   H.-P. Dommel et al.: "Floor Control for Multimedia Conferencing and Collaboration". In *ACM Journal on MM Sys.*, Vol. 5, No. 1, January 1997.

[Ell91]   C. A. Ellis et al.: "Groupware - Some Issues and Experiences". In *Communications of the ACM*, Vol. 34, No. 1, p. 38-58, 1991.

[Flu95]   F. Fluckiger: "Understanding Networked Multimedia Applications and Technology". Prentice Hall, New York, 1995.

[Gey97]   W. Geyer et al.: "Multimedia-Technolgie zur Unterstützung der Lehre an Hochschulen". In *Multimediales Lernen in der Beruflichen Bildung*, BW - Bildung und Wissen, Nürnberg, 1997.

[Gut95]   T. F. Gutekunst: "Shared Window Systems". Dissertation ETH No. 11120, Swiss Federal Institute of Technology, Zurich, 1995.

[Han95]   M. Handley et al.: "SDP: Session Description Protocol". Internet-Draft, IETF, draft-ietf-mmusic-sdp-01.ps, November 1995.

[Han96]   M. Handley et al.: "The Internet Multimedia Conferencing Architecture". Internet-Draft, draft-ietf-mmusic-confarch-00.ps, IETF, February 1996.

[Hil96]   V. Hilt: "Modellierung und Implementierung von Protokollen und Mechanismen zur Unterstützung von kooperativem Lernen in einer TeleTeaching-Umgebung". Master's thesis, University of Mannheim, Germany, 1996.

[ITU96]   International Telecommunication Union: Recommendation T.120 - Data protocols for multimedia conferencing, July 1996.

[MMC96]   "Multimedia Collaboration". User's & Installation Guide, A.01.00, http://www.prz.tu-berlin.de/docs/html/MMC/manual/index.html, Hewlett-Packard Company, June 1996.

[MMU96]   "Multiparty Multimedia Session Control (MMUSIC) Charter". URL: http://sauce.mmlab.uninett.no/ietf/mmusic-charter.html, 1996.

[Nak93]   A. Nakajima: "Telepointing Issues in Desktop Conferencing Systems". In *Computer Communications*, Vol. 16, No. 9, 1993.

[She95]   S. Shenker et al.: "Managing Shared Ephemeral Teleconferencing State: Policy and Mechanism". Internet draft, IETF, July 1995.

Using Distributed Multimedia Infrastructures for Advanced Teleteaching Applications[1]

Olaf Neumann, Sandra Rennecke, Alexander Schill

Dresden University of Technology, Department of Computer Science, 01062 Dresden

Freiberg University of Mining and Technology

schill/rennecke/neumann@ibdr.inf.tu-dresden.de, http://wwwrn.inf.tu-dresden.de

Abstract

Distributed multimedia systems and advanced network and Internet technologies enable novel application areas such as teleteaching. This paper first presents an experimental distributed infrastructure for such teleteaching applications and discusses recent experiences with various distributed lectures and exercises. In addition to key technological issues such as multimedia information management and distribution, aspects of user interaction, content provision, and acceptance of given solutions are also discussed.

Based on the implementation and evaluation results of the first project phase, an extended support platform using the Java programming language has been designed. Its major architectural components and their integration are presented in the second part of the paper. Special issues addressed are management of animated lecturing material, user administration, workgroup support, and evaluation of the overall learning process.

Keywords: Multimedia, Teleteaching, Distributed Applications, Internet, Java

1. Introduction

With the evolution of Internet services and distributed multimedia technology, new tele-education facilities become feasible. This enables numerous new opportunities for students and universities: Courses can be designed in a more interactive way, lectures of specialists from abroad can be integrated into a regular curriculum, solutions to exercises can be submitted and be corrected online, and interdisciplinary work can be stimulated.

This paper first presents experiences with a teleteaching project between the Universities of Technology in Dresden and Freiberg in Saxony, Germany. Major scenarios discussed in section 2 are video broadcast of lectures and Computer Supported Cooperative Work (CSCW) in small seminars. Important technical issues focus on the use of distributed multimedia infrastructures for these scenarios. In particular, design decisions with respect to multimedia information management and distribution, to server and network capacities and to application categories are discussed. Moreover, experiences with the practical validation and acceptance of the infrastructure during recent semesters are reviewed.

Based on these experiences, an integrated architecture for more systematic development and runtime support of such scenarios based on Java has been developed recently. In its second part, the paper presents the major components and design issues

---

[1] This work was supported by funds from the BMBF.

of this environment in section 3 (JaTeK - Java Based Teleteaching Kit). The main JaTeK component implements user administration, management of user groups, integration of multimedia lecturing material, and several auxiliary tools. JaWoS (Java-based Workgroup Support) offers various multimedia-based CSCW tools for interaction within small learning groups. Finally, JaVal (Java-based Evaluation) implements tools and techniques for evaluating learning success, especially with respect to online exercises.

Section 4 reviews and compares related approaches in the teleteaching and distributed multimedia area. Briefly spoken, our work differs from similar teleteaching approaches such as the Project TeleTeaching Heidelberg-Mannheim /EFF95/ by emphasizing not only video/audio distribution of lectures and multimedia-enhanced lecturing material in the World Wide Web, but also integration of interactive exercises, use of CSCW techniques for distributed interaction among learning groups and built in evaluation of access and acceptance. From a technological point of view, we are employing existing distributed multimedia technology on a day-to-day basis for implementing these scenarios. However, we are also working on technological advances in this area, for example concerning new resource reservation techniques for multimedia data streams in heterogeneous networks.

2. Infrastructure and Experiences

2.1. Background

Our teleteaching approach is mainly characterized by the following technical background and scenarios:

- *Use of the Internet:* Due to its broad availability, the Internet serves as the major networking basis of the project. Our current access rate is 34 Mbit/s and is envisaged to be upgraded to 155 Mbit/s. Locally between major campus sites, a 155 Mbit/s ATM network infrastructure is operational. All lecturing materials are offered via WWW servers using HTML and postscript formats. Internet access has been enabled for several dormitories in Dresden, making the teleteaching infrastructure accessible from the students' home. Most client-site applications are running on relatively unexpensive PCs under Windows NT, Windows 95, and Linux.

- *Variety of scenarios:* To enable flexible teleteaching support, three major scenarios are being implemented (1) Broadcast of lectures: First, lectures can be broadcast online or batch-oriented to remote locations via the Internet. In addition to audio and video of the lecturer, the teaching materials are displayed via shared applications. Internet tools such as mbone, and wb (whiteboard) serve as an experimental basis. (2) Interactive exercises: Secondly, exercises can be accessed via WWW in an asynchronous way, and solutions can be sent back by the students (also from dormitories) via HTML forms, recently also augmented with animations based on Java applets. Some exercises can be corrected automatically via solution masks (e.g. for multiple choice questions), but most are corrected interactively. Feedback is given to the student by MIME mail. (3) CSCW in small groups: Thirdly, we offer seminars to small groups of students. The students cooperate concerning special exercises, for example designing web pages for a virtual company.

- *Interdisciplinary approach:* While most teleteaching courses are offered in the area of Computer Science, other faculties are also participating. In particular, the Psychology department contributes to issues of the psychology of teaching and learning, the Economics and Business Administration department contributes to an office communication course, and other departments, especially at our partner site in Freiberg are importing Computer Science classes to enhance their curriculum.

In the following subsections, details concerning the various scenarios and the recent experiences will be reported.

2.2. Online Lectures

The first lectures supported by the project were "Office Communication", a course for Computer Science majors held in Dresden and exported to Freiberg in the summer semester '96 (SS 96). Moreover, a course on "Computer Networks" was also exported and partially integrated into "Computer Science I and II" at the remote location. In the winter semester '96 (WS 96), these lectures were augmented by an online course in Psychology, emphasizing the interdisciplinary nature of our project. Other Talks held in Freiberg about "Modelling and Efficiency Measurement of Distributed Systems" were used in Dresden. Further courses held by the participating faculties will also be supported.

The lecturing materials used are mainly based on large collections of animated Powerpoint foils. They are presented remotely based on wb (whiteboard). Although this works fine for purely displaying the material, it is not suitable for communicating the behavior of complex processes and procedures to the students; this rather requires WWW and Java-based techniques discussed later. As the implemented lectures mainly cover technological aspects themselves (Java, HTML, EDIFACT, RPC etc.), there is a close correlation between the contents and the technology. This has facilitated the introduction of teleteaching and has led to an increased acceptance as also outlined later in more detail.

Video transmission is based on mbone. Even between the main campus and selected dormitories, an mbone tunnel has been configured so that lectures can be viewed remotely in the students' rooms. Several specific problems had to be solved within this context: For example, the Linux system used in the dormitory is not multicast-enabled and therefore had to be replaced by a BSD-Unix system. The video frame rate had to be kept much lower in this case as compared to the intra-campus communication. This is not that much a problem as the importance of video seems to be much lower than the relevance of high-quality audio and lecturing material due to our recent polls among students.

2.3. Evaluation and Experiences

The technical infrastructure has been operational for about one year. Several experiences with the use of the tools mentioned above have been made. Experiences with Internet access from dormitories are very positive, with a very intensive use of these facilities by a large number of students. In special lectures such as computer networks, positive experiences with online animations (using PowerPoint and additional tools) were made, for example for illustrating the behaviour of transparent bridges in networks.

Digital videos containing important statements made in the lecture can be retrieved by students from the pages. The WWW pages for the affected classes are created while the project is running, in cooperation with the involved psychologists. Special effort is put into making them effective for learning and acceptable by the students.

Digital video sequences of ten to twenty minutes duration with accompanying Powerpoint-slides have also been shown in lectures, using large scale projection equipment. The importance of very good audio quality was severely underestimated in the beginning. Improving audio with using higher sampling rates, professional microphones and more powerful speakers in the auditorium has led to much higher acceptance. Video is a problem, as high quality digitized sequences are still expensive, time- and loadwise, to create, transmit, store and play. Despite those technical problems, the projection into lectures has been found both fitting in topic and difficulty as well as interesting by most students. Much positive feedback in the free-from „remarks" section of the questionnaire encourages us to continue using video/slide inserts by experts in lectures.

Accompanying the courses is an evaluation of the materials offered by teleteaching, results of which flow back into the project. Return of questionnaires was voluntary, of about 80 students present in the lectures, 50 to 60 questionaires were returned.

The comments we received in the free form section of the questionaire were mostly positive („good idea", „interesting"). Concerning specific criteria, we evaluated the difficulty of the course materials, the organization of the slides, the synchronization between video/audio and slides mentioned already, and the audio quality. An example result is summarized and interpreted in fig. 1.

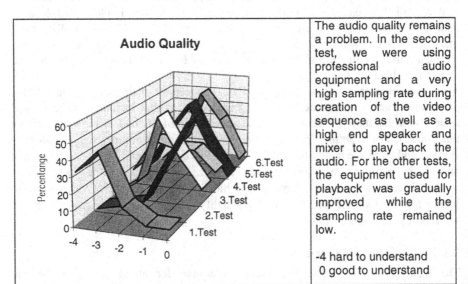

The audio quality remains a problem. In the second test, we were using professional audio equipment and a very high sampling rate during creation of the video sequence as well as a high end speaker and mixer to play back the audio. For the other tests, the equipment used for playback was gradually improved while the sampling rate remained low.

-4 hard to understand
0 good to understand

Figure 1: Example evaluation result

3. Advanced Support Platform: JaTeK

3.1. Architecture

The programming language Java offers the means to use the Internet for interaction /ABD96/. A systematic and open architecture for all levels, from creation to usage through several loops of evaluation and redesign of the courseware, remains to be implemented to achieve the true potential of distributed multimedia teaching and learning in the Internet. The proposed architecture is a modular, object-oriented and distributed design that can encompass future enhancements to existing creation, presentation and interaction methods. The implementation will be accompanied by usage of the existing modules, so that practical feedback and evaluation can be incorporated into the ongoing design process and examples exist in all stages of the project. Similar to Tim Berners-Lee's description of the Wold-Wide Web /BER94/, JaTeK can be associated with the following concepts, from lowest to higest level:

- the program module JaTeK and its applet-clients
- a set of distributed Java programs and applets and communication protocols between those programs, divided into several modules
- a collection of services to create and run Internet-supported tutorials and lectures
- a consistent WWW/Java-based environment for learning, containing information and interaction parts.

Basically, three levels of users will work with JaTeK:
- programmers that extend the functionality and integrate new tools in the environment
- information providers (lecturers) that use JaTeK to hold courses and evaluate learning success
- information retrievers (students) that use the system in their classwork.

The boundaries between these user levels are not static, a programmer might also be an information provider if she is holding lectures, and an information retriever might also want to offer information. The aim of the architecture is to remain as flexible and open as possible to leave room for possible later needs of the different user groups. On a system level, this means modularization and breaking down the complex task of teleteaching into small, manageable applications that can be exchanged for newer versions without requiring recompilation/reinstallation of the whole system. A consistent interface on the provider and retriever level must be guaranteed at all times to restrict the time needed to learn interaction with JaTeK to a minimum.

At the present stage, JaTeK consists of three larger modules that are independently developed, all consisting of sets of applets and programs themselves. These modules will be described in more detail in the next sections, briefly they are:

- JaTeK, the main module, capable of having connection oriented communication with its clients using HTTP, handling user authentication, client communication, communication with the other modules and offering interactive courseware
- JaWoS (Javabased Workgroup Support), a component handling interaction and cooperation of several users

- and JaVal (Javabased Evaluation System), a component to monitor statistical access data for sets of information and to present the data to the provider.

The components of JaTeK make use of an applet/application pool management to distribute new tools between servers and from server to client. The Java classes will be transmitted using object serialization, that is supported in the newly released JDK 1.1. This dynamic handling of software components ensures that JaTeK can be extended during runtime. The description of applets and applications in the pool allows an application to request task specific programs without knowing their actual names. To implement the server side pool management component, the use of a CORBA-compliant middleware component or of a Java enabled database are under consideration.

3.2. Teleteaching Kit

The central component, JaTeK, consists of a system allowing to create WWW-sites containing courseware. The material can be classified and integrated into the learning environment with special tools. A user administration component allows linking personal data, i.e. comments or solutions to exercises, within the environment. Moreover, login and authentication are supported. For designing exercises, templates

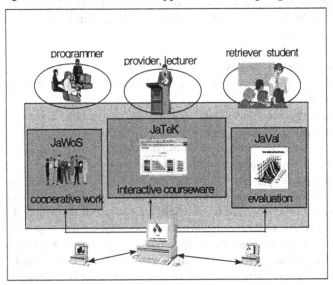

Figure 2 Overview of the JaTeK system

for multiple choice, free text entry and others are offered and can be adapted in order to generate specific exercises by the lecturer.

The web pages are managed by a database system. Relevant data, that is user data, directory data, data pertaining to web pages and media specific data are being stored on a SQL-server. This allows the dynamic creation of web pages from the stored data. The contents of the server can also be accessed by applets used in tutorials. Special tools allow to manipulate directory entries and enable the authoring of exercises.

3.3. Workgroup Support

Designing and implementing teaching materials that use communication and interaction capabilities to support cooperative learning and teaching is the aim of the component for Java-based Workgroup Support (JaWoS). JaWoS offers services and examples to make teaching materials support multiple interacting users. Through modules that offer communication in groups, as well as applets and tools that can be manipulated by several users simultaneously, asynchronous and synchronous cooperative work-scenarios in a JaTeK-environment are made simple to create. Group- and role information for human users as well as for involved computer systems are managed by JaWoS, so the system is extended by roles offering flexibility and ease of use as described in /GOT96/. A JaWoS-server encapsulates these services and information and offers client access though a well-defined interface. It also offers Java libraries and applets that can be used by clients to create an interactive component. The JaWoS server usually runs on the same host as the JaTeK component that manages most of the users registered with it. Roles and groups can be edited from any host on the network by a qualified user with Java applets. It is not planned to implement all applets distributed by JaWoS from scratch in the course of the project. The aim is rather to use existing free applets and code and create a calling interface where possible.

3.4. Evaluation Component

To evaluate the quality of the courseware created with JaTeK, data about tool- and information usage is collected within the JaVal component . To this end, access times and frequency are gathered in a database. Scores reached in exercises and comments, help requests or note marks about the usage of the environment are also collected based on a learning progress control and task evaluation. Data is stored anonymously.

After statistical evaluation of the data, conclusions for improving the learning environment can be drawn. For example, it was found that fonts with serifs are hard to read when projected with a beamer. Also, audio- and video components were rarely used by students with low-bandwidth access, as file sizes of several megabytes are needed to achieve sufficient quality.

4. Related Approaches

Looking at related approaches, we can mainly distinguish between teleteaching projects and platforms on one hand and basic technology of distributed multimedia systems on the other. Both areas are important with respect to our presented approach and are briefly reviewed and compared to our work below.

General surveys about requirements and approaches in the teleteaching area are found in /DP94, BW95/. These papers illustrate that videoconferencing, application sharing via whiteboards, the use of hypermedia, animated lecturing materials, support for the authoring process and a powerful networking infrastructure are key components of an overall solution for computer-based training and teleteaching. Although technically the usage of modern multimedia materials in regular university courses is certainly possible, applying them on a day to day basis still demands a lot of effort. Experiences in the last year have shown that many of the time consuming tasks in producing, editing, offering and even retrieving digital course enhancements can be simplified once a pattern of work is established and examples exist.

Several projects are using videoconferencing facilities and have reported about the experiences with their use in distance lecturing; examples are /WG96, GL96/. In /SRH95/, the implementation of a virtual classroom as an extension of these concepts is described. /TDA95/ discusses the implementation of virtual scientific conferences based on Internet tools and reports about experiences of an extensive trial. Related projects with similar goals are /TTT95/ (video-enhanced lecturing material offered via WWW)or /HB96/ (cooperative learning using a WWW interface). Several projects also employ database technology for maintaining and offering lecturing material, sometimes in direct combination with WWW services; an example is presented in /AS93/. Our SQL server approach uses similar concepts.

With respect to distributed multimedia systems /NAS95, STN95, VKB95/, we are mainly using existing services and facilities such as mbone, whiteboard, advanced WWW techniques such as forms and applets etc. on top of conventional networks (such as IP over Ethernet or ATM). Our particular emphasis has been on gathering real-world experiences with such infrastructures and on developing integrated higher-level services for direct support of teleteaching.

However, within related projects, we are also investigating the use and extension of advanced protocols for multimedia communication /ZDE93, DHH94/, especially in the context of an RSVP /BRZ96 / and IPng implementation over ATM /SKB96/. At the higher, application-related layers, several approaches in the area of quality of service (QoS) research have been considered within the design of our infrastructure and shall also be incorporated at the implementation level. Important examples are the Cinema system /BDF95, RBH94/ with configurable components for implementing QoS-aware distributed multimedia applications, the QoS-Architecture (QoS-A) /CCN94/ with its layered abstraction facilities for QoS specification and mapping, and operating system and scheduling approaches for guaranteeing quality of service at the end system level /GOP96/.

Within our own implementation work, we also developed a QoS-aware video transmission system named Xnetvideo /HHS96/ with integrated video scaling facilities in order to adapt multimedia streams to the available bandwidth in heterogeneous networks. Similar work is going on concerning the extension of the Berkom MMC (Multimedia Collaboration) system (cf. /ADH93/) with video scaling mechanisms. We plan to take advantage of these initially separate developments by incorporating these services into our teleteaching environment.

5. Conclusions and Future Work

This paper has described our recent experiences with the use of distributed multimedia infrastructures for advanced teleteaching applications and has also presented an emerging Java-based platform for systematically supporting such applications. Most important, it should have become obvious that today's multimedia facilities, network infrastructures and Internet services already enable rather sophisticated, interdisciplinary teleteaching scenarios. In particular, we often discovered that didactic questions and issues of presentation and user interaction are as challenging as the basic technological problems.

However, we also got aware of the direct correlation between available bandwidth and possible scenarios - for example, asynchronous interaction within exercises demand much less network capacities than synchronous audio- and application sharing interactions or even video-based interaction. For technological reasons, but mainly for financial reasons (e.g. concerning dorm networking), bandwidth in the wide area range will remain scarce within even the mid- or long-term future. Therefore, it is crucial for strong acceptance of teleteaching facilities to offer various heterogeneous mechanisms and scenarios adapted to the given bandwidth.

We also recognized the important potential of user interaction within exercises and practical online courses. The use of Java applets to implement such interactions turned out to be rather effective. This guided our decision to implement a more systematic and sophisticated support platform for teleteaching based on Java. Nevertheless, there are still significant problems concerning the rapid release of new versions of Java JDK and of limited stability of the Java platform.

Our future work will focus on the full implementation and evaluation of this platform. Particular emphasis is put on further supporting collaborative exercises and evaluation of learning progress. Moreover, the integration of QoS-aware media transmission facilities and of resource reservation techniques at the network layer will also be important issues.

Acknowledgements

This project is supported by BMBF (Bundesministerium für Bildung, Forschung und Technologie) via DFN (Deutsches Forschungsnetz). We would like to thank all colleagues and students who contributed to the results presented in this paper. In particular, Katrin Franze provided a lot of support by reviewing and classifying related work.

References

/ABD96/ Abdel-Wahab, H., Kvande, B., Nanjangud, S., Kim, O., Favreau, J.P.: Using Java for Multimedia Collaborative Applications, Proceedings of the 3rd International Workshop on Protocols for Multimedia Systems (PROMS'96), October 1996, Madrid

/ADH93/ Altenhofen, M., Dittrich, J., Hammerschmidt, R., Käppner, T., Kruschel, C., Kückes, A., Steinig, T.: The BERKOM Multimedia Collaboration Service; ACM Conf. on Multimedia, 1993

/AS93/ Dr. Sloane, A.: Aspects of multimedia database technology for teleteaching; School of Computing and Information Technology, University of Wolverhampton, 1993 (http://www.wlv.ac.uk/~cm9037/as1993.html)

/BDF95/ Barth, I., Dermler, G., Fiederer, W.: Levels of Quality of Service in CINEMA, GI-SI Jahrestagung, Zürich, Schweiz, 18.-20. September, 1995

/BER94/ Berners-Lee, T. and others: The World-Wide Web; Communications of the ACM, Vol. 37, No.8, 1994

/BRZ96/ Braden, R., Zhang, L.: Resource ReSerVation Protocol; Version 1, Functional Specification, Work in Progress, 1996

/BW95/ Dr. Willis, B.: Distance Education at a Glance... ; Engineering Outreach University of Idaho, Moscow, 1995 (http://www.uidaho.edu/evo/distglan.html)

/CCN94/ Campbell, A., Coulson, G., Hutchison, D.: A Quality of Service Architecture; Computer Communication Review, Vol. 1, No. 2, April 1994, pp. 6-27

/DHH94/ Delgrossi, L., Herrtwich, R.G., Hoffmann, F.: An Implementation of ST-II for the Heidelberg Transport System; Internetworking: Research and Experience, Vol. 5, 1994, pp. 43-69

/DP94/ Peraya, D.: Distance Education and the WWW; WWW Conference Workshop: Teaching & Learning with the Web, Genf, May 25-27, 1994 (http://tecfa.unige.ch/edu-ws94/contrib/peraya.fm.html)

/EFF95/ Effelsberg, W.: Das Projekt TeleTeaching der Universitäten Heidelberg und Mannheim, PIK 18 (4), 1995 http://www.informatik.uni-mannheim.de/informatik/pi4/projects/teleTeaching/index.html

/GL96/ Grebner, R., Langenbach, C.: Dokumentation des Projektes „Multimediaunterstützte Dezentralisierung interdisziplinärer Lehre"; Leibnitz-Rechenzentrum, München, 1996, http://www.wi2.uni-erlangen.de/project/RTB-312/doc/index-d.html

/GOP96/ Gopalakrishnan, R., Parulkar, G.: Bringing Real-Time Scheduling Theory and Practice Closer for Multimedia Computing; SIGMETRICS 96-5/96 Philadelphia, PA, USA

/GOT96/ Gottlob, G., Schrefl, M., Röck, B.: Extending Object-Oriented Systems with Roles, ACM Transactions on Information Systems, Vol. 14, No. 3, July 1996, pp. 268-296

/HB96/ Hoogstoel, F., Bourguin, G.: Using a WWW Server to Access and Manage a Gemstone Smalltalk Server Supporting a Virtual Collaborative Learning Organisation; WebNet 96 - San Francisco, CA - October 15-19, 1996, http://rom.informatik.uni-freiburg.de/webnet96/Html/471.htm;internal&sk=021F6C44

/HHS96/ Hess, R., Hutschenreuther, T., Schill, A.: Video Communication and Scaling System Xnetvideo: Design and Implementation; European Workshop on Interactive Distributed Multimedia Systems and Services, LNCS 1045, Springer-Verlag, Berlin, 1996, pp. 195-210

/NAS95/ Nahrstedt, K., Steinmetz, R.: Resource Management in Networked Multimedia Systems; IEEE Computer, Vol. 28, No. 5, May 1995, pp. 52-63

/RBH94/ Rothermel, K., Barth, I., Helbig, T.: CINEMA - An Architecture for Distributed Multimedia Applications; in: Architecture and Protocols for High Speed Networks, Amsterdam, Kluwer Academic Publishers, 1994, pp. 253-271

/SKB96/ Schill, A., Kühn, S., Breiter, F.: Internetworking over ATM: Experiences with IP/IPng and RSVP; Computer Networks and ISDN Systems, Vol. 28, 1996, pp. 1915-1927

/SRH95/ Hiltz, S. R.: Teaching in a virtual classroom; International Conference on Computer Assisted Instruction, National Chiao Tung University Hsinchu, March 7-10, 1995 (http://www.njit.edu/Department/CCCC/VC/Papers/ Teaching.html)

/STN95/ Steinmetz, R., Nahrstedt, K.: Multimedia Computing, Communications and Applications; Prentice-Hall, Englewood Cliffs, NJ, 1995

/TDA95/ Anderson, T. D.: The Virtual Conference: Extending Professional Education in Cyberspace; International Journal of Educational Telecommunications, 1995 ,http://www.atl.ualberta.ca/papers/vircon_menu.html

/TTT95/ Dokumentation zum Projekt Teleteaching and -training; Technische Hochschule Darmstadt, 1995, http://www.informatik.th-darmstadt.de/PI/TTT/ Welcome.html

/VKB95/ Vogel, A., Kerherve, B., von Bochmann, G., Gecsei, J., Distributed Multimedia and QoS: A Survey, IEEE Multimedia, Vol. 2, No. 2, 1995, pp. 10-19

/ZDE93/ Zhang, L., Deering, S., Estrin, D., Shenker, S., Zappala, D.: RSVP: A New Resource Reservation Protocol; IEEE Network, Sept. 1993, pp. 8-18

A Scalable Scheme to Access Multimedia Documents with Quality of Service Guarantees

Abdelhakim Hafid and Raouf Boutaba

Computer Research Institute of Montreal
1801, McGill College avenue, #800
Montreal, Canada, H3A 2N4
http://www.crim.ca/~ahafid

Abstract: This paper describes a scalable scheme, we call SAS, that allows users to access multimedia documents in a distributed system. SAS assumes a system which consists of a set of client machines, a set of server machines, and networks connecting these machines. *Popular documents* are replicated and stored on the servers; this replication is realistic and required since popular documents are accessed by most users. A user can ask to play a document with the desired quality of service (QoS) from any server; SAS allows to select the *best* server which is able to deliver the document to the user. This means that (1) this server stores the requested document; (2) has enough available resources to deliver the document with the desired QoS; preferably it is the least loaded server; and (3) there are enough network resources to support the transfer of the document from the server to the user. SAS consists of high level interactions between some management entities and thus may be applied to an arbitrary system with different types of commercial and research multimedia-on-demand servers; it is essentially independent of the technologies used to deliver multimedia data.

SAS allows to minimize the blocking probability as much as possible; a user request is rejected only if all servers (storing the requested document) are loaded at maximum or the communication system is overloaded; it also allows to recover automatically, if this is possible, from QoS degradations during the presentation of the document. Last and not least, SAS allows to mask server failures.

Keywords: scalable, multimedia server, quality of service, adaptation

1. Introduction

Multimedia-on-demand systems, such as video-on-demand and news-on-demand, allow users to retrieve MM information in real-time over a broadband network onto their display devices. These systems should be able to support location transparent and concurrent access to hundreds or thousands of users; this is of great importance for their viability and their acceptance by potential users.

Many research groups are developing multimedia-on-demand servers, particularly video-on-demand servers [Gem 95]; however, most of current approaches used in designing and implementing these servers have concentrated on the performance of the continuous media file servers in terms of seek time overhead, and real-time disk scheduling. Examples are the continuous media file server at Lancaster [Lou 93], the UCSD video-on-demand server [Ran 93], and the multimedia storage server at Washington University [Bud 96]. Upon receipt of a user request to play a document, the server checks whether it has enough resources, e.g., CPU slots, to support the request; if the response is yes, the document presentation starts; otherwise a simple reject message is sent to the user even if available resources, e.g., another server, exist somewhere in the system. This means that the current systems focus on some local resources (server's resources) reservation issues and do not take benefit from the system distribution.

In this paper we propose a scheme to access MM documents which scales well the characteristics and requirements of MM systems. The here described scheme, we call SAS, assumes a system which consists of a set of client machines, a set of server machines, and networks connecting these machines. The servers in use can be of any type as they provide facilities to play MM documents; SAS is essentially independent of the servers technologies and the software they use. Popular documents are replicated and stored on the servers; this replication is realistic and required since popular documents are accessed by most users. A user can ask to play a document with the desired quality of service (QoS) from any server; it is the task of SAS to select a *best* server which is able to deliver the document to the user.

In a distributed system, generally some servers are overloaded while others are idle. Users usually request services from the same sub-set of servers all the time; they would always access their favorite (well-known) server (s). If this server is down, e.g. overloaded, they cannot get the requested services. SAS allows to find a server that (1) has the requested document; (2) has enough available resources to deliver the document with the desired QoS; preferably it is the least loaded server; and (3) there are enough network resources to support the transfer of the document from the server to the user. The objective of SAS is to minimize the blocking probability as much as possible; a user request is rejected only if all servers (storing the requested document) are loaded at maximum or the communication system is overloaded. This is in opposition with current multimedia-on-demand systems which concentrate to check the capacity of a *single* server (which may consist of a cluster of computing nodes) to play the requested document. Furthermore, SAS allows an automatic adaptation to react to QoS degradations without the direct intervention by the user/application. That is, during the play-out of the document, if the network and/or the server machine become congested thus leading to lower presentation quality, SAS selects another suitable server (which is able to deliver the requested document) and automatically (user-transparently) performs a transition from the current server (in difficulty) to the new one. More specifically, SAS stops the presentation of the document after having obtained the current position of the document, and restarts the presentation (using the alternate server) from the position parameter determined earlier. This transition procedure has been implemented in the context of the CITR news-on-demand prototype [Won 97]. Last but not least, SAS allows to mask server failure; this is done by detecting failures and reconfiguring the system such that the failed server(s) is not considered when processing a new user request.

It is worth noting that the emphasis of SAS is on the rapid access to a small number of MM documents by a large number of users; this is in opposition to systems with an emphasis in providing access to a large number of videos by a smaller number of users, such as the Berkeley video-on-demand system [Bru 96].

The rest of the paper is organized as follows. Section 2 presents an overview of SAS and describes protocols that implement it In Section 3 we qualitatively evaluate SAS performance through simulations. Section 4 concludes the paper.

2. Scalable Access Scheme (SAS)

Given a user request to play a MM document with a desired QoS, SAS allows to locate a MM server which is able to support the delivery of the document while satisfying the user requirements. In order to improve the overall system performance, and thus its scalability, SAS operates a dynamic server localisation according to several criteria that are: (1) MM server's load; (2) the QoS characteristics of the MM document stored in the MM server in question; and (3) the user QoS requirements.

The operation of SAS can be summarised as follows: (1) identify the MM servers which contain a variant of the MM document to play; this gives the list of potential MM server candidates; (2) select the MM server which contains the document variant which satisfies the most the user request, and which has the lightest load; (3) check whether there are enough local and network resources to deliver the document from the MM server (selected in (2)) to the user's host; (4) if the response is yes, start the presentation of the document; otherwise go to step (2).

We decided to use hierarchical protocols to implement SAS; this seems to be natural thanks to the hierarchical structure of distributed systems [Bou 95]. For the sake of simplicity and clarity of the description of the protocols that implement SAS, we consider a two-level hierarchy system as shown in Figure 1. However, the proposed protocols can be used in a system with an arbitrary hierarchical structure.

We identified mainly two approaches for implementing SAS: a hierarchical protocol based on a top-down approach (as described below), and a hierarchical protocol based a bottom-up approach. The operation of the bottom-up approach is similar to some extent to the operation of the top-down approach; a more detailed description of the bottom-up approach can be found in [Haf 97c].

The top-down approach is used if we assume that the server manager receives directly requests from the users. Concretely, this is done via the user agent; the user asks to play a document with the desired QoS via a user-interface provided by the user agent. Just after this operation, the user agent directs the user request to the server manager.

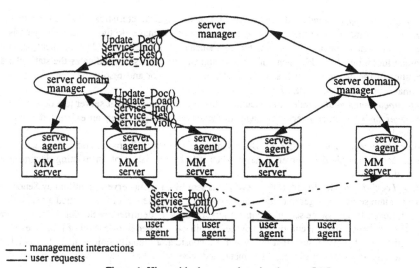

Figure 1. Hierarchical protocols to implement SAS

When the server manager receives the user request, it determines the domain which contains the document to play and which is the least loaded domain. It is up to the domain server manager to select (or to find) a server which is able to support the delivery of the requested document. If this activity succeeds, then the presentation of the document can start; otherwise, a reject message is sent back to the server manager; the latter selects another domain which contains the document to play and forwards the user request to it. If all domain server managers answer with a reject message, the server manager sends a reject message to the user.

Signals Description

We define the following signals:

- *Service_Inq signal* (Sender_Id, User_Id, Document_Id, QoS, Host_Constraints): It is sent by the user, the server manager, a domain manager, or a server agent; it contains five parameters: (1) Sender_Id: indicates the identifier of the sender; (2) User_Id: indicates the IP address of the user's host and the port number in use; (3) Document_Id: is the unique identifier of document to play; (4) QoS: indicates the QoS requirements of the user; they are described in terms of parameters that are directly expressed in terms of human-perceptible quantities [Haf 96]; (5) Host_Constraints: indicates the static constraints of the user's host, such as the monitor size and resolution and the compression schemes which are supported.

- *Service_Res signal* (User_Id, Document_Id, Status): It is sent by an agent server manager or a domain server manager in response to Service_Inq(); it contains three parameters: (1) User_Id; (2) Document_Id; and (3) Status: indicates whether the user request can be supported or not; Status can assume two values: ACCEPT or REJECT.

- *Service_Conf* (User_Id, Document_Id, Status, Proposition): It is initially sent by an agent server manager to the user; it contains four parameters: (1) User_Id; (2) Document_Id; (3) Status: indicates whether the user accepts Proposition or not; Status can assume two values: ACCEPT or REJECT; and (4) Proposition: indicates the way (e.g., QoS characteristics) the server agent will play the document, identified by Document_Id.

- *Service_Viol signal* (Sender_Id, User_Id, Document_Id, QoS, Host_Constraints): It is sent by (1) a server agent when the server cannot maintain the delivery of the document (identified by Document_Id) to the user (identified by User_Id) while satisfying the user QoS requirements QoS; or (2) the user, identified by User_Id, when he/she notices a QoS degradation of the presentation of the document identified by Document_Id.

- *Service_Alive signal* (Sender_Id): It is sent by a server agent, identified by Sender_Id, to its domain server manager; the signal is used to detect server failures.

- *Add_document signal* (Sender_Id, Document_Id, Document_Characteristics): It is sent by an agent server to a domain server manager or by a domain server manager to the server manager when a new document is created; it contains three parameters: (1) Sender_Id; (2) Document_id: is the unique identifier of the document; and (3) Document_characteristics: indicates the static characteristics of the document, such as QoS constraints (e.g., color and resolution), and compression format (e.g., MPEG or MJPEG).
- *Delete_document signal* (Sender_Id, Document_Id): It is sent by an agent server to a domain server manager or by a domain server manager to the server manager when an existing document is deleted;
- *Update_document signal* (Sender_Id, Document_Id, Document_Characteristics): It is sent by an agent server to a domain server manager when the characteristics of an existing document are altered;
- *Update_Load signal* (Sender_Id, Load_Level): It is sent by an agent server, identified by Sender_Id, to a domain server manager to report that the current load of the server has reached a load level of Load_Level. It can also be sent by a domain server manager, identified by Sender_Id, to the server manager to report that the current load of the domain has reached a load level of Load_Level;

In a distributed systems such that the system we consider, a globally unique identifier should be used to identify any entity in use (e.g., documents and servers).

2.1. Description of the Server Manager

The task of the server manager is to redirect user requests to the appropriate domain of servers. For this purpose, the server manager maintains two state variables M_Document_List and M_Load_List as shown in Figure 2.

M_Document_List	
Document_Id	List_of_Domains [Domain_Id, ...]
I	I
I	I
I	I

M_Load_List	
Domain_Id	Load
I	I
I	I
I	I

Figure 2. State variables of the server manager

Variables Description

We define the following variables:
V_List_of_Domains: a variable which indicates a list of identifiers of a subset of the system domains.
D: a variable which indicates a domain identifier

Operation

- When a *Service_Inq (User_Id, User_Id, Document_Id, QoS, Host_Constraints) signal* is received: The server manager performs the following operations:

  - sets V_List_of_Domains to List_of_domains which is associated with the entry which corresponds to *Document_Id*.

  - orders the elements of V_List_of_Domains in terms of their load; ideally, the first element of V_List_of_Domains would represent the least loaded domain.

  - considers the first element, D, of V_List_of_domains and sends Service_Inq() to D; D is the current domain.

- When a *Service_Res (User_Id, Document_Id, Status) signal* is received:
  If *Status* is equal to REJECT, then if the current domain is not the last element of V_List_of_domains, then the server manager considers the next element, D, of V_List_of_domains and sends Service_Inq() to D (D becomes the current domain); otherwise it sends Service_res() signal to the user.

- When a *Service_Viol (Sender_Id, User_Id, Document_Id, QoS, Host_Constraints) signal* is received:

If the sender is the user (*Sender_Id* is equal to *User_Id*) then the Service_Viol() signal is processed in the same way as a Service_Inq() signal; otherwise, if the sender is a domain server manager, then the Service_Viol() signal is processed as Service_Inq() signal with slight modifications: the operation (V_List_of_Domains:=V_List_of_Domains - [*Domain_Id*]) should be performed before the ordering of V_List_of_Domains.

- When a *Add_document (Domain_Id, Document_Id, Document_Characteristics) signal* is received:

If an entry for *Document_Id* exists in M_Document_List then the server manager simply adds the element Domain_Id to its List_of_Domains; otherwise it creates a new entry for *Document_Id* and sets the corresponding List_of_domains to [*Domain_Id*].

- When a *Delete_document (Domain_Id, Document_Id) signal* is received:

If List_of_Domains which associated with the entry which corresponds to *Document_Id* is equal to [*Domain_id*] then the server manager deletes this entry; otherwise it deletes *Domain_Id* from List_of_Domains (List_of_Domains:=List_of_Domains - [*Domain_Id*]).

- When a *Update_Load (Domain_Id, Load_Level) signal* is received:

The server manager sets Load, which is associated to the entry which corresponds to *Domain_Id*, to *Load_Level*.

2.2. Description of the Domain Server Manager

The task of a domain server manager is to redirect user requests to the appropriate server; a domain server manager knows some details about the servers of the domain, but knows nothing about the servers of other domains. The domain server manager maintains two state variables D_Document_List and D_Load_List as described in Figure 3.

D_Document_List			D_Load_List	
Document_Id	List_of_Server_Doc [(Server_Id, Doc_Char), ...]		Server_Id	Load
¦	¦		¦	¦
¦	¦		¦	¦
¦	¦		¦	¦

Figure 3. State variables of the domain server manager

Variables Description

We define the following variables:

V_List_of_Servers: a variable which indicates a list of identifiers of a subset of the system servers.

S: a variable which indicates a server identifier

Operation

- When a *Service_Inq (Sender_Id, User_Id, Document_Id, QoS, Host_Constraints) signal* is received:

The domain server manager performs the following operations:

- puts in V_List_of_Servers the identifier of any server which contains the document, identified by *Document_Id*, such that the characteristics of this copy of the document satisfy the user's host constraints *Host_Constraints*. A detailed description of this activity in the context of news-on-demand systems can be found in [Haf 97a]. It is worth noting that V_List_of_Servers is built based on the information contained in List_of_Server_Doc which is associated with the entry which corresponds to *Document_Id*.

- orders the elements of V_List_of_Server_Doc in terms of two parameters: (1) the degree of satisfaction of the user QoS requirements by a server [Haf 96]; and (2) the load of the server in question. The ordering of the list will be primarily based on the first parameter.

- considers the first element, S, of V_List_of_Servers and sends Service_Inq() to S; S is the current server.

- When a *Service_Res (User_Id, Document_Id, Status) signal* is received:

If Status is equal to ACCEPT or the current server is the last element of V_List_of_Servers then the domain server manager sends Service_Res() signal to the server manager; otherwise it considers the next element. S, of V_List_of_Servers and sends Service_Inq() signal to S; S becomes the current server.

- When a *Service_Viol (Sender_Id, User_Id, Document_Id, QoS, Host_Constraints) signal* is received:

If the sender is the server manager (*Sender_Id* is equal to the server manager identifier) then the Service_Viol() signal is processed in the same way as a Service_Inq() signal; otherwise, if the sender is a server agent, then the Service_Viol() signal is processed as Service_Inq() signal with slight modifications: the operation (V_List_of_Servers:= V_List_of_Servers - [*Sender_Id*]) should be performed before the ordering of V_List_of_Servers.

- When a *Add_document (Server_Id, Document_Id, Document_Characteristics) signal* is received:

If an entry for *Document_Id* exists in D_Document_List then the domain server manager simply adds the tuple *(Domain_Id, Document_Characteristics)* to the corresponding List_of_Server_Doc; otherwise it creates a new entry for *Document_Id* and sets the corresponding List_of_Server_Doc to [*(Domain_Id, Document_Characteristics)*]

- When a *Delete_document (Server_Id, Document_Id) signal* is received:

If List_of_Server_Doc which associated to the entry which corresponds to *Document_Id* is equal to [*(Server_id,)*] then the domain server manager deletes this entry; otherwise it extracts *(Server_Id,)* from List_of_Server_Doc (List_of_Server_Doc:= List_of_Server_Doc - [*(Server_Id,)*]).

- When a *Update_Load (Server_Id, Load_Level) signal* is received:

The domain server manager sets Load, which is associated to the entry which corresponds to Server_Id, to Load_Level.

- When the domain server manager does not receive k successive Service_Alive(Server_Id) signals from a server agent (identified by Server_Id):

The domain server manager simply sets Load of the entry which corresponds to Server_Id in D_Load_List to OVERLOADED (this value indicates that the server is overloaded); this means that the domain server manager will not send Service_Inq() signal to the server agent, identified by Server_Id, until this value changes. Thus, the server failure will be masked; users will not see the failure.

The value of k should be carefully chosen.

2.3. Description of the Server Agent

A server agent is responsible for monitoring the load of its MM server. The decisions of what server to select in order to serve the user request will be mainly based on the monitored server load. Periodically, server agents measure their load and send the results to their domain server managers. No load message (Update_Load() signal) is sent if the load has not changed. A detailed description of this activity can be found in [Haf 97b].

To report that it is still able to server service requests (still alive), a server agent sends periodically a Service_Alive() signal to its domain server manager. The period value depends on many factors, such as the reliability of the server in question, and can be computed based on some statistics on the past behaviors of the server. It is worth noting that the period values of different servers do not need to be equal. When repaired after a failure a server agent has only to send a Update_Load() signal to its domain server manager; this will allow the latter to consider the server again as a potential server able to process Service_Inq() signals.

Variables Description

We define the following variables:

V_Status: a variable which can assume ACCEPT or REJECT.

V_Proposition: a variable which indicates a QoS offer. Definitions and examples of QoS offers can be found in [Haf 96].

Operation

- When a *Service_Inq (Domain_Id, User_Id, Document_Id, QoS, Host_Constraints) signal* is received:

If we assume that the load value stored in D_Load_List is consistent with the actual load of the server then the server agent reserves successfully the local resources required to support the service request; otherwise, it checks whether there are enough available resources to support the requested service. If the response is yes, then the server agent initiates a network resource reservation protocol, such as RSVP [Zha 95] or the Tenet suite protocols [Fer 95], to reserve the network resources required to support the delivery of the document from the server to the user's host. If this operation succeeds, then the server agent sets V_Status to ACCEPT; otherwise it sets V_Status to REJECT. Then, the server agent performs the following operations:

- sets V_Proposition to the offer the server and the network can provide; this means that V_Proposition may or may not satisfy the user requirements QoS. For example, if the user asks to play a document with (color, TV resolution, 25 frames/s) as QoS, V_Proposition may be (color, TV resolution, 25 frames/s), (black&white, TV resolution, 25 frames/s), or (color, TV resolution, 15 frames/s).

- sends Service_Res (*User_Id, Document_Id*, V_Status) to its domain server manager.
- if V_Status is equal to ACCEPT, two policies may be used:

Policy 1: the server starts immediately the delivery of the document;

Policy 2: the server sends Service_Conf (*User_Id, Document_Id*, V_Status, V_Proposition) signal to the user (via the user agent) who may accept Proposition or not (in this case he/she updates V_Status to REJECT). Then, the user sends Service_Conf (*User_Id, Document_Id*, V_Status,) signal to the server agent. If V_Status is equal to ACCEPT, then the server starts the delivery of the document; otherwise it deallocates the (local and network) resources that have been reserved to support the user request.

- When a new document, identified by Document_Id, is stored in the server:

The server agent sends Add_document (Server_Id, Document_Id, Document_Characteristics) signal to its domain server manager(s).

- When an existing document, identified by Document_Id, is deleted from the server:

The server agent sends Delete_document (Server_Id, Document_Id) signal to its domain server manager(s).

- When the server load reaches a certain load level Load_Level:

The server agent sends Update_Load signal (Server_Id, Load_Level) to its domain server manager(s).

- When the server cannot maintain the delivery of the document identified by Document_Id:

The server agent sends Service_Viol (Server_Id, User_Id, Document_Id, QoS, Host_Constraints) signal to its domain server manager.

2.4. Description of the User Operation

The user agent maintains and updates a state variable, U_Document_List, as shown in Figure 4; this variable is used to identify the server (server_Id) involved in the delivery of a document (Document_Id) when generating a Service_Viol() signal.

U_Document_List

Document_Id	Server_Id

Figure 4. State variable of the user agent

During the presentation of the document, if some QoS degradations occur, a notification should be sent to some management entity. In the context of our work, depending on the policy in use either the user (user-oriented policy) or the server (server-oriented policy) generates a Service_Viol() signal; that is, when the user notices an unacceptable presentation quality and a user-oriented policy is in use he/she may send (via the user agent) a Service_Viol() signal to the server manager; when the server agent cannot maintain the delivery of the document and a server-oriented policy is used, it sends a Service_Viol() signal to its domain server manager.

- When a *Service_Conf signal (Server_Id, User_Id, Document_Id, Status, Proposition)* is received:

If the user accepts *Proposition* then the user agent sets *Status* to ACCEPT; otherwise it sets *Status* to REJECT. Then it sends Service_Conf (, *User_Id, Document_Id, Status,*) signal to the server identified by *Server_Id.*

It is worth noting that metadata describing the MM documents (location and characteristics) might reside in a centralized or distributed repository. This means that the server manager and the domain server manager do not need to handle the state variable M_Document_List and D_Document_List respectively; instead, they can ask metadata directly from the repository. To use a repository or state variables to handle metadata depends on many factors, such as the response time to get metadata (short when state variable are used), and the memory amount to store metadata (minimized when a repository is used).

3. SAS Evaluation

We performed a number of simulation experiments to evaluate the gain obtained while using SAS. First we ran simulations of service requests without using SAS, that is the server response is either an acceptance or a rejection. Second we run simulations of service requests using SAS.

The simulation is parameterized by the following:

- *Number of users making requests:* the number of users making requests over a given period of time (6 hours is selected).
- *User Request pattern in time:* we consider a duration of 6 hours divided into time intervals. The service request probability is equal for all slots.
- *Length of the requested documents (LSD):* this process is used to model the generation of the lengths, *length*, of the document requested. We assume that the length of a requested document is generated randomly between 60 min. and 90 min.
- *Server capacity:* we assume that the server provides only one QoS class. The server capacity indicates the maximum number of documents which the server can deliver concurrently.
- *Access pattern of users to different servers:* most users asks services from the well-known servers (e.g., reliable and high-performance servers).
- *Load levels:* the load levels which the server agent should communicate to its domain server manager, e.g, a load level of 100% means that the server is loaded at maximum and cannot serve any new request.

The main metric we adopted for evaluation and comparison was the *blocking probability:*
blocking probability= (number of rejections)/(number of service requests)

A rejection corresponds to a rejection initiated by a server agent or a rejection initiated by the user who does not accept the proposition of the server agent. We do not consider rejections because of network resources shortage; we assume that we have enough network resources. This assumption does not affect the accuracy of the experiments, since it is used for both approaches (classical and SAS).

Simulation Hypothesis

We assume that there are six servers in the system: two high-performance servers (S1 and S2) with a capacity of 50 and four servers (S3, S4, S5, and S6) with a capacity of 20. S1, S3 and S4 (respectively, S2, S5 and S6) belong to the same domain. Initially (just before the experiments), the servers are loaded at 0%.

Results

Figure 5shows that the number of service requests rejected using a classical access approach increases more rapidly. In this experiment we assume that 80% (resp. 20%) of user requests are sent to the high-performance servers (resp. to the others).

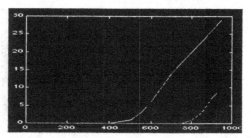

Figure 5. Blocking probability Vs number of user requests

Figure 6 shows that the variation of *access pattern of users to different servers* parameter has an important impact when using a classical access scheme, while it has not impact when using SAS. For this experiment we assume that the system receives 1000 requests and we vary the percentage of requests sent to high -performance and low-performance servers (from 0% to 100%).

Instead of trying to estimate (e.g. to statistically compute) the accurate user pattern, which continuously changes, it is better to use SAS which does not make any assumption on the user's access pattern.

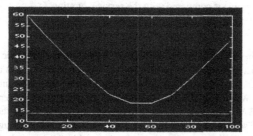

Figure 6. Blocking probability Vs. user's access pattern

Figure 7(resp. Figure 8) shows the distribution of requests served by S1 and S3 using SAS (resp. using a classical approach). In this experiment we assume that: (1) 1000 user requests are issued for the duration of the experiments; and (2) 80% (resp. 20%) of user requests are sent to the high-performance servers (resp. to the others). It is clear that the load is better balanced using SAS than classical approaches.

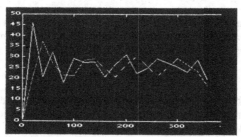

Figure 7. Proportion of requests started by S1 and S3 using SAS
in the last 20 minutes Vs. time

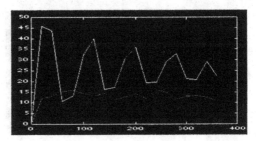

Figure 8. Proportion of requests started by S1 and S3
in the last 20 minutes Vs. time

The experiments show that we can service more users when using SAS (e.g, see Figure 5); this allows us to create a system that can service a large number of users for reasonable cost. Furthermore it does allow to better balance the load among available servers. We can state that SAS is more efficient and flexible than classical approaches to access MM documents. It is obvious that SAS requires a longer period of time (response time) to answer the user request than the existing approaches. However, we believe that a long response time (we are currently working to evaluate this) is not a problem since it will be necessary negligible if compared with the service duration.

4. Conclusion

This paper describes a new scheme, we call SAS, that allows scalable access to multimedia documents in a distributed system. SAS assumes a system which consists of a set of client machines, a set of server machines, and networks connecting these machines; popular documents are replicated and stored on the servers. When the user asks (from any server) to play a document with the desired quality of service (QoS) SAS selects a suitable server which is able to deliver the document to the user. This means that the server has the requested document, and there enough (local and network) resources to support the delivery of the document; preferably this server is the least loaded. During the presentation of the document, SAS allows to recover automatically from any QoS degradation that might occur. Last and not least, SAS allows to mask server failures.

SAS consists of high level interactions between management entities and thus may be applied to an arbitrary system with different types of commercial and research multimedia-on-demand servers; it is essentially independent of the technologies used to deliver multimedia data. A number of implementations of SAS may be developed based on the characteristics of the environment (e.g., topology and structure).

SAS allows to minimize the blocking probability as much as possible; a user request is rejected only if all servers (storing the requested document) are loaded at maximum or the communication system is overloaded. It also allows to balance the load among servers.

References

[Bou 95]   R. Boutaba and S. Znaty, An Architectural Approach for Integrated Network and Systems Management, In ACM-Sigcom Computer Communication Review, vol. 25, no. 5, P. 13-39. ACM Press. October, 1995.

[Bru 96]   D.W. Brubeck and L.A. Rowe. *Hierarchical Storage Management in a Distributed Video-On-Demand System*. IEEE Multimedia, Vol. 3, No. 3, pp. 37-47 (1996).

[Bud 96]   M.Buddhikot, G.Parulkar and J.Cox, *Design of a Large Scale Multimedia Storage Server*, Journal of Computer Networks and ISDN Systems, Elsevier (North Holland), December 1994.

[Fer 95]   D.Ferrari, *The Tenet Experience and the Design of Protocols for Integrated Services Internetworks*, Multimedia Systems Journal, November 1995

[Gem 95]   D.Gemmel et al., *Multimedia Storage Servers: A Tutorial*, IEEE Computer, Vol. 28, No. 5, 1995

[Haf 96]   A.Hafid, G.v.Bochmann and B.Kerherve, *A QoS Negotiation Procedure for Distributed MM Presentational Applications*, In Proceedings of the Fifth IEEE International High Speed Distributed Computing (HPDC-5), Syracuse, New York, 1996

[Haf 97a]   A.Hafid and G.v.Bochmann, *Quality of Service Adaptation in Distributed Multimedia Applications*, ACM Multimedia Systems Journal, 1997 (to appear) (http://www.crim.ca/~ahafid/papers/MMSystems.ps)

[Haf 97b]   A.Hafid and R.Boutaba, *A Scalable Video-on-Demand System: Architecture and Implementation*, In Proceedings of ICCC'97, Cannes, France, 1997 (to appear)

[Haf 97c]   A.Hafid and R.Boutaba, *A Scalable and Highly Available Video-on-Demand System*, Telecommunication Systems Journal, Special Issue on Multimedia, 1997 (submitted)

[Lou 93]   P.Lougher, *The Design of a Storage Server for Continuous Media*, Ph.D. Thesis, Lancaster University, 1993

[Ran 93]   V.Rangan, *Architecture and Algorithms for Digital Multimedia On-Demand Servers*, Performance 93 and Sigmetrics 93, L.Donatiello and R.Nelson eds., 1993

[Wan 85]   Y. Wang, *Load Sharing in Distributed Systems*, IEEE Transactions on Computers, March 1985

[Won 97]   J.Wong, K.Lyons, R.Velthuys, G.Bochmann, E.Dubois, N.Georganas, G.Neufeld, T.Ozsu, J.Brinskelle, D.Evans, A.Hafid, N.Hutchinson, P.Inglinski, B.Kerherve, L.Lamont, D.Makaroff, and D.Szafron, *Enabling Technology for Distributed Multimedia Applications*, IBM Systems Journal, 1997 (to appear)

[Zha 95]   L.Zhang, et al., *RSVP Functional Specification*, Working Draft, draft-ietf-rsvp-spec-07.ps, 1995

A Client-Controlled Adaptation Framework for Multimedia Database Systems*

Silvia Hollfelder[1], Achim Kraiss[2], and Thomas C. Rakow[3]

[1] GMD – Integrated Publication
and Information Systems Institute (IPSI)
Dolivostraße 15, 64293 Darmstadt
Germany
hollfeld@darmstadt.gmd.de

[2] University of the Saarland,
Dept. of Computer Science
P.O. Box 151150
66041 Saarbrücken, Germany
kraiss@cs.uni-sb.de

[3] Springer-Verlag,
Tiergartenstr. 17
69121 Heidelberg
Germany
rakow@springer.de

Abstract

Multimedia applications require continuous delivery of time-dependent data. In best-effort systems where no resource reservation is possible, adaptation strategies are necessary in case of high system load. We propose an adaptation framework for a client-pull architecture where continuous data are requested and buffered by the presenting client. The current client buffer utilization is used as an indicator for performance bottlenecks and determines which adaptation strategy the client invokes. The aim of the invoked adaptation is to reduce the data volume which has to be transferred to the client in a controlled manner leading to a graceful degradation of presentation quality. Frequent switches between different qualities are avoided by using a hysteresis and by specifying smooth transitions between different degrees of data dropping. As we consider adaptation control only at the client side we believe that our approach can easily be used for achieving intra-media synchronized presentations in most open and heterogeneous environments.

1 Introduction

Multimedia presentations must obey time constraints to achieve synchronized presentations. In environments where no guarantees for continuous data transport can be given hiccups may occur during the presentation of continuous media. These hiccups violate intra-media synchronization of a continuous object where presentation units have to be displayed with the same temporal requirements as they were fixed at recording time. Violating intra-media synchronization constraints usually leads to unacceptable presentation quality and to complicated inter-media synchronization [BS96, STE96]. Therefore, multimedia applications require a "just in time" transport of data via networks, that is, the data unit to be presented has to be loaded into the client buffer just before it will be requested by the presenting application. One way to support the transport of continuous data in such environments is *resource reservation*. This means that resources being involved in the presentation, like processor, disk, and network, are requested by the client and dedicated to the presenter at presentation time. The main advantage of this approach is that, due to the static resource preallocation, the requested quality of service can be guaranteed throughout the presentation. Concurrent resource requests by multiple users are subject to an admission control mechanism. If a single presentation does not get enough resources at start time, the presentation has to be delayed or rejected. The drawback is that in such systems the resources have to be exclusively controlled which is unrealistic for commonly used open distributed environments like the Internet and non-real-time operating systems. Another disadvantage is that resources are wasted if user interactions like pause and switches to slower presentation speeds occur. In this case concurrent presentations are rejected although they could

---

* This work has partly been funded by the European Union within the framework of the ESPRIT Long-Term Research Project HERMES, No. 9141 [http://www.ced.tuc.gr/hermes/hermes.html].

get enough resources to start, which reduces the overall throughput of presentations. In order to increase the number of parallel presentations the quality guarantees for each presentation have to be reduced. This can be managed by specifying quality requirements based on probabilistic parameters [VGG94]. Although this increases the degree of parallel multimedia presentations, the system again has to guarantee minimum performance and therefore has to control involved components.

The simplest realization of multimedia presentations in open distributed environments is to accept and serve presentations by *best-effort*. Each application is allowed to start presentations at any point in time without exclusively allocating resources. This allows a higher number of accepted applications compared to the usual resource reservation approach, and no specialized network protocols or system software have to be used. However, the drawback is that no guarantees for the timeliness of data delivery can be given and that temporary load peaks may delay the presentation. In order to keep up the intra-media synchronization of a presentation in best-effort systems under varying available resources, *adaptation* techniques, also called scaling methods or flow management, have to be used [DHH93]. These techniques try to dynamically change the quality of a presentation after the system has detected bottlenecks in data delivery. Reducing the presentation quality leads to a reduced data volume to be transported from the server to the client and the presentation expects to keep up its intra-media synchronization by reducing disk utilization, memory consumption and used network bandwith. In general, adaptation strategies have to be invoked, if it can be foreseen that a running presentation cannot keep up intra-media synchronization assuming that the resource consumption remains constant.

Based on the system components that control the data transfer from the server to the presenting client, we have to distinguish between *server-push* and *client-pull* architectures [RVT96]. In a server-push architecture data are retrieved by the server and transmitted continuously to a requesting client without any further client requests. In a client-pull architecture clients request data from the server at the time they are needed. Typically, video servers are server-push architectures, which serve the clients in periodic rounds. The reason is that having the timing of data delivery under the control of the server gives potential for optimizing the scheduling of resources like disks and the usage of server buffers. Database management systems are usually based on client-pull architectures, where data are requested on demand by passing queries from the client application to the server. The advantage of client-pull architectures is that the client can immediately react to interactions as well as to performance bottlenecks in order to keep its presentation intra-media synchronized [TK96].

In this paper, we present an adaptation framework we have implemented in the *AMOS* (Active Media Object Store) prototype of a multimedia database management system (MM-DBMS) developed at GMD-IPSI [RKN96a, RNL95]. Our approach is based on a client-pull architecture where continuous data are delivered in best-effort. The client-pull architecture gives the client the possibility to invoke adaptation strategies as soon as it detects bottlenecks. An adaptation framework is supported for the combination of temporal and resolution adaptations, which allows to achieve a multi-staged change of presentation qualities. We introduce a *smooth adaptation technique* which assigns different degrees of adaptation to specific intervals of client buffer utilization. The lower the buffer utilization is the more likely are hiccups and the more effective the adaptation strategy has to be. Frequent and abrupt changes of presentation quality are avoided by overlapping the buffer utilization intervals and by specifying constraints on temporal adaptation, e.g., how much presentation units can be dropped. The framework provides media-specific specifications for each continuous object which are kept in the database as metadata. Having media-specific specifications is necessary as the usage of adaptation strategies depends on the temporal, auditory, and visual importance of each presentation unit [AFK95]. The adaptation strategy may furthermore depend on the contents of the object. These are the reasons why it must be possible to specify different adaptation strategies with different degrees of adaptations for each single continuous object. Our experiences show that our adaptation framework is a good means to react to resource consumption in a MM-DBMS. Earlier work on this topic was pre-published in [HKR96].

The remainder of the paper is organized as follows. Section 2 describes the architecture of the AMOS prototype. In Section 3, we introduce our adaptation framework and in Section 4 the adaptation feedback mechanism is given as it is realized in the prototype. Section 5 gives some preliminary experimental results. In Section 6, we compare our approach with existing work. Finally, Section 7 concludes the paper and gives an outlook on further work.

2 System Architecture

The AMOS prototype is based on a client/server architecture where the server is responsible for the storage of continuous and discrete data. The client is responsible for fetching the data from the server, for presenting them and for handling interactions with the user. An application program only has to invoke the presentation of data from a remote database, parametrized with specific QoS parameters. The MM-DBMS then retrieves the media data and sets up a continuous data connection from the database server to the client utilizing an appropriate network.

In addition to the tasks of traditional object management, in a MM-DBMS the *continuous delivery* of data like audio and video "just in time" has to be supported. It enables an application to access data by loading objects from persistent storage into the buffer. Because of the large volume of time-dependent data they usually cannot be entirely loaded into the buffer. These objects are therefore segmented into small blocks which are preloaded into the buffer. We call our generic data type for multimedia objects *Continuous Long Fields* [RKN96b]. A Continuous Long Field differs from conventional long fields by the additional specification of QoS parameters and potentically used adaptation schemes. The atomic unit used for presentation is called *Continuous Object Presentation Unit (COPU)*. A Continuous Long Field consists of a sequence of *COPU*s. Furthermore, metadata like format and recording parameters are stored with every field. Preloading and replacement of COPUs in the client buffer is managed by using an algorithm called *Least/Most Relevant for Presentation* (L/MRP) [MKK95]. The algorithm does not only preload those COPUs needed for a continuous presentation of a Continuous Long Field but also considers the current presentation state and likely interactions when deciding which COPU should be replaced from buffer.

3 The Adaptation Framework

In the following, we give an overview of the general requirements of intra- and inter-media synchronization and show how our system supports intra-media synchronization using adaptation. Furthermore, we describe how different adaptation strategies can be specified in the AMOS prototype.

3.1 Synchronization by Adaptation

The encoding process of a time-dependent object determines the time when a COPU has to be presented relative to the beginning of its playback. Usually, the timing information of COPUs is given by the display rate, e.g., a video is specified by a number of frames presented per second. To achieve *intra-media* synchronization, the presenter has to keep up this requirement, i.e., the display rate has to be the same as the recording rate and the duration of a playback should exactly correspond to that of the original determined at recording time. During the display small deviations in the time of COPU presentations are acceptable. The quality of intra-media synchronization only depends on the deviations between required and presented data. To achieve inter-media synchronization, we do not specify any inter-media synchronization model, like in [QN97], but use the following approach: with intra-media synchronized continuous objects, the problem of *inter-media* synchronization can be solved by combining multiple intra-media synchronized Continuous Long Fields in order to get inter-media synchronized data [BS96]. Possible interactions make an inter-media synchronization more complicated, but a simultaneous control of the presentation can be maintained by control messages between the presentation processes [TK96a].

To achieve the synchronization requirements in distributed non real-time systems, it has to be ensured that the COPUs needed for presentation are already residing in the client buffer at the time they are requested. To keep the buffer always filled with the required COPUs, the client has to prefetch the needed COPUs from the server into the client buffer. A decreasing buffer utilization reflects a bottleneck in COPU delivery, that is, the consumption of COPUs in the client buffer is faster than the transfer of COPUs from server to client. The aim of an adaptation strategy is to avoid that the buffer gets completely empty, as in this case the presentation would have to wait for the next COPU to be transferred from the server, possibly violating intra-media synchronization constraints. Clearly, the adaptation strategy has to decrease the data transfer rate from server to client.

3.2 Adaptation Techniques

In the following, we summarize some adaptation techniques for the most important compression formats. For videos, the single frame-encoded formats based on JPEG [WAL91], called Motion-JPEG, as well as the ISO standards MPEG-1 [LEG91], and MPEG-2 [MPE2] are widely used. The MPEG standard uses redundant information of subsequent frames for coding only differences to preceding (P-frames), or preceding and succeeding frames (B-frames). Due to the differential encoding of the video, some frames can be decoded only if other frames have been decoded before.

Adaptation for multimedia objects can be classified along two dimensions (Fig. 1): (1) the method used for data reduction and (2) the effect the adaptation has on the presentation. In field one, a video stream is adapted by dropping single frames. Clearly, using this adaptation leads to a reduced rate of presented frames. The synchronization requirements are met by presenting the previously presented frame each time the presentation detects a dropped frame in the stream. However, presentation quality is reduced because the smoothness of movements in the video could be lost. It's easy to drop

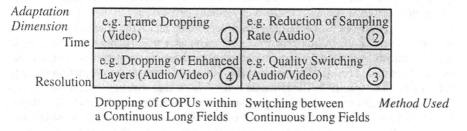

Figure 1: Classification of Adaptation Mechanism

frames in Motion-JPEG videos because the frames are independent of each other, but the dropping of frames in MPEG videos must consider interdependencies between the frames. In this case, the B-frames have to be dropped first and if it is not sufficient, P-frames are dropped in the next stage. For audio, dropping single samples is usually not suitable since this leads to significant disturbances which the user will perceive.

In field two, the rate of presented COPUs is reduced by switching to another stream. For example, the rate of samples presented in an audio stream can be reduced by switching to an audio stream of a lower sampling rate. Obviously, this kind of adaptation is suitable for video streams as well.

Field three uses switching to another data stream for reducing the quality of single COPUs, but keeping the original display rate. Quality switching can be done in two ways. First, Continuous Long Fields can be converted at runtime to get COPUs with lower quality. Second, multiple Continuous Long Fields with the same contents but with different qualities can be stored at the server. In the former case, the Continuous Long Field is stored in best quality. At high system load a lower quality is dynamically generated at the server for being presented at the client. The MPEG-2 standard supports this strategy by using a hierarchical compression scheme that provides the scalability of frames in different substreams with additional quality. Some JPEG implementations for hierarchical JPEG coding exist, too. In the latter case, the Continuous Long Field is stored in different qualities or sizes which leads to redundant storage of the same object in the database.

Finally, field four shows how the dropping of COPUs could lead to a reduction of the quality of single COPUs. If the raw information of a continuous object is stored in more than one stream, the COPUs of the basic stream can be loaded first. If the system has enough time to load other streams, the next incremental stream can be transferred to the client. Otherwise, the incremental streams are dropped and the COPU quality will not be increased. An example for such an adaptation is an audio which is stored in different sample sizes, e.g., 16 Bit or 8 Bit samples.

Any kind of adaptation reduces the required data rate and, as a consequence, the load of system components like server and network, but it is obvious that not all adaptation strategies can be used for all media. Rather the used adaptation strategy is highly media-specific and may furthermore depend on the contents of the object.

3.3 Specification of Adaptation Strategies

As outlined in the introduction, our client-controlled adaptation mechanism is designed for client-pull systems without resource reservation (best-effort systems). Each presentation starts with a filled client buffer and continuously requests COPUs at the client. While COPUs are presented, a preloader asynchronously prefetches the next COPUs from the server into the client buffer. Because of the client-pull architecture a buffer overflow never occurs. If prefetching from the server is faster than or as fast as the presentation of COPUs, the buffer utilization should never decrease. But if server CPU, server disk or network bandwith have temporary bottlenecks, the clients buffer utilization starts to decrease. In this case, the client starts to invoke adaptation techniques in order to gracefully degrade presentation quality but keeping the presentation intra-media synchronized. Therefore, for each buffer utilization interval the adaptation strategy to be used can be specified. The rationale behind this is that with high buffer utilizations the danger of getting hiccups in the presentation is low and that only moderate adaptations may be necessary. But as with low buffer utilizations the danger of loosing intra-media synchronization increases, more effective adaptation techniques have to be invoked by the client in order to fill up the buffer more quickly with inferior quality data.

In order to ensure that the dropping of COPUs leads to an increased buffer utilization, we logically insert *empty COPUs* for each dropped COPU.

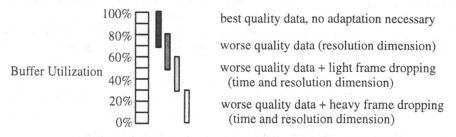

Buffer Utilization

100% — best quality data, no adaptation necessary
80% — worse quality data (resolution dimension)
60% —
40% — worse quality data + light frame dropping (time and resolution dimension)
20% —
0% — worse quality data + heavy frame dropping (time and resolution dimension)

Figure 2: Adaptation Specification Example

Figure 2 shows an example of a specified adaptation strategy for videos. Having the buffer completely filled means that the preloader seems to have no bottleneck. Thus it continues to request best-quality COPUs from the server. If buffer utilization goes down below 70 percent, the sample specification leads to the request of worse quality COPUs from the server. If the reduced amount of requested data leads to reduced resource consumption and/or the overall system load decreases, the buffer utilization may reach 80 percent after some time. In this case the preloader switches back to best quality COPUs. If the buffer fill level continues to decrease, a more effective adaptation will have to be initiated. In our example, the preloader starts to drop COPUs, if the buffer utilization reaches 50 percent. The dropping of COPUs can be specified in different degrees. On the one hand, the more COPUs are dropped per unit of time, the less data have to be requested from the server. Dropping a lot of COPUs means that the buffer utilization quickly increases because we handle our dropped COPUs as empty COPUs in the buffer, but they are not actually in the buffer, of course. But if, on the other hand, more empty COPUs are in the buffer, the presentation quality gets worse. This justifies the usage of different buffer intervals for dropping. In our example we modeled two intervals of adaptation in time dimension. We try to avoid frequent presentation quality switches induced by short-term variances in data delivery by using a hysteresis. Therefore, the intervals of different adaptation strategies overlap. Only in case of "long term" performance improvements or degradations of the system, another quality will be requested. The higher these short-term variances in data delivery are and the jerkier a quality switch between two intervals is, the more these intervals should overlap. As it is possible to specify a multistaged dropping of COPUs, the intervals need not overlap for adaptations using COPU dropping.

Our concept allows controlled dropping by specifying the minimum number of non-empty COPUs in a sequence of COPUs and the maximal number of empty COPUs in this sequence. Figure 3 shows two sample sequences of COPUs loaded into buffer for different specifications of COPU dropping. The constraints are specified by simple integer values d and s. The parameter d specifies the minimal distance (in COPUs) between two dropped COPUs. If the value of d is zero, the preloader is allowed to insert successive empty COPUs into the buffer. In order to limit the size of this sequence, the parameter s specifies the maximum sequence size of empty COPUs. Note that if s equals zero no empty COPU is allowed to be inserted and thus no COPU dropping will be invoked. The two parameters d and s together control the distribution of empty COPUs within a continuous stream implying different presentation qualities. The first specification in Figure 3 requires that there must be at least three COPUs between two empty

COPUs whereas in the second example the sequence size of empty COPUs is limited to four. It is obvious that small d values lead to a lower presentation quality. Thus such rigid adaptation in temporal dimension should be used only for buffer utilization intervals with a low upper bound. The lower the buffer utilization value is the more important is it to fill the buffer up with a growing number of empty COPUs. Otherwise the buffer will get completely empty leading to a COPU-fault at COPU request time.

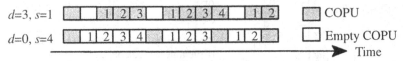

Figure 3: Examples for Adaptation in Temporal Dimension (Dropping)

4 Adaptation Feedback Mechanism

In Figure 4 the basic adaptation feedback mechanism is illustrated as it is realized in the AMOS prototype. The presentation process, called *Single Media Presenter (SMP)*, continuously requests COPUs, presents them in an intra-media synchronized way, and handles user interactions. In case of user interactions, like playing fast forward, fast rewind or rewind, the SMP requests the corresponding COPUs. The *Continuous Object Manager (COM)* provides COPUs for a Single Media Presenter under the guarantee that at any time the consumer requests a COPU for presentation, this COPU is available in the buffer. An instance of the adaptation specification is given to the COM when the corresponding presentation is started. The COM permanently controls the buffer utilization and, based on the adaptation specification, determines which quality has to be used. Based on the chosen strategy, the preloader manages the request of COPUs from the server and the dropping of COPUs.

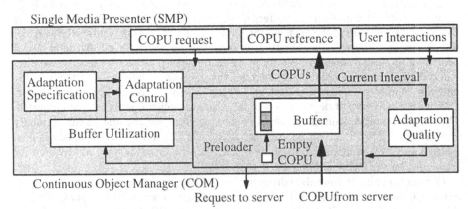

Figure 4: Adaptation Feedback Mechanism

The algorithm for adaptation is given in Figure 5. The *producer* process implements the preloading of COPUs and the adaptation strategies. The input value *adaptation_specification* represents the modeled buffer intervals and the corresponding adaptation strategies. The buffer is filled with best-quality COPUs before the presentation starts and the producer asyncronously preloads further COPUs. The preloading basically consists of a loop starting with the first COPU to be preloaded and ending with the last one. For each COPU to be inserted into the buffer it determines the current buffer fill level and the current interval by calling the *control* function with the input parame-

PROCESS *producer* (*adaptation_specification*)
BEGIN
current_interval := *best_quality_interval;*   *// current buffer interval, we start with best quality COPUs*
⟨fill buffer with best-quality COPUs⟩;   *// start up phase*
copuNr: = ⟨first COPU to present⟩;
WHILE ⟨presentation active⟩ **DO**
    buffer_utilization := *calculate_BufferUtilization*();
    current_interval := *control*(*buffer_utilization, current_interval*);   *// set current interval*
    current_strategy := *adaptation_strategy*(*current_interval*);   *// get current adaptation strategy*
    IF NOT *bufferFull*() **THEN**
        IF ⟨resolution adaptation specified in current_strategy⟩ **THEN**
            IF *copuNr* **NOT IN** *buffer* **THEN** *load_from_server* (*copuNr, qualityId*(*current_strategy*));
        END IF;
        ELSE   *// no resolution adaptation*
            IF *copuNr* **NOT IN** *buffer* **THEN**   *// maybe we can drop the COPU*
                IF *dropping_allowed*(*current_strategy*) **THEN** ⟨insert empty COPU into buffer⟩;
                ELSE *load_from_server*(*copuNr, qualityId*(*current_strategy*)); *//COPU must be loaded*
                END IF;
            END IF;
        copuNr: = ⟨nextCOPU to be presented⟩;
    ELSE *// buffer full, benefit from idle time*
        ⟨replace lower quality COPU in buffer by its best quality COPU⟩;
    END IF;
END WHILE;
END PROCESS;

Figure 5: Algorithm for the Continuous Object Manager

ters current value of buffer utilization and the current adaptation interval. Based on these values it will be tested if another quality has to be invoked. This is done by checking if the current buffer utilization is still in the adaptation interval of the currently active adaptation strategy. If the value is out of the range of this interval the new adaptation strategy is determined by calculating a new interval that contains the current buffer utilization. If resolution adaptation is specified, the producer has to request different streams from the server (*qualityId* in Fig. 5). If temporal adaptation by frame dropping is specified, the producer has to check whether frame dropping is allowed for the requested COPU. This is realized by means of the function *dropping_allowed*() which controls the sequence size of loaded non-empty COPUs and inserted empty COPUs. The COPU is forced to be loaded from the server, if the minimum number of non-empty COPUs has not yet been loaded or if the maximum number of empty COPUs in sequence is reached. Otherwise the COPU can be dropped.

When the buffer is full, no more COPUs have to be preloaded and the producer process will have some idle time. This may be the case, if the consuming process stops or slows down presentation because of a user interaction. Another reason for increased loading speed could be that in the meantime additional resources got free. The producer process does not spend idle time waiting for decreasing buffer utilizations but starts to replace lower-quality and empty COPUs with the highest-quality COPUs. As these CO-PUs are preloaded for the current presentation, the presentation benefits from this action by getting best quality COPUs at its subsequent requests.

5 Implementation and Experiences

We have implemented the algorithm and the adaptation specifications using Continuous Long Fields in the AMOS MM-DBMS. Buffer loading and replacement run under control of the L/MRP algorithm.

We made some measurements of our adaptation, using the video data type for Motion-JPEG. We decided to implement it for the video format Motion-JPEG rather than using the more prominent video format MPEG because the implementation of the adaptation for Motion-JPEG is more efficient and all concepts can be shown. An implementation of MPEG videos has to consider the dependencies between frames. Further, because of the smaller sizes of B- and P-Frames, the dropping of B- and P-Frames will not reduce the system load to the same extent as dropping frames in the Motion-JPEG implementation does. Nevertheless, the results should be applicable to video presentations in general. We performed tests with different specifications for temporal and spatial dimensions of adaptation, respectively. The quality of a presentation using frame dropping is characterized by the frequency distribution of empty frames within COPU sequences. An important characteristic of a presentation which switches between different qualities is the frequency of these switches. Each time the presentation requests a COPU for presentation, we recorded whether the returned COPU was empty or the quality of the returned COPU was reduced. The results were averaged over several runs. A constant workload was put on the network connection between one client and the server during the presentation of a synchronized video and audio. Note, that in all compared presentations only the adaptation specifications were changed while all other parameters remained constant.

In our first experiment, we compared the presentation of a video without any adaptation with a presentation where frame dropping was specified. We denote the adaptation of the former presentation as uncontrolled frame dropping and the latter presentation as controlled frame dropping. Note that the presentation with uncontrolled frame dropping gets empty frames only when the buffer is empty. Thus, empty COPUs lead to hiccups in the presentation as it has to wait until the server delivers the next COPU. By means of controlled frame dropping the client starts earlier with dropping frames, exactly, when the buffer fill level corresponds to the adaptation interval. In Figure 6, the frequency distribution of the distances between two empty frames is given. The greater the distances between empty COPUs are, the better is the presentation quality. For example, in the diagram, the value pair (2, 13) means that 13 percent of all successive empty frames have two non-empty frames in between them. The diagram shows that with uncontrolled frame dropping an empty frame was very often (53%) preceded by only one or no non-empty frame. As expected, using controlled frame dropping leads to lower densities of empty frames throughout the video presentation.

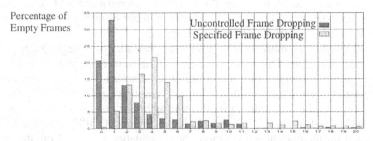

Figure 6: Frequency Distribution of Empty Frames

In Figure 7, the occurrence of empty frames over the runtime of the two presentations is given. For each presented frame, the number of empty frames was determined within a one second interval around the frame. The rate of each presentation was 25 frames per second. Both presentations started with a filled buffer. Figure 7 (a) shows that the presentation without frame dropping begins to insert empty COPUs as soon as the initially filled buffer is consumed (after second 120). Once the buffer becomes empty the density of empty COPUs significantly decreases leading to hiccups of uncontrolled length. On the other hand, Figure 7(b) shows that the presentation using controlled frame dropping based on the adaptation specifications starts inserting empty frames early, in order to compensate the decreasing buffer utilization (before second 80). Thus, it can keep up a better presentation quality with lower rates of empty frames over a long presentation time.

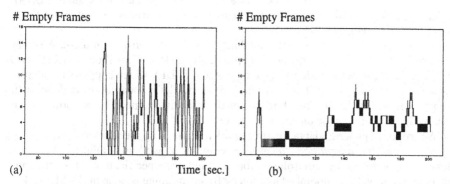

Figure 7: Insertion of empty Frames over Presentation Time

In another experiment, we investigated the effects of overlapping intervals for adaptations based on switches between streams of different frame qualities. As a result, quality changes occur very (!) frequently with a non-overlapping interval specification (Fig. 8 (a)) but only seldom with overlapping intervals, in spite of the extreme chosen workload. (Fig. 8 (b)). This indicates that the usage of a hysteresis is indeed suitable in order to ensure that the adaptation strategy does not react on short-term variances in system loads too early.

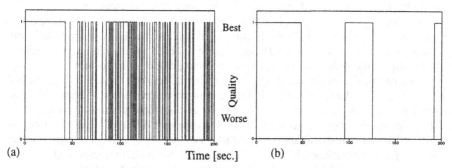

Figure 8: Time-line Distribution of Quality Switching

Finally, we investigated different adaptation specifications in some visual experiments. Single switches between different qualities could surely be detected by the user but did not entail to unacceptable presentation quality. In case of video clips with a high movement rate switching to streams with low quality frames is a good solution while frame dropping is often bothering. In general, switching to streams with lower qualities yields better presentations than dropping frames. As expected, a frequent change in spatial adaptation is very annoying and should be avoided. These first rough experiences coincide with similar results reported in [AFK95].

6 Related Work

In the area of adaptation, a lot of work has already been done. In order to classify our work more clearly, we will discuss the following aspects with regard to adaptation: *multi user access,* available *network protocols, implementation* of the framework, and *adaptation dimensions.* The most relevant differences result from *monitoring* system resources to detect bottlenecks and *adaptation control* which reacts on bottlenecks.

We realized the *multi user access* as a best-effort scheme. The system VOSAIC [CTC95] uses an admission control mechanism which determines or estimates the requirements of a request.

In our approach, we assume the availability of a reliable *network protocol.* We do not require any multimedia-specific network protocols like Real Time Protocol (RTP). VOSAIC [CTC95] uses the Video Datagram Protocol (VDP), which supports an adaptive flow control. In the Heidelberg Transport System (HeiTs) adaptation is done at the transport level [DHH93]. Therefore, the media encoding is modified according to the bandwidth available in the underlying network.

We propose a generic adaptation framework that can be applied for any kind of media by supporting the *adaptation dimensions* temporal as well as resolution. We try to achieve *smoothness* by controlled dropping and the specification of overlapping adaptation intervals. Temporal adaptation by frame dropping is done in [RPSL94]. Furthermore, [BGB96], [QN97], and [CPS95] employ temporal adaptation, but no adaptation in the resolution dimension. Controlled frame dropping is done in VOSAIC [CTC95]. Adaptation in the resolution dimension for video flows like MPEG-2 is proposed in the complex dynamic QoS management in [CH95].

A general adaptation is specified for all streams in the same way while in a *stream specific* scheme stream nongeneric adaptation is allowed for each. Our framework allows the *implementation* of a *stream specific adaptation specification*, using temporal as well as resolution adaptation. This implies that for each stream an individual adaptation specification is possible, reflecting QoS-parameters, which are specified by the user. To the best of our knowledge, only [NQ96] considers this aspect, too.

By *monitoring* only the client buffer we try to detect bottlenecks. [CPS95] have developed a feedback mechanism for client/server synchronization, dynamic QoS control, and system adaptiveness. Bottlenecks are monitored by the client, too. Further, [QN97] inspect the difference between the recording rate of a media and the rate supported by the system on the client side. Monitoring of network capacities is described for example by [EA95]. In case of limited transport service capabilities, dynamic adaptation is used to reduce the video data rate of the stored continuous data before sending the stream over the network. With this approach different network protocols can be considered. The disadvantage is that no server or client bottlenecks are regarded. More complex approaches which inspect more than one system component are proposed in [CH95] and [KW94]. In [CH95] a framework is developed which uses QoS adaptors, QoS filters, and QoS groups. In [KW94] resources are monitored on the server and the network side, e.g., CPU, buffer space, and network resources are inspected.

In our approach the *adaptation control* is done by the client by modified requests. Therefore, we model buffer level intervals as adaptation marks. In [NHK96] a probe-based algorithm determines a possible QoS for the presentation and calculates a degradation point as an adaptation mark, assuming that the client is the bottleneck. We do not use pre-calculated values for our adaptation marks, because we assume that the server and network are the bottleneck and their load varies during the presentation. We only consider short-term disturbances of bottlenecks by inspecting the buffer fill level at each request. This implies the possibility of fast buffer control and fast reaction on inter-actions. In [CPS95] detected bottlenecks are handled by the server. In [BGB96] a syn-chronization scheme of inter- and intra-streams in client-pull architectures is proposed. They use filters to differentiate long-term and short-term bottleneck deviations on the client and send control messages to the server which reacts to these messages. In VOSA-IC [CTC95] the server adapts the video frame rate after a client feedback.

7 Conclusion and Further Work

In this paper, we presented a method to achieve synchronization in best-effort systems by reducing the system load in case of bottlenecks. We proposed an adaptation algo-rithm that inspects the buffer utilization and invokes an adaptation strategy on the client side. We consider the temporal as well as the resolution adaptation dimensions includ-ing controlled dropping. The adaptation specification for a continuous medium is mod-eled by overlapping intervals which realizes smooth changes in presentation quality. Further, individual adaptation strategies for each continuous medium are considered.

We plan to investigate how the buffer utilization intervals of an adaptation strategy could be adjusted automatically during the runtime of a presentation. For example, fre-quently occuring, large variances of system load cause frequent quality changes, al-though the buffer space could be large enough to compensate these load variances. In this case, the buffer utilization intervals should overlap to a larger extent in order to pro-vide best quality COPUs, as long as possible.

Furthermore, comprehensive experimental evaluations are subject of future work. The combination of admission control techniques to guarantee a minimum quality for the presentation is another important topic. It seems to be a good idea to study the re-quirements of complex presentations, and adapt our buffer adaptation mechanism to them. Related work on adaptation can be done on other system layers like storage man-agement. Currently, a combined MPEG video and audio data type is under development for the AMOS MM-DBMS.

Acknowledgement

We gratefully acknowledge the contribution of our student colleague *Jürgen Spieker* for the implementation and measurements of our adaptation framework. *Michael Löhr* and *Michalis Vazirgiannis* (Nat. Technical University of Athens) gave helpful comments on an earlier version of this paper.

References

[ACH96] Christina Aurrecoechea, Andrew T. Campbell and Linda Hauw: A Survey of QoS Architectures. In: *Multimedia Systems Journal, Special Issue on QoS Architecture*, 1997 (to appear).

[AFK95] Ronnie T. Apteker, James A. Fischer, Valentin S. Kisimov and Hanoch Neishlos: Video Acceptability and Frame Rate. In *IEEE Multimedia*, pages 32–40, Fall 1995.

[BGB96] Ernst Biersack, Werner Geyer, and Christoph Bernhardt: Intra- and Inter-Stream Synchronization for Stored Multimedia Streams. In *Proc. of the International Conference on Multimedia Computing and Systems*, pages 372–381, 1996.

[BR94] Klemens Böhm and Thomas C. Rakow: Metadata for Multimedia Documents. In *ACM SIGMOD Re-cord Special Issue on Metadata for Digital Media*, 23(4), pages 21–26, 1994.

[BS96] Gerold Blakowski and Ralf Steinmetz: A Media Synchronization Survey: Reference Model, Specifica-tion, and Case Studies. In *IEEE Journal on selected Areas in Communications*, 14 (1), pages 5–35, Jan. 1996.

[CH95] Andrew Campbell and David Hutchison: Dynamic QoS Management for Scalable Video Flows. In: *Proc. Fifth International Workshop on Network and Operating System Support for Digital Audio and Video (NOSSDAV '95)*, pages 101–112, Apr. 1995.

[CPS95] Shanwei Cen, Calton Pu, Richard Staehli, Crispin Cowan and Jonathan Walpole: A Distributed Real-Time MPEG Video Audio Player. In *Fifth International Workshop on Network and Operating System Support for Digital Audio and Video (NOSSDAV '95)*, pages 142–153, Apr. 1995.

[CS95] Saurav Chatterjee and Jay Strosnider: A Generalized Admission Control Strategy for Heterogeneous, Distributed Multimedia Systems. In *Proc. ACM Multimedia 95*, pages 345–356, 1995.

[CTC95] Zhigang Chen, See-Mong Tan, Roy H. Campbell, and Yongcheng Li: Real Time Video and Audio in the World Wide Web. In *World Wide Web Journal*, Volume 1, Number 1, pages 333–348, Dec. 1995.

[DHH93] Luca Delgrossi, Christian Halstrick, Dietmar Hehmann, Ralf Guido Herrtwich, Oliver Krone, Jochen Sandvoss, Carsten Vogt: Media Scaling for Audiovisual Communication with the Heidelberg Transport System. In *Proc. ACM Multimedia 93*, pages 99–104, 1993.

[EA95] Alexandros Eleftheriadis and Dimitris Anastassiou: Meeting Arbitrary QoS Constraints Using Dynamic Rate Shaping of Coded Digital Video. In *Fifth International Workshop on Network and Operating System Support for Digital Audio and Video (NOSSDAV '95)*, pages 89–100, Apr. 1995.

[HKR96] Silvia Hollfelder, Achim Kraiss, and Thomas Rakow: A Client-Controlled Adaptation Framework for Multimedia Database Systems. In: *GMD Technical Report* (Arbeitspapiere der GMD), No. 1002, *Sankt Augustin*, Germany, June 1996.

[IT95] Y. Ishibashi and S. Tasaka. A Synchronization Mechanism for Continuous Media in Multimedia Communications. In *IEEE Infocom'95*, volume 3, Boston, Massachusetts, pages 1010–1019, April 1995.

[KW94] Thomas Käppner and Lars C. Wolf: Media Scaling in Distributed Multimedia Object Services. In *Multimedia: Advanced Teleservices and High-Speed Communication Architectures, Proc. of Second International Workshop, IWACA '94*, pages 34–43, Sept. 1994.

[LEG91] D. LeGall: MPEG: A Video Compression Standard for Multimedia Applications. In: *Communications of the ACM*, Vol. 34, No. 4, pages 45–68, Apr. 1991.

[MKK95] Frank Moser, Achim Kraiss and Wolfgang Klas. L/MRP: A Buffer Management Strategy for Interactive Continuous Data Flows in a Multimedia DBMS. In *Proc. Int. Conf. of Very Large Data Bases 1995 (VLDB)*, pages 275–286, Sept. 1995.

[MPE2] Generic Coding of Moving Pictures and Associated Audio (MPEG-2), ISO/IEC 13818-2 International Standard, May 1996.

[NHK96] Klara Nahrstedt, Ashfaq Hossain, and Sung–Mo Kang: A Probe-based Algorithm for QoS specification and Adaptation. In *Int. Workshop on QoS*, Paris, France, pages 89–100, March 1996.

[NQ96] Klara Nahrstedt and Lintian Qiao: A Tuning System for Distributed Multimedia Applications. In *Report* No. UIUCDCS-R-96–1958, UILU-ENG-96-1721, *University of Illinois*, May 1996.

[NY96] Raymond T. Ng and Jinhai Yang: An Analysis of Buffer Sharing and Prefetching Techniques for Multimedia Systems. In *Multimedia Systems*, No. 4, pages 55–69, 1996.

[QN97] Lintian Qiao and Klara Nahrstedt: Lip Synchronization within an Adaptive VOD System. In *Int. Conference on Multimedia Computing and Networking*, Feb. 1997, San Jose.

[RNL95] Thomas C. Rakow, Erich J. Neuhold and Michael Löhr: Multimedia Database Systems - The Notations and Issues. In Georg Lausen (Ed.): *Datenbanksysteme in Büro, Technik und Wissenschaft (BTW), GI-Fachtagung, Dresden, März 1995*, S. 1–29, Springer, Reihe Informatik Aktuell, 1995.

[RKN96a] Thomas C. Rakow, Wolfgang Klas, and Erich J. Neuhold: Research on Multimedia Database Systems at GMD–IPSI. In *IEEE Computer Society Multimedia Newsletter*, Vol. 4, No. 1, pp. 41–46, April 1996.

[RKN96b] Thomas C. Rakow, Wolfgang Klas, and Erich J. Neuhold: Abstractions for Multimedia Database Systems. *Proc. 2nd Int. Workshop on Multimedia Information Systems*, Sept. 26–28, 1996. West Point, New York, USA.

[RPSL94] L.A. Rowe, K. Patel, B.C. Smith and K. Liu: MPEG Video in Software: Representation, Transmission and Playback. In *Proc. of SPIE - The International Society for Optical Engineering 1994*, Vol. 2188, pages 134–144, Feb. 1994.

[RVT96] Siram S. Roa, Harrick M. Vin and Asis Tarafdar: Comparative Evaluation of Server-push and Client-pull Architectures for Multimedia Servers. In *Nossdav 96*, pages 45–48, 1996.

[STE96] Ralf Steinmetz: Human Perception of Jitter and Media Synchronization. In *IEEE Journal on selected Areas in Communications*, Vol. 14, No. 1, pages 61–72, Jan. 1996.

[TK96] Heiko Thimm and Wolfgang Klas: Delta-Sets for Optimized Reactive Adaptive Playout Management. In *Proc. 12th Int. Conf. on Data Engineering (ICDE)*, pages 584–592, Feb. 1996.

[TK96a] Heiko Thimm and Wolfgang Klas: Playout Management in Multimedia Database Systems. In Kingsley C. Nwozu, P. Bruce Berra, and Bhavani Thuraisingham (Eds.), *Design and Implementation of Multimedia Database Management Systems*. Kluwer Academic Publishers, pages 318-376, 1996.

[VGG94] Harrick M. Vin, Pawan Goyal, Alok Goyal, and Anshuman Goyal: A Statistical Admission Control Algorithm for Multimedia Servers. In: *Proc. ACM Multimedia 94*, pages 33–40, 1994.

[WAL91] G. Wallace: The JPEG Still Picture Compression Standard. In *Communications of the ACM*, 34 (4), pages 30–44, Apr. 1991.

Interactive Multimedia Communications at the Presentation Layer*

Chung-Ming Huang and Chian Wang

Laboratory of Multimedia Networking
Institute of Information Engineering
National Cheng Kung University
Tainan, Taiwan 70101, R. O. C.
Correspondence: huangcm@locust.iie.ncku.edu.tw

Abstract. A main feature of the next generation's computing software is supporting convenient user interactions. Interactive multimedia presentations that are executed in distributed environments raise some new computer communication behaviors, which we call "interactive multimedia communications". In this paper, we (1) classify user interaction types, (2) point out the main problems in interactive multimedia communications, (3) distinguish two approaches for solving these main problems, (4) identify the possible control architectures for processing user interactions, and (5) propose some methods and control schemes accordingly.

1 Introduction

To have flexible multimedia presentations, user interactions are becoming more popular and required. In true interactive systems, users are allowed to take over the execution control during presentations to modify some execution configurations, e.g, speed up/slow down the presentation speed. Synchronization control of interactive multimedia presentations for continuous media in distributed environments becomes much more complicated because of (i) the continuous media's nature, i.e., a sequence of media units should be continuously supplied to the display devices, (ii) the distributed nature, i.e., the client display site and the server source site are geographically separated, and (iii) the user interactions' random nature, i.e., user interactions can be issued dynamically and unpredictably, and the interval between the interrupt action and the resume action is random. For example, a user can issue a pause action, wait for 0.5 second, 5 seconds, or even 1 minute, and then issue the restart action to resume the presentation.

In addition to intra-medium synchronization and inter-media synchronization [2, 8], we classify the third synchronization type, which is called *interaction-based synchronization*, for providing interactive multimedia services. When a user issues an interaction at the client display site, the presentation of the associated media streams may not be synchronous due to the jitter phenomenon; media units that are currently stored in the client buffer and that are flowing in the network may become useless and redundant in order to satisfy the features and modifications that are associated with the issued interaction; the source server site is still transmitting some media units,

---

* The research is supported by the National Science Council of the Republic of China under the grant NSC 86-2213-E-006-071.

which may also be useless and redundant, before an interrupt message is received. Interaction-based synchronization mainly deals with the issues that result from the processing of re-synchronization and the process of useless and redundant media units. In this paper, we concentrate on the interaction-based synchronization issue.

Based on the continuous media's nature, the distributed nature, and the user interactions' random nature, interactive multimedia presentations that are executed in distributed environments raise some new computer communication behaviors, which we call "interactive multimedia communications". Some related work for processing user interactions has been reported [5, 6, 7, 9], which are either for the centralized environment, or for one or some specific user interactions that are executed in the distributed environment. In this paper, we provide a comprehensive study about interactive multimedia communications at the presentation layer. We (1) classify user interaction types, (2) point out the main problems in interactive multimedia communications, (3) distinguish two approaches for solving these main problems, (4) identify the possible control architectures for processing user interactions, and (5) propose some methods and control schemes accordingly.

2 Classification of User Interactions

User interactions can be classified as non-temporal-related ones and temporal-related ones. Non-temporal-related user interactions, e.g., adjust audio volume, don't affect media's flow sequences. That is, non-temporal-related user interactions don't affect both media transmission at the server site and media reception at the client site. On the other hand, temporal-related user interactions, e.g., reverse the presentation direction, affect media's flow sequence. When the user issues a user interaction, current media presentation and transmission should be interrupted and stop temporarily. Presentation and transmission are continued when the processing of user interactions is over. Since our goal is for temporal behaviors of user interactions, we concentrate on the temporal-related user interactions. For convenience, temporal-related user interactions that are exemplified in this paper are *reverse*, *skip*, *freeze-restart*, *fast-forward*, *fast-backward*, *scale*, *step*, and *play*.

With these interactions, users are allowed to (1) *reverse* the direction of the presentation flow, (2) *skip*, either forwardly or backwardly, to a new presentation point, (3) *fast-forward/fast-backward* presentation with a default speed, e.g., 3 times/$\frac{1}{3}$ time of the normal presentation speed, when the presentation is in forward/backward presentation direction, (4) (re-)*scale* the presentation speed, either speed-up or slow-down, that is specified by users' choices, (5) *freeze* the presentation, and then *restart* the presentation after some time units that are decided by users, (6) *step* the presentation with y frames each time for n times, in which y and n are decided by users, and (7) *play* the presentation with the default (normal) configuration, i.e., forward with the normal speed.

User interactions can be classified in many ways. Depending on the number of user interrupts that is needed to achieve the corresponding user interaction, temporal-related user interactions can be classified as 1-phase and m-phase types. The 1-phase type can be achieved by one user interrupt, while the m-phase type can be achieved by multiple user interrupts[1]. For m-phase user interactions, e.g., freeze-restart, media transmission at the server site is suspended for k time units between two consecutive user interrupts, where k is decided by users. Among the above eight user interactions, 1-phase user interactions include *reverse, skip, scale, fast-forward, fast-backward,* and *play*; m-phase user interactions include *freeze-restart* and *step*.

Depending on the association with zero or some parameters, temporal user interactions can be classified as 0-parameter type and m-parameter type. Among the above eight user interactions, reverse, freeze-restart, fast-forward, fast-backward, and play are 0-parameter ones; skip, scale, and step are m-parameter ones. For example, let the currently displayed video frame be the xth, and a step(y) user interaction be issued. The next frame to be displayed is the $(x+y)^{th}$ video frame.

Depending on the number of operations that users should apply using the provided user interface, e.g., the number of buttons and/or keyboard that a user should click, temporal-related user interactions can be classified as 1-operation type and m-operation type. Users should click one button (multiple buttons) to achieve 1-operation (m-operation) user interactions. Among the above eight user interactions, 1-operation user interactions include reverse, fast-forward, fast-backward, and play; m-operation user interactions include skip, scale, freeze-restart, and step. For example, when a user U wants to trigger the skip (scale) user interaction, user U should (1) click a "skip" ("scale") button, (2) enter the value of the destination point (the new speed) parameter, and (3) click the "apply" button. Using some system-supported routines such that a sequence of multiple operations can be achieved automatically by systems, some user interactions that originally belong to the m-operation type can become the 1-operation type. For example, fast-forward and fast-backward are in fact pre-setup m-operation ones, in which the speed and direction are automatically adjusted to the default values by systems.

Temporal-related user interactions can also be classified as primitive type and compound type, in which a compound user interaction is either a variation of a primitive one with some default values for the associated parameters, or is composed of k primitive ones. Among the above eight user interactions, reverse, skip, freeze-restart, and scale, belong to the primitive type; play, step, fast-forward and fast-backward belong to the compound type. Depending on the provided functions, compound user interactions can be achieved automatically and/or manually. The execution of the composed primitive user interactions or the entrance of the associated parameters' values is achieved by systems for automatic compound user interactions. For example, fast-forward and fast-backward belong to automatic compound user interactions, in which the corresponding value of the speed parameter is automatically set by systems. The execution of manual compound user interactions are achieved by users manually during the run time, if the system is with the capability. For example, after a user U issuing a *freeze* interrupt, U also wants to reverse the presentation direction when the presentation is resumed. Under this situation, U can issue *reverse* instead of issuing *restart* for the second interrupt. Moreover, if systems can provide the programming capability, users are allowed to program some automatic compound user interactions themselves. For example, let the start (end) video frame of a "loop" user interaction be the X_s^{th} (X_e^{th}) video frame, a "loop(k)" user interaction, in which k is the execution number for the loop, can be programmed as follows: "loop(k) = skip(X_s), freeze when

---

[1]For convenience, the non-temporal-related user interactions are called 0-phase type.

the X_e^{th} video frame is reached, skip(X_e)."

In the extreme classification case, freeze-restart can be the only primitive one and the other user interactions belong to the compound ones. The execution sequence of the compound user interactions is as follows: (1) freeze, (2) modify the configuration, i.e., change the values of direction, speed, resumed presentation point, etc., and (3) restart. For example, the execution sequence of the reverse user interaction is as follows: freeze the current presentation, change the direction to the opposite one, and then restart the presentation. Among the remaining user interactions, reverse, fast-forward, fast-backward, and play are associated with pre-setup configurations, i.e., the system can have the execution sequence fully automatically; skip, scale, and step are associated with some user-input parameters, i.e., users should supply some parameters' values during the execution sequence.

3 Temporal Relationships with User Interactions

Temporal relationships among media streams can be represented by *temporal intervals*. Given any two intervals, there are thirteen different ways in which the two intervals may be related [1]. Fig. 1 depicts seven of the thirteen relationships; the remaining ones are the inverse of relations (2)...(7) that are depicted in Fig. 1.

With the provision of the eight user interactions discussed in Section 2, users are allowed to change the presentation flow direction or speed, or specify a new starting point of the presentation. Because user interactions are issued by users dynamically and unpredictably during the presentation, the temporal relationships of two related media streams may be changed, e.g., from *overlap* to *start*, when the presentation is resumed. That is, at the commencement time of the resumed presentation, the temporal relationships of two media objects may become different from the originally specified ones. The resume point of the issued user interaction determines the changes of temporal relationships that are associated with the user interaction.

Fig. 2 depicts the possible changes of temporal relationships for the reverse user interaction. Fig. 3 depicts the possible changes of temporal relationships for skip, freeze-restart, fast-forward, fast-backward, scale, and step. For the play user interaction, if the presentation direction is from backward (forward) to forward (backward), the resulted changes of temporal relationships is the same as that is depicted in Fig. 2 (3). Table 1 has the summary of the possible changes of temporal relationships caused by the eight user interactions.

4 Main Issues

There are two problems in interaction-based synchronization that is executed in distributed environments. The first problem is as follows. Since each medium has its own property and its own Quality-Of-Service (QOS) requirement, an individual communication channel between the client site and the server site is established for each medium stream. Because of (i) the different communication situations in communication channels, i.e., jitters occur at different time in different communication channels, and (ii) the adopted synchronization policies, e.g., the blocking policy is suitable for audio and the nonblocking policy is suitable for video [3, 8], the displaying media units of media streams in different communication channels may not be the originally

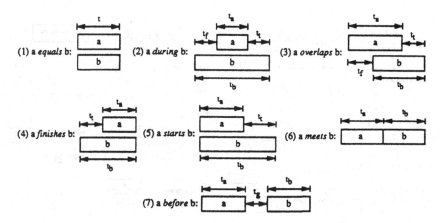

Fig. 1. Possible temporal relationships between two intervals.

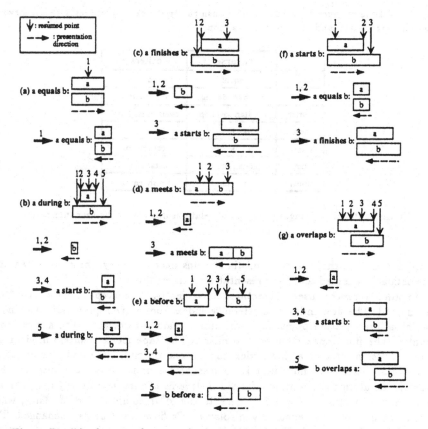

Fig. 2. Possible changes of temporal relationships for the reverse user interaction.

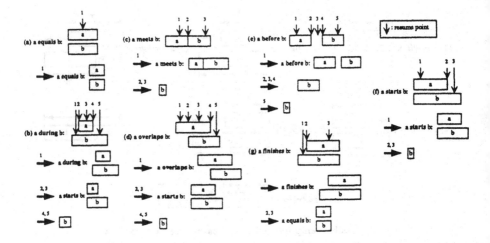

Fig. 3. Possible changes of temporal relationships for the skip, freeze-restart, fast-forward, fast-backward, scale, and step user interactions.

	reverse	others
equal	equal	equal
during	no, start, during	during, start, no
overlap	no, start, overlap	overlap, start, no
meet	no, meet	meet, no
before	no, before	before, no
start	equal, finish	start, no
finish	no, start	finish, equal

Table 1. Possible changes of temporal relationships caused by user interactions.

coupled ones, i.e., the display of media streams may not be synchronous, when user interactions are issued and the presentation is interrupted.

In order to process user interactions conveniently, the concept of "flow index" can be adopted. The flow index is a reference index from a given presentation point to the initial presentation point. Each medium stream is associated with a flow index counter. The flow index counter is similar to the tape index counters used in our home VCRs. A unit of a flow index can be one or multiple atomic media units, in which an atomic medium unit is an inseparable unit of the video medium. For example, an atomic medium unit for video streams is a frame. In this paper, the xth flow index of medium stream S represents the xth medium unit of S. Thus, when a medium unit of S is repeatedly displayed, S's flow index keeps unchanged. The display length of a flow index unit is the time duration of a video frame. The time length of a medium's flow index unit is equal to the others', but the data size may be different from others'. Let a presentation contain video, audio, and slide streams. Due to the jitter phenomenon, flow indices of video, audio, and slide streams be x, y, and z respectively when the user issues a user interaction. After processing the user interaction, resume points of all media streams should be the same, e.g., w, which is derived by using a re-synchronization function $F_I(x, y, z)$, where I is the interaction type. Thus, the presentation will be resumed synchronously.

The second problem is that when the presentation is paused at the client site to process user interactions, processes at the server site are still transmitting media units to the client site, which may be useless/redundant for the following resumed presentations, before interrupt messages are received. That is, some useless/redundant media units are stored in the buffer at the client site and are flowing in the networks. Let a presentation contain n streams and $_cI_i$ ($_sI_i$), $i=1..n$, be the last displayed (transmitted) medium unit in stream i at the client (server) site. When server process SA_i receives an interrupt message and then pauses transmission, client process CA_i has to discard k_i media units, i.e., $(_cI_i+1)^{th}$, $(_cI_i+2)^{th}$, ..., $(_cI_i+k_i)^{th}$, in its buffer, and discard the media units $(_cI_i+k_i+1)^{th}$ to $(_sI_i)^{th}$ that are in medium channel i.

The above two issues should be resolved to have correct and synchronous display when the presentation is resumed. (1) The re-synchronization issue: how to identify from which media units the associated processes at the server site should transmit when the presentation is resumed? (2) the identification issue: how many media units should be dropped and from which incoming media units the resumed presentation should start at the client site?

The re-synchronization issue is resolved by adopting different re-synchronization functions F_I, where I is skip, reverse, scale, fast-forward, fast-backward, freeze-restart, step, or play. Let medium stream i be in flow index x_i, $i=1..n$, when the presentation is temporarily paused. For the skip user interaction, since the destination point is specified by the user, e.g., skip to the xth medium unit, the presentation is resumed synchronously from medium unit x, no matter what the values of x_1, x_2, ..., x_n are. For the other user interactions, F_I can be (1) the maximum or minimum value of $(x_1, x_2, ..., x_n)$ [4], (2) the current flow index of a master stream x_i, or (3) others.

Two approaches that can be adopted to resolve the identification problem are the query-based approach and the token-based approach. We have proposed a query-based approach for interactive multimedia presentations in [4]. Through the cooperative messages exchanging between related control processes at the client site and at the server site, (1) the media units from which the server source site should transmit and (2) the number of media units to be dropped at the client display site, can be calculated. To shorten the calculation overhead and to have better response time, one can adopt the token-based approach. Let the resumed point be the xth medium unit, which is derived by using the appropriate re-synchronization function. The token-based approach is based on inserting a synchronization token in front of the xth medium unit in each medium stream. The client display site discards incoming media units until the synchronization token is received. In this way, the computation overhead spent in deriving the number of dropped media units, which is required in the query-based approach, can be saved.

The main concern of adopting the token-based approach is how to correctly deliver the synchronization token to the client site. Since the main characteristic of transmitting continuous media is maintaining real-time requirement with some error tolerance, it is possible that the synchronization token may be lost or garbled. There are two methods that can guarantee the correct delivery of the synchronization token. If the underlying networks can provide dynamic QOS, the synchronization token can be transmitted by associating with the "can't be lost/garbled" attribute. Although the dynamic QOS approach is the simplest way, it is not very available in practical networks. We propose the following timer-based control scheme.

At the client site, a timer is set up with an appropriate timeout interval T when the corresponding interrupt message is sent to the server site. T is derived according to the round trip transmission time between the client site and the server site and the associated overhead for processing interrupt messages at client and server sites. The synchronization token is also associated with the Time-To-Live (TTL) attribute. That is, if the synchronization token can't reach the client site by the TTL interval, the synchronization token is removed by networks. The time interval of TTL should be equal to or less than T–(the transmission time of the interrupt message from the client site to the server site + the associated overhead for processing interrupt messages at client and server sites). The client site (i) discards all of the received media units until the synchronization token is received, and (ii) re-transmits a resume interrupt message when time-out occurs in any one of media channels. In the case that some media channels' synchronization tokens are correctly received but the others are not, all media streams should re-transmit again.

5 Control Architectures and Methods

Suitable synchronization control mechanisms should be adopted for interactive multimedia presentations in distributed environments. Fig. 4 depicts the abstract synchronization control architecture. In Fig. 4, (i) each client medium process i, $i=1..n$, handles the receiving of media units and the display of media units to the corresponding display device for stream i and handles intra-medium synchronization, (ii) each server medium process j, $j=1..n$, handles the capture of media units from a medium base and the transmission of media units to network for stream j, (iii) synchronization controllers handle inter-media synchronization issues. For convenience, synchronization controller is called Synchronizer and medium process is called Actor. Media channels between associated Actors deliver media units. Signal channels, which are associated with reliable and error-free transmission, deliver interrupt and control messages for processing user interactions. Synchronizer at the client site is in charge of accepting user interactions from the corresponding user input channel.

The execution of re-synchronization, e.g., the calculation of the re-synchronization function F_I, is done by the cooperation of Synchronizers and Actors. When the new start point, i.e., the value of z that is derived by applying the suitable re-synchronization function F_I, is derived, (i) each Actor at the server site resumes its transmission from the zth medium unit, and (ii) each Actor at the client site discards received media units until the synchronization token arrives, and then resumes the presentation from the zth medium unit that arrives following the synchronization token. Depending on the re-synchronization processing site being either at the client site

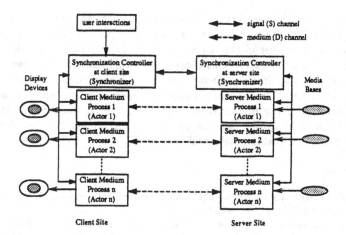

Fig. 4. The abstract synchronization control architecture in distributed environments.

or the server site, there are two approaches for interaction-based synchronization resolution schemes. Due to the space limitation, the details of these two control schemes
are not presented. Interested readers can get the complete version of this paper by
visiting the following web site: *http://bear.iie.ncku.edu.tw/English.html/miscellanea*.

6 Discussion and Conclusion

One of the main goal for processing user interactions is to shorten the response time.
Adopting some buffering schemes may be able to improve the response time for some
user interactions. The buffering schemes are especially applicable for m-phase user
interactions because there are some time gap between the t^h and the $(i+1)^{th}$ user
interrupts. The system can utilize the time gap to pre-fetch some media units from
the server site to the client site for the following resumed presentation. For example,
if the step value is fixed as one, i.e., the value of y in "step(y)" is 1, the client
site can pre-fetch multiple media units such that the following presentation can be
executed immediately when the "step" interrupt is issued at the client site. Through
some sophisticated calculation, freeze-restart can also be improved by pre-fetch k
media units during the idle time between the "freeze" interrupt and the "restart"
interrupt. If the presentation interval of these k media units is enough to compensate
the transmission delay of (i) the "restart" interrupt message sent from the client site
to the server site, (ii) the computation overhead for restarting media transmission,
and (iii) the transmission delay of the first medium that is sent from the server site
to the client site, then the presentation can be immediately resumed at the client site
when the "restart" interrupt is issued.

When a user issues a user interaction, the media units that are stored at the client
buffer and that are flowing in the network may still be useful after the presentation
is resumed. Depending on the usefulness of the media units that are (i) stored at
the client buffer and (ii) flowing in the network when the user interaction is issued,
temporal user interactions can be classified as the bn-useful type and the bn-useless

type. Among the eight user interactions, reverse, skip, and play belong to the bn-useless type, and the others belong to the bn-useful type. The play user interaction can be classified as the bn-useful (bn-useless) type if the original presentation is forward (backward). To have more strict classification, we still classify the play user interaction as the bn-useless type. However, not all of the buffered and flowing media units are useful. For example, if the speedup ratio is 3, then only 1 of every 3 media units becomes useful after the presentation is resumed. Thus, more complicated control schemes should be proposed for the bn-useful user interactions, i.e., fast-forward, fast-backward, freeze-restart, step, and scale.

References

1. J. F. Allen, "Maintaining Knowledge about Temporal Intervals," *Communications of the ACM*, VOL. 26, NO. 11, pp. 832-843, 1983.
2. G. Blakowski and R. Steinmetz, "A Media Synchronization Survey: Reference Model, Specification, and Case Studies," *IEEE Journal on Selected Areas in Communications*, VOL. 14, NO. 1, pp. 5-35, 1996.
3. C. M. Huang and C. M. Lo, "An EFSM-based Multimedia Synchronization Model and Authoring System," *IEEE Journal on Selected Areas in Communications*, VOL. 14, NO. 1, pp. 138-152, 1996.
4. C. M. Huang, C. Wang, J. S. Chen, and C. H. Lin, " Synchronization Architectures and Control Schemes for Interactive Multimedia Presentations," *IEEE Transactions on Consumer Electronics*, VOL. 42, NO. 3, pp. 546-556, 1996.
5. L. Lamont and N. D. Georganas, "Synchronization Architecture and Protocols for a Multimedia News Service Application," *Proceedings of the 1st IEEE International Conference on Multimedia Computing and Systems*, pp. 3-8, 1994.
6. L. Li, A. Karmouch and N. D. Georganas, "Multimedia Teleorchestra with Independent Sources: Part 2 - Synchronization Algorithms," *ACM Multimedia Systems*, VOL. 1, NO. 4, pp. 154-165, 1994.
7. J. Schnepf, J. A. Konstan, and D. H. C. Du, "Doing FLIPS: FLexible Interactive Presentation Synchronization," *IEEE Journal on Selected Areas in Communications*, VOL. 14, NO. 1, pp. 114-125, 1996.
8. R. Steinmetz, "Synchronization Properties in Multimedia Systems," *IEEE Journal on Selected Areas in Communications*, VOL. 8, NO. 3, pp. 401-412, 1990.
9. P. S. Yu, J. L. Wolf, and H. Shachnai, "Design and Analysis of a Look-ahead Scheduling Scheme to Support Pause-resume for Video-on-demand Applications," *ACM Multimedia Systems*, VOL. 3, NO. 4, pp. 137-149, 1995.

Modeling of Adaptable Multimedia Documents

Stefan Wirag

University of Stuttgart
Institute of Parallel and Distributed High-Performance Systems (IPVR)
Breitwiesenstr. 20-22
D-70565 Stuttgart
wirag@informatik.uni-stuttgart.de

Abstract. Multimedia presentations are applicable in various domains such as advertising, commercial presentations or education. If multimedia documents which describe multimedia presentations can be accessed on-line via different network types and be presented on various types of terminals different amounts of resources may be available at presentation time. Hence, it can happen that there are not enough resources to render a multimedia document according to the specification. For usual multimedia documents resource scarcity implies an arbitrary reduced presentation quality. To handle resource scarcity in a better way, multimedia documents can be made adaptable to different resource situations. The presented temporal model provides abstractions to specify multimedia documents with alternative presentation parts. Further on, the presentation behavior of media objects can vary within specified limits. Hence, the presented temporal model allows to compose presentations which have a defined behavior when resource restrictions occur and a resource scarcity need not result in an arbitrary reduced presentation quality.

1 Introduction

Multimedia documents present discrete media objects, such as graphic or text, and continuous media objects, such as video, audio or animation, in an orchestrated manner to the user. In a distributed environment, the structural description of a multimedia document can be stored apart from media object content or even the structural description can be stored in distributed fashion. Hence, the presentation of a document requires not only processing cycles and buffer space but also bandwidth.

For the presentation of a given document various amounts of resources may be available, depending on the type of the presentation terminal, the underlying network as well as the system load at presentation time. For example, consider multimedia documents stored in so-called digital libraries. Ideally, those documents can be accessed via different network types and be displayed at various types of terminals, such as PCs, Set-Top-Units or even PDAs. Even if we consider one type of terminal and one type of network, different load conditions may cause different amounts of resources to be available. If the underlying system provides for resource reservation, the resources needed to present a (single) media object (e.g. a video clip) can be determined and reserved prior to its presentation. Without reservation the resource situation may change even while the presentation of the object is in progress. However, reservation in general cannot prevent resource shortages to occur during the presentation of multi-object documents. Especially in the case of interactive documents it is not feasible or even possible to reserve all resources before presentation starts.

When a resource shortage occurs there are basically three ways to react to it. First, just ignore it and accept the quality of the presentation to be decreased *somehow*. Second, abort the presentation, and third, adapt the presentation in a *user-controlled manner* to the given resource situation. Of course, the third approach requires an appropriate document model as well as an adaptive scheduling mechanism. In this paper, we are going to propose extensions to the temporal model *Tiempo*[1] [7, 8] to support this type of adaptability.

The Tiempo model is hierarchical. Temporal dependencies between media objects are defined by so called interval operators. Interaction is modeled by reaction relations. The desired adaptability is achieved by selection groups and Quality of Service ranges. *Selection groups* allow to define alternative presentation parts representing the same information in different form. *Quality of Service ranges* allow to specify alternative presentation options for media objects. These abstractions can be applied in combination to achieve a high degree of adaptability. Thus, resource shortages need not result in a reduced presentation quality of Tiempo documents.

The remainder of the paper is structured as follows: In Section 2, we present the basic concepts of the Tiempo model. In Section 3, the means that make specifications adaptable are described. Then, we take a glance at the scheduling of Tiempo documents. Related work is considered in Section 5. Finally, we summarize our results.

2   Basic Concepts of the Tiempo Model

In Tiempo, documents consist of composite media objects, such as scenes, which are composed of single media objects, such as videos or a texts. A media object is modeled by a temporal space, a presentation interval and a projection. The *temporal space (TS)* represents the information of the media object. The *presentation interval* represents the period the media object is presented. It is described by its extent l. The projection describes how information from the TS is presented in the presentation interval. Hence, media objects can be presented with other than recording-time properties. Interaction possibilities are described by *interaction intervals*. Each such interval represents the period in which a particular user-interaction (e.g. click with mouse, insertion of text) is possible with a media object. An interaction interval is described by its extent and a state that changes when the user triggers the associated interaction.

2.1 Temporal Spaces

A TS of a single media object is called *temporal data space (TDS)*. It is defined by the temporally ordered data units of the object, e.g. the TDS of a video object is built by its frames. A TDS is described by a *data interval* with the length n/m where m is the re-

---

1. Temporal integrated model to present multimedia-objects

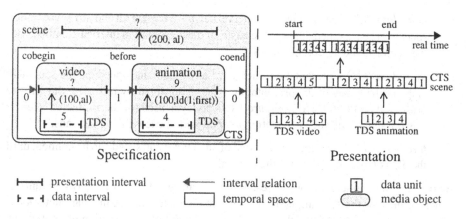

Fig. 1. Specification example

cording speed of the object and *n* the number of data units. As a discrete media object consists only of one data unit, its TDS is by default described by an interval of length 1.

A TS of a composite media object is called *composite temporal space (CTS)*. It is composed by single or composite media objects. The CTS of the top composite object represents the whole document and its presentation interval the document presentation period. To define the temporal layout of a CTS, presentation intervals of contained media objects are arranged by *interval operators* [7] relative to each other or the CTS.

The specification example in Figure 1 describes a scene where an animation is presented after a video-clip. Here, the interval operator "before" specifies that the animation has to be started 1 second after the video has ended. The "cobegin" operator describes that the CTS starts with the video and the "coend" operator describes that the CTS ends with the animation. To define the temporal layout of interaction possibilities, interaction intervals are related to other intervals by interval operators.

2.2 Projection of Temporal Spaces

A projection from a TS to a presentation interval is described by a tupel *(v,s)*. *v* defines how many data units of the TS are presented per second within the presentation interval. A presentation speed of 100 implies playout with normal speed. For discrete media objects *v* is set to 0, which implies that during the presentation always the same data unit is presented. In Figure 1 the presentation speed of the scene is set to 200. Hence, the video and the animation are presented with the double recording speed.

Parameter *s* represents an *alignment policy* that determines which information of the TS is presented in the presentation interval. Tiempo offers four alignment policies:

- *align-length (al)*: The whole TS is presented in the presentation interval. As the extent of the TS is known, it is sufficient to specify parameter *v* or the extent of the presentation interval *l*. In Figure 1 the video and the scene apply the "al" policy.

- *hold-data (hd):* The presentation interval can have any extent and any presentation speed can be specified. Gaps at the edges of the presentation interval are filled with the last or first data unit of the TS.

- *loop-data (ld):* A presentation interval of any length and speed can be specified. Gaps at the edges of the presentation interval are filled by repeating the already presented part of the TS. In Figure 1 this policy is used for the animation. Hence, it is possible to fill the presentation interval with the length of 9 seconds.

- *force-end (fe):* A presentation interval of any length and speed can be defined, until the TS is large enough to fill the interval.

As the last three policies allow to present only a part of a TS, they have two parameter which define how the TS is aligned in the presentation interval. The first parameter identifies an instant i of the TS and the second parameter defines the position of i within the presentation interval. i can be mapped to the *first*, to the *last* or to the *center* instant of the presentation interval. According to this relation and the defined speed, all instants of the TS can be associated with an instant in the presentation interval. With discrete media objects, the policy *hd(1,first)* is used. The right side of Figure 1 shows how a presentation is created by projection according to the specification on the left side. It is assumed that the video consists of 5 frames, the animation consists of 4 frames and both media objects were recorded with one frame per second.

2.3 Interaction Relations

In Tiempo, an interaction is described as a so-called *reaction relation* between an interaction interval and affected elements. A reaction relation consists of a trigger-condition and actions. The actions describe behavior modification of projections, presentation or interaction intervals. Actions can start and stop intervals, pause, freeze and speed up the playout [8]. The trigger-condition describes how the state of the interaction interval must change (e.g. a state change of an interaction interval of a button from not-pressed to pressed) before the actions are executed.

3 Modeling of Adaptability in Tiempo

To be able to adapt documents to different resource situations flexible specifications are needed, since flexibility enables the presentation system to create alternative presentations from a document. Considering the structure of a document, flexibility is possible on two levels. Adaptable documents can contain alternative media objects or even scenes which represent the same information in different form. As different information forms have various resource requirements, the presentation system can handle different resource situations. Adaptable documents can contain totally different arrangements of the same media objects. For example, sometimes it makes no difference whether an advertising animation is presented before or after a video-clip. If there are not enough resources to present the media object in one position, it might be possible to present it in another position. In Tiempo, this kind of flexibility is supported by *selection groups*.

Adaptable documents can contain alternative presentation options for the same media objects. For example, it could be specified that a video which is normally presented with 20 frames per second, can alternatively be presented with 10 frames per second. Further, sometimes the starting or terminating instant of a media object can vary within certain limits whether the arrangement to other media objects is not disrupted. To model such flexibility, we introduced *Quality of Service (QoS) ranges*. Selection groups and QoS ranges can be combined to achieve an optimal adaptability of documents.

3.1 Selection Groups

A selection group consists of an arbitrary number of *presentation alternatives*. It has the semantics that exactly one of the alternatives has to be implemented by the presentation system. Each presentation alternative can consist of several media objects and interval operators. Hence, it can represent any part of a specification. Selection groups can be nested to specify selection possibilities on different levels.

As multimedia presentations should be attractive for users, they often contain media objects which contribute little to the information content, e.g. a company logo animation. The playout of such information can be omitted when a resource scarcity occurs. Even some temporal relations can be less relevant. If they are omitted the temporal layout might change so that a resource shortage can be compensated. Scenarios where it is allowed to omit specification parts are modeled by selection groups with an empty presentation alternative.

To limit specification overhead, interval operators relating media objects which are contained in a presentation alternative need not be part of the presentation alternative. The presentation system can detect these interval operators automatically, because if a media object is not presented, interval operators relating the media object to other media objects have no meaning and have to be omitted.

Objects of presentation alternatives of a selection group can be located in different CTSs. In Tiempo, media objects and interval operators are specified within CTSs of composite media objects. Hence, a presentation alternative contains only references to objects.

A multimedia document may contain several selection groups consisting of multiple presentation alternatives. Hence, the presentation system could have various possibilities to adapt a specification. In Tiempo, priorities can be assigned to presentation alternatives that indicate which alternative should be preferred when more than one can be implemented. The presentation system associates with higher priorities a higher relevance respectively presentation quality. When adapting a presentation, the presentation system will try to implement a combination of presentation alternatives so that the sum of priorities is as high as possible.

So far it is not defined when the presentation system can choose a presentation alternative of a selection group, e.g. sometimes it can be wished that the presentation system continues with the playout of subtitles when the playout of audio is no longer possible and sometimes not. Especially in interactive documents, scenes can be rendered more than once. It might be desirable that even scenes with presentation alternatives present the same arrangement of media objects in each rendition. In Tiempo, the selection of presentation alternatives is controlled by the following *selection policies*:

- *init*: The presentation system can only once in a document presentation make a selection. If the document part with the selection group is rendered more than once, always the same presentation alternative is implemented.

- *static*: The presentation system can only make a selection at the start of the presentation part which contains the selection group. If the document part is rendered more than once, it can select a different presentation alternative in each rendition.

- *dynamic*: The presentation system can change the implemented presentation alternative during its presentation.

3.2 Quality of Service (QoS) Ranges

A QoS range is a value range where a priority is associated with each value. In the Tiempo model, QoS range arguments can be used to specify the extent of presentation and interaction intervals, the presentation speed and the delays implied by interval operators. If applied, a presentation system can select extent, speed or delay values out of QoS ranges with regard to the resource situation.

A multimedia presentation consist of several media objects arranged by interval operators. Hence, a presentation system can have different options to adapt a specification, e.g. to reduce either the speed of a video or an animation. To indicate which values of QoS ranges are preferred, priorities are assigned to the contained values. A higher priority represents a better presentation quality. When adapting a scene of a presentation, the presentation system should try to select a combination of QoS range values of all objects so that the sum of priorities is as high as possible.

The priority structure of a QoS range is defined by arbitrary anchor points. The QoS range in Figure 2 may define the extent of a presentation interval, which can be between 10 sec and 55 sec. In the example, an extent of 10 sec has the priority 30, an extent of 55 sec has the priority 70 and an extent of 35 sec has a priority of 100. With the extents between 10 and 35 linear increasing priorities from 30 to 100 are associated and with the extents between 35 and 55 linear decreasing priorities from 100 to 70 are associated. Here, the presentation system should implement an extent of 35 sec for an optimal presentation quality.

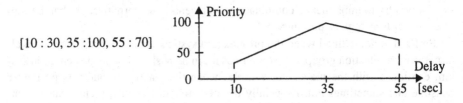

Fig. 2. QoS range semantics

If the continuous playout of a media object is a quality criterion, the presentation speed should not change during its presentation. But it can be allowed to choose a particular speed when the media object is started. Further, if scenes are rendered more than once, it might be desirable that some object behavior is equal in each rendition. To define when a value from a QoS range can be chosen, theoretically the same selection policies can be assigned to QoS ranges as to selection groups. To distinguish between static and dynamic selection is only reasonable for speed values, since the user will recognize a dynamic speed change. For interval extents and delays implied by interval operators a dynamic selection will cause a modified presentation behavior in the future. As the user does not know the future, he cannot distinguish between a static and a dynamic selection policy. Hence, to QoS ranges that define interval extents and delays implied by interval operators only the selection policies init and dynamic can be assigned.

3.3 Specification Example

Figure 3 shows the temporal specification of an adaptable multimedia document which represents a tutorial on a technical process. Selection group *sg1* describes that the presentation can start with a technical animation, a textual description or directly with the video. The presentation alternative with the technical animation contains a reference to selection group *sg2*. Hence, if the technical animation is presented, simultaneously a speech sequence or subtitles have to be presented. As the selection policy of *sg2* is dynamic, the presentation system can switch between subtitles and speech during the presentation. The selection policy of *sg1* is static. This means, it is not allowed to stop the playout of a presentation alternative and continue with another. During the whole scene an animation of the company logo is presented. As the logo animation has a low relevance, selection group *sg3* is specified with an empty presentation alternative and the selection policy dynamic. Thus, the logo animation can be omitted or if already started, its playout can be terminated. The assigned priorities indicate that in *sg1* the playout of the technical animation either with the speech or the subtitles is preferred. In *sg2* the playout of the speech is preferred. The low priority of the presentation alternative with the logo animation in *sg3* cause that the logo animation is terminated first when there is a resource shortage which makes it impossible to present all objects.

The speed of the video and the logo animation are specified flexible. The speed of the video and the animation can be between 90% and 100% of its original speed. The priorities of the video speed values are higher than the priorities of the logo animation speed values. Thus, the presentation system will reduce first the speed of the logo animation when there is a resource shortage. The delay between the end of the text or technical animation and start of the video is also specified flexible. It can be between 1 and 3 seconds.

4 Adaptive Scheduling of Tiempo Documents

When a multimedia document should be presented, a presentation schedule has to be created from the specification. Then, the presentation can be performed according to the

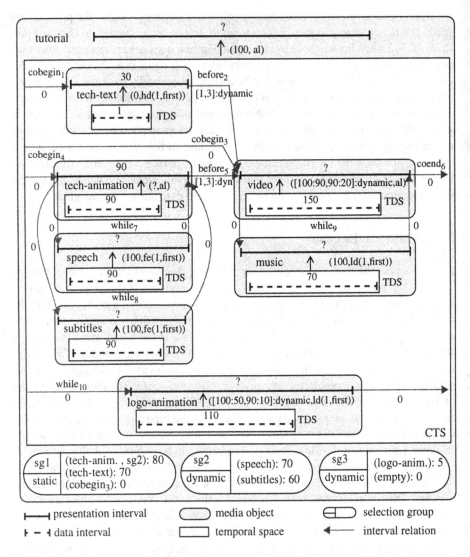

Fig. 3. Adaptable tutorial specification

schedule. If user interaction occurs, the schedule has to be modified appropriately. With flexible Tiempo documents, the schedule is additionally adapted periodically to the available resources.

To be able to adapt a specification effectively, information about resources which are required to present media objects are needed. Such data can partly be computed from the specification, partly it have to be determined experimentally. Further, information about the resource situation are necessary. Such information can be delivered from resource monitoring components, a management system or components which perform the playout of media objects. To increase the probability for a good adaptation, the presentation system should plan the presentation in advance for the whole scene. Because if the future would not be considered, it might happen that it is impossible to present all remaining objects of the scene because of an earlier adaptation decision. Hence, a profile of available resources in the future has to be determined from the delivered data by approximation.

Based on this information, the presentation system has to create a schedule so that in each instant in the future less resources are required than available. It starts with the best combination of presentation alternatives of selection groups and tries to find some start and end instant for these objects according to the specification. If this is not possible without violating the specification, the presentation system repeats the same procedure for the next best combination of presentation alternatives and so on. If a possible combination is found, it will try to find speed values, start and end instants for media objects from QoS ranges so that the sum of priorities is as high as possible. If a part of the scene was already presented, the described algorithm is only applied for the not presented part of the scene and selection restrictions implied by selection policies of already running objects must be considered. If all combinations require more resources than available, the combination with the lowest violation can be implemented or the presentation can be terminated.

5   Related Work

Various temporal models and synchronization concepts have been developed for multimedia presentations. Multiple variations of Petri Net models such as TSPN [5] exist which have a strong expressive power according to the specification of synchronization aspects. However, these models do not adapt documents with regard to the resource situation.

Besides Petri Net-based models, several high-level temporal models have been proposed. Models that base on the time-line concept, such as MAEestro [2], integrate no flexibility. Hence, it is not possible to compose adaptable specifications. Some models that describe temporal relations between media objects explicitly support indeterminism, e.g. Mbuild [3] and FLIPS [6]. In these models the contained indeterminism is not applied to adapt the specification, it helps either to ease the composition of presentations or to synchronize presentations where interaction is possible.

The only model that has the potential to specify adaptable documents is Firefly [1]. Firefly integrates mechanisms to specify flexible presentations. It is event-based, which means that the start and end of media objects are modeled as instants. The delay between related events is described by a minimum, optimal and maximum value. Additionally, costs are defined for shrinking the extent of an object to minimum or extending it to maximum. To automatically compute the optimal schedule out of these values, a linear programming algorithm has been proposed. To gain adaptable specifications, the cost values could be defined with regard to the relevance or resource requirements of media objects. However, the authors of Firefly do not exactly define how cost values are determined and the presentation system does not consider the resource situation when the presentation schedule is created.

6 Conclusion

To perform multimedia presentations under different resource situations, adaptable documents models are required. We have identified various forms of flexibility which make multimedia documents temporally adaptable. The presented Tiempo model offers selection groups to represent alternative presentation parts consisting of media objects and interval operators. With QoS ranges alternative presentation options of media objects or interval operators can be defined. Assigned priorities define which alternatives of selection groups or QoS ranges are preferred by the author. Hence, Tiempo conform multimedia documents can integrate a high degree of flexibility and resource shortages need not result in a reduced presentation quality. Based on the described concepts, a presentation system is being developed on top of the CINEMA-platform [4]. We currently aim at developing adaptive scheduling algorithms for different system configurations.

7 References

[1]   M. Cecelia Buchanan and Polle T. Zellweger. "Scheduling Multimedia Documents Using Temporal Constraints", in Proc. of 3rd International Workshop on Network and Operating System Support for Digital Audio and Video, San Diego, USA, pp. 223–235, 1992.

[2]   George D. Drapeau. "Synchronization in the MAEstro Multimedia Authoring Environment", in Proc. of ACM 1st Intl. Conference on Multimedia, Anaheim, USA, pp. 331 – 340, 1993.

[3]   Rei Hamakwa, Hidekazu Sakagami, and Jun Rekimoto. "Mbuild - Multimedia Data Builder with Box and Glue", in Proc. of IEEE 1st Intl. Conference on Multimedia Computing and Systems, Boston, USA, pp. 520–525, 1994.

[4]   Kurt Rothermel, Ingo Barth, and Tobias Helbig. "CINEMA - an architecture for distributed multimedia applications", in Architecture and Protocols for High Speed Networks. Norwell, MA: Kluwer, 1994, pp. 253-271.

[5]   P. Senac, Michel Diaz, Alain Leger, and Pierre de Saqui-Sannes. "Modeling Logical and Temporal Synchronization in Hypermedia Systems", IEEE Journal on Selected Areas in Communications, vol. 14, no. 1, pp. 84-104, 1996

[6]   J. Schnepf, J. A. Konstan, and D. H.-C. Du. "Doing FLIPS: FLexible Interactive Presentation Synchronization", IEEE Jounal on Selected Areas in Communications, vol. 14, no. 1, pp. 114-125, 1996.

[7]   Thomas Wahl and Kurt Rothermel. "Representing Time in Multimedia Systems", in Proc. of IEEE 1st Intl. Conference on Multimedia Computing and Systems, Boston, USA, pp. 538–543, 1994.

[8]   Thomas Wahl, Stefan Wirag and Kurt Rothermel. "TIEMPO: Temporal Modeling and Authoring of Interactive Multimedia", in Proc. of IEEE 2nd Intl. Conference on Multimedia Computing and Systems, pp. 274-277, Washington DC, 5 1995

QoS Aware Browsing in Distributed Multimedia Systems

Georg Michelitsch, Max Ott, Daniel Reininger, Girish Welling

C&C Research Laboratories, NEC USA, Inc.

Abstract : It has been widely recognized that QoS is important in designing
distributed, interactive multimedia systems. However, although much has been
said about QoS in the networking domain, and to a lesser degree in processor
scheduling, there has hardly been any research activity concerning QoS in the
user interface community. We show how the use of 3D graphics techniques in
the user interface can lead to a natural way of implicitly specifying and presenting QoS to the end-user. Further, we introduce the concept of generic QoS control tools that allow users to explicitly control and monitor quality of service
across different media types. In order to support such a novel user interface, a
mechanism is needed to communicate these quality requirements to lower level
system components. Although there exists a plethora of QoS architectures that
define the semantics and interface of every component in many different ways,
none offers a definitive way of structuring QoS aware systems. We instead propose a generalized, abstract concept of QoS for all layers of a software architecture. Each layer in a software system deals with QoS at its appropriate level of
abstraction using a generic API for communicating QoS parameters and values
to layers above and below. We call the aggregation of these parameters and values a "service contract". This abstract concept can be applied recursively to
build hierarchies of services. This paper describes our framework for building
QoS aware software systems and explains in detail the user interface for a multimedia browser as an example application.

1 Introduction

Distributed computing is often expected to be an extension of personal computing,
where user commands are immediately serviced on a best effort basis. However, this
expectation is unrealistic when the source of information is moved from the local hard
disk to a server located anywhere in the world. The expectation becomes impractical
with multimedia content, like motion video and high quality audio. There is clearly a
need for a system with quality of service (QoS) support. We believe that for effectiveness, QoS control must be naturally and unobtrusively exposed to the end-user, with
different levels of detail.

Consider an application scenario where a financial analyst follows an inherently multi-
modal work style. He monitors and interacts with different sources of information,
real-time financial data in numerical and graphical form, articles from on-line news

feeds, and a variety of broadcast audio and video services. Each of these information sources compete with each other for limited screen real estate, and the attention of the analyst.

During a typical session, an important story from a news report showing in a small window on the computer screen might catch the attention of our analyst. He focuses on this report for a short while, ignoring everything else on the screen. He then searches for related background information, and finally looks at how the financial market has reacted to the report. During this session, the analyst has moved his attention from one source of information to another, demanding different levels of detail in the data presented.

At the user interface level our system allows the user to place objects on an information landscape in 3D space. Less important things are pushed back towards the horizon, more important ones are pulled forward for closer inspection. This information about the level of importance the user attaches to an object is mapped into a service contract with the underlying software responsible for presenting the data. That component in turn sub-contracts its work to other components with different service contracts building a hierarchy of services in the process. As an example of a contract one can think of a request to a video server to play a particular movie at a certain frame rate with a given resolution.

So far we have only discussed the scenario of a user requesting different levels of QoS from the system. However, it is as likely to have components throughout the system not being able to maintain the level of quality of service demanded by their service contracts. As an example, let us go back to our financial analyst. Imagine him commuting to his office by train. While riding the train he will have a different network connection from what he can expect at his office with frequent hand-off operations taking place in between. It is difficult to reserve resources for mobile users and maintain the desired QoS for each of them and at the same time achieve high utilization of the network. Adaptive mobile multimedia applications improve their chance for successful communication since they can continue to operate with the available resources as the user roams among microcells.

In order to build such systems we provide a framework within which multimedia application with awareness of QoS changes can easily be constructed. We first present a definition of our QoS framework and then describe our approach to a QoS aware user interface based on that framework.

2 QoS Framework

The multimedia research community has been concerned for quite some time with the question of how to build multimedia systems based on platforms which were not originally designed to handle real-time data. As a result the notion of QoS, which originated in the networking community and became a central research issue due to the surge in interest in ATM technology, had to be extended to the domain of distributed computing.

The article by Andrew Campbell et al [4] offers a comprehensive overview of QoS architectures. In the tradition of the OSI network reference model, the architectures described in this survey attempt to lay out the functionality and interfaces between every architectural component in the system. The QoS broker [3] for example, defines an architectural entity that functions as the intermediary between system components. This entity brokers a deal between these components in order to ensure a certain end-to-end system behavior. This architecture defines the protocols between the entities, the semantics of QoS parameters, and the proper place for each component within a system.

Our work, however, does not attempt to define the exact semantics and layering of every component in a multimedia system. Since we believe that these questions are still open research issues, we propose an architecture for building QoS aware systems, where we only define the mechanisms for negotiating and monitoring QoS contracts between software components. The semantics of such contracts is left to the parties concerned. In doing so, we allow software designers to recursively apply this concept in order to allow different levels of abstraction for QoS.

2.1 Service Contract

We express the characteristics of a service in terms of parameters, which have a name, a type, and a value (see table 1). These parameters can either be used as requirements that a software component imposes on a service, or as a measure of compliance of a service with its requirements. The aggregation of requirement and compliance parameters for a given service is called a service contract.

Table 1: QoS parameter examples

Name	Type	Value
CurrentFrameRate	IntegerValue	30
FrameRateRange	IntegerRange	[1 .. 30]
Cost	IntegerTargetRange	[100 (150) 200]
Sensitivity	OrderedSet	{low, medium, high}

2.2 Service Abstraction

A service can itself enlist the service of other software components creating a hierarchy of services. There can be a distinct service contract at each level using the same generic interface for contract negotiation.

2.3 Generic Interface for Contract Negotiation

There are three primitives necessary for service contract negotiations:

- Request: presents a contract to the service provider specifying QoS values and value ranges to be maintained.

- Notify: calls the client in case a target value for a parameter committed to in the service contract cannot be met by the service provider.

- Status: queries the current status of parameters defined in the contract.

Figure 1 shows the relationship between client and service provider expressed through a service contract, the generic API, and the recursive application of that principle.

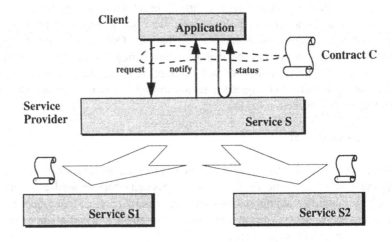

Fig.1. Service Hierarchy

The service provider offers service "S" to a client according to contract "C". The service provider decides on whether a contract can be satisfied by the service provider itself or has to be implemented by requesting the services of other service providers (services S2 and S3 in figure 3). The service provider monitors compliance of its sub-contractors with contracts negotiated at that level and will try to switch to different service providers in case the terms of the contracts are not met anymore. If this attempt fails the service provider will send an alert to its client passing control to the next level up in the service hierarchy.

3 User Interface

Though user interface design issues are important for any interactive system there are few results reported in the research community about user interface issues in relation with QoS. Work such as [1] and [2] discusses the relation between user perception and video playout quality. The survey in [5] describes an approach where users select QoS parameters on the basis of example displays. We are exploring the opportunities to blend new interaction techniques and alternative user interface concepts with the notion of QoS exposed to the user of an interactive multimedia system. For that purpose we extended *CockpitView* [6], an experimental user interface developed at the C&C Research Laboratories in Princeton which offers the following key features:

- Support for large number of objects on the screen that can be viewed and manipulated concurrently through the use of 3D techniques, surpassing current windowing user interfaces, which have to rely on icons to reduce the consumption of screen real estate.

- Use of user interaction techniques that work equally well on a office computer with a mouse attached, and on a portable device with a touch panel.

- Provision for non-expert users to work with the system "out of the box" without having to read an extensive manual first.

Fig.2. Screen Shot of *CockpitView* User Interface

Figure 2 shows a snapshot of the user interface in action. A compound document (NEC Times) representing a service provider is shown on the right. Two embedded video clips have been dragged off the document and placed on the data landscape for viewing, one at the bottom left playing at high resolution and one near the horizon at quarter size.

Also visible are the video control tool to the upper left of the screen and the service meter tool to the right which offers a control portion and a speed meter for feedback.

3.1 3D Data Landscape

Based on the fact that humans are generally very good at remembering objects by their location, we chose a spatial metaphor for organizing information entities in one global virtual landscape. Recent advances in low-cost 3D graphics hardware make this approach feasible, but in contrast to the video game industry, which emphasis realistic rendering of 3D objects, our goal is to optimize the use of limited screen real estate while at the same time provide users with the context they need [7, 8].

However, we do not believe that ordinary users will want to use special equipment for 3D input and output. We want to be able to manipulate objects directly on a flat screen with either a pen or our fingers. In order to do so we restrict the degrees of freedom an object has in our information landscape. By grabbing the projected image of an object and dragging it on the screen, we move the object on the data landscape along a path on the surface of that landscape as shown in figure 3.

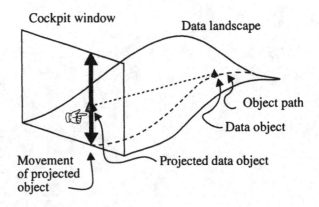

Fig.3. Interaction with Object in 3D Landscape

The user looks at the landscape as if through the window of a cockpit (hence the name *CockpitView*). He can reach out of the cockpit, grab an object on the landscape, and move it around. When he drags an object closer to the cockpit window, it will expand revealing more of its content. If pushed back towards the horizon an object will shrink. However, all objects will keep their full functionality no matter at what scale factor they are displayed.

3.2 Active tools

Every object on the screen is an active object that reacts to user generated events as well as to other objects. When the user drops an object onto another one and both objects are compatible, a compound object will be created. If one of these objects is what we call an Active Tool, the attachment will also cause the tool object to perform an operation on the other object.

This is a less abstract, more real-world like way of manipulating objects (see R. B. Smith [9]). The active tool concept also allows for both an "object-verb" approach, where the user first identifies the target of an operation and then the command, as well as for a "verb-object" approach, where the user starts with selecting a command first and then specifies the target object(s).

The familiar set of "pickup", "drag", and "drop" operations is all the user has to master in order to be able to operate the entire user interface. The possibility to group, modify, and therefore customize commands simply by assembling existing active tools enables users to perform tasks that previously required some form of programming, or the use of a dedicated user interface builder.

3.3 Service Meter: An example of an active tool

We designed a generic widget for users that allow them to specify QoS contracts in a consistent way across different types of media objects. The ServiceMeter changes its appearance and the semantics of controls based on the type of object it is attached to. For that purpose it uses the common convention of active tools and the inherent mechanisms for building compound objects defined in the CockpitView framework.

The value range of each parameter or parameter pair in the contract is projected to a different side of a cube. At the highest level of abstraction we start with a single parameter which expresses quality in terms from low to high. At the next level this parameter is mapped to a different set of parameters, in case of video this could be frame rate and detail, where detail then again maps to resolution and quantization.

Figure 4 shows the ServiceMeter clip in three different stages, first with the control portion hidden, then revealing the slider for the quality parameter, and finally showing the two-dimensional slider for controlling detail over frame rate. The user moves from one level of abstraction to another one by clicking on a particular area on the cube which causes it to rotate to the next side.

This mechanism of mapping one parameter to a set of less abstract parameters recursively is a direct application of the service contract concept described in the previous section. The user interface visualizes the service hierarchy through progressively revealing levels of detail in the parameters making up these contracts.

3.4 Explicit versus implicit QoS control

The ServiceMeter gives the user explicit control over the service contract for each media object. At the same time we can use information derived from the placement of objects in the data landscape to formulate a service contract. Objects placed close to the horizon can reduce their requirements for certain parameters. In case of video for example both the resolution and to a lesser degree the frame rate can be reduced without changing the user's perception of the video played.

The strategy we use in our system is to employ implicit QoS control for all objects with no ServiceMeter attached. As soon as the user attaches the meter tool to an object, the ServiceMeter updates its parameter settings and displays reflecting the current setting maintained by that media object. As long as the ServiceMeter is attached the QoS contract as shown on that tool is maintained regardless of the placement of the media object, effectively overriding the implicit mechanism.

Thus we can leverage the cost saving potential (assuming that cost is related to bandwidth requirements from the network) of our QoS approach without even having to introduce the concept to end-users by using the implicit strategy. However, the advanced or simply curious user can easily override this behavior by using the service meter tool.

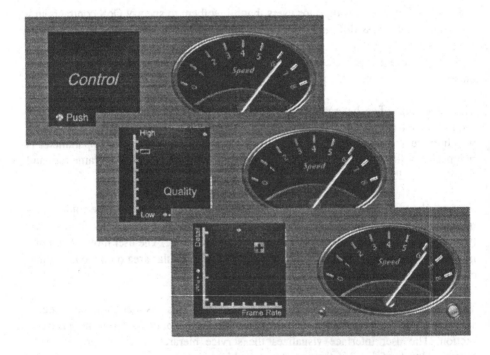

Fig.4. ServiceMeter showing QoS parameters at different levels of abstraction.

4 System Implementation

Our system is implemented on a test-bed which consists of an Ultra-SPARC 1 acting as a video server, another Ultra-SPARC 1 with Creator 3D graphics for the client side and an ATM switch by NEC.

All videos are encoded with M-JPEG and stored on disks local to the video server. The video and audio data is streamed over the ATM network using either a native ATM interface or RTP/IP over ATM.

The CockpitView user interface library is implemented in C++ relying on the X-window system for low level graphics operations and event handling. In addition, the current implementation uses the XIL library from Sun Microsystems for image scaling and decompression of JPEG (or MPEG-1) encoded video in order to achieve the desired performance.

The multimedia middleware [11] abstracts media processing such as transport and display of video and audio in a distributed environment including synchronization and scheduling. It consists of a set of libraries written in C++ which are in use on both Solaris and Linux platforms.

Among system components making use of our generalized QoS framework we have finished the dynamic, network bandwidth allocation portion. As described in [10] this architecture allows clients to specify a QoS contract that reflects the application's requirements and the terminal's capabilities. Servers dynamically negotiate the required bandwidth with the network and regulate the source's rate while renegotiation is in progress [12]. The network model considered, extends the broadband ATM service to support bandwidth renegotiation. A new service class, called VBR+[13], is proposed to solve the practical difficulties experienced in maintaining QoS and achieving significant statistical multiplexing gains for multimedia traffic.

5 Conclusion

Based on our generalized QoS framework we built a multimedia content delivery system that adapts quality of service to both user demands and fluctuations in system resources. Throughout the system, quality of service is managed using the single, generic API of our framework. At each level, quality of service is specified by a contract which is established between the client and the service provider component at that level. The recursive application of this mechanism creates a service hierarchy which constitutes QoS for the system as a whole.

At the user interface level we showed how limited screen real estate and the shifting focus during a typical machine-user-interaction offer ample opportunities for media tasks to change their resource demands. The placement of multimedia objects in a 3D data landscape naturally mirrors the different levels of quality requested from services by users. This, in combination with powerful direct manipulation techniques and the concept of active tools, results in a comprehensive set of interaction mechanisms for users of multimedia systems.

References

1. R. Steinmetz and C. Engler. "Human Perception of Media Synchronization". Tech. Report 43.9310, IBM European Networking Center, Heidelberg, 1993.
2. R.T. Apteker et al. "Distributed Multimedia: User Perception and Dynamic QoS". In Proceedings of IS&T/SPIE Symposium on Electronic Imaging: Science & Technology, Workshop on High-speed Networking and Multimedia Computing, SPIE, New York, 1994.
3. Klara Nahrstedt and Jonathan Smith, "The QoS Broker". In the IEEE Multimedia Magazine, Spring 1995, Vol. 2, No. 1.
4. Andrew Campbell, Cristina Aurrecoechea, and Linda Hauw. "A Review of QoS Architecture". In the ACM Multimedia Systems Journal, 1996.
5. Andreas Vogel, Brigitte Kerherve, Gregor von Bochman, and Jan Gecsei. "Distributed Multimedia and QoS: A Survey". In the IEEE Multimedia Magazine, Summer 1995, Vol. 2, No 2.
6. G. Michelitsch. "CockpitView: A User Interface Framework for Future Network Terminals". In conference companion proceedings of CHI'96, Vancouver, BC, Canada, ACM, 1996.
7. G. G. Robertson, S.K. Card, and J.D. Mackinlay. "Information Visualization using 3D Interactive Animation". In the Communication of the ACM, 36(4), 1993.
8. L. Staples. "Representation in Virtual Space: Visual Convention in the Graphical User Interface". In proceedings of INTERCHI'93, ACM, 1993.
9. R. B. Smith. "The Alternate Reality Kit: An Animated Environment for Creating Interactive Simulations". In proceedings of the 1986 IEEE Computer Society Workshop on Visual Languages, 1986.
10. M. Ott, D. Reininger, and W. Luo. "Adaptive and Scalable QoS for Multimedia using Hierarchical Contracts". In proceedings of the Fourth ACM International Multimedia Conference, Boston, MA, 1996.
11. M. Ott, D. Reininger, G. Michelitsch, V. Bansal, R.J. Siracusa, D. Raychaudhuri. "Heidi-II: A Software Architecture for ATM Network Based Distributed Multimedia Systems". In Proceedings of European Workshop on Interactive Distributed Multimedia Systems and Services (IDMS'96), Berlin, March 1996.
12. D. Reininger, M. Ott, G. Michelitsch, and G. Welling. "Scalable QoS Control for VBR Video Servers". In Proceedings of First IEEE Conference on Multimedia Signal Processing, Princeton University, Princeton, NJ, June 1997.
13. D. Reininger, D.Raychaudhuri, and J. Hui. "Dynamic Bandwidth Allocation for VBR Video over ATM Networks". In IEEE Journal on Selected Areas in Communications Special Issue on Video delivery to the home, Vol. 14, No 6, August 1996.

An Object-Oriented Client/Server Architecture for Video-on-Demand Applications

J. Deicke, U. Mayer and M. Glesner

Darmstadt University of Technology
Institute of Microelectronic Systems
E-Mail: {deicke|mayer|glesner}@mes.th-darmstadt.de

Abstract. One of the most important standardization efforts in the broad field of audiovisual coding is MPEG-4 that introduces objects as smallest accessible units inside frames. This creates new possibilities in scalable scene compositions where more interesting foreground objects (e. g. a news speaker) can be lossy encoded with a better quality than a less interesting background object. This leads to new degrees of freedom in achieving high compression rates. Furthermore, this new audiovisual coding paradigm provides new functionalities in interactive audiovisual presentations. In this paper, we present a prototype realization of a video-on-demand application that enables distributed event handling of object-related events created by a user. Our system consists of a client/server architecture, where the client side handles events like zooming or moving single video objects at presentation time, while the server side considers user control concerning the quality of a lossy encoding of the single video objects. The term object-oriented stands for the access of objects rather than frames in our coding scheme as well as the object-oriented design of our software system by means of aggregation, association or inheritance. This paper describes the architecture of our system by explaining how the desired distributed functionality can be achieved. The software is entirely written in the Java language. Therefore, platform independence is achieved by using the Java Virtual Machine. Java's multithreading capabilities are used to design a highly extendible and flexible system that can be dynamically configured at runtime. The overall functionality of the system is explained with OMT class diagrams, block diagrams and state diagrams describing a protocol exchange by client and server, including the influence of the user interaction on the quality of the transmitted video objects.

1 Introduction

Since 1988, the expanding field of audiovisual communication has been addressed by the MPEG[1] standardization group that is now developing the new MPEG-4 standard. MPEG-4 is much more than just an improved coding scheme that wants to outperform existing coding standards like H.263 [1] or MPEG-2 [2].

---

[1] MPEG = Moving Pictures Expert Group is a working group of ISO. Its official name is ISO/IEC JTC1/SC29/WG11

Originally, MPEG-4 addressed audiovisual communication via low bitrate channels (up to 64 kbits/s), but during the last two years it became clear that MPEG-4 will mainly focus on new functionalities. When the committee realized that a new compression paradigm was needed and they introduced *objects* in *frames* as smallest accessible units [3], the idea of interaction with those objects arose. So MPEG-4 started to turn out as the first audiovisual coding standard considering real user interaction. The idea of *distributed event handling*, i. e. saving bandwidth by handling user events at the client side, if possible, and passing server-related events to the server side leads to new and complex problems in distributed system design.

One of the central ideas of MPEG-4 is the idea of *scalability*. Reducing spatial redundancy (intra-frame coding) as well as temporal redundancy (inter-frame coding or simply skipping frames of objects of minor interest) can be achieved during runtime by user interaction. The typical example is a news speaker and the background in a common head-and-shoulders scene where bandwidth can be saved by skipping some frames of the background object, or in other words, transmitting the more interesting foreground object with a higher frame rate than the more static background object.

In the following, the frequently cited new functionalities of MPEG-4 [4] like, e. g. content-based interactivity or content-based scalability, shall be presented in a prototype realization of a client/server system that demonstrates some of those functionalities in a scalable video-on-demand application.

In this context, *object-oriented* relates to accessible video objects or VOPs[2] in a scene as well as to an object-oriented approach in the software design of the system.

Section 2 describes the realized functionalities in detail. In section 3 the overall architecture of the whole system is introduced. Finally, section 4 contains some concluding remarks about our system and a description of the current and future work. At the end of this paper, a screen shot of the client front end is added that shows the functionalities realized so far.

2 Distributed Functionality

Designing distributed systems means deciding which functionalities belong to which part of the system. Starting from a user-created event, it has to be clarified, whether the client or the server is more suitable to handle this event. In the latest verification model of MSDL[3] [5], some core experiments are proposed, especially one called *client-side interactivity* which offers a guideline for the design of event distribution. User-requested operations concerning the sole presentation of audiovisual objects like zooming, rotating or mirroring single objects can easily be handled on the client side. Requests about the QoS[4] like frame rate or quality parameters in a compression scheme which affect the data stream have to be handled by the server. User requests have to be treated properly during

---

[2] VOP = Video Object Plane
[3] MSDL = MPEG-4 Syntax and Systems Description Language
[4] QoS = Quality of Service

presentation time and show effect as soon as possible. This holds true for both client and server, of course depending on the reliability and real time functionality of the used network.

Client-side actions that are implemented in our system, are

- zooming and
- moving

of single video objects. To the server side, the following requests can be sent:

- controlling the number of frames to skip for single video objects and
- setting the QoS of single video objects by changing a parameter that controls the compression rate of the single video objects.

Up to now, our system has no real time capabilities. This would require an underlying real time operating system and a real time protocol Therefore, we use a number of frames to skip instead of a *frame rate*. In our description, the term *number of frames to skip* means how many frames are skipped sequently (that is 3 means every fourth frame is transmitted).

Currently, the encoding of the single video objects is achieved by a DPCM[5] procedure followed by an Huffman encoding scheme. The amount of grey levels or colors, respectively, can be controlled by user interaction: the range is from lossless to a very poor quality (providing a very high compression rate) in ten steps, arbitrarily chosen. The used video objects come from the presegmented sequence *Hall Monitor*, where a segmentation of each frame is available by the COST211[ter] simulation subgroup[6]. For extraction of video objects from existing video sequences, we have proposed different approaches (e.g. [6]).

In the following, a coarse description of the functionality of the single modules is given: The client contains a user front end and, exploiting Java's benefit of platform independence, it can be started on any operating system, a Java Virtual Machine has been ported to (e. g. Sun Solaris or Window 95/NT). Besides, we have developed two different versions of the Client class, so it can be run as a standalone application invoked from a shell like environment or be executed as an applet within a Java capable browser. For more information about Java, refer to [7] or [8].

The whole system has been designed under consideration of multithreading aspects. Java's multithreading capabilities, it's object-oriented nature and the MPEG-4 Systems group's choice of Java as the MSDL basic language reason our choice for Java as the implementation language.

3 Description of the Prototype

Our model allows to connect multiple clients to one server simultaneously as well as the connection of a single client to multiple servers. For sake of simplicity, we describe only the one-client one-server connection procedure.

---

[5] DPCM = Differential Pulse Code Modulation
[6] COST = European Cooperation in the field of Scientific and Technical Research

3.1 Overview

For demonstration of a data flow example, a simplified block structure diagram of the whole system is shown in figure 1.

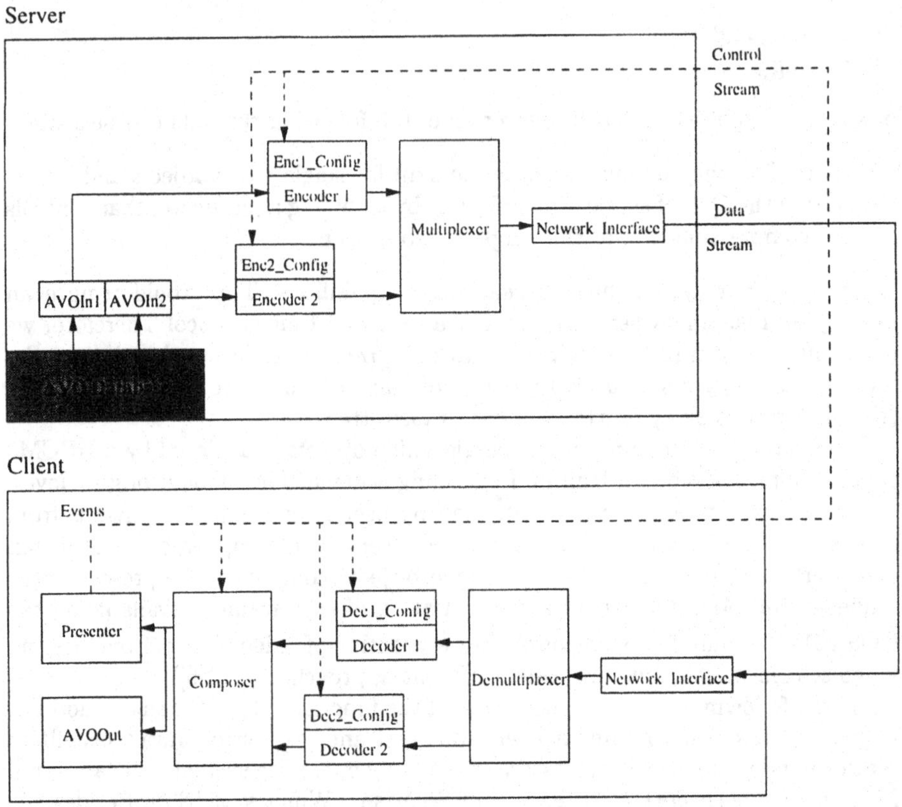

Server

Fig. 1. Block Structure of the Client/Server System. Data and Control Flow in an Example with two Encoder and two Decoder Entities

The server contains an AVO database[7]. After the request of the client, the transmission of a demanded video sequence begins. The server starts the transmission of all AVOs composing the sequence. Through AVOInBuffers, the AVOs enter the multithreaded environment and are separately encoded in their individual encoders. Then the encoded streams are multiplexed and sent downstream by a network interface (e. g. a TCP/IP network interface, that we used in our experiments). It has to be pointed out that the data stream drawn in figure 1 is actually bidirectional, because the configuration and initialization parts of the protocol are exchanged through that channel. Figure 1 depicts only the real transmission phase.

At the client side, the data stream is picked up by a peer network interface, passed to a demultiplexer and divided to the single encoded elementary streams. These are further processed by their individual decoders and passed to an entity

---

[7] AVO = Audiovisual Object. A 2D AVO with no sound is a VOP.

called Compositor that builds a scene from the incoming video objects. This is the place, where the local coordinate systems of the single video objects are transformed in the world coordinate system of the scene. For a more detailed description, look at the latest MPEG-4 system descriptions [9], [5].

Afterwards, the built scene is sent to the ClientConnectionWindow and, optionally, to an entity called AVOOutBuffer that saves the sequence to a storage medium on the client side. Finally, the ClientConnectionWindow shows the built scene to the user and awaits the user's interaction (see also the attached screen shot of the ClientConnectionWindow at the end of this paper). We remark that our class ClientConnectionWindow plays the role of the Presenter class in the official MPEG-4 systems description. The caught events in the ClientConnectionWindow are then adequately distributed like described in section 2 and depicted in figure 1. Therefore, the unidirectional upstream called control stream is needed.

3.2 OMT Notation

The overview in figure 2 and figure 3 uses the widely accepted OMT[8] notation [10]. Solid lines show a simple association relation, the diamond symbol illustrates an aggregation and the triangle stands for inheritance. The solid ball refers to possibly multiple relations (zero or more). We extended the OMT notation by boxes with a shaded border, that represent classes, that inherit from a thread superclass. Performance handicaps of the whole system due to the interpreted character of the Java language will probably be overcome by code optimizing and the awaited Java just-in-time compiler or Java chips, respectively.

The instantiated objects of the classes ClientConnectionManager, Client-Connection, Demultiplexer, Decoder and AVOOutBuffer of the client side and AVOInBuffer, Encoder, Multiplexer, ServerConnectionManager and ServerConnection at the server side run in separate threads to provide the possibility of establishing multiple client/server connections. The corresponding data processing modules work on the elementary streams on each side of the connection (= streamed representation of VOPs) in separate threads. This fact gives our system an enormous flexibility in the handling of a varying number of elementary streams. Each video object can have it's individually optimized elementary stream encoding entity, where optimization could mean the best of a fixed amount of basic tools.

3.3 Initialization and Data Transmission

In the following, the dynamic behaviour of our system is shown by description of the instantiation process, when the server and the client are started. Firstly, the Server object has to be started, instantiating a threaded object of the class ServerConnectionManager that itself instantiates a ServerConnectionMan-agerWindow object and its related ServerEventListener object, that handles

---

[8] OMT = Object Modeling Technique

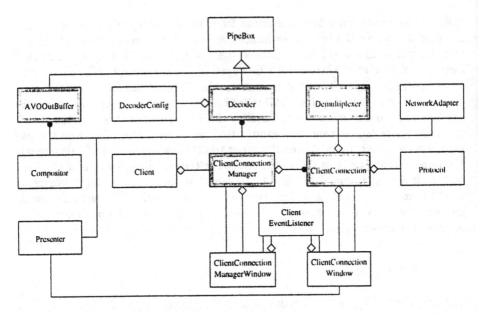

Fig. 2. Main Classes of the Client Side

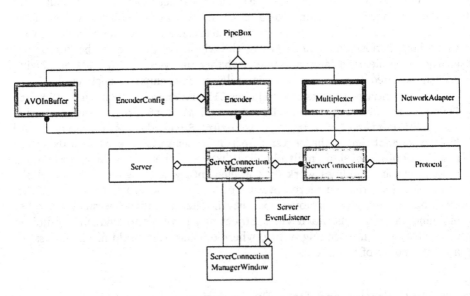

Fig. 3. Main Classes of the Server Side

GUI events as it is foreseen in the Java Development Kit 1.1. Then the **Server-ConnectionManager** listens at a port for clients.

If a user starts a **Client** object, a **ClientConnectionManager** thread is instantiated and started, presenting a GUI with the **ClientConnectionManager-Window** that enables the user to connect to one or more of the offered servers by simply pressing a connect button. Then the connection to a **ServerConnection-Manager** object is built which now instantiates a **ServerConnection** object. The peer objects `ClientConnection` and `ServerConnection` handle the following Protocol message exchange, instantiate the necessary amount of the objects `NetworkInterface`, `Demultiplexer`, `Decoder`, `DecoderConfig`, `AVOOutBuffer` or `NetworkInterface`, `Multiplexer`, `Encoder`, `EncoderConfig` or `AVOInBuffer`, respectively, in order to create the required encoding and decoding structures. After the `ClientConnection` thread has been started, the `ClientConnection-Window` is presented to the user enabling the dialogue with the server: choosing a video sequence (our prototype is limited to simple video QCIF sequences without sound), configuring a default QoS and starting the transmission. While transmitting, the user on the client side can control the presentation through various widgets on the `ClientConnectionWindow`. The instances of `ClientConnectionMan-agerWindow` and `ClientConnectionWindow` await user events that are handled by the instantiated `ClientEventListener` object. So, as proposed in the MSDL Verification Model [5], several client-side events like choosing a video object, zooming it in or out, etc. are implemented in our system. Two server-related requests are realized: The user can control the number of frames to skip of single video objects separately and manipulate the quality of single video objects by changing a *quality parameter* of the chosen encoding scheme (DPCM and Huffman Coding). Here, the easiest way to achieve a scalable lossy compression scheme is to limit the size of the Huffman code book.

For an intuitive understanding of the control possibilities, a screen shot of the `ClientConnectionWindow` like it looks in the data receiving phase is added at the end of this paper.

An important fact about our proposed system is the extendible pipelining concept. Java uses so called `PipedStreams` (input and output) for communication between threads. Because the computing-time expensive modules like encoder or multiplexer should run in separate threads, we have subclassed them from a common super class called `PipeBox`, that delivers basic connection of PipedInputStreams to PipedOutputStreams through a buffer. Due to the reusability idea of object-oriented design, our system is extendible to a high degree concerning new modules that can easily be added in the data flow chain.

3.4 Protocol

In this section, a coarse overview of the used protocol is given. The used terminology on the high level is leaned on the terminology of the MPEG-2 [2] and MPEG-4 [5] standards, procedure names are partly taken from the definitions in the ITU-T[9] standard H.245 [11].

The overall structure of the protocol is modeled for the client side as a state diagram and is depicted in figure 4. In the following, a description of the main steps of the protocol is given (for the client side): After the program start of the client, a connection phase has to be passed, in which client and server introduce themselves as MPEG-4 conform client and server by sending identification messages.

---

[9] ITU-T = International Telecommunication Union - Telecommunication Standardization Sector

The client leaves the state *Start* and changes to the state *Connected*, i. e. the fundamental connection messages have been exchanged successfully. After that, a capability exchange has to occur, especially the client has to let the server know, which decoding and demultiplexing algorithms it has implemented. Besides, the client should inform the server, how many sequences at a time, how many frames, and how many video objects per frame it can process. Also information about the possible receiving file formats should be exchanged here (e. g. QCIF, 4:2:0 color sub-sampling). So the client's state changes to *CapabilitesExchanged* and the server knows the necessary information about the clients capabilities for the following protocol steps. Immediately after the capability exchange, the server sends the suitable data archive to the client, offering a choice of audiovisual information. At that point, the server already sends only the choice of sequences, the client can handle. In our prototype realization this offer is limited to QCIF sequences without sound. The client has changed to the state *DataArchiveReceived* Now the first real user interaction after the establishing of the connection has to be done: the user chooses the sequence she or he is interested in. The client changes it's state to *ReceivingData Structure*. This is a very complex part of the protocol, because now the client has to instantiate all the necessary data processing modules such as Decoder and related objects it needs to handle the following elementary streams. A *DataStructureBuilt* message to the server starts the actual data transmission phase and the client changes to the state *ReceivingData*. After the transmission of the whole sequence, the user can decide, whether she or he wants to exit or to go back to the *DataArchiveReceived* state.

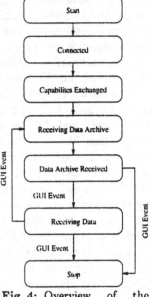

Fig. 4: Overview of the Client Side Protocol

The user influence on the data quality has not to be considered by the protocol, because it doesn't lead to a new state. Actually, the user events generated by the graphical user interface are sent to the server through an additional control channel. It is important to point out, that two different communication channels exist between client and server: a bidirectional *data channel*, which is used for the whole protocol exchange described in this section and the unidirectional *control channel* (only upstream), which is used for the exchange of the quality parameters during the transmission. So the *Receiving Data* state could be further subdivided by distinguishing between the states of the control channel; we didn't do so to keep the overview character of figure 4.

3.5 Encoder and Multiplexer

Each VOP frame builds an ES-DU[10]. The encoder adds a header with three elements to the encoded data: The length of the ES-DU, a four bit compression

---

[10] ES-DU = Elementary Stream Data Unit

scheme tag and a 4 bit compression quality parameter. In our case, the compression scheme tag has always a fixed value indicating our used DPCM and Huffman coding scheme. The compression quality parameter is a normalized value, providing the decoder with the achieved quality, i. e. the size of the Huffman code book. The Huffman code book can be seen as belonging to the payload of the elementary stream per frame and is transmitted every frame independently. The ES-DU is shown in figure 5.

Total Length (Header and Payload)	Compression Scheme (4 bit)	Compr. Quality Parameter (4 bit)	Payload .....

Fig. 5. An ES-DU

Up to now, the Multiplexer has no adequate scheduler but simply multiplexes the elementary streams data units (or single video objects) alternating one after each other to so called MUX-PDUs[11] [12] (see figure 6). The transmitted MUX-

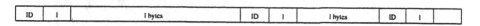

Fig. 6. A MUX-PDU

PDUs have a fixed length. If an ES-DU is longer than the remaining part of a MUX-PDU, it will be transmitted at the beginning of the next MUX-PDU. The remainder of the current MUX-PDU is filled with stuffing bits (ones).

The MUX-PDUs are not error corrected; this can be understood like transmitting program streams, opposite to transport streams according to the definitions in MPEG-2.

The MUX-PDUs simply consist of an elementary stream identification number, a length indicator and the following payload. This is the minimum information to provide the possibility of skipping frames in a non-scheduled multiplex procedure.

4 Conclusions and Future Work

We have proposed an object-oriented client/server architecture for the connection of multiple clients and multiple servers in a TCP/IP network. As an application, a video-on-demand server has been implemented. The software system is written multithreaded and offers a great flexibility in opening and closing connections and adding new modules because of a consequent use of object-oriented design methods. The necessary client modules are instantiated at runtime before the actual data transmission begins, dynamically configured by the server that sends the necessary configuration data on request of the client. So, the system demonstrates some of the main aspects (e. g. flexibility and extensibility) frequently mentioned in the context of MPEG-4 systems. Content-based MPEG-4 functionalities as scalability of single video objects and control of frame transmission rate at transmission time are realized. A distribution of user generated

---

[11] MUX-PDU = Multiplexer Protocol Data Unit

events enables a highly interactive user front end without wasting bandwidth and server performance by wrongly delegated event handling. Real-time protocol extensions of the system are partly prepared, the transmission and adjusting of time base information between client and server will be added soon.

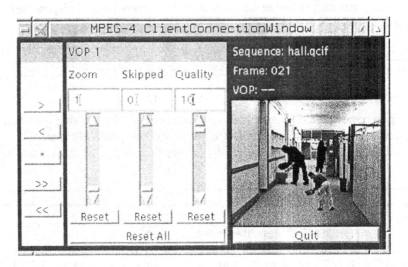

Fig. 7. The ClientConnectionWindow during data transmission of a QCIF sequence

References

1. ITU-T Recommendation H.263. Video Coding for Low Bitrate Communication, July 1995.
2. ISO/IEC JTC1/SC29/WG11. IS 13818: Generic Coding of Moving Pictures and Associated Audio, November 1994.
3. F. Pereira. MPEG-4: A New Challenge for the Representation of Audio-Visual Information. In *Picture Coding Symposium*, Melbourne, Australia, 1996.
4. ISO/IEC JTC1/SC29/WG11. MPEG-4 Proposal Package Description (PPD), July 1995.
5. ISO/IEC JTC1/SC29/WG11. N1484 MSDL VM, November 1996.
6. Alexander Steudel, Jürgen Deicke and Manfred Glesner. Unscharfe Segmentierung mit dem Bereichswachstumsverfahren zur Codierung in MPEG-4. In *9. Aachener Kolloquium "Signaltheorie"*, Aachen, 1997.
7. David Flanagan. *Java in a Nutshell*. O'Reilly, Sebastopol, 1996.
8. Ken Arnold and James Gosling. *The Java Programming Language*. Addison-Wesley, Reading, 1996.
9. ISO/IEC JTC1/SC29/WG11. N1483 Systems Working Draft, November 1996.
10. James Rumbaugh, Michael Blaha et al. *Object-Oriented Modeling and Design*. Prentice-Hall Inc., Englewood Cliffs, 1991.
11. ITU-T Recommendation H.245. Control Protocol for Multimedia Communciation, May 1996.
12. ITU-T Recommendation H.223. Multiplexing Protocol for Low Bitrate Multimedia Applications, 1996.

Interactive Remote Recording and Playback of Multicast Videoconferences

Wieland Holfelder

Praktische Informatik IV
University of Mannheim, Germany
whd@pi4.informatik.uni-mannheim.de

Abstract

One of the most exciting technologies in today's Internet is the MBone. MBone stands for Multicast Backbone; it provides the infrastructure for efficient multipoint packet delivery in the Internet. The most popular scenario is worldwide audio-video conferencing. However, so far there are no satisfactory solutions available on how to archive multimedia data streams of multicast videoconferences, and how to make them accessible for remote sites (i.e. how to remotely record *and* remotely play back MBone conferences). In this paper we present architectural considerations, design issues and describe a prototypical implementation of a **Video Conference Recording on Demand (VCRoD)** service for the MBone. Our system is called **MBone VCRoD** Service (MVoD) and is a client-server based architecture for interactive remote recording and playback of MBone sessions. MVoD is implemented based on open standards (e.g. CORBA), making it possible for other applications to interface it. Since the MVoD client application is implemented using JAVA, it will be possible to access the service from almost any platform.

1 Introduction

Popular MBone applications like vat [15], rat [9], nevot [28], vic [21] or ivs [30] support audio-video conferencing over the Internet using IP-multicast technology [2]. The MBone VCR project ([10]) presented one of the first applications that could interactively record and play back multicast multimedia data streams from and to these applications. However, the MBone VCR tool was designed as a single-user application and cannot be accessed from remote sites, thus making it inapplicable as a VCR on Demand system. After the release of the MBone VCR software to the Internet community in 1995 ([11]), we received very valuable feedback and suggestions on how to further improve the application. One of the most frequently addressed issues was the request for a distributed version of the MBone VCR. Since the original design and architecture was for a different scenario (local, single- user application) we could not simply add the necessary code for interactive distributed recording and playback. Therefore we developed a new architecture, taking the new requirements into account. The basic functionality of recording and playing back of MBone sessions, providing random access within these sessions and many other features have already been successfully implemented in the MBone VCR; we were able to reuse major parts of the former MBone VCR core code.

In this paper we present architectural considerations and address design issues that led to the architecture of our new MBone VCR on Demand Service (MVoD), and we describe a prototypical implementation. MVoD consists of three major components: an

MVoD Manager, an MVoD VideoPump, and an MVoD Client, accompanied by a number of newly developed MVoD protocols. Among other features, MVoD supports personalized access (i.e. users log into the service with userid and password), scheduling of recording- and playback requests on a per user basis for later fulfillment, and multiple active sessions per client. MVoD closes the gap between traditional VoD systems and traditional videoconferencing systems. We believe that this architecture will enable new usage scenarios of the MBone infrastructure.

The remainder of the paper is structured as follows: after a brief overview of related projects, chapter 3 motivates the MVoD architecture with some example scenarios. Chapter 4 discusses some design considerations including relevant standards for distributed systems and video servers. Chapter 5 then describes the architecture of the MVoD project and its components. A description of the MVoD protocols follows in chapter 6 before we conclude the paper in chapter 7.

2 Related Work

Though the very first applications developed for the MBone were mainly for audio/video conferencing or shared editing, the need for applications that would also support recording and playback of MBone sessions was soon evident. In this section we briefly describe projects that to our knowledge address the area of MBone recording and playback.

First Generation MBone Recording: The first generation of recording tools for the MBone was a set of small, command-line oriented tools developed by Anders Klemets. With these tools one could record exactly one data stream of a particular medium from the network and dump the packets to the local file system. For playback, the tools just read the packet stream from the disk and re-sent it into the network. No intra- or inter-media synchronization was provided and there was no interactive control.

Interactive MBone Playback: The next generation of recording tools for the MBone introduced some interactivity. Klemets' Media on Demand System for WWW ([17]) provided a form-based access to recorded MBone sessions. However, interactive recording was not possible, and since the interaction between WWW clients and servers was handled by the HyperText Transfer Protocol (HTTP) real interactivity could not be achieved (see [17] for details).

Interactive MBone Recording & Playback: The MBone VCR ([10]) was the first application that supported interactive, synchronized recording and playback of MBone sessions. For the first time random access, fast-forward and rewind within sessions was possible. Other features of the MBone VCR included a so-called blank-skip mechanism where parts of a session with no audio could be skipped automatically. In addition an automatic recording timeout was provided when no data was received for a certain amount of time, with an automatic restart of the recording when data started arriving again. Additional features included indexing, muting of individual media streams, and a primitive macro language. Even though the MBone VCR solved a number of problems inherent to earlier approaches, it had the disadvantage that it was not designed as a distributed application.

Distributed Interactive MBone Playback: Peter Parnes addresses some of the drawbacks of the MBone VCR in his mMOD - multicast Media-on-Demand system [25].

mMOD is an on-demand system that offers user-friendly remote access to previously recorded MBone material. It also provides many of the features of the MBone VCR in a distributed fashion. However, the mMOD system is still what we consider a VoD system: it does not support remote, interactive recording; hence, it does not fall into the category of VCRoD systems.

3 Example Scenarios

In this section we will motivate the MVoD service by giving several example scenarios.

3.1 On-line Videoconferencing

Consider a scenario where users from all over Europe take part in an MBone conference discussing a particular problem. In such a scenario it would often be helpful if participants could send not only live audio-video streams from their desktop microphone or camera to the other conference participants but could also bring archived material such as digital video-clips into the ongoing conference. Offering this feature, the MVoD service can enhance on-line videoconferencing significantly.

3.2 Off-line Videoconferencing

Imagine that a user could not participate in the conference mentioned above. Considering the worldwide scope of the MBone, this might be because he/she lives in a different time zone (e.g. the U.S.) and could not be in the office at the time the conference took place. In this case it would be desirable to be able to record the on-line videoconference in order to retransmit it off-line at a later point in time. Optionally this retransmission could be sent to a different multicast address, could use a different scope (i.e. using a different IP time-to-live value), or could even be retransmitted to a unicast address only. We call this feature off-line videoconferencing because the user was off-line when the MVoD service was recording the conference but can still follow the conference later just as if he or she had been a passive participant in the original transmission. In principle, this could be solved using the former MBone VCR, however, to achieve the best quality possible, it is desirable to record the conference as close to the sender(s) as possible. In the example above, this can be achieved by using an MVoD server located in Europe.

3.3 Tele-Teaching and Home-Learning

Interesting and useful applications for the MVoD Service can be found furthermore in education. Many so-called teleteaching projects (e.g. [5]) transmit lessons on a regular basis using the MBone technology. Students may participate in the session at a remote site (teleteaching) or may access archived lessons from their home computer (home-learning). The teleteaching scenario can be improved by making use of the on-line videoconference enhancements as described above, whereas the latter home-learning scenario can only be offered to the students if the institute of instruction is able to archive sessions with a service like MVoD.

3.4 Remote Recording

One can also think of two other scenarios where a) a session is transmitted over the MBone using a scope that is too small to reach a person interested in this topic or b) a person would like to record a session but he or she lacks direct access to the MBone. In these scenarios remote recording on a server that is connected to the MBone and located within the regional scope of the session will solve either problem. The users can connect to the server and schedule the recording for later playback. In scenario b) the playback could then be point-to-point.

4 Design Considerations

Having motivated the MVoD service with the above example scenarios, in the following chapter we give an overview of our design goals and address some crucial design issues, after which we describe standards that influenced the architecture of the MVoD service.

4.1 Design Goals

Before going into the details of the MVoD architecture, we want to give an overview of the basic design goals we had in mind during the development process. In particular we found the following features important for the MVoD service:

- interactive distributed client-server architecture where multiple clients can request recordings and playbacks of MBone sessions from multiple servers.
- multi-user capabilities with personalized access (user, group, others paradigm).
- off-line VCRoD services where users can schedule requests and must not be connected to the server when the requests are actually fulfilled (e.g. a professor records a lesson and schedules multiple playbacks for later fulfillment when he/she is off-line).
- near VCRoD features and floor control mechanisms for users connecting to sessions that are already accessed by other users.
- connection oriented client-server communication to allow for a state-based design and an event interface from MVoD servers back to the MVoD clients.
- open, standard-based client-server communication offering a standardized interface to the service. This makes it possible for other applications to easily access the service as well.
- platform-independent, intuitive and easy-to-use graphical user interface for the MVoD client.
- flexible and scalable implementation of the MBone VCRoD service and protocols, making it possible to serve multiple clients and multiple sessions simultaneously.

4.2 Design Issues

In addition to the major design goals mentioned above, we found further design issues to be important for the development of the MBone VCRoD service. The MVoD service is an interactive system. In principle there are two basic modes for presenting information to human users, passive and interactive. Passive presentations are those where users have no control over the presented data (e.g. a television broadcast). Interactive presentations, such as those provided by a VoD system, are presentations where the users should have control at least over the following "degrees of customization" [6]:

- the time when the presentation starts
- the order in which the various information items are presented
- the speed at which they are displayed
- the form of presentation.

However, since a VCRoD system offers more than a VoD system we needed to define some additional criteria. A VCRoD system not only allows for interactive playback of a presentation but also for interactive recording. Furthermore both playback and recording take place in a networked environment, and may be initiated from various remote clients. Hence, we require that the users should also have control over the following additional degrees of customization:

- the location in the network where the recording or playback takes place
- the regional scope for recordings and playbacks (e.g. as defined as the "time to live" value in IP packets)
- the transmission mode for a playback (i.e. unicast or multicast)
- the users and groups that may access a recorded session and the way they may access it (read, write, etc.).

Besides these basic design issues, we found other, MBone-specific design issues that we derived from the overall MBone scenario and infrastructure ([18], [19], [26]). They are influenced by existing MBone standards like RTP [27], SAP [7] or SDP [8] and existing MBone applications (e.g. [9], [15], [21], [28], [30]).

The MVoD service, in our opinion, should not necessarily need to know any application-specific details of the data stream, such as format, color or brightness of a video stream, format or volume of an audio stream etc., nor should it need to take care of the presentation of the recorded data during playback (rather this should be done by the original application). This implies that both recording and playback should be transparent for existing and future MBone applications.

The Real-Time Transport Protocol RTP [27] provides a level of abstraction that offers the application independence described above. Among other things, the timestamps in RTP data-packets and RTCP control-packets allow a VCRoD service to do inter-media (e.g. two audio stream) and intra-media (e.g. one audio and one video streams) synchronization without the need to know about the actual payload of the RTP data packets. Therefore the RTP protocol is crucial to achieve application independence as claimed above.

We also believe that an interface to the Session Announcement Protocol (SAP) [7] and the Session Description Protocol (SDP) [8] is useful in order to be able to retrieve information about scheduled or ongoing MBone sessions, and to be able to announce sessions that are to be played back by such a service. Furthermore, an efficient announcement mechanism for the MVoD service itself is desirable. This is to enable interested MVoD clients to learn about the existence and location of active MVoD servers and to obtain detailed information about particular servers.

To achieve compatibility and portability we also had to take current standards for video servers and distributed systems into account. The following sections will therefore address some of these standards in more detail and discuss which standards (or parts thereof) influenced the MVoD architecture.

4.3 DSM-CC

Digital Storage Media - Command and Control (DSM-CC) is an ISO/IEC standard developed for the delivery of multimedia broadband services ([1]). It is defined in terms of a simple functional reference model (Fig. 1) consisting of client and server entities (jointly called users) that use a network to communicate with each other.

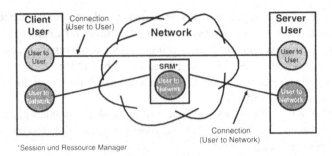

Fig. 1:DSM-CC reference model

DSM-CC covers many different functions needed for interoperability between service providers and service consumers of broadband services in general. In the MBone VCRoD system we incorporated some of the ideas of the DSM-CC media-stream interface where VCR-like control of video streams is described ([13]). The stream interface is part of the user-to-user interface specification (within DSM-CC both server and clients are called users), the user-to-network interface specification covers other issues like network session and resource control. The complete DSM-CC standard covers much more than we need within the MVoD project; on the other hand, DSM-CC neither covers remote recording nor multicast. Therefore, our design and architecture is influenced by DSM-CC as far as the VoD features are concerned, however, it will also include features that are not part of DSM-CC.

4.4 DAVIC

The Digital Audio Video Council (DAVIC), a non-profit association with a membership over 200 corporations from over 20 countries is defining one of the initial uses of DSM-CC. The purpose of DAVIC is to favor the success of emerging digital audio-visual applications and services, by the timely availability of internationally agreed upon specifications of open interfaces and protocols that maximize interoperability across countries and applications/services ([3]). As shown in Fig. 2 the DAVIC system consists of three basic components: a Content Provider System (CPS), a Service Provider System (SPS) and a Service Consumer System (SCS). Located between these systems is a CPS-SPS and a SPS-SCS Delivery System to distributed the multimedia content. The current DAVIC 1.0 version of specifications allows the deployment of systems that support initial applications such as TV distribution, near video on demand, video on demand, and some basic forms of teleshopping. The MBone VCRoD system uses some basic ideas of the DAVIC specification (e.g. CPS-, SPS-, and SCS-like components), however, its functionality can only cover a very small area of the comprehensive DAVIC specifications. On the other hand, similar to DSM-CC, the current DAVIC specification does not address the issues of interactive remote recording and multicast that we need for the MVoD service.

4.5 CORBA

DSM-CC uses the Interface Definition Language IDL ([22]) to specify its interfaces; it does not specify which Remote Procedure Call (RPC) scheme is used. However, its aim is to permit interoperability with CORBA 2.0, the Common Object Request Broker Ar-

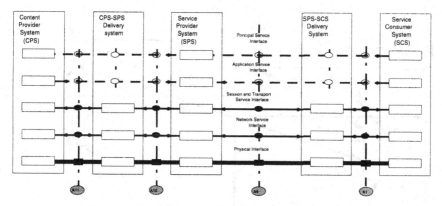

Fig. 2:Structure of a DAVIC system

chitecture and Specification, Revision 2.0 ([23]). Consequently, DAVIC has chosen CORBA 2.0 RPC (Universal Networked Object - UNO) and encoding (Common Data Representation - CDR). The idea of CORBA is to allow for applications to communicate with one another, no matter where they are located or who has designed them. The main advantages and reasons behind our decision to use a CORBA-compliant system within the MBone VCRoD project are:

- CORBA supports distribution and object orientation (MVoD is a distributed system and has an object-oriented design)
- CORBA supports many existing programming languages (including JAVA and C++, which are the languages used in the MVoD project)
- CORBA provides a high degree of interoperability (which will enable other applications to easily access the MVoD service).

5 MVoD Architecture

The MVoD service consists of the following three basic components (see Fig. 3):
- MVoD Manager
- MVoD VideoPump
- MVoD Client

(where Manager and VideoPump together form the logical unit of the MVoD Server), and the following three protocols for the communication between these components:
- VCR Service Access Protocol (VCRSAP)
- VCR Stream Control Protocol (VCRSCP)
- VCR Announcement Protocol (VCRAP)

5.1 MVoD Server

The MVoD Server consists of two modules, the MVoD Manager ([29]) and the MVoD Video-Pump. As an illustration, within the DAVIC specification, the MVoD Manager could be called a Service Provider System (SPS) and the MVoD VideoPump could be called a Content Provider System (CPS).

MVoD Manager: The MVoD Manager is the core of the MVoD service. It provides the service access for MVoD clients, it announces the MVoD Service over the VCRAP (see below), and it interfaces SAP and SDP to obtain and distribute session information. The

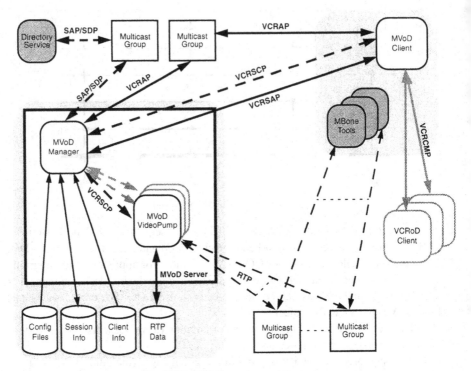

Fig. 3:The MVoD service architecture

MVoD Manager itself is subdivided into different threads responsible for specific tasks. Besides the main process handling all the request from MVoD clients, it has additional special-purpose threads: an announcement-receiver thread responsible for the reception of SAP/SDP packets, an announcement-sender thread responsible for announcing played sessions, and a VCRAP-sender thread responsible for the distribution of VCRAP-packets. The MVoD Manager does not handle any real-time data itself; instead, for each MBone session it creates a new MVoD VideoPump thread responsible for that task. The MVoD Manager is completely implemented in JAVA. The interfaces are specified using the standardized Interface Definition Language (IDL) from the Object Management Group (OMG) [22], and the remote objects are made available to the clients using Sun Microsystem's JAVA-IDL system ([14]).

MVoD VideoPump: An MVoD VideoPump runs as a separate thread and is responsible for the handling of the real-time data of one particular MBone session (possibly consisting of multiple media streams). It is created on demand and controlled by the MVoD Manager. The VideoPump can either receive RTP and RTCP packets as specified in RFC 1889 [27] from the network, synchronize them and store them on the local file system, or it can read already recorded RTP data from the file system and send it out to the network again. The MVoD VideoPump is implemented in C++, using the native method concept of JAVA. This was done mainly for two reasons:

- the VideoPump handles time critical data and C++ has a better run-time performance than JAVA, and
- it reuses code from the MBone VCR which was implemented in C++.

To keep the interfaces between the MVoD Manager and the MVoD VideoPump as simple as possible, a lean and flexible VideoPump interface in JAVA is provided. This interface is then mapped to the native C++ code.

5.2 MVoD Client

The MVoD Client ([16]) is an example of an application with which a user can interact with the MVoD service. In the DAVIC specification, the host running the MVoD client and the MBone tools would be called the Service Consumer System (SCS). Since the interfaces of the MVoD service are specified using the standardized Interface Definition Language (IDL), every client application that implements this interface could access the service as well.

The MVoD client is completely implemented in JAVA and can be started either as a JAVA applet within a JAVA-enabled Web browser, or as a stand-alone JAVA application. It has an easy-to-use graphical user interface, quite similar to that of the MBone VCR (see Fig. 4).

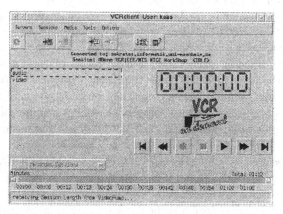

Fig. 4: Graphical User Interface of an MVoD client

6 MVoD Protocols and Interfaces

Within the MVoD service we specified a number of new protocols, and we interfaced to some already defined Internet protocols.

6.1 VCR Service Access Protocol

The VCR Service Access Protocol (VCRSAP) offers the basic service access for MVoD Clients. It is implemented using CORBA objects and itself is split into three interfaces:

- a Service-Access-Interface
- a Service-Interface, and
- a Session-Interface.

Service Access Interface: The Service Access Interface provides the initial access point for MVoD Clients. It is a distributed CORBA object that an MVoD Client can access through a so-called door-handle. A door-handle is a URL-like identifier of a server object in the JAVA IDL system. Information about the door-handle is distributed by the

MVoD Manager through the VCRAP (see below). Through the Service Access Interface clients can only retrieve information about the server or log in to the MVoD service. There is only one Service Access Interface object per MVoD Manager. After a successful login a client will have access to a Service Interface.

Service Interface: The Service Interface is again a distributed CORBA object. For each client a new Service Interface object is created by the manager and provides a person: Through this interface clients have access to private sessions, sessions of their group and sessions that can be accessed by everybody. These sessions are managed and made available to the client in four different lists. A list of recorded sessions on this server, a list of sessions that are currently announced through SAP/SDP (see below), a list of all active sessions[1] on this server and a list of all active session of this client on this server. A client can select a particular session either for playback if it is in the recorded list, or for recording if it is in the announcement list, or for joining if it is an active session. In principal, a user can have three different roles; he/she can either be:

- the *proprietor* of a session (if he/she recorded the session)
- the *owner* of a session (if he/she has control over the session for playback)
- a *listener* of a session (if he/she is a passive participant of the session).

If a user joins an active session he/she either becomes a passive member of the session if the session is owned by another user, or regains control over the session if he or she is the owner. The latter case may for example occur if a user connects to the MVoD service, selects a session for playback (thus becoming the owner of the session), starts playing the session, disconnects from the service without stopping the playback and later reconnects again while the session is still active. A selected session will be made accessible to the client through a Session Interface.

Session Interface: the Session Interface is also a distributed CORBA object to which a client application gets access after it has selected a session through the Service Interface. Per selected session a new Session Interface object is created. Through the Session Interface the client can control a particular session (e.g. invoke stream control commands) or receive events. The protocol used to control a session is the VCR Stream Control Protocol (VCRSCP).

6.2 VCR Stream Control Protocol

Access to the VCR Stream Control Protocol (VCRSCP) is provided by the *Session Interface* of the VCRSAP. It is used to control the real-time data of a selected session. It is basically a communication protocol between a client and a Video-Pump. However, MVoD Clients do not know about MVoD VideoPumps, and therefore the protocol is handled by the MVoD Manager, who will forward requests to the appropriate MVoD VideoPump.

The interface of the VCRSCP is a *One-Call Interface*, which means that all VCRSCP requests are sent using only one method of the Session Interface object. This method expects a stream control command (e.g. stop, play, rec, ff, rew, etc.) as parameter, which may optionally have a *start time* (to specify when to start the command), a *duration* (to specify how long the command should last), and a list of *free parameters* depending on the type of the command. The One-Call Interface method returns with a

---

1. An active session is either currently playing or recording, or scheduled for playing or recording.

```
v=     <protocol version>
i=     <identifier>
n=     <name>
d=     <description>
u=     <URL>
e=     <email address>
p=     <phone number>
ss=    <server sending>
ttl=   <TTL>
ms=    <max sessions>
mu=    <max users>
nuc=   <number users currently connected>
nsr=   <number sessions recorded>
nsa=   <number sessions active>
```

Fig. 5:A VCRAP packet

return code which indicates either the successful completion of the command or that an error occurred, a *state* representing the state of the session after the command was executed (e.g. idle, playing, recording, etc.), an optional *message* further explaining the return code and a free parameter list depending on the type of the stream control command.

6.3 VCR Announcement Protocol

Within the MVoD service we also specified the so-called VCR Announcement Protocol (VCRAP). Through this protocol, MVoD Managers regularly send VCRAP announcement packets to a well- known multicast address[1] in order to allow clients to learn about currently active servers and to obtain specific information about these servers. This information could be the current load, topics the server offers, number of currently logged-in users, etc. The VCRAP is a text-oriented protocol similar to SDP. It uses the known format: <type>=<value> for each line of text in a VCRAP packet. The information in one VCRAP packet must be transmitted exactly in the order specified in Fig. 5 with all fields mandatory.

The current protocol version v is 1, the identifier i is the door-handle already mentioned in section 6.1, n is any name of the server, d can give further information about the content on this server, e and p are email address and phone number of the administrator of the server, ss is the time when the server process was started, ttl is the maximum ttl that the server will use to transmit data as well as VCRAP packets, ms is the maximum number of sessions that this server will hold, mu is the maximum number of users that are allowed to be connected to the server at the same time, nuc is the number of users currently connected, nsr is the number of recorded sessions on this server and finally, nsa is the number of currently active sessions.

---

1. The address is 224.0.1.52 and is registered under the name *mbone-vcr-directory* by the Internet Assigned Numbers Authority (IANA) [12].

6.4   VCR Client Message Protocol

The last protocol specified in the MVoD service is the VCR Client Message Protocol (VCRCMP). This protocol is used by MVoD clients to communicate with each other (e.g. to implement a floor control mechanism or to indicate state transitions in shared sessions). Even though Fig. 3 shows VCRCMP as if it were handled directly between MVoD clients, there is in fact no direct communication. Instead, since only the MVoD Manager holds a list of active participants, clients send VCRCMP messages to the manager, who then will forward the message to all other participants. However, since an MVoD Manager only forwards but does not process the VCRCMP messages, the protocol is virtually handled between clients.

As an illustration, consider a client b that connects to a session already owned by another client a, making it impossible for client b to get full control over the session (i.e. to become owner of the session). However, b could still join the session, either as a passive participant receiving all the state information through the VCRCMP from client a, or b could negotiate over the VCRCMP to assume session control from client a.

The basic mechanism to enable and implement floor control is provided by VCRCMP, however, the actual floor-control policy (e.g. first-come-first-serve, priority-based) is not yet implemented.

6.5   SAP/SDP Interface

In order to be able to receive information needed to record sessions, the VCRoD Manager interfaces SAP/SDP. Additionally, the VCRoD Manager also announces sessions currently played back by the server or scheduled for later playback through SAP/SDP. This allows for other users to join the session and avoids conflicts with the allocation of multicast addresses or network bandwidth.

7   Status, Conclusion and Outlook

Whereas earlier work has focused on interactive remote playback of already recorded MBone sessions, this paper presented architectural considerations and design issues for interactive remote recording *and* playback of such multicast videoconferences which lead us to the architecture of the MBone VCRoD Service (MVoD).

All components and protocols described in this paper are implemented as prototypes on various platforms. MVoD Manager and MVoD Client are implemented in Java using JDK 1.02 and the alpha version 2.2 of SUN Microsystem's JAVA IDL. Both client and manager were successfully tested under SUN Solaris 2.5, DEC OSF 4.0 and Linux (kernel 2.0), furthermore the client also runs under Microsoft Windows '95. The MVoD VideoPump is implemented in C++ using TCL's C-library [24] and invoked from JAVA code as native methods. The Video-Pump is successfully tested under SUN Solaris 2.5, DEC OSF 4.0 and Linux (kernel 2.0). We are now evaluating and testing the system to obtain some performance results.

A critical look at the selection of recorded material of most of the MBone VoD systems available today shows that the amount of data typically is not very large nor up to date. One of the problems might be that nobody other than the administrator of the system can add new, interesting material. We believe that with the MVoD service this situation can be changed dramatically.

References

[1] Balabanian V. et. al.: **An Introduction to Digital Storage Media-Command and Control (DSM-CC)**, Online document: http://www.cselt.stet.it/ufv/leonardo/mpeg/documents/dsmcc.htm

[2] Deering, S. E.: **Multicast Routing in a Datagram Internetwork**, PhD thesis, Stanford University, 1991.

[3] Digital Audio-Video Council (DAVIC): **What is DAVIC**, Online document: http://www.davic.org

[4] Digital Audio-Video Council (DAVIC): **DAVIC 1.0 Specification Part 02, System Reference Models and Scenarios**, Geneva Switzerland, 1996.

[5] Eckert, A., Geyer, W., Effelsberg, W.: **A Distance Learning System for Higher Education Based on Telecommunication and Multimedia**, accepted paper at ED-MEDIA/ED-TELECOM'97, June 1997.

[6] Fluckinger, F.: **Understanding networked multimedia**, Prentice Hall, 1995

[7] Handley, M.: **SAP: Session Announcement Protocol**, Internet Draft, Internet Engineering Task Force, Multiparty Multimedia Session Control Working Group, Nov. 1996, Draft expires: 25 May 1997.

[8] Handley, M., Jacobsen, V.: **SDP: Session Description Protocol**, Internet Draft, Internet Engineering Task Force, Multiparty Multimedia Session Control Working Group, Mar. 1997, Draft expires: 26 Sept. 1997.

[9] Hardman, V., Sasse, A., Handley, M., Watson, A.: **Reliable Audio for Use over the Internet**, in Proceedings of Inet'95, Honolulu, Hawaii, June 1995.

[10] Holfelder, W.: **MBone VCR - Video Conference Recording on the MBone**, Proceedings of ACM Multimedia '95, ACM, pp 237-238, Oct. 1995.

[11] Holfelder, W.: **The MBone VCR**, Software online: ftp://pi4.informatik.uni-mannheim.de/pub/mbone/vcr

[12] IANA - Internet Assigned Numbers Authority: **Assignments, General Assignments, multicast-adresses,** Online document: ftp://ftp.isi.edu/in-notes/iana/assignments/multicast-addresses

[13] ISO/IEC JTC1/SC29/WG11: **Information Technology - Generic Coding of Moving Pictures and Associated Audio: Digital Storage Media - Command and Control -** ISO/IEC 13818-6 International Standard, July 1996.

[14] JavaSoft: **Java IDL,** Online document: http://splash.javasoft.com/JavaIDL/pages/index.html, Sun Microsystems Inc., 1996.

[15] Jacobsen, V., McCanne, S.: **Visual Audio Tool (vat)**, Lawrence Berkeley Laboratory, Software online: ftp://ftp.ee.lbl.gov/conferencing/vat

[16] Kaas, M.: **Entwurf und Implementierung einer Clientkomponente für einen MBone VCR on Demand Service**, Diploma Thesis (in German), Lehrstuhl für praktische Informatik IV, University of Mannheim, Jan. 1996.

[17] Klemets, A.: **The Design and Implementation of a Media on Demand System for WWW**, Proceedings of the First International Conference on WWW, Geneva, May 1994.

[18] Kumar, V.: **MBone: Interactive Multimedia on the Internet**, New Riders Publishing, 1996.

[19] Macedonia, M., Brutzman, D.: **MBone provides audio and video accross the Internet**, IEEE Computer, pp 30-36, Apr. 1994.

[20] McCanne, S.: **Scalable Compression and Transmission of Internet Multicast Video**, Dissertation, Computer Science Division (EECS), University of California at Berkeley, 1996.

[21] McCanne, S., Jacobsen, V.: **vic: a flexible framework for packet video**, Proceedings of ACM Multimedia '95, ACM, pp 511-522, Oct. 1995.

[22] Object Management Group (OMG): **Interface Definition Language (IDL)**, ISO/IEC 14750 International Standard

[23] Object Management Group (OMG): **CORBA 2.0, the Common Object Request Broker Architecture and Specification Revision 2.0**, July 1995.

[24] Ousterhout, J.K.: **Tcl and Tk Toolkit**, Addison-Wesley, 1994.

[25] Parnes, P.: **mMOD: the multicast Media-on-Demand system**, paper submitted to NOSSDAV'97, http://ctrl.cdt.luth.se/~peppar/progs/mMOD/

[26] Savetz, K., Randall, N., Lepage, Y.: **MBone: Multicasting Tomorrow's Internet**, IDG Books WorldWide, 1996.

[27] Schulzrinne, H., Casner, S., Frederick, R., Jacobsen, V.: **RTP: A Transport Protocol for Real-Time Applications**, Internet Request for Comments 1889, Internet Engineering Task Force, Audio-Video Transport Working Group, Jan. 1996.

[28] Schulzrinne, H.: **Voice Communication accross the Internet: A network voice terminal**, Technical Report TR 92-50, Department of Computer Science, University of Massachusetts, Amherst, Massachusetts, July 1992.

[29] Steinacker, A.: **Entwurf und Implementierung einer Managementkomponente für einen MBone VCR on Demand Service**, Diploma Thesis (in German), Lehrstuhl für praktische Informatik IV, University of Mannheim, Jan. 1996.

[30] Turletti, T.: **INRIA Video Conferencing System (ivs)**, Institut National de Recherche en Informatique et en Automatique, Software online: http://www.inria.fr/rodeo/ivs.html

[31] Zhang, L., Deering, S., Estrin, D., Shenker, S., Zappala, D.: **RSVP: A new resource reservation protocol**, IEEE Network 7, pp. 8-18, Sept. 1993.

Author Index

Springer
and the
environment

At Springer we firmly believe that an
international science publisher has a
special obligation to the environment,
and our corporate policies consistently
reflect this conviction.
We also expect our business partners –
paper mills, printers, packaging
manufacturers, etc. – to commit
themselves to using materials and
production processes that do not harm
the environment. The paper in this
book is made from low- or no-chlorine
pulp and is acid free, in conformance
with international standards for paper
permanency.

Lecture Notes in Computer Science

For information about Vols. 1–1234

please contact your bookseller or Springer-Verlag

Vol. 1272: F. Dehne, A. Rau-Chaplin, J.-R. Sack, R. Tamassia (Eds.), Algorithms and Data Structures. Proceedings, 1997. X, 476 pages. 1997.

Vol. 1273: P. Antsaklis, W. Kohn, A. Nerode, S. Sastry (Eds.), Hybrid Systems IV. X, 405 pages. 1997.

Vol. 1274: T. Masuda, Y. Masunaga, M. Tsukamoto (Eds.), Worldwide Computing and Its Applications. Proceedings, 1997. XVI, 443 pages. 1997.

Vol. 1275: E.L. Gunter, A. Felty (Eds.), Theorem Proving in Higher Order Logics. Proceedings, 1997. VIII, 339 pages. 1997.

Vol. 1276: T. Jiang, D.T. Lee (Eds.), Computing and Combinatorics. Proceedings, 1997. XI, 522 pages. 1997.

Vol. 1277: V. Malyshkin (Ed.), Parallel Computing Technologies. Proceedings, 1997. XII, 455 pages. 1997.

Vol. 1278: R. Hofestädt, T. Lengauer, M. Löffler, D. Schomburg (Eds.), Bioinformatics. Proceedings, 1996. XI, 222 pages. 1997.

Vol. 1279: B. S. Chlebus, L. Czaja (Eds.), Fundamentals of Computation Theory. Proceedings, 1997. XI, 475 pages. 1997.

Vol. 1280: X. Liu, P. Cohen, M. Berthold (Eds.), Advances in Intelligent Data Analysis. Proceedings, 1997. XII, 621 pages. 1997.

Vol. 1281: M. Abadi, T. Ito (Eds.), Theoretical Aspects of Computer Software. Proceedings, 1997. XI, 639 pages. 1997.

Vol. 1282: D. Garlan, D. Le Métayer (Eds.), Coordination Languages and Models. Proceedings, 1997. X, 435 pages. 1997.

Vol. 1283: M. Müller-Olm, Modular Compiler Verification. XV, 250 pages. 1997.

Vol. 1284: R. Burkard, G. Woeginger (Eds.), Algorithms — ESA '97. Proceedings, 1997. XI, 515 pages. 1997.

Vol. 1285: X. Jao, J.-H. Kim, T. Furuhashi (Eds.), Simulated Evolution and Learning. Proceedings, 1996. VIII, 231 pages. 1997. (Subseries LNAI).

Vol. 1286: C. Zhang, D. Lukose (Eds.), Multi-Agent Systems. Proceedings, 1996. VII, 195 pages. 1997. (Subseries LNAI).

Vol. 1287: T. Kropf (Ed.), Formal Hardware Verification. XII, 367 pages. 1997.

Vol. 1288: M. Schneider, Spatial Data Types for Database Systems. XIII, 275 pages. 1997.

Vol. 1289: G. Gottlob, A. Leitsch, D. Mundici (Eds.), Computational Logic and Proof Theory. Proceedings, 1997. VIII, 348 pages. 1997.

Vol. 1290: E. Moggi, G. Rosolini (Eds.), Category Theory and Computer Science. Proceedings, 1997. VII, 313 pages. 1997.

Vol. 1291: D.G. Feitelson, L. Rudolph (Eds.), Job Scheduling Strategies for Parallel Processing. Proceedings, 1997. VII, 299 pages. 1997.

Vol. 1292: H. Glaser, P. Hartel, H. Kuchen (Eds.), Programming Languages: Implementations, Logigs, and Programs. Proceedings, 1997. XI, 425 pages. 1997.

Vol. 1294: B.S. Kaliski Jr. (Ed.), Advances in Cryptology — CRYPTO '97. Proceedings, 1997. XII, 539 pages. 1997.

Vol. 1295: I. Prívara, P. Ružička (Eds.), Mathematical Foundations of Computer Science 1997. Proceedings, 1997. X, 519 pages. 1997.

Vol. 1296: G. Sommer, K. Daniilidis, J. Pauli (Eds.), Computer Analysis of Images and Patterns. Proceedings, 1997. XIII, 737 pages. 1997.

Vol. 1297: N. Lavrač, S. Džeroski (Eds.), Inductive Logic Programming. Proceedings, 1997. VIII, 309 pages. 1997. (Subseries LNAI).

Vol. 1298: M. Hanus, J. Heering, K. Meinke (Eds.), Algebraic and Logic Programming. Proceedings, 1997. X, 286 pages. 1997.

Vol. 1299: M.T. Pazienza (Ed.), Information Extraction. Proceedings, 1997. IX, 213 pages. 1997. (Subseries LNAI).

Vol. 1300: C. Lengauer, M. Griebl, S. Gorlatch (Eds.), Euro-Par'97 Parallel Processing. Proceedings, 1997. XXX, 1379 pages. 1997.

Vol. 1301: M. Jazayeri (Ed.), Software Engineering - ESEC/FSE'97. Proceedings, 1997. XIII, 532 pages. 1997.

Vol. 1302: P. Van Hentenryck (Ed.), Static Analysis. Proceedings, 1997. X, 413 pages. 1997.

Vol. 1303: G. Brewka, C. Habel, B. Nebel (Eds.), KI-97: Advances in Artificial Intelligence. Proceedings, 1997. XI, 413 pages. 1997. (Subseries LNAI).

Vol. 1304: W. Luk, P.Y.K. Cheung, M. Glesner (Eds.), Field-Programmable Logic and Applications. Proceedings, 1997. XI, 503 pages. 1997.

Vol. 1305: D. Corne, J.L. Shapiro (Eds.), Evolutionary Computing. Proceedings, 1997. X, 313 pages. 1997.

Vol. 1308: A. Hameurlain, A M. Tjoa (Eds.), Database and Expert Systems Applications. Proceedings, 1997. XVII, 688 pages. 1997.

Vol. 1309: R. Steinmetz, L.C. Wolf (Eds.), Interactive Distributed Multimedia Systems and Telecommunication Services. Proceedings, 1997. XIII, 466 pages. 1997.

Vol. 1310: A. Del Bimbo (Ed.), Image Analysis and Processing. Proceedings, 1997. Volume I. XXI, 722 pages. 1997.

Vol. 1311: A. Del Bimbo (Ed.), Image Analysis and Processing. Proceedings, 1997. Volume II. XXII, 794 pages. 1997.

Vol. 1312: A. Geppert, M. Berndtsson (Eds.), Rules in Database Systems. Proceedings, 1997. VII, 213 pages. 1997.

Vol. 1313: J. Fitzgerald, C.B. Jones, P. Lucas (Eds.), FME '97: Industrial Applications and Strengthened Foundations of Formal Methods. Proceedings, 1997. XIII, 685 pages. 1997.

Vol. 1314: S. Muggleton (Ed.), Inductive Logic Programming. Proceedings, 1996. VIII, 397 pages. 1997. (Subseries LNAI).

Vol. 1315: G. Sommer, J.J. Koenderink (Eds.), Algebraic Frames for the Perception-Action Cycle. Proceedings, 1997. VIII, 395 pages. 1997.

Vol. 1317: M. Leman (Ed.), Music, Gestalt, and Computing. IX, 524 pages. 1997. (Subseries LNAI).

Vol. 1324: C. Peters, C. Thanos (Ed.), Research and Advanced Technology for Digital Libraries. Proceedings, 1997. X, 423 pages. 1997.